SCHAUM'S OUTLINE OF

THEORY AND PROBLEMS

OF

BASIC MATHEMATICS

with Applications to
Science and Technology

Second Edition

•

HAYM KRUGLAK, Ph.D.
Professor Emeritus of Physics
Western Michigan University

JOHN T. MOORE, Ph.D.
Former Professor of Mathematics
University of Western Ontario

RAMON A. MATA-TOLEDO, Ph.D.
Associate Professor of Computer Science
James Madison University

SCHAUM'S OUTLINE SERIES
McGRAW-HILL
New York   San Francisco   Washington, D.C.   Auckland   Bogotá
Caracas   Lisbon   London   Madrid   Mexico City
Milan   Montreal   New Delhi   San Juan   Singapore   Sydney
Tokyo   Toronto

Dedicated to the Lion of Judah for his many blessings, opportunities and the wonderful family that he has given me.
RAMT

Dr. Haym Kruglak is a Professor Emeritus of Physics at Western Michigan University.

Dr. John T. Moore was a former professor of Mathematics at the University of Western Ontario.

Dr. Ramon A. Mata-Toledo is a tenured Associate Professor of Computer Science at James Madison University. He holds a Ph.D. from Kansas State University in Computer Science. He earned his M.S and M.B.A from the Florida Institute of Technology. His bachelor degree with a double major in Mathematics and Physics is from the Instituto Pedagogico de Caracas (Venezuela). Dr. Mata-Toledo's main areas of interest are databases, natural language processing, and applied mathematics. He is the author of numerous papers in professional magazines, national and international congresses. Dr. Mata-Toledo can be reached at mata1ra@jmu.edu.

Schaum's Outline of Theory and Problems of
BASIC MATHEMATICS

9 10 11 12 13 14 15 16 17 18 19 20 VFM VFM 9 8 7 6 5

ISBN 0-07-037132-6

Sponsoring Editor: Barbara Gilson
Production Supervisor: Clara Stanley
Editing Supervisor: Maureen B. Walker

Library of Congress Cataloging-in-Publication Data

Kruglak, Haym.
    Schaum's outline of basic mathematics with applications to science
and technology. — 2nd ed. / Haym Kruglak, John T. Moore, Ramon A.
Mata-Toledo.
        p. cm. — (Schaum's outline series)
    Rev. ed. of: Schaum's outline of theory and problems of basic
mathematics with applications to science and technology / Haym
Kruglak, John T. Moore. c1973.
    Includes index.
    ISBN 0-07-037182-6 (pbk.)
    1. Mathematics. I. Moore, John T. II. Mata-Toledo, Ramon A.
III. Kruglak, Haym. Schaum's outline of theory and problems of
basic mathematics with applications to science and technology.
IV. Title.
QA39.2.K77   1998
510—dc21   98-13992
                                                           CIP

McGraw-Hill

A Division of The McGraw-Hill Companies

Preface

This book is designed for the many individuals who have difficulties in *applying* mathematics. The topics were selected primarily because of their USEFULNESS. Thus, the emphasis throughout the book is on the formulation and solution of problems from the physical world.

It is assumed that the user of this book has been exposed to some high school mathematics. The selected principles, techniques, and examples were developed to aid:

1. Those who have an average background in high school mathematics but have not used this discipline for several years. Veterans and other adults returning to school frequently need not only a review of the mathematics fundamentals, but also an exposure to applications which may be new to them.

2. Those who have a good background in mathematics but need a handy "refresher," as well as a source book of unfamiliar concepts, techniques, and applications.

3. Those who have a poor background in high school mathematics. These students need a compact reference source of mathematical concepts immediately applicable to their mathematics, science, and technology courses.

Basic Mathematics may also be used successfully as a classroom text (1) in community and technical junior colleges as a first course in mathematics, (2) in colleges and universities for non-credit remedial and compensatory courses, (3) in high schools for special senior-year review and preparatory courses, (4) in vocational schools for technical mathematics courses, (5) in adult evening schools for refresher and terminal courses.

Many concepts and applications not usually taught in high school or college courses are included in this book; it should thus become a valuable supplementary text for high school and college courses in physical science, chemistry, physics, astronomy and computer science.

Each chapter contains a summary of basic definitions, principles, and techniques, with illustrative examples. The accompanying Solved Problems, drawn largely from the elementary sciences and technology, are provided with step-by-step solutions.

The sets of Supplementary Problems concluding the chapters are selected to reinforce the understanding of each concept and skill. Answers are given for all the problems.

If the users of this book find it helpful in increasing their mathematical mastery, the authors will feel richly rewarded.

HAYM KRUGLAK
RAMON A. MATA-TOLEDO

How to Use This Book

Mathematics is indispensable to the understanding and application of the basic laws of the physical world. The principles, examples, exercises, and illustrations in this book have been especially selected so as to be of maximum value to you in using them effectively.

If you are not sure what to study, ask your science instructor or curriculum adviser to point out the topics in this book which are most essential for your needs.

We believe that the following suggestions will enable you to learn more efficiently the desired mathematical skills.

1. Get to know what topics are covered in this book by scanning the table of contents and index. Skim the pages from beginning to end.

2. Have paper and pencils handy. Use standard size paper or a notebook. Nothing is as subject to error as work done on small odd scraps of paper.

3. Learn the terms and definitions.

4. Reproduce in *writing* all the steps of the solved examples and problems.

5. Read most carefully the statement of the problem.

6. Work out all the steps of the supplementary problems in a systematic fashion. Check your solution before looking up the answer.

7. Use the index and follow up all the cross-references for a given topic.

8. Obtain, if necessary, additional information from high school and college textbooks, available in school and public libraries.

9. Read carefully the manual of your calculator. When working out the problems, understand that there may be differences in accuracy and precision between the different models and brands of calculators.

10. The following conventions have been adopted to explain the use of the graphing calculators: Function keys are enclosed in curly brackets in CAPS, for example, {ENTER}. This instructs the user to press the key labeled ENTER. Options on screen menus are shown underlined. For example, for the graphing calculator HP-38G, the sequence, {LIB}Function {ENTER}, will indicate that Function is an option of the screen menu obtained after pressing the key {LIB}. Comments about the use of function keys or options are enclosed in double quotes. For example, "Press the shift key (the turquoise key) first and then press the PLOT key in the SETUP Menu."

Contents

Chapter 1

Decimal Fractions

The Meaning of Decimals

1.1 THE DECIMAL SYSTEM

Our number system is called the *decimal system* because it is based on the number 10.

The value of a digit in a number depends on its position. For example, in 528 the digit 5 has a value of 500 or 5×100; the digit 2 stands for 20 or 2×10.

1.2 DECIMAL FRACTIONS

The positional system of writing numbers can be extended so that the digits represent tenths, hundredths, thousandths, etc. A number with one or more digits beyond the units place is called a *decimal fraction* or simply a *decimal*. The integer digits are separated from the decimal digits by a period called the *decimal point*.

A number which has digits on both sides of the decimal point is called a *mixed decimal*.

EXAMPLE 1.1. The number 528.35 means $(5 \times 100) + (2 \times 10) + (8 \times 1) + (3 \times 1/10) + (5 \times 1/100)$.

1.3 POSITION DIAGRAM FOR DECIMALS

The positional values of some of the digits in the decimal system are shown in Fig. 1-1.

Fig. 1-1. Positional values in the decimal system

1.4 IMPORTANCE OF DECIMAL FRACTIONS

Calculations are greatly simplified by the use of decimal fractions. Scientists use the metric system, which is decimal, and it is very likely that this system will become universal.

Reading and Writing Decimals

1.5 READING NUMBERS WITH DECIMAL FRACTIONS

To read a mixed decimal, first read the part to the left of the decimal point. The decimal point is read as "and". Then read the digits after the decimal point and add the place name of the last digit.

EXAMPLE 1.2.   273.16 is read as two hundred seventy-three and sixteen hundredths.

EXAMPLE 1.3.   1.609 is read as one and six hundred nine thousandths.

If the decimal has no integral part, then read the digits after the decimal and add the place name of the last digit.

EXAMPLE 1.4.   0.39 is read as thirty-nine hundredths.

EXAMPLE 1.5.   0.00245 is read as two hundred forty-five one hundred-thousandths.

1.6   WRITING NUMBERS WITH DECIMAL FRACTIONS

The writing of decimal numbers is based on the knowledge of the positional value of the digits as shown in Fig. 1-1.

EXAMPLE 1.6.   Expressed in decimal form, the number thirty-seven and four tenths is 37.4.

EXAMPLE 1.7.   Expressed in decimal form, the number four hundred twenty-five and seventy-three thousandths is 425.073.

Operations with Decimals

1.7   THE BASIC LAWS OF ARITHMETIC

Decimal numbers like all real numbers obey the three basic laws of arithmetic.

I.   *Commutative laws*

EXAMPLE 1.8.
$$4.51 + 3.75 = 3.75 + 4.51$$
$$(4.51)(3.75) = (3.75)(4.51)$$

II.   *Associative laws*

EXAMPLE 1.9.
$$(7.03 + 58.6) + 1.20 = 7.03 + (58.6 + 1.20)$$
$$(7.03 \times 58.6) \times 1.20 = 7.03 \times (58.6 \times 1.20)$$

III. *Distributive laws*

EXAMPLE 1.10.
$$3.95(7.03 + 1.20) = (3.95 \times 7.03) + (3.95 \times 1.20)$$

1.8   ADDITION OF DECIMALS

The procedure is to write the numbers to be added so that their decimal points are aligned vertically.   Then the addition is carried out as with integers, but with the decimal point in the same position.

EXAMPLE 1.11.   Determine the sum of 0.1243, 21.3123, and 1.9362.

$$
\begin{array}{r}
0.1243 \\
+\ 21.3123 \\
+\ \ 1.9362 \\
\hline
23.3728 \\
\end{array}
$$

If the numbers do not have the same number of decimal places, assume that the empty spaces may be filled with zeros.

EXAMPLE 1.12. Add: 0.014, 1.0056, and 745.7.

$$
\begin{array}{r}
0.0140 \\
+ \quad 1.0056 \\
+ 745.7000 \\
\hline
746.7196
\end{array}
$$

1.9 SUBTRACTION OF DECIMALS

As in addition, write one number above the other with the decimal points aligned.   Then subtract and place the decimal point between the units and tenths.

EXAMPLE 1.13.   Subtract: 1.8264 from 23.3728.

$$
\begin{array}{r}
23.3728 \\
- \quad 1.8264 \\
\hline
21.5464
\end{array}
$$

EXAMPLE 1.14.   Subtract: 3.75 from 282.9.

$$
\begin{array}{r}
282.90 \\
- \quad 3.75 \\
\hline
279.15
\end{array}
$$

1.10 MULTIPLICATION OF DECIMALS

For the multiplication of two numbers in decimal form, first ignore the decimal points and perform the multiplications as if the numbers were integers.   Then insert the decimal point in the product so that the number of decimal places in this number is equal to the sum of the numbers of decimal places in the factors.

EXAMPLE 1.15.   Multiply: 0.214 by 1.93.

The product of 214 by 193 is 41,302.   There are three decimal places in 0.214 and two in 1.93, so there must be five decimal places in the final answer.   The desired product is then 0.41302.

EXAMPLE 1.16.   Multipy: 1.21 by 0.0056.

The product of 121 by 56 is 6776.   The number 1.21 has two decimal places while 0.0056 has four, so that the product has six decimal places and must be 0.006776.

Notice in Example 1.16 that is was necessary to insert two zeros after the decimal point in order to obtain six decimal digits.

Multiplying a decimal by 10 will move the decimal point one place to the right; multiplication by 100 will move it two places to the right; etc.   Thus,

$$1.25 \times 10 = 12.5 \qquad 3.14 \times 100 = 314 \qquad 0.0065 \times 1000 = 6.5$$

1.11   DIVISION OF DECIMALS

If one decimal number is to be divided by another, it is desirable to make the divisor an integer by multiplying (if necessary) both numbers by a sufficiently high power of 10.   This does not affect the value of the quotient.   The procedure is then to divide as in ordinary long division, placing a decimal point in the answer after the last digit before the decimal point in the dividend has been brought down.

EXAMPLE 1.17.   Divide 0.1547 by 0.014.

Move the decimal point 3 places to the right in each of the numbers to make the divisor an integer.

The desired quotient 0.1547/0.014 is equivalent to 154.7/14, which results, under ordinary division, in 11.05.

$$
\begin{array}{r}
11.05 \\
14\overline{)154.7} \\
\underline{14} \\
14 \\
\underline{14} \\
70 \\
70
\end{array}
$$

It is common practice to place the decimal point of the quotient above the decimal point of the dividend, as shown in the computation above.

Dividing a number by 10 will move the decimal point one place to the left; division by 100 will move it 2 places to the left; etc.   Thus,

$$39.37 \div 10 = 3.937 \qquad 42 \div 100 = 0.42 \qquad 72.5 \div 1000 = 0.0725$$

1.12   ROUNDING OFF DECIMALS

In some cases it is desirable to express a decimal to fewer places.   This type of approximation is called *rounding off*.

There are four commonly used rules for rounding off decimals:

Rule 1.   If the first of the digits to be dropped is less than 5, then the last retained digit is unchanged.

EXAMPLE 1.18.   Round off 3.1416 to two decimal places.

Since 4 is the last digit to be retained and the next digit to its right is less than 5, the rounded number becomes 3.14.

Rule 2.   If the first of the digits to be dropped is greater than 5, then the last retained digit is increased by 1.

EXAMPLE 1.19.   Round off 0.4536 to three decimal places.

Since 6 is the digit to be dropped and it is greater than 5, the last retained digit is changed from 3 to 4.   The rounded number is 0.454.

Rule 3.   If the first digit to be dropped is 5 followed by zeros, then the last retained digit is kept unchanged if it is even.   If the last digit to be kept is odd, then it is increased by 1.

EXAMPLE 1.20.   Round off 16.450 to one decimal place.

The first digit to be dropped is 5 followed by a zero.   Since the last digit to be kept is even, it is left unchanged.   The rounded number is 16.4.

EXAMPLE 1.21.   Round off 193.73500 to two decimal places.

The first digit to be dropped is 5 followed by two zeros.   The last digit to be kept is 3, which is increased to 4.   The rounded number is 193.74.

Rule 4. If the first digit to be dropped is 5 followed by a nonzero digit, then the last retained digit is increased by 1.

EXAMPLE 1.22. Round off 2.8530 to one decimal place.

The first digit to be dropped is 5 followed by 3. The last digit to be kept is 8; it is increased to 9. The rounded number is 2.9.

Solved Problems

1.1. Read the following numbers: (*a*) 24.7, (*b*) 0.285, (*c*) 6.03, (*d*) 0.0015, (*e*) 0.000437.

　　Using the positional values of the digits, the numbers are read as: (*a*) twenty-four and seven tenths, (*b*) two hundred eighty-five thousandths, (*c*) six and three hundredths, (*d*) fifteen ten-thousandths, (*e*) four hundred thirty-seven millionths.

1.2. Write the following numbers in decimal form: (*a*) forty-five and sixty-three hundreths, (*b*) one and thirteen thousandths, (*c*) ninety-five hundredths, (*d*) three and one hundred twenty-five ten-thousandths, (*e*) nine hundred eighty and three tenths.

　　From the values of the places in the decimal sequence, the numbers are written as: (*a*) 45.63, (*b*) 1.013, (*c*) 0.95, (*d*) 3.0125, (*e*) 980.3.

1.3. Add the following numbers: (*a*) 97, 364.23, and 0.759; (*b*) 23.5, 816.07, 8.62, and 0.233; (*c*) 7.001, 24.9, 96.93, and 0.682; (*d*) 7.58, 94.6, and 4.989.

　　Set up the numbers so that their decimal points are in a vertical line, and add the digits in the usual manner. The decimal point of the answer is in line with the other decimal points.

(*a*)	(*b*)	(*c*)	(*d*)
$\quad$ 97.000	$\quad$ 23.500	$\quad$ 7.001	$\quad$ 7.580
+ 364.230	+ 816.070	+ 24.900	+ 94.600
+ $\quad$ 0.759	+ $\quad$ 8.620	+ 96.930	+ $\quad$ 4.989
$\overline{461.989}$	+ $\quad$ 0.233	+ $\quad$ 0.682	$\overline{107.169}$
	$\overline{848.423}$	$\overline{129.513}$	

1.4. Subtract:

　　(*a*) 24.61 from 393.5　　　　(*c*) 362.78 from 457.06

　　(*b*) 0.917 from 1.165　　　　(*d*) 14.758 from 100.39.

　　Set up the numbers so that the larger is above the smaller with their decimal points lined up vertically. Subtract in the same way as with integers. The decimal point of the answer is in line with the other decimal points.

(*a*)	(*b*)	(*c*)	(*d*)
$\quad$ 393.50	1.165	$\quad$ 457.06	$\quad$ 100.390
− $\quad$ 24.61	− 0.917	− 362.78	− $\quad$ 14.758
$\overline{368.89}$	$\overline{0.248}$	$\overline{94.28}$	$\overline{85.632}$

1.5. Without multiplying longhand, write out the result if the number 1.602 is multiplied by: (*a*) 100, (*b*) 100,000, (*c*) 10,000,000, (*d*) 1,000,000.

　　For each power of 10 the decimal point has to be moved one place to the right.

　　(*a*) $1.602 \times 100 = 1.602 \times 10 \times 10 = 160.2$

(*b*) $1.602 \times 100,000 = 1.602 \times 10 \times 10 \times 10 \times 10 \times 10 = 160,200$

(*c*) $1.602 \times 10,000,000 = 16,020,000$

(*d*) $1.602 \times 1,000,000 = 1,602,000$

1.6. Without dividing longhand, write out the result if the number 2.540 is divided by: (*a*) 10, (*b*) 10,000, (*c*) 100,000, (*d*) 10,000,000.

Division by each factor of 10 moves the decimal point one place to the left.

(*a*) $2.540 \div 10 = 0.2540$

(*b*) $2.540 \div 10,000 = 2.540 \div (10 \times 10 \times 10 \times 10) = 0.000254$

(*c*) $2.540 \div 100,000 = 0.0000254$

(*d*) $2.540 \div 10,000,000 = 0.000000254$

1.7. Multiply: (*a*) 4.5 by 9.72, (*b*) 57.8 by 0.023, (*c*) 9389 by 0.52, (*d*) 4.186 by 1.03.

The multiplication process is shown below.

(*a*)   4.5   The product has 3 decimal places since the sum of
         9.72   the numbers of places in the two given numbers is $1 + 2 = 3$.
         ────
          90
         315
         405
         ────
       43.740 *Ans.*

(*b*) 57.8   There are four decimal places in the product, since the
       0.023   sum of the numbers of places in the two given numbers is
       ───── $1 + 3 = 4$.
        1734
        1156
       ──────
       1.3294 *Ans.*

(*c*) 9389   (*d*) 4.186
      0.52   1.03
     ──────   ──────
     18778   12558
     46945   4186
    ───────   ───────
    4882.28 *Ans.*   4.31158 *Ans.*

1.8. Divide to three decimal places and round off the quotient to two places: (*a*) 22 by 26, (*b*) 2.913 by 0.37, (*c*) 50.219 by 9.35, (*d*) 6.75 by 106.

Move the decimal point in the divisor and dividend so that divisor becomes an integer. Place the decimal point of the quotient above the changed decimal point of the dividend.

```
            0.846                          7.872
(a)  26)22.000              (b)  0.37.)2.91.30
         220                           ‿     ‿
        ─────
         208                            259
        ─────                           323
         120                            296
         104                           ─────
        ─────                           270
         160                            259
         156                           ─────
        ─────                           110
           4                             74
                                        ────
    Ans.  0.85                           36

                         Ans.   7.87
```

```
                5.371                        0.063
 (c) 9.35.)50.21.9              (d) 106)6.75
      ⌣      ⌣                          6 36
                                          390
           4675                           318
           3469                            72
           2805
           6640              Ans.  0.06
           6545
            950
            935
            150
      Ans.  5.37
```

1.9. Round off the following decimals to the indicated number of places.

(a) 1.609 to two decimal places   (d) 752.6412 to three decimal places

(b) 10.650 to one decimal place   (e) 17.3652 to two decimal places

(c) 0.8750 to two decimal places

Using the rules for rounding off:

(a) 1.61, since the first digit to be dropped is greater than 5.

(b) 10.6, since the first digit to be dropped is 5 followed by zero and the last digit to be kept is even.

(c) 0.88, since the first digit to be dropped is 5 followed by zero and the last digit to be kept is initially odd.

(d) 752.641, since the first digit to be dropped is smaller than 5.

(e) 17.37, since the first digit to be dropped is 5 followed by a nonzero digit.

Supplementary Problems

The answers to this set of problems are at the end of this chapter.

1.10. Write the following numbers in decimal form: (a) six and sixty-seven hundredths,   (b) nine hundred thirty-one and two tenths,   (c) one hundred and fifty-five thousandths,   (d) one thousand eighty-three ten-thousandths.

1.11. Read the following: (a) 0.103,   (b) 752.05,   (c) 36.708,   (d) 2.0419.

1.12. Without multiplying longhand, write out the result if the number 4.19 is multiplied by: (a) 10, (b) 10,000,   (c) 1,000,000,   (d) 0.01.

1.13. Without performing longhand division, write out the result when the number 6.02 is divided by: (a) 100,   (b) 1000,   (c) 100,000,   (d) 10,000,000.

1.14. Find the sum of the decimal numbers in each of the following sets: (a) 12.314, 4.5932, 5.416, 2.59;   (b) 0.0012, 0.0145, 4.65;   (c) 0.05467, 4.5862, 6.496;   (d) 12.6, 0.135, 3.56, 6.013.

1.15. Find the product of the numbers in each of the following sets to four rounded decimal places: (a) 3.567, 45.65;   (b) 1.498, 0.0065;   (c) 56.2, 0.015;   (d) 0.62, 0.047.

1.16. Divide the first of each of the following pairs of numbers by the second to four rounded decimal places: (*a*) 4.567, 34.46; (*b*) 4.006, 0.0063; (*c*) 45.02, 0.047; (*d*) 70, 1.609.

1.17. Round off the following to the specified number of places: (*a*) 54.109 to two places, (*b*) 8.53 to one place, (*c*) 762.50 to an integer, (*d*) 0.27750 to three places, (*e*) 4.0251 to two places.

Answers to Supplementary Problems

1.10. (*a*) 6.67   (*b*) 931.2   (*c*) 100.055   (*d*) 0.1083

1.11. (*a*) One hundred three thousandths,   (*b*) seven hundred fifty-two and five hundredths,
(*c*) thirty-six and seven hundred eight thousandths,   (*d*) two and four hundred nineteen ten-thousandths.

1.12. (*a*) 41.9   (*b*) 41,900   (*c*) 4,190,000   (*d*) 0.0419

1.13. (*a*) 0.0602   (*b*) 0.00602   (*c*) 0.0000602   (*d*) 0.000000602

1.14. (*a*) 24.9132   (*b*) 4.6657   (*c*) 11.13687   (*d*) 22.308

1.15. (*a*) 162.8336   (*b*) 0.0097   (*c*) 0.8430   (*d*) 0.0291

1.16. (*a*) 0.1325   (*b*) 635.8730   (*c*) 957.8723   (*d*) 43.5053

1.17. (*a*) 54.11   (*b*) 8.5   (*c*) 762   (*d*) 0.278   (*e*) 4.03

Chapter 2

Measurement and Scientific Notation

Measurement

2.1 BASIC CONCEPTS

To measure a physical quantity means to compare it in size (magnitude) with a similar standard quantity called a *unit*. For example, the measurement 4.35 m for the length of a room indicates that a meter was used as a unit of length and that the length of the room was 4.35 times larger than the length of the meterstick.

A *measurement* is the ratio of the magnitude of any physical quantity to that of a standard.

Mensuration is the process of making a measurement. Mensuration involves counting either whole units or their fractions. The choice of units depends on convenience, tradition, or law.

Standard Unit

A measure of a physical quantity with which other units are compared is called a *standard unit*. All standards of measure are defined by some legal authority or by a conference of scientists. For example, the universally accepted standard of mass is the International Prototype Kilogram.

There are many different devices for taking measurements of physical quantities: rules, balances, stopwatches, voltmeters, thermometers, barometers, etc. Each one of the above instruments has a scale with divisions. Basically, mensuration by means of an instrument requires an estimation of the distance on the scale between the initial and final positions of the instrument pointer, which may be a needle or the top of a mercury column.

Exactness in mensuration is a relative term; even an experimental genius using the most delicate known instrument cannot take a perfect measurement of a continuous quantity. Thus, the record of every measurement must consist of a number, an estimate of its uncertainty, and a unit.

2.2 EXPERIMENTAL ERRORS OR UNCERTAINTIES

One of the axioms of experimental science is: No measurement of a continuous physical quantity is absolutely exact. Even if it were possible to find the accurate value of a measure, there is no way to prove that this value is absolutely true or correct. Thus, every measurement of a continuous quantity is only an approximation to the true or absolute value of that measure. The inherent uncertainties in mensuration are called *experimental errors* or *uncertainties*. There are of two types: systematic and random.

Systematic Errors

These errors have the same algebraic sign, thus making the measurement either too large or too small. Systematic errors are also called *constant*.

Worn or corroded weights, electrical instruments with needles off the zero mark, clocks which gain or lose time are examples of faulty measuring devices which may be responsible for systematic uncertainties. Heat leakage, friction, barometric pressure, temperature, and humidity are some of the external conditions that may cause systematic errors.

EXAMPLE 2.1. The slider of a laboratory balance must be moved to the 0.2-g position on the beam to produce equilibrium when there is no load on the weighing pan. This means that every weight measurement will be 0.2 g too high. To correct for this error, the adjusting screw must be turned until the balance is in equilibrium with the slider on the 0.0 division; the alternative is to correct each weighing for the "zero error" by subtracting 0.2 g from the indicated weight measurement.

Random Errors

Random errors in the measurement of a physical quantity result from the chance variations in it or in the measuring devices. Random errors are generally small and have an equal probability of being positive and negative. These errors are unavoidable. Random errors are also called accidental.

Random errors can be caused by changes in pressure and temperature, by small irregularities in the object being measured, by building vibrations, by the variations in the observer's reading of a scale, and by many other small fluctuations.

EXAMPLE 2.2. The diameter of a metal rod is measured with a micrometer which can be read to 0.0001 cm. The results of three measurements are recorded in Table 2.1.

Table 2.1

Trial Number	Micrometer Reading, cm	Deviation from the Average, cm
1	0.0596	− 0.0001
2	0.0598	+ 0.0001
3	0.0597	0
Average	0.0597	

The deviations from the average value are small, and do not appear to vary in any systematic fashion. The random errors in this set of measurements might have been caused by the variations in the manufacture of the rod, small differences in its wear, the variation in pressure when the micrometer jaws were closed, or the changes in the observer's estimate of the scale reading.

The effect of random errors can be minimized by taking a large number of observations under the same conditions and averaging the measurements.

Mistakes and Personal Bias

Arithmetical mistakes are not considered to be experimental errors. There is need for alertness in performing and checking calculations so as to avoid arithmetical mistakes. The improper use of instruments is a personal mistake and can lead to worthless results. The personal bias against certain digits or colors can sometimes be discovered by checking against a standard or by comparing the observations of several experimenters.

2.3 ACCURACY

Accuracy refers to the degree of agreement between an experimental value of a physical quantity and its "true" or correct value. Since it is impossible to eliminate all experimental errors, absolute accuracy can never be attained. Usually, an "accurate" or "accepted" value is the result of many measurements by a number of experts in the field.

Absolute Error

The difference between an experimental value of a physical quantity and the accepted value is called an absolute error or absolute uncertainty. This term is somewhat misleading because it suggests that the "true" value is known exactly.

EXAMPLE 2.3. The acceleration of gravity g is measured in the laboratory. The experimental value of g is 962 cm/sec^2. The accepted value of g is 980 cm/sec^2. How large is the absolute error?

$$\text{absolute error} = \text{experimental value} - \text{accepted value}$$
$$= 962 - 980 \text{ cm/sec}^2$$
$$= -18 \text{ cm/sec}^2$$

The minus sign means that the experimental value is lower than the accepted value.

Relative Error

The relative error or relative uncertainty of an experimental value is the ratio of the absolute error to the accepted value.

EXAMPLE 2.4. Using the data of Example 2.3, determine the relative error.

$$\text{relative error} = \frac{\text{absolute error}}{\text{accepted value}}$$
$$= \frac{-18 \text{ cm/sec}^2}{980 \text{ cm/sec}^2}$$
$$\approx -0.0184$$

Since the two quantities in the ratio have identical units, the units cancel out.

2.4 PRECISION

The degree of consistency or reproducibility of a measurement is called precision. The more precise the measurement, the less difference there is between two observations of the same event.

EXAMPLE 2.5. A sample is weighed on a balance having 0.1 g as the smallest subdivision. Even the most skilled observer could not estimate the balance reading to better than 0.1 of the smallest subdivision, or $0.1 \times 0.1 \text{ g} = 0.01 \text{ g}$. Therefore, the maximum or "upper bound" precision of a measurement with this balance is $\pm 0.01 \text{ g}$.

High precision does not of itself guarantee high accuracy. For example, one could determine the surface tension of a liquid with a precision of ± 0.01 dyne/cm. Yet this experimental value could be 2 dyne/cm lower than the value obtained in another trial with the same apparatus because the purity of the liquid was not rigidly controlled. The presence of a small amount of impurity radically changes the value of surface tension.

A measurement has high precision only if the random errors are very small. Precision is indicated by the use of significant figures, instrument reading uncertainties, and by statistical treatment of data.

Significant Figures

2.5   DEFINITION

The digits in a measurement which a scientist reads and estimates on a scale are called *significant figures*. These include all the certain digits and one additional doubtful digit based on the observer's estimate of a fractional part of the smallest scale subdivision.

EXAMPLE 2.6. Let the length of an object be measured with a ruler whose smallest subdivision is 1 cm (Fig. 2-1). Since the end of the object falls between the 10- and 11-cm marks, we are sure of the first two digits (1 and 0) but are doubtful of the third digit.

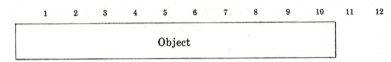

Fig. 2-1. Length of an object to three significant figures

A reasonable estimate of the object's end position might be 0.6 of the distance between the 10- and 11-cm marks. Therefore, the length of the object to three significant figures is 10.6 cm. The third digit might well be any number between 5 and 7, but not even the most experienced observer would dare to estimate that fraction to 0.01 cm. Thus, the length of the object should be recorded to three significant figures as 10.6 cm. It is also necessary to indicate the reading uncertainty of the scale. The recorded value is then 10.6 ± 0.1 cm.

EXAMPLE 2.7. The object of Example 2.6 is remeasured with a ruler with 0.1 cm subdivisions. The end of the object and a portion of the enlarged scale are shown in Fig. 2-2.

Fig. 2-2. Length of an object to four significant figures

The certain digits are 10.6. The fourth digit is doubtful, but is estimated as 4. The measurement is recorded to *four* significant figures as 10.64 ± 0.01 cm.

Sometimes a scale can not be read to better than one-half of the smallest scale subdivision. Example 2.7, the measurement would then be recorded as 10.65 ± 0.05 cm, because the end of the object is closest to the middle of the smallest subdivision. If the end of the object were near the 10.7 mark the measurement would be recorded as 10.70 ± 0.05 cm.

2.6   READING SIGNIFICANT FIGURES

The number of significant figures in a measurement can be ascertained by the use of the following rules:

1. The first significant figure in a measurement is the first digit other than zero, counting from left to right. In the measurement 29.84 ml, the first significant figure is 2.

2. Zeros to the left of the first nonzero digit are not significant. In the measurement 0.002335 g/cm^3, the zeros to the left of 2 are not significant. There are four significant figures in this measurement.

3. Zeros which occur between two significant digits are significant since they are part of the measurement. In the measurement 30.0809 g, the zeros are significant figures because they occur between the digits 3 and 9. There are thus six significant figures in this measurement.

4. Final zeros in measurements containing decimal fractions are significant. Thus in the measurement 7.0 sec the final zero is significant; it indicates that the measurement was precise to one-tenth of a second and has two significant figures. The measurement 7.00 sec is more precise and has three significant figures.

5. The number of significant figures is independent of the measurement unit. The measurement of the object in Fig. 2-2 could be expressed as 10.6 cm, 106 mm, 0.106 m, or 0.000106 km. In each case the number of significant figures is three.

2.7   OPERATIONS WITH SIGNIFICANT FIGURES

(a) Rounding Off Significant Figures

If the first digit to the right of the last significant figure is less than 5, it is dropped; if the first digit to the right of the last significant figure is more than 5, then the last significant figure is increased by one and retained. If the first digit to be dropped is 5 followed by zeros, then Rule 3 on page 4 is applicable.

EXAMPLE 2.8.   42.54 ml to three significant figures is 42.5 ml; 42.5 ml to two significant figures is 42 ml; 425.6 ml to three significant figures is 426 ml.

(b) Addition and Subtraction with Significant Figures

When measured values are to be added or subtracted, they should be rounded off to the precision of the least precise measurement.

EXAMPLE 2.9.   A speed of 331.46 m/sec is increased by 14.9 m/sec. What is the resultant speed?

Before combining the two measurements, round off 331.46 to the nearest tenth. The addition gives $331.5 + 14.9 = 346.4$. The sum of the two measurements is 346.4 m/sec.

EXAMPLE 2.10.   The original temperature of a liquid is 99.27 °C; its final temperature is 22.5 °C. What is the temperature change?

Round off 99.27 to 99.3 and perform the subtraction: $99.3 - 22.5 = 76.8$. The temperature change is 76.8 °C.

(c) Multiplication and Division with Significant Figures

In the multiplication or division of measurements, the number of significant figures in the result is no greater than the number of significant figures in the measurement with the fewest significant figures.

EXAMPLE 2.11.   The dimensions of a plate are: length $= 13.2$ cm; width $= 4.8$ cm. Find the area of the plate.

Multiply 13.2 by 4.8. Longhand multiplication gives 63.36. The answer should be rounded off to 63 cm^2 because 4.8 cm has only two significant figures.

EXAMPLE 2.12.   An object weighing 120.7 g absorbs 3.78 cal. Determine the number of calories absorbed by 1 g.

Performing the division, $3.78/120.7 = 0.03132$. The answer is given to three significant figures is 0.0313 cal/g because 3.78 cal has three and 120.7 g has four significant figures.

(d) Operations with Pure Numbers and Defined Ratios

Natural Numbers are the sequence of whole numbers starting with 1. They are the numbers that we use when counting with our fingers. A natural number like 6 can be written as 6 or 6.0, 6.00, 6.000, etc.—i.e., with an arbitrary number of zeros. Therefore multiplication and division of measurements by natural numbers preserve the number of significant figures. For example, if the measured radius of a circle is 1.96 cm, then the diameter of the same circle is 2×1.96 cm $= 3.92$ cm. The number of significant figures remains three.

The above discussion applies also to quantities whose relationship is fixed by definition. Thus, since 1 in. $= 2.5400$ cm, the conversion of centimeters to inches, or vice versa, will result in as many significant figures as there are in the *measured* quantity. A measured length of 4.3 cm is $4.3/2.5400 \approx 1.7$ in.

Scientific Notation

2.8   DEFINITION

A number is said to be in *scientific notation* if it is expressed as the product of a number between 1 and 10 and some integral power of 10. For example, 1000 is written in scientific notation as 1×10^3 or, more simply, 10^3; 0.0001 is written as 10^{-4}; $1247 = 1.247 \times 10^3$; $0.000186 = 1.86 \times 10^{-4}$.

Another name for this method is *powers-of-ten notation*.

2.9   ADVANTAGES OF SCIENTIFIC NOTATION

There are several advantages of expressing numbers in this notation.

1. Very large and very small numbers can be written more compactly. The arithmetical operations with them are greatly simplified.

2. An estimate of the result of an involved computation can be obtained quickly.

3. Scientific notation is used widely by scientists to indicate the precision of measurements. The measurements in this form are recorded with the correct number of significant figures.

2.10   CONVERSION TO AND FROM SCIENTIFIC NOTATION

Since the decimal point of any number can be shifted at will to the left or to the right by multiplying by an appropriate power 10, any number can be expressed in scientific notation. In general, if N is any number, we write $N = a \times 10^z$, where a is a number between 1 and 10 and z is an integer, positive or negative.

The writing of numbers in this notation is based on the effect produced by moving the decimal point. Thus, if the decimal point is moved one place to the right, the number is multiplied by 10. Moving the decimal point one place to the left divides the number by 10.

Rule.   To convert a number to scientific notation, move the decimal point so that it appears after the first significant figure. If the decimal point was moved to the left, multiply the resulting number by 10 raised to a power equal to the number of places moved. If the decimal point was moved to the right, the power of 10 is negative and numerically equal to the number of places moved.

A few powers of ten are shown in Table 2.2. The power of ten is also called the *exponent* of ten.

Table 2.2.   Powers of Ten

$10^0 = 1$	
$10^1 = 10$	$10^{-1} = \dfrac{1}{10^1} = \dfrac{1}{10} = 0.1$
$10^2 = 100$	$10^{-2} = \dfrac{1}{10^2} = \dfrac{1}{100} = 0.01$
$10^3 = 1000$	$10^{-3} = \dfrac{1}{10^3} = \dfrac{1}{1000} = 0.001$
$10^4 = 10,000$	$10^{-4} = \dfrac{1}{10^4} = \dfrac{1}{10,000} = 0.0001$
$10^5 = 100,000$	$10^{-5} = \dfrac{1}{10^5} = \dfrac{1}{100,000} = 0.00001$
$10^6 = 1,000,000$	$10^{-6} = \dfrac{1}{10^6} = \dfrac{1}{1,000,000} = 0.000001$
$10^9 = 1,000,000,000$	$10^{-9} = \dfrac{1}{10^9} = \dfrac{1}{1,000,000,000} = 0.000000001$

EXAMPLE 2.13.   Write each of the following numbers in scientific notation:

$$(a)\ 10,000,000 \qquad (b)\ 84,562 \qquad (c)\ 0.005239$$

(a) The decimal point in 10,000,000 is after the last zero.   The first significance digit is 1.   If the decimal point is moved to the left seven places, it is located after 1, or 1.000 000 0.   Multiplying by 10^7 gives the number in scientific notation as 1×10^7, usually written 10^7.

(b) The decimal point is moved to the left four places.   The number becomes 8.4562.   Multiplying by 10^4 expresses the number in scientific notation as 8.4562×10^4.

(c) The decimal point is moved to the right three places.   The resulting number is multiplied by 10^{-3}.   Thus, 0.005329 in scientific notation is 5.239×10^{-3}.

EXAMPLE 2.14.   Express the following numbers in ordinary decimal notation:

$$(a)\ 5.95 \times 10^4 \qquad (b)\ 3.46 \times 10^{-2} \qquad (c)\ 0.107397 \times 10^5$$

The rule is applied in reverse.

(a) Move the decimal point four places to the right: $5.95 \times 10^4 = 59,500$.

(b) Move the decimal point two places to the left: $3.46 \times 10^{-2} = 0.0346$.

(c) Move the decimal point five places to the right: $0.107397 \times 10^5 = 10,739.7$.

EXAMPLE 2.15.   Write $v = 3500$ m/sec in scientific notation.

Final zeros in measurements represented by integral numbers may or may not be significant.   In some cases the last significant zero is underlined.   Thus $v = 3500$ m/sec has three significant figures.   To avoid misunderstanding, it is best to express all measurements in scientific notation.   3500 m/sec is written as 3.5×10^3 m/sec if there are two significant figures in the result.   For three significant figures one would record 3.50×10^3 m/sec; for four significant figures the value should be written 3.500×10^3 m/sec.

2.11   OPERATIONS WITH SCIENTIFIC NOTATION

The parts of the numbers preceding the powers of ten obey all the rules of decimals given in Chapter 1.

(a) Addition and Subtraction

Rule. Before adding or subtracting two numbers expressed in scientific notation, their powers of ten must be equal.

EXAMPLE 2.16. Add 1.63×10^4 and 2.8×10^3.

To convert 2.8×10^3 to $N \times 10^4$, move the decimal point one place to the left $2.8 \times 10^3 = 0.28 \times 10^4$. Now add the two decimal parts in the usual manner.

$$
\begin{array}{r}
1.63 \\
+\ 0.28 \\
\hline
1.91
\end{array}
$$

The sum is 1.91×10^4.

EXAMPLE 2.17. Subtract 5.2×10^{-5} from 9.347×10^{-3}.

To change 5.2×10^{-5} to $N \times 10^{-3}$, move the decimal point two places to the left: $5.2 \times 10^{-5} = 0.052 \times 10^{-3}$.

Subtracting the two decimal parts of the given numbers,

$$
\begin{array}{r}
9.347 \\
0.052 \\
\hline
9.295
\end{array}
$$

The difference is 9.295×10^{-3}.

(b) Multiplication and Division

Rule. To multiply two powers of ten, add their exponents.

EXAMPLE 2.18. Multiply 100 by 1000.

$$100 = 10 \times 10 = 10^2 \qquad 1000 = 10 \times 10 \times 10 = 10^3$$

Therefore, $100 \times 1000 = 10 \times 10 \times 10 \times 10 \times 10 = 10^2 \times 10^3 = 10^{2+3} = 10^5$.

Rule. To divide two powers of ten, subtract the exponent of the divisor from the exponent of the dividend.

EXAMPLE 2.19. Divide 1,000,000 by 10,000.

$$1,000,000 = 10 \times 10 \times 10 \times 10 \times 10 \times 10 = 10^6$$
$$10,000 = 10 \times 10 \times 10 \times 10 = 10^4$$

Therefore, $\dfrac{1,000,000}{10,000} = \dfrac{10^6}{10^4} = 10^{6-4} = 10^2$ or   100.

The above result could have been obtained by cancellation of four zeros, which is the equivalent of subtracting 4 from 6 in the exponent of the dividend.

Rule. To multiply two numbers expressed in scientific notation, multiply separately their decimal parts and their powers of ten.

EXAMPLE 2.20. Multiply 1.245×10^3 by 2.70×10^2.

$$(1.245 \times 2.70) \times 10^{3+2} = 3.3615 \times 10^5 \approx 3.36 \times 10^5$$

Rule. To divide two numbers expressed in scientific notation, divide separately their decimal parts and their powers of ten.

EXAMPLE 2.21.   Divide 6.75×10^5 by 1.5×10^3.

$$\left(\frac{6.75}{1.5}\right) \times \left(\frac{10^5}{10^3}\right) = 4.5 \times 10^{5-3} = 4.5 \times 10^2$$

EXAMPLE 2.22.   Divide 8.5×10 by 1.25×10^{-2}.

$$\left(\frac{8.5}{1.25}\right) \times \left(\frac{10}{10^{-2}}\right) = 6.8 \times 10^{1-(-2)} = 6.8 \times 10^{1+2} = 6.8 \times 10^3$$

Note that a negative exponent in the divisor has its sign changed and is then added to the exponent of the dividend, as in subtraction of negative numbers.

2.12   APPROXIMATE COMPUTATIONS WITH SCIENTIFIC NOTATION

Scientific notation simplifies calculations with very large or very small numbers.

EXAMPLE 2.23.   Find the approximate product: $26,900 \times 0.00934$, to one significant figure.

Express the numbers in scientific notation:

$$2.6990 \times 10^4 \times 9.34 \times 10^{-3} \approx ?$$

Round off each number to the nearest integer.   Use the previously given rules for rounding off:

$$3 \times 10^4 \times 9 \times 10^{-3} \approx ?$$

Multiply the integers and the powers of 10; round off to one significant figure:

$$3 \times 9 \times 10^4 \times 10^{-3} = 27 \times 10^1 \approx 30 \times 10^1$$

Express the answer in scientific notation:

$$30 \times 10^1 = 3 \times 10^2$$

Longhand calculation gives $26,990 \times 0.00934 = 2.520866 \times 10^2$.

EXAMPLE 2.24.   Find the approximate quotient of $0.00969 \div 107,100$.

Express the numbers in scientific notation:

$$(9.69 \times 10^{-3}) \div 1.07100 \times 10^5$$

Round off the numbers to the nearest integer:

$$(10 \times 10^{-3}) \div (1 \times 10^5)$$

Divide the integers and the powers of 10:

$$\frac{10}{1} \times \frac{10^{-3}}{10^5} = 10 \times 10^{-8} = 10^{-7}$$

Longhand calculation gives 9.047619×10^{-8}.

2.13   ORDER OF MAGNITUDE

Sometimes we are not interested in the exact value of a measure but an approximation to it.   *The order of magnitude of a measure is an approximation to an integer power of ten.*   It answers the question: What is the closest power of ten to the given measure?   The following examples will illustrate this concept.

EXAMPLE 2.25   Given a measure of 2.47×10^8 km, what is its order of magnitude?

To calculate the order of magnitude of this measure, let us concentrate on the number accompanying the power of ten.   Observe that this number has an integer and a fractional part.   Ignoring the fractional part, notice that the number

2 is closer to 1 than it is to 10.   Replacing the decimal and fractional part of this number by 1, we can rewrite the number as 1×10^8 km.   Therefore, we can say that the order of magnitude of the measure is 10^8 km.   That is, the given measure is close to 10^8 km.

EXAMPLE 2.26.   Given the measure 8.35×10^4 miles, what is its order of magnitude?

Concentrating on the number 8.35 and ignoring its fractional part, we observe that 8 is closer to 10 than it is to 1.   Replacing the decimal and fractional part of the number by 10, we can rewrite the number as 10×10^4 miles.   Therefore, we can say that the order of magnitude of the measure is 10^5 miles.   That is, this measure is close to 10^5 miles.

The preceding two examples illustrate a simple rule for calculating the order of magnitude assuming that the number has been expressed as a number multiplied by a convenient power of ten.   If the number has an integer and a fractional part, ignore the fractional part and concentrate on the integer part.   If this integer part is less than 5, the order of magnitude is the power of ten given with the number.   If the integer part is 5 or greater, the order of magnitude is 10 times the power given with the number.

EXAMPLE 2.27.   Given a measure of 5.74×10^{-9} miles, what is its order of magnitude?

Since the number preceding the decimal point is 5, according to the rule stated above the order of magnitude is 10 times the power of ten given with the number.   That is, $10 \times 10^{-9} = 10^{-8}$.   Therefore the order of magnitude is 10^{-8} miles.

When comparing similar quantities, a factor of 10 is called *one order of magnitude*.   This factor is an indication, using powers of ten, of the relative size of the two quantities.   When we say that quantity A is two orders of magnitude bigger than quantity B, what we really say is that the quantity A is 100 times, or 10×10, bigger than the quantity B.   The following example illustrates this concept.

EXAMPLE 2.28.   If 1 fg $= 10^{-15}$ g and 1 µm $= 10^{-6}$ g, how many orders of magnitude is 1 µm greater than 1 fg?

Dividing 10^{-6} (the largest quantity) by 10^{-15} (the smallest quantity) we obtain 10^9.   Therefore, 1 µm is 9 orders of magnitude greater than 1 fg.

2.14   CONVERSION OF UNITS

Units of measurement in science are treated as algebraic quantities.   For example, if an object covers a distance d of 15 meters in a time t of 3.0 seconds, then the object's speed v is

$$v = \frac{d}{t} = \frac{15}{3.0} = 5.0 \quad \text{or} \quad 5.0 \,\frac{\text{meter}}{\text{second}} \quad \text{or} \quad 5.0 \text{ m/sec}$$

The units of speed are meters per second.   The term *per* means that the first measured quantity is divided by the second.

The area of a square $3.0 \text{ cm} \times 3.0 \text{ cm}$ is 9.0 cm^2.   The unit of the area is read "square centimeters."

A measurement in one system of units can be expressed in a different system by means of relationships between the two systems.   Such defined relationships are called *conversion factors*.   For example, 1 meter $= 39.37$ inch is a conversion factor for length between the metric and the British systems.   This conversion factor may be written as

$$\frac{1 \text{ meter}}{39.37 \text{ inch}} = 1,$$

since the dividend and the divisor are equal by definition.   We could also write this factor as

$$\frac{39.37 \text{ inch}}{1 \text{ meter}} = 1.$$

The conversion of the units of a given physical quantity is based on the fact that multiplication by a conversion factor does not change the value of the quantity.   This is because every conversion factor is equal to 1, and multiplication of a number by unity leaves the number unchanged.

EXAMPLE 2.29.   Convert $15 \frac{\text{meter}}{\text{second}}$ to $\frac{\text{inch}}{\text{second}}$.

The "meter" above the division line in the units is to be replaced by "inch."   The conversion factor is $\frac{1}{39.37} \frac{\text{meter}}{\text{inch}}$ or $\frac{39.37}{1} \frac{\text{inch}}{\text{meter}}$.

The second factor is to be used so that the "meter" above the line will cancel the "meter" below the line:

$$15 \frac{\cancel{\text{meter}}}{\text{second}} \times 39.37 \frac{\text{inch}}{\cancel{\text{meter}}} \approx 5.9 \times 10^2 \frac{\text{inch}}{\text{second}} \quad \text{or} \quad 5.9 \times 10^2 \text{ in./sec}$$

EXAMPLE 2.30.   Express $1.20 \times 10^2 \frac{\text{kilometer}}{\text{hour}}$ in feet per second.

The conversion factors are 1 km = 1000 m; 1 m = 3.281 ft; 1 hour = 3600 sec.

Each of the conversion factors is written in a form that will multiply 120 km/hr by 1 and replace the given units by the desired ones.

$$120 \frac{\cancel{\text{km}}}{\cancel{\text{hr}}} \times \frac{1 \cancel{\text{hr}}}{3600 \text{ sec}} \times \frac{1000 \cancel{\text{m}}}{1 \cancel{\text{km}}} \times \frac{3.281 \text{ ft}}{1 \cancel{\text{m}}} \approx 1.09 \times 10^2 \text{ ft/sec}$$

2.15   THE INTERNATIONAL SYSTEM OF UNITS

The International System of Units (Le Système International d'Unités) is the modern version of the metric system.   The 11th General Conference on Weights and Measures adopted it in 1960.   It is generally known as, "SI", the abbreviated form of its French name.

The SI begins with a set of seven base units as shown in Table 2.3.

Derived Units

In the International System of Units, appropriate combinations of the base units represent all physical quantities.   These combinations are called *derived units*.   For example, speed, which is the ratio of distance to elapsed time, has the SI unit of meter per second, or m/s.   Likewise, the density of an object, the ratio of its mass to its volume, has the SI unit of kilogram per cubic meter, or kg/m^3.

EXAMPLE 2.31.   If the area of an object is calculated by multiplying its length by its width, what is the derived SI unit of the area of a rectangular object?

The SI unit for both length and width is the meter (m).   Therefore, to obtain the SI unit of the area of the object, we multiply the unit of length by the unit of width.

$$\text{SI unit of area} = (\text{SI unit of length}) \times (\text{SI unit of width})$$
$$= (\text{meter}) \times (\text{meter})$$
$$= \text{meter}^2 \text{ or } m^2$$

Table 2.3.   The SI Base Units

Quantity	Unit	Symbol
Length	meter	m
Mass	kilogram	kg
Time	second	s
Electric current	ampere	A
Temperature	kelvin	K
Amount of substance	mole	mol
Luminous intensity	candela	cd

EXAMPLE 2.32.   The kinetic energy of an object, or the energy of the object while in motion, is given by the equation $KE = \frac{1}{2}mv^2$, where m is the mass of the object and v is its speed.   If the SI unit for mass is the kilogram (kg) and the SI derived unit for speed is meter/second (m/s), what is the derived SI unit for kinetic energy?

$$\text{SI unit of kinetic energy} = (\text{SI unit of mass}) \times (\text{SI unit of speed})$$
$$= (\text{kg}) \times (\text{m/s})$$
$$= \text{kg} \cdot \text{m/s}$$

2.16   PREFIXES AND DECIMAL MULTIPLIERS

Sometimes the basic units are not convenient for expressing the dimension of an object.   For instance, the distance between cities is better expressed in kilometers rather than meters.   Likewise, the dimensions of small objects such as the length of bacteria cannot be adequately expressed in meters.   To solve this problem, the SI allows the use of prefixes and decimal multipliers (see Table 2.4) to express quantities in more convenient units.   Whenever the name of a unit is preceded by one of these prefixes, the size of the unit is obtained by multiplying it by its corresponding multiplication factor.

EXAMPLE 2.33.   The radius of the earth is approximately 6.4 Mm.   Express this distance in meters.

Since a megameter is 10^6 meters, we simply substitute the prefix mega by its multiplying factor.

Therefore, 6.4 Mm $= 6.4 \times 10^6$ meters.

EXAMPLE 2.34.   The radius of a proton is about 10^{-15} meters.   Express this distance in femtometers.

Since the multiplying factor for femto is 10^{-15}, we can say that the radius of a proton is 1 fm.

EXAMPLE 2.35.   The average life span of our sun is estimated to be around 10^{18} seconds.   Express this time in exaseconds.

Since an exasecond is 10^{18} seconds, we can say that the average life of the sun is about 1 Es.

EXAMPLE 2.36.   Express 0.235 meters in millimeters.

From the table we can see that 1 mm $= 10^{-3}$ m.   In consequence, 1 m $= 10^3$ mm.

Therefore, 0.235 m $= 10^3 \times 0.235$ mm $= 235$ mm.

EXAMPLE 2.37.   The wavelength of infrared rays is sometimes measured in non-SI units such as the angström (Å), the micron (μ) and the millimicron (mμ).   The metric equivalent of these units is given by the following conversion

Table 2.4. Prefixes and Multiplying Factors of the SI

Prefix	Symbol	Multiplication Factor
exa	E	10^{18}
peta	P	10^{15}
tera	T	10^{12}
giga	G	10^{9}
mega	M	10^{6}
kilo	k	10^{3}
hecto	h	10^{2}
deka	da	10^{1}
deci	d	10^{-1}
centi	c	10^{-2}
milli	m	10^{-3}
micro	μ	10^{-6}
nano	n	10^{-9}
pico	p	10^{-12}
femto	f	10^{-15}
atto	a	10^{-18}

factors: $1 \text{ Å} = 10^{-8}$ cm, $1 \text{ m}\mu = 10 \text{ Å}$, and $1 \mu = 10^{3} \text{ m}\mu$. If the wavelength of an infrared ray is 7.7×10^{-5} cm, express this length in angstroms, microns, and millimicrons units.

Since $1 \text{ Å} = 10^{-8}$ cm, we need to multiply the given measure by a convenient power to express it as a multiple of 10^{-8}. Knowing that $10^{-8} = 10^{-5} \times 10^{-3}$, we can rewrite the given measure as

7.7×10^{-5} cm $= 7.7 \times 10^{-5}$ cm $\times (10^{-3}/10^{-3})$ ← Notice that we are multiplying and dividing the given measure by the same amount. This does not alter the numerical value of the expression.

7.7×10^{-5} cm $= 7.7/10^{-3} \times 10^{-5} \times 10^{-3}$ cm ← Grouping the powers

7.7×10^{-5} cm $= 7.7 \times 10^{3} \times 10^{-8}$ cm

7.7×10^{-5} cm $= 7700 \text{ Å}$ ← Remember that $1 \text{ Å} = 10^{-8}$ cm.

Since $1 \text{ m}\mu = 10 \text{ Å}$, we have that 7.7×10^{-5} cm $= 770 \text{ m}\mu$.

Finally, since $1 \mu = 10^{3} \text{ m}\mu$ we have that 7.7×10^{-5} cm $= 0.77 \mu$.

An easier way of converting a given measure (the existing measure) to an equivalent measure (the desired measure) in the metric system is by means of a *conversion factor*. The value of this factor *is obtained by dividing the multiplying factor of the existing measure by the multiplying factor of the desired measure*. The number of desired units is obtained by multiplying the existing units by the conversion factor. The following example will illustrate this.

EXAMPLE 2.38. Convert 3.5 milligrams (mg) to kilograms.

The existing measure is 3.5 milligrams. The multiplying factor of *milli*grams is 10^{-3}. Since we want to convert milligrams to *kilo*grams (the desired measure), the multiplying factor of this desired measure is 10^{3}. Dividing these multiplying factors according to the rule stated above, we have

$$10^{-3} \div 10^{3} = 10^{-3-(3)} = 10^{-6}$$

The conversion factor is 10^{-6}.

Therefore, 3.5 milligrams $\times (10^{-6}$ kilogram/milligram$) = 3.5 \times 10^{-6}$ kilograms.

In the computer field, the term "kilo" is used with a slightly different meaning. Instead of standing for 10^3, it stands for 2^{10}. This difference is significant when considering multiples of the basic storage unit, the byte. Here, 1 Kilobyte (1 Kb or Kbytes) stands for 2^{10} or 1024 bytes. Notice that when used this way, the word "kilo" is a misnomer. However, this terminology has become popular and is widely used throughout the world.

EXAMPLE 2.39. An advertisement announces that a computer program uses 256 Kb of memory to run. What is the exact number of bytes used by this program?

Since 1 Kbyte $= 1024$ bytes, the program uses 256×1024 bytes $= 262,144$ bytes.

EXAMPLE 2.40. Computer data is stored in "files" which may have several thousand bytes. One of the most popular units for measuring the size of a file is a Megabyte (Mb or Mbytes), where 1 Mb $= 10^3$ Kbytes. If a data file occupies 52 Mb of storage, how many bytes exactly does it have?

$$52 \text{ Mb} = 52 \times 10^3 \text{ Kb} = 52 \times 10^3 \times 1024 \text{ bytes} = 53,248,000 \text{ bytes}.$$

EXAMPLE 2.41. A hard disk is a large capacity computer storage device. The capacity of a hard disk is measured sometimes in gigabytes (Gb or Gbytes), where 1 Gb $= 10^3$ Mb. If a particular brand of computer has a hard disk with a capacity of 3.5 Gb, what is the exact capacity of the hard disk in bytes?

$$3.5 \text{ Gb} = 3.5 \times 10^3 \text{ Mb}$$
$$= 3.5 \times 10^3 \times 10^3 \text{ Kb} \leftarrow \text{ Remember that } 1 \text{ Mb} = 10^3 \text{ Kb}$$
$$= 3.5 \times 10^3 \times 10^3 \times 1024 \text{ bytes} \leftarrow 1 \text{ Kb} = 1024 \text{ bytes}$$
$$= 3,584,000,000 \text{ bytes}.$$

Solved Problems

2.1. Which one of the following statements concerning random errors is *incorrect*? (1) They are variable. (2) They cannot be controlled. (3) They are accidental errors. (4) They are caused by faulty apparatus. (5) They can be minimized by taking the arithmetical average of several observations.

Ans. (4)

2.2. Which one of the following statements is *true* when speaking of systematic errors? (1) These errors are never the same. (2) They are caused by incorrect method, defective apparatus, or incomplete theory. (3) They can be reduced by taking and averaging a large number of observations. (4) They are called accidental errors. (5) They can be eliminated by the use of a single mathematical formula.

Ans. (2)

2.3. A motorist wishes to check the fuel consumption per mile of his car. He divides the distance traveled by the number of gallons used on the trip, but makes a mistake in division. This error is: (1) random, (2) systematic, (3) either random or systematic, (4) neither random nor systematic, (5) both systematic and random.

Ans. (4)

2.4. The speed of sound in air at 25 °C has a "handbook" value of 346 m/sec. A laboratory measurement at the same temperature results in a value of 342 m/sec. Calculate   (a) the absolute error,   (b) the relative error.

(a) absolute error = measured value − handbook value

$$= 342 \text{ m/sec} - 346 \text{ m/sec} = -4 \text{ m/sec}$$

The negative value indicates that the measured value is smaller than the accepted or "true" or handbook value of the speed.

(b) relative error $= \dfrac{\text{absolute error}}{\text{handbook value}} = \dfrac{-4 \text{ m/sec}}{346 \text{ m/sec}} \approx -0.01$

2.5. The specific gravity of a solution is found by two methods: (1) specific bottle;   (2) buoyant effect on a solid of known weight suspended in the solution.   The two experimental values are 0.848 and 0.834, respectively.

(a) Calculate the absolute error or discrepancy between the two values.

(b) Compute the relative error or relative discrepancy between the two values.

Since it is not known which one of the two values is the more accurate one, the average of the two is taken as the base for calculation.

$$\text{average value of s.g.} = \frac{0.848 + 0.834}{2} = 0.841$$

(a) absolute error = 0.848 − 0.834 = 0.014.

(b) relative error $= \dfrac{\text{absolute error}}{\text{average value}} = \dfrac{0.848 - 0.834}{0.841} \approx 0.017$

2.6. The circumference C and the diameter D of a penny were measured with a metric ruler to be $C = 6.07$ cm, $D = 1.83$ cm.

(a) Divide C by D to the correct number of significant figures.

(b) Find the discrepancy (absolute error) between the experimental value C/D in (a) and its theoretical value, 3.142.

(c) Calculate the relative error in (b).

(a) $\dfrac{C}{D} = \dfrac{6.07 \text{ cm}}{1.83 \text{ cm}} = 3.317$

Since the measured values have three significant figures each, the quotient should be expressed to the same precision. The division is carried out to one more significant figure than the least precise measurement and the last digit is rounded off. The value of C/D is 3.32.

(b) absolute error = experimental value − theoretical value

$$= 3.32 - 3.14 = 0.18$$

(c) relative error $= \dfrac{\text{absolute error}}{\text{theoretical value}} = \dfrac{0.18}{3.14} \approx 0.057$

Since the absolute error has two significant figures, the relative error should be calculated to three figures and then rounded off to two.

2.7. How many significant figures are there in the following measured quantities?   (*a*) 43.20 cm, (*b*) 0.005 m,   (*c*) 0.001117 kg,   (*d*) 1.6×10^{11} cm/sec,   (*e*) 6.0507 g.

Applying the rules for reading significant figures, the answers are: (*a*) 4, (*b*) 1, (*c*) 4, (*d*) 2, (*e*) 5.

2.8. Convert   the   following   measurements   to   scientific   notation: (*a*) 1038 cm/sec, (*b*) 980.7 cm/sec², (*c*)865,000 mi,   (*d*) 0.0009004 g/cm³,   (*e*) 30.06 AU.

For numbers greater than 1, move the decimal point to the left to the first digit; for numbers smaller than 1, move the decimal point to the right past the first digit.   The answers are:

(*a*) 1.038×10^3 cm/sec   (*c*) 8.650×10^5 mi   (*e*) 3.006×10 AU

(*b*) 9.807×10^2 cm/sec   (*d*) 9.004×10^{-4} g/cm²

In part (*c*) the underlined zero is significant and must be written in the decimal part.

2.9. Round off the decimal parts to answers in Problem 2.8 to three significant figures.

Applying the rules for rounding off significant figures, the answers are: (*a*) 1.04, (*b*) 9.81, (*c*) 8.65, (*d*) 9.00, (*e*) 3.01.

2.10. Add the following measurements, keeping the correct number of significant figures: 20.6 cm, 7.350 cm, 0.450 cm.

The least precise measurement is 20.6 cm, since the last digit is uncertain to about 0.1 cm.   Rounding off the other measurements to the nearest 0.1 cm and adding:

$$\begin{array}{r} 20.6 \\ +\ \ 7.4 \\ 0.4 \\ \hline 28.4 \end{array}$$

The sum of the given measurements is 28.4 cm.

2.11. Subtract 9.26×10^{-4} kg from 4.575×10^{-2} kg and give the answer to the correct number of significant figures.

First express 9.26×10^{-4} to the same power of ten as 4.575×10^{-2}:

$$9.26 \times 10^{-4} = 0.0926 \times 10^{-2}$$

Since 4.575 has an uncertainty in the third decimal place, round off 0.0926 to the nearest 0.001, or 0.093.   Subtracting:

$$\begin{array}{r} 4.575 \times 10^{-2} \\ -\ 0.093 \times 10^{-2} \\ \hline 4.482 \times 10^{-2} \end{array}$$

The difference in the two measurements is 4.482×10^{-2} kg.

2.12. Multiply 12.45 m by 18.3 m and express the product to the correct number of significant figures.

Multiplying: $\qquad\qquad\qquad 12.45 \times 18.3 \approx 227.8$

Rounding off the result to three significant figures, the product is 228 m.

2.13. Divide to the correct number of significant figures 1.98 km by 59 sec.

Carry out the division to three significant figures and round off to two.

$$
\begin{array}{r}
0.0335 \\
59\overline{)1.98} \\
177 \\
\hline
210 \\
177 \\
\hline
330 \\
295 \\
\hline
\end{array}
$$

The quotient of the two measurements is 0.034 km/sec.

2.14. Express the following numbers in ordinary decimal notation: (a) 4.5023×10^5, (b) 6.8×10^{-3}, (c) 2.796×10^3, (d) 4.186×10^7, (e) 10^{-8}.

For positive powers of ten move the decimal to the right as many places as the exponent of ten; for negative powers of ten move the decimal point to the left. The answers are: (a) 450,230, (b) 0.0068, (c) 2796, (d) 41,860,000, (e) 0.00000001.

2.15. Multiply 2.4×10^6 by 6.25×10^{-4}.

Multiply separately the two parts:

$$2.4 \times 6.25 = 15.0 = 1.50 \times 10^1$$
$$10^6 \times 10^{-4} = 10^{6-4} = 10^2$$

The product is $1.50 \times 10^1 \times 10^2 = 1.50 \times 10^{1+2} = 1.50 \times 10^3$.

If the two numbers represent measured quantities, then the answer should be written 1.5×10^3.

2.16. Calculate $(3 \times 10^8)^2$.

$$(3 \times 10^8)^2 = 3 \times 10^8 \times 3 \times 10^8 = 9 \times 10^{8+8} = 9 \times 10^{16}$$

2.17. Compute $\dfrac{1.56 \times 10^2 \times 2.0 \times 10^8}{5.2 \times 10^5}$.

$$\frac{1.56 \times 2.0}{5.2} \times 10^{2+8-5} = 0.60 \times 10^5 = 6.0 \times 10^{-1} \times 10^5 = 6.0 \times 10^{5-1} = 6.0 \times 10^4$$

2.18. Calculate $\dfrac{1}{4.0 \times 10^{11} \times 5.25 \times 10^{-5}}$.

$$4.0 \times 10^{11} \times 5.25 \times 10^{-5} = 21 \times 10^{11-5} = 2.1 \times 10^1 \times 10^6 = 2.1 \times 10^7$$

$$\frac{1}{2.1 \times 10^7} = \frac{1}{2.1} \times \frac{1}{10^7} \approx 0.476 \times 10^{-7} \approx 4.76 \times 10^{-1} \times 10^{-7} \approx 4.76 \times 10^{-8}$$

If the two quantities were measurements, the answer would be 4.8×10^{-8} in appropriate units.

2.19. Calculate the approximate value of $\dfrac{2 \times 746}{38.5 \times 440}$.

Express the numbers in scientific notation:

$$\frac{2 \times 7.46 \times 10^2}{3.85 \times 10^1 \times 4.40 \times 10^2}$$

Round off the numbers to the nearest integer:

$$\frac{2 \times 7 \times 10^2}{4 \times 10^1 \times 4 \times 10^2}$$

Do the indicated operations with the integers and the powers of ten:

$$\frac{2 \times 7}{4 \times 4} \times \frac{10^2}{10^1 \times 10^2} = \frac{14}{16} \times 10^{-1} \approx 1 \times 10^{-1} \approx 0.1$$

Longhand calculations to two significant figures give 0.088.

2.20. One light-minute is the distance light travels in one minute. Calculate the approximate number of miles in one light-minute. Use scientific notation for calculations and the answer to one significant figure.

$$\text{speed of light} = 186,000 \text{ mi/sec,} \qquad 1 \text{ min} = 60 \text{ sec}$$

$$\text{distance traveled by light} = \text{speed} \times \text{time}$$

$$\text{distance traveled in one minute} = 186,000 \ \frac{\text{mi}}{\cancel{\text{sec}}} \times 60 \ \cancel{\text{sec}}$$

$$\text{distance} = 1.86 \times 10^5 \times 6 \times 10^1$$

$$\approx 2 \times 10^5 \times 6 \times 10^1 \approx 12 \times 10^6 \approx 1.2 \times 10^7 \text{ mi} \approx 10^7 \text{ mi}$$

2.21. Calculate the approximate value of $\dfrac{0.00469 \times 35.8 \times 0.742}{0.01362 \times 833}$.

Express the numbers in scientific notation:

$$\frac{4.69 \times 10^{-3} \times 3.58 \times 10^1 \times 7.42 \times 10^{-1}}{1.362 \times 10^{-2} \times 8.33 \times 10^2}$$

Round off the numbers to the nearest integer:

$$\frac{5 \times 10^{-3} \times 4 \times 10^1 \times 7 \times 10^{-1}}{1 \times 10^{-2} \times 8 \times 10^2}$$

Do the indicated operations with the integers and the powers of 10. Round off to one significant figure.

$$\frac{5 \times 4 \times 7}{8} \times \frac{10^{-3} \times 10^1 \times 10^{-1}}{10^{-2} \times 10^2} = \frac{140}{8} \times 10^{-3} \approx \frac{100}{10} \times 10^{-3}$$
$$\approx 10 \times 10^{-3} \approx 10^{-2}$$

Longhand calculations give an answer of 1.098109×10^{-2}.

2.22. The hardness of a sample of water is measured to be 17 parts per million. (*a*) Express this concentration in scientific notation. (*b*) Write the answer in ordinary decimal notation.

(*a*) 1 million $= 10^6$.

$$\frac{17}{10^6} = 1.7 \times 10^1 \times 10^{-6} = 1.7 \times 10^{1-6} = 1.7 \times 10^{-5}$$

(*b*) $1.7 \times 10^{-5} = 0.000017$.

2.23. The total automotive horsepower in the U.S.A. was 14.306 billion in 1965. (*a*) Express this horsepower in scientific notation. (*b*) Round off the value in (*a*) to three significant figures.

(*a*) 1 billion = 1 thousand million $= 10^3 \times 10^6 = 10^9$. Thus:

$$14.306 \times 10^9 = 1.4306 \times 10^1 \times 10^9 = 1.4306 \times 10^{1+9} = 1.4306 \times 10^{10}$$

(*b*) 1.43×10^{10}.

2.24. The U.S. population in 1965 was estimated to be one hundred ninety-four million five hundred ninety-two thousand. (*a*) Express this number in scientific notation to three significant figures. (*b*) Using the data from Problem 2.23 compute the automotive horsepower per capita in 1965.

(*a*) $194{,}592{,}000 = 1.94592 \times 10^8 \approx 1.95 \times 10^8$.

(*b*) $\dfrac{1.43 \times 10^{10}}{1.95 \times 10^8} = \dfrac{1.43}{1.95} \times 10^{10-8} \approx 0.7333 \times 10^2 \approx 73.3$ horsepower per capita

2.25. Convert 14,000 BTU to calories, given that 1 BTU $= 2.52 \times 10^2$ calorie.

Expressing the given value in scientific notation, 14,000 BTU $= 1.4 \times 10^4$ BTU. The conversion factor has to be written so that BTU appears below the division line. Then the BTU will cancel out. Round off 2.52 to two significant figures.

$$1.4 \times 10^4 \; \cancel{\text{BTU}} \times 2.5 \times 10^2 \; \frac{\text{calorie}}{1 \; \cancel{\text{BTU}}} = (1.4 \times 2.5) \times 10^{4+2} = 3.5 \times 10^6 \text{ calorie}$$

2.26. Convert 1.6 g/cm^3 to kg/m^3, given that 1 kg $= 1000$ g, 1 m $= 100$ cm.

$$1 \text{ kg} = 10^3 \text{ g} \qquad 1 \text{ m} = 10^2 \text{ cm} \qquad 1 \text{ m}^3 = (100 \text{ cm})^3 = 10^6 \text{ cm}^3$$

Multiply by the conversion factors so as to cancel out the units g and cm^3.

$$1.6 \; \frac{g}{cm^3} \times \frac{1 \text{ kg}}{10^3 \text{ g}} \times \frac{10^6 \text{ cm}^3}{1 \text{ m}^3} = 1.6 \times 10^3 \text{ kg/m}^3$$

2.27. Convert 45.0 mi/hr to m/min.

$$1 \text{ mi} = 1.609 \text{ km} \qquad 1 \text{ km} = 1000 \text{ m} = 10^3 \text{ m} \qquad 1 \text{ hr} = 60 \text{ min}$$

Rearrange the conversion factors so as to replace mi by m and hr by min:

$$45 \; \frac{\text{mi}}{\text{hr}} \frac{10^3 \text{ m}}{1 \text{ km}} \times \frac{1.609 \text{ km}}{1 \text{ mi}} \frac{1 \text{ hr}}{60 \text{ min}} = \frac{45 \times 1.609}{60} \times 10^3 \text{ m/min} \approx 1.207 \times 10^3 \text{ m/min}$$

Rounded off to three significant figures the answer is 1.21×10^3 m/min.

2.28. Evaluate $\dfrac{(9.0 \times 10^9) \times (1.6 \times 10^{-19})^2}{(10^{-15})^2}$ to two significant figures.

$$(9.0 \times 1.6^2) \times \frac{10^9 \times (10^{-19})^2}{(10^{-15})^2} = (9.0 \times 2.56) \times \frac{10^9 \times 10^{-38}}{10^{-30}}$$

$$\approx 23 \times 10^{9-38+30} \approx 23 \times 10^1 \approx 230$$

2.29. The age of the earth is estimated to be 1.3×10^{17} seconds.   What is the order of magnitude of this estimation?

Since the digit before the decimal period of the given quantity is less than 5, the order of magnitude is 10^{17} seconds.

2.30. The time it takes the earth to complete a revolution about its axis is 8.6×10^4 seconds.   What is the order of magnitude of this measure?

Since the digit before the decimal period is greater than 5, the order of magnitude is 10 times the power of ten of the given number.   Therefore, the order of magnitude is 10^5.

2.31. Astronomical distances are generally measured in light years (9.5×10^{15} m) and astronomical units (1.5×10^{11} m).   How many orders of magnitude is a light year greater than an astronomical unit?

The order of magnitude of a light year is 10^{16}.   The order of magnitude of an astronomical unit is 10^{11}.   Dividing 10^{16} by 10^{11}, we obtain 10^5.   Therefore, a light year is 5 orders of magnitude greater than an astronomical unit.

2.32. The proton rest mass (M_p) is 1.673×10^{-27} kg.   What is the order of magnitude of this measure?

Replacing the decimal and fractional part of the number by 1, the number can be rewritten as 1×10^{-27} kg.   Therefore, the order of magnitude is 10^{-27}.

2.33. Convert 0.9 pg to fg.

The given measure is 0.9 pg. The multiplying factor for pico is 10^{-12}. Since we want to convert it to femtograms, the multiplying factor of the desired measure is 10^{-15}. Dividing the multiplying factor of the exiting measure by that of the desired measure, we have that $10^{-12} \div 10^{-15} = 10^3$. Therefore, $0.9 \text{ pg} = 0.9 \times 10^{-12} \times 10^3 \text{ fg} = 0.9 \times 10^{-9} \text{ fg}$.

2.34. Convert 2800 μm to mm.

The given measure is 2800 μm. The multiplying factor for micro is 10^{-6}. We want to convert this measure to mm. The multiplying factor of the desired measure, *milli*meter, is 10^{-3}. Dividing the multiplying factor of the existing measure by that of the desired measure, we obtain $10^{-6} \div 10^{-3} = 10^{-3}$. Therefore, we have that $2800 \text{ μm} = 2800 \times 10^{-6} \times 10^3 \text{ mm} = 2.8 \text{ mm}$.

2.35. Express 125,000 m in km.

The given quantity 125,000 m can be rewritten as follows: $125,000 \text{ m} = 125 \times 10^3 \text{ m}$. Since the multiplying factor for kilo is 10^3, we can say then that $125,000 \text{ m} = 125 \times 10^3 \text{ m} = 125 \text{ km}$.

2.36. Magnetic tapes are used extensively in the computer industry as an inexpensive means of storing data for backup purposes. Tapes are characterized by their density (D) which is defined as the number of characters per inch of tape. If L is the length of a tape in inches, the capacity (C) of a tape is defined by $D \times L$. If the length of tape is 2400 feet and its density is 1600 bytes per inch, what is the capacity of the tape in bytes? Express the result in kilobytes and megabytes.

$$C = D \times L = 1600 \text{ bytes/inch} \times 2400 \text{ feet} \times 12 \text{ inches/feet} = 46,080,000 \text{ bytes}.$$

Since 1 kilobyte = 1024 bytes, the capacity of the tape in Kb = 45,000 Kb or 45 Mb.

2.37. The capacity (C) of a hard disk is defined as the number of bytes that can be stored in it. This capacity is given by the following formula: $C = NC \times TPC \times TC$ where, NC is the number of cylinders of a disk, TPC the number of tracks per cylinders is represented and TC the track capacity (the number of bytes that can be stored in a track). If a disk has 555 cylinders, 39 tracks per cylinder, and a track capacity of 19,254 bytes, what is the capacity of the disk in megabytes?

In this case, $NC = 555$ cylinders, $TPC = 39$ tracks/cylinder, $TC = 19,254$ byte/track. The disk capacity is given by $C = 555 \text{ cylinders} \times 39 \text{ tracks/cylinder} \times 19,254 \text{ byte/track} = 416,752,830 \text{ bytes}$.
Since 1 Kb = 1024 bytes, the capacity of the disk in Kb is approximately 406,985.186 Kb. Knowing that 1 Mb = 1000 Kb, we can say that the disk capacity is about 407 Mb.

Supplementary Problems

The answers to this set of problems are at the end of this chapter.

2.38. Evaluate the main type of error by placing in the blank spaces below:

R for random errors

S for systematic errors

E for either random or systematic

N for neither random nor systematic

B for both random and systematic

(a) In all measurements the observer records 4 as a 5 and 5 as a 4.   _____

(b) A speedometer which is accurate on a level road is read on a rough road.   _____

(c) An object is weighed many times with a chipped or worn weight.   _____

(d) A clock which is losing time at a constant rate is used to time a chemical reaction.   _____

(e) A farmer looks at a field and says its area is 40 acres.   Mensuration shows the field to be
    35 acres.   _____

(f) A metal tape measure correct at 68 °F is used at 85 °F.   _____

(g) An unknown volume is measured several times with an accurate graduate cylinder (neglect
    water which adheres to the walls of both vessels) with the eyes in line with the liquid level. _____

(h) A pint of water is used to check the accuracy of the 1-lb division on a balance.   _____

(i) An ammeter is connected as a voltmeter.   _____

(j) A spring scale is read in a building located on a busy truck highway.   _____

2.39. Which one of the following measurements involves primarily random errors?

(a) Measuring the length of a long hallway with a good 12-in. steel ruler.

(b) Weighing objects with a set of scales that do not balance for zero weight.

(c) Reading a thermometer on which the glass tube has been displaced relative to the scale.

(d) Measuring class periods with a clock that runs "fast".

2.40. Which one of the following is correct? One might make systematic errors if we were:

(a) Allowing a meter stick to slip when making a measurement.

(b) Reading a meter stick from various angles and positions.

(c) Estimating to the nearest 1/2 scale division.

(d) Taking measurements with a tape measure that had stretched.

(e) Consistently using the value of 6 multiplied by 7 as 45.

2.41. The freezing-point depression of a 1-molar solution has an accepted value of 1.86 °C. A student determines this constant to be 1.90 °C. (a) Find the absolute error. (b) Calculate the relative error.

2.42. The reading of a burette is given as 12.59 ± 0.02 ml. Find the relative uncertainty in making the measurement.

2.43. A resistor is measured by the voltmeter-ammeter method and by using a Wheatstone bridge. $R_{va} = 22.3$ ohm; $R_w = 22.2$ ohm.

(a) Calculate the absolute and relative uncertainty between the two methods.

(b) It is possible to tell from the data which of the two methods is more accurate?

2.44. A voltmeter has two scales: 0 to 5 and 0 to 50. The reading uncertainty of the 0- to 5-volt scale is 0.05 volt; the uncertainty of the 0- to 50-volt scale is 0.2 volt. If a reading of 2.5 volt is taken with each of the two scales, what are the two relative uncertainties?

2.45. The period of a 250-g mass oscillating on a spring is found by substituting experimental values of the mass and the spring constant into a formula. The actual period is determined by timing 100 oscillations with a stopwatch. T (formula) $= 1.15$ sec; T (observation) $= 1.14$ sec. Find the relative uncertainty between the two values.

2.46. A flask is marked 500 ml. Its volume is compared with a standard flask and is found to be 499 ml. (*a*) What is the absolute error in the flask calibration? (*b*) What is the relative error?

2.47. A meterstick is compared with a standard meter. If the meterstick is 0.5 mm too short, what is the relative error? (1 meter $= 1000$ mm.)

2.48. How many significant figures are there in the following measured quantities?

(*a*) 30.7 °C   (*c*) 0.30 ml   (*e*) 1703.5 mm   (*g*) 1.81×10^{-8} m/sec

(*b*) 0.2028 g   (*d*) 0.0146 in³   (*f*) 0.0100 volt

2.49. Convert the following measurements to scientific notation with three significant figures:

(*a*) 0.000972 cm   (*e*) 109,737 cm⁻¹

(*b*) 1038 cps   (*f*) 23.9632 g

(*c*) 299,776 km/sec   (*g*) 34.02 ml

(*d*) 4.1847 joule/cal   (*h*) 149.9 micropoise

2.50. A block has the dimensions 153.1 cm × 18.6 cm × 7.5 cm. (*a*) Find the volume of the block to the proper number of significant figures. (*b*) Express the volume in scientific notation.

2.51. Add the measurements to the correct number of significant figures: 331.46 g, 18.9 g, 5.93 g, 0.247 g.

2.52. (*a*) A force of 225 g-wt elongates a spring 37 cm. Find the elongation per gram to the correct number of significant figures. (*b*) Express the answer in scientific notation.

2.53. (*a*) Find the cross-sectional area of a wire whose diameter is 0.0597 cm to the correct number of significant figures ($A = \pi D^2/4$, where $\pi = 3.142$). (*b*) Express the answer in scientific notation.

2.54. Calculate the following value of t to the correct number of significant figures from the measurements with a gas thermometer:

$$t = \frac{75.94 - 69.84}{69.84} \times \frac{1}{0.00358} \ °C$$

2.55. (*a*) What are the relative instrument reading uncertainties to two nonzero digits in the following measurements?

(1) 6.5 ± 0.1 cm   (3) 528 ± 1 g   (5) 321.3 ± 0.1 sec

(2) 0.80 ± 0.005 amp   (4) 3.27 ± 0.02 ml

(*b*) Express the reading uncertainties in scientific notation to one significant figure.

2.56. Express the following numbers in scientific notation:

(*a*) 22,400   (*d*) 0.0000862   (*g*) 299,793

(b) 0.001293   (e) 41,850,000   (h) 965.19

(c) 980.2   (f) 273.17   (i) 0.0002389

2.57. Express the following in ordinary decimal notation:

(a) 4.5023×10^5 (c) 3.937×10^4 (e) 8.25×10^{-9} (g) 2.8979×10^{-3}

(b) 1.75890×10^7 (d) 8.31696×10^7 (f) 5.6686×10^{-8} (h) 1.450×10^{-4}

2.58. Perform the indicated operations and express the answers in scientific notation:

(a) $2.5 \times 10^{-2} \times 6.2 \times 10^{12}$ (c) $(3 \times 10^{10})^2$ (e) $\dfrac{6.62 \times 10^{-34} \times 3 \times 10^8}{1.602 \times 10^{-19} \times 10^{-9}}$

(b) $\dfrac{75 \times 10^8}{5.0 \times 10^5}$ (d) $\dfrac{6 \times 10^{-14}}{9 \times 10^{16}}$

2.59. Find approximate results of the following indicated computations (to one significant figure):

(a) $\dfrac{62 \times 5150}{96,500}$ (c) $\dfrac{1,013,000 \times 22,400}{32 \times 273}$ (e) $\dfrac{0.585 \times 4}{3.14 \times 16 \times 0.0024}$

(b) $\dfrac{743}{760} \times \dfrac{273}{308} \times 19.7$ (d) $\dfrac{300 \times 4}{3.14 \times (0.135)^2}$

2.60. The chlorine concentration in tap water is 0.15 parts per million. Express this concentration in scientific notation.

2.61. The mean distance of the planet Saturn from the sun is 1,430,000,000 km. If the velocity of light is 300,000 km/sec, how long does it take light to travel from the sun to Saturn? Use scientific notation for calculations and for the final answer.

2.62. The charge of an electron is 1.60×10^{-19} coulomb. How many electrons are needed for a total charge of 1 coulomb?

2.63. One light-year is the distance light travels in a year. Compute the approximate number of miles in a light-year. Velocity of light = 186,000 mile/sec.

2.64. The mass of 6.02×10^{23} atoms of aluminum is 27.0 g. Determine the mass of a single atom of aluminum.

2.65. For an electron $e/m = 1.76 \times 10^{11}$ coulomb/kg. If $e = 1.60 \times 10^{-19}$ coulomb, determine the value of m.

2.66. Find an approximation for $\dfrac{(8.4 \times 10^{10})^2 \times 2.2}{6.7 \times 10^{-11}}$.

2.67. (a) The electric energy generated in 1960 in the United States was 842 billion kilowatt-hours. Express this quantity in scientific notation.

(b) The United States population in 1960 was 179.3 million. Express this number in scientific notation.

(c) Calculate the electric energy generated per capita in the United States in 1960, expressing it in scientific notation.

2.68. One microcurie (μc) is one-millionth of a curie; one micro-microcurie (μμc) is one-millionth of a millionth of a curie. Express 120 μμc in scientific notation.

2.69. (a) Convert 115 ft/sec to m/min to three significant figures. 1 m ≈ 3.28 ft. (b) Write the answer in scientific notation.

2.70. A rectangular plate has the dimensions 4.50 in. × 18.35 in.

     (*a*) Calculate the area of the plate in scientific notation to three significant figures.

     (*b*) Express this area in scientific notation in cm^2 and m^2.

$$1 \text{ in.} = 2.540 \text{ cm} \qquad 1 \text{ m} = 100 \text{ cm}$$

2.71. Convert 9.80 m/sec^2 to km/hr/sec (kilometers per hour per second). $1 \text{ km} = 1000 \text{ m}$. Express the answer in scientific notation to three significant figures.

2.72. (*a*) Convert 14.7 lb/in^2 to $newton/m^2$.   (*b*) Express the answer in scientific notation to three significant figures.

$$1 \text{ lb} \approx 4.45 \text{ newton} \qquad 1 \text{ in.} = 2.540 \text{ cm} \qquad 1 \text{ m} = 100 \text{ cm}$$

2.73. The diameter of the moon is 3.5×10^6 m and the diameter of the sun is 1.4×10^9 m. How many orders of magnitude is the diameter of the sun greater than the diameter of the moon?

2.74. The mass of an electron is estimated to be about 9.10×10^{-28} g. What is its order of magnitude?

2.75. The number of atoms in 1.000 g of oxygen is estimated to be 3.77×10^{22}. What is the order of magnitude of this measure?

2.76. The period of an infrared radiation has an order of magnitude of 10^{-13} sec and the period of an X-ray is 10^{-18} sec. How many orders of magnitude is the period of an infrared radiation greater than that of an X-ray radiation?

2.77. The velocity of light in vacuum is 2.998×10^8 m/s. What is the order of magnitude of this measure?

2.78. The gravitational constant (G) is $6.673 \times 10^{-11} \text{ N} \cdot \text{m}^2/\text{kg}^2$. What is the order of magnitude of this constant?

Answers to Supplementary Problems

2.38. (*a*) N, (*b*) R, (*c*) B (*d*) S, (*e*) N, (*f*) S, (*g*) R, (*h*) S, (*i*) N, (*j*) R

2.39. (*a*)

2.40. (*d*)

2.41. (*a*) 0.04 °C (*b*) 0.02

2.42. $0.0016 \approx 0.002$

2.43. (*a*) $0.1; 0.0045 \approx 0.004$ (*b*) No

2.44. 0.02 for the 0–5 voltmeter; 0.08 for the 0–50 voltmeter

2.45. $0.0087 \approx 0.01$

2.46. (a) 1 ml   (b) 0.002

2.47. -0.0005

2.48. (a) 3   (b) 4   (c) 2   (d) 3   (e) 5   (f) 3   (g) 3

2.49. (a) 9.72×10^{-4} cm   (e) 1.10×10^5 cm^{-1}
(b) 1.04×10^3 cps   (f) 2.40×10 g
(c) 3.00×10^5 km/sec   (g) 3.40×10 ml
(d) 4.18 joule/cal   (h) 1.50×10^2 micropoise

2.50. (a) 21,000 cm^3 (b) 2.1×10^4 cm^3

2.51. 356.5 g

2.52. (a) 0.15 cm/g-wt   (b) 1.5×10^{-1} cm/g-wt

2.53. (a) 0.00280 cm^2 (b) 2.80×10^{-3} cm^2

2.54. 24.4 °C

2.55. (a) (1) 0.015;   (2) 0.0063;   (3) 0.0019;   (4) 0.0061;   (5) 0.00031
(b) (1) 2×10^{-2};   (2) 6×10^{-3};   (3) 2×10^{-3};   (4) 6×10^{-3};   (5) 3×10^{-4}

2.56. (a) 2.24×10^4 (d) 8.62×10^{-5} (g) 2.99793×10^5
(b) 1.293×10^{-3} (e) 4.185×10^7 (h) 9.6519×10^2
(c) 9.802×10^2 (f) 2.7317×10^2 (i) 2.389×10^{-4}

2.57. (a) 450,230   (c) 39,370   (e) 0.00000000825   (g) 0.0028979
(b) 17,589,000   (d) 83,169,600   (f) 0.000000056686   (h) 0.0001450

2.58. (a) 1.55×10^{11} (c) 9×10^{20} (e) 1.24×10^3
(b) 1.5×10^4 (d) 6.7×10^{-31}

2.59. (a) 3   (b) 20   (c) 2,000,000   (d) 40,000   (e) 200

2.60. 1.5×10^{-7}

2.61. 4.77×10^3 sec

2.62. 6×10^{18}

2.63. 6×10^{12} mile

2.64. 4.49×10^{-23} g

2.65. 9.09×10^{-31} kg

2.66. 2×10^{32}

2.67. (a) 8.42×10^{11} (b) 1.793×10^8 (c) 47$\underline{00}$ kw-hr $\approx 4.70 \times 10^3$ kw-hr per capita

2.68. 1.2×10^{-10} curie

2.69. (a) 21$\underline{00}$ m/min   (b) 2.10×10^3 m/min

2.70. (*a*) 8.26×10^1 in^2　　(*b*) 5.33×10^2 cm^2; 5.33×10^{-2} m^2

2.71. 35.3 km/hr/sec

2.72. 1.01×10^5 newton/m^2

2.73. 3 orders of magnitude

2.74. 10^{-27}

2.75. 10^{22}

2.76. 5 orders of magnitude

2.77. 10^8

2.78. 10^{-10}

Chapter 3

Common Fractions

Basic Concepts

3.1 FRACTIONS AND MEASUREMENT

The need for greater precision in measurement led to the concept of fractions. For example, "the thickness is $\frac{3}{4}$ in." is a more precise statement than "the thickness is between 0 and 1 in." In the first measurement, the spaces between the inch marks on the scale were likely subdivided into quarters; on the second scale, there were no subdivisions. In the metric system, the subdivisions are multiples of 10 and all measurements are expressed as *decimal* fractions. In the British system, the subdivisions are *not* multiples of 10 and the measurements are usually recorded as common fractions. The universal use of the metric system would greatly simplify the making and recording of measurements. However, common fractions would still be necessary for algebraic operations.

3.2 TERMS

Fraction: numbers of the form $\frac{3}{4}, \frac{1}{2}, \frac{6}{5}, \frac{7}{10}$ are called fractions. The line separating the two integers indicates division.

A *numerator* (or dividend) is the integer above a fraction line.

A *denominator* (or divisor) is the integer below a fraction line. The denominator in a measurement may show the number of subdivisions in a unit.

A *common fraction* may have for its denominator numbers other than 10, 100, 1000, etc. Other names for it are simple fraction and vulgar fraction.

EXAMPLES: $\frac{1}{125}, \frac{2}{7}, \frac{15}{32}$

A *decimal fraction* has for its denominator 10 to some power. Decimal fractions were discussed in Chapter 1.

The *terms of a fraction* are its numerator and denominator.

A *mixed number* is a combination of an integer and a fraction. For example, the mixed number $3\frac{2}{5}$ indicates the addition of an integer and a fraction: $3 + \frac{2}{5}$.

3.3 READING AND WRITING FRACTIONS

To read a common fraction read the numerator as an integer with the denominator read as a fraction of a unit. To read a mixed number read the integer, include "and", then read the fraction.

EXAMPLE 3.1. $\frac{1}{2}$ is read: one half.

$\dfrac{3}{4}$ is read: three fourths or three quarters.

$\dfrac{5}{8}$ is read: five eighths.

$\dfrac{24}{125}$ is read: twenty-four one hundred twenty-fifths.

$14\frac{2}{5}$ is read: fourteen and two fifths.

EXAMPLE 3.2.   Three halves is written $\dfrac{3}{2}$.

Five sixths is written $\dfrac{5}{6}$.

One thirty-second is written $\dfrac{1}{32}$.

Twelve three hundred sixty-sixths is written $\dfrac{12}{366}$.

Ninety-five and four eighty-thirds is written $95\dfrac{4}{83}$.

Since a fraction represents a quotient, $\dfrac{1}{0}$ is not defined or has no meaning.   The fraction line is frequently represented by /.   Thus $3/4 = \dfrac{3}{4} = 3 \div 4$.

3.4   BASIC PRINCIPLE

A fundamental principle used in work with fractions is: If both the numerator and denominator of a fraction are multiplied or divided by the same nonzero number, the value of the fraction is unchanged.

Another way of expressing this rule is: If a fraction is multiplied by 1, the value of the fraction remains unchanged.

EXAMPLE 3.3.

$$\frac{2}{3} = \frac{2 \cdot 2}{3 \cdot 2} = \frac{4}{6} \qquad \frac{2}{3} = \frac{2 \cdot 5}{3 \cdot 5} = \frac{10}{15} \qquad \frac{2}{3} = \frac{2 \cdot 6}{3 \cdot 6} = \frac{12}{18}$$

The numerator and denominator of the fraction $\dfrac{2}{3}$ were multiplied by 2, 5, and 6, respectively. The fractions $\dfrac{2}{3}, \dfrac{4}{6}, \dfrac{10}{15}, \dfrac{12}{18}$ are equal and are said to be in *equivalent forms*.

It can be seen that in the above example the change of a fraction to an equivalent form implies that the fraction was multiplied by 1.   Thus, the multipliers $\dfrac{2}{2}, \dfrac{5}{5}, \dfrac{6}{6}$ of the fraction $\dfrac{2}{3}$ in this case are each equal to 1.

3.5   REDUCTION TO LOWEST TERMS

The basic principle given above allows us to simplify fractions by dividing out any factors which the numerator and denominator of a fraction may have in common.   When this has been done, the fraction is in reduced form, or reduced to its lowest terms.   This is a simpler and more convenient form for fractional answers.   The process of reduction is also called cancellation.

EXAMPLE 3.4.   Since $\frac{18}{30} = (2 \cdot 3 \cdot 3)/(2 \cdot 3 \cdot 5)$, the numerator and denominator can be divided by the common factors 2 and 3.   The resulting fraction is $\frac{3}{5}$, which is the reduced form of the fraction $\frac{18}{30}$.

Factoring

The process of factoring is very useful in operations involving fractions.

If an integer greater than 1 is not divisible by any positive integer except itself and 1, the number is said to be prime.   Thus, 2, 3, 5, 7, etc., are prime numbers, the number 2 being the only even prime number.

If a number is expressed as the product of certain of its divisors, these divisors are known as factors of the representation.   The prime factors of small numbers are easily found by inspection.   Thus, the prime factors of 30 are 2, 3, 5.

The following example illustrates a system which can be used to find the prime factors of a large number.

EXAMPLE 3.5.   Find the prime factors of 1386.

Try to divide 1386 by each of the small prime numbers, beginning with 2.   Thus, $1386 \div 2 = 693$.   Since 693 is not divisible by 2, try 3 as a divisor: $693 \div 3 = 231$.   Try 3 again: $231 \div 3 = 77$.   Try 3 again; it is not a divisor of 77, and neither is 5.   However, 7 is a divisor, for $77 \div 7 = 11$.   The factorization is complete since 11 is a prime number.   Thus, $1386 = 2 \cdot 3 \cdot 3 \cdot 7 \cdot 11$.   The results of the successive divisions might be kept in a compact table as shown below.

Table 3.1.

Dividends	1386	693	231	77	11
Factors	2	3	3	7	11

The following divisibility rules simplify factoring:

Rule 1.   A number is divisible by 2 if its last digit is even.

EXAMPLE 3.6.   The numbers 64, 132, 390 are each exactly divisible by 2.

Rule 2.   A number is divisible by 3 if the sum of its digits is divisible by 3.

EXAMPLE 3.7.   Consider the numbers 270, 321, 498.   The sums 9, 6, 21 of the digits are divisible by 3. Therefore, the given numbers are exactly divisible by 3.

Rule 3.   A number is divisible by 5 if its last digit is 5 or zero.

EXAMPLE 3.8.   The numbers 75, 135, 980 are each divisible by 5.

Rule 4.   A number is divisible by 9 if the sum of its digits is divisible by 9.

EXAMPLE 3.9. The numbers 432, 1386, and 4977 are exactly divisible by 9 since the sums of their digits are 9, 18, 27, and these are divisible by 9.

3.6 COMPARISON OF FRACTIONS

It is not always easy to compare by inspection the magnitudes of two common fractions. For example, it is not immediately obvious which is larger, $\frac{3}{4}$ or $\frac{8}{11}$. If the two fractions have equal denominators, then the fraction with the larger numerator is the larger. Thus, $\frac{5}{8}$ is larger than $\frac{4}{8}$. The *least common multiple* (LCM) of a set of integers is the smallest integer that is divisible by each member of the set. The LCM is found by using each prime factor the largest number of times that it occurs in any of the given numbers.

EXAMPLE 3.10. Determine the LCM of the numbers 180, 360, and 450.

We write
$$180 = 2 \cdot 2 \cdot 3 \cdot 3 \cdot 5$$
$$360 = 2 \cdot 2 \cdot 2 \cdot 3 \cdot 3 \cdot 5$$
$$450 = 2 \cdot 3 \cdot 3 \cdot 5 \cdot 5$$

Reference to the factorization of these numbers shows that the LCM is $2 \cdot 2 \cdot 2 \cdot 3 \cdot 3 \cdot 5 \cdot 5$ or 1800.

The *least common denominator* (LCD) of two or more fractions is the least common multiple of their denominators.

EXAMPLE 3.11. Write the fractions $\frac{3}{8}, \frac{1}{2}$, and $\frac{5}{18}$ with a least common denominator.

Write the denominators as the product of their prime factors, omitting 1.
$$8 = 2 \cdot 2 \cdot 2 \qquad 2 = 2 \qquad 18 = 2 \cdot 3 \cdot 3$$

The factors occurring the largest number of times are $2 \cdot 2 \cdot 2$ in 8 and $3 \cdot 3$ in 18. Therefore the LCD $= 2 \cdot 2 \cdot 2 \cdot 3 \cdot 3 = 72$.

Therefore, the three fractions may be written in an equivalent form by using the basic principle.
$$\frac{3 \cdot 9}{8 \cdot 9} = \frac{27}{72} \qquad \frac{1 \cdot 36}{2 \cdot 36} = \frac{36}{72} \qquad \frac{5 \cdot 4}{18 \cdot 4} = \frac{20}{72}$$

Note that the number by which the numerator and denominator were multiplied is obtained by dividing the original denominator into the LCD. For example, in the fraction $\frac{3}{8}$, the multiplier 9 was obtained as the quotient of $72 \div 8$.

EXAMPLE 3.12. Compare the magnitude of the fractions $\frac{4}{9}$ and $\frac{7}{15}$.

Factoring the denominators, we get
$$9 = 3 \cdot 3 \qquad 15 = 3 \cdot 5$$

Therefore, LCD $= 3 \cdot 3 \cdot 5 = 45$. The fractions can be written in equivalent form as
$$\frac{4 \cdot 5}{9 \cdot 5} = \frac{20}{45} \qquad \text{and} \qquad \frac{7 \cdot 3}{15 \cdot 3} = \frac{21}{45}$$

Since $\frac{21}{45}$ is larger than $\frac{20}{45}, \frac{7}{15}$ is larger than $\frac{4}{9}$.

Operations with Fractions

3.7 ADDITION OF FRACTIONS

To add fractions which have a common denominator, add their numerators and keep the same common denominator.

EXAMPLE 3.13. Determine the sum of $\frac{1}{4}+\frac{5}{4}+\frac{3}{4}$.

Adding the numerators, $1+5+3=9$.

Keeping the same common denominator 4, the desired result is $\frac{9}{4}$ or $2\frac{1}{4}$.

To add fractions with unlike denominators, determine the least common denominator. Express each fraction in equivalent form with the LCD. Then perform the addition.

EXAMPLE 3.14. Determine $\frac{1}{2}+\frac{3}{4}+\frac{2}{8}$.

The LCD is 8. The sum with fractions in equivalent form is $\frac{4}{8}+\frac{6}{8}+\frac{2}{8}$. Adding the numerators and keeping the same LCD, the result is $\frac{12}{8}$ or $\frac{3}{2}$.

To add mixed numbers, calculate the sum of the integers separately from the sum of the fractions. Then add the two sums.

EXAMPLE 3.15. Add the mixed numbers $7\frac{1}{6}$ and $2\frac{3}{4}$.

First we add the integers: $7+2=9$. Adding the fractions, we get $\frac{1}{6}+\frac{3}{4}=\frac{2}{12}+\frac{9}{12}=\frac{11}{12}$. Combining the two sums, the answer we obtain is $9\frac{11}{12}$.

3.8 SUBTRACTION OF FRACTIONS

As with addition, before two fractions can be subtracted they must have a common denominator.

EXAMPLE 3.16. Subtract $\frac{2}{3}$ from $\frac{3}{4}$.

The LCD is 12. The subtraction with the fractions in equivalent form gives

$$\frac{3\cdot3}{4\cdot3}-\frac{2\cdot4}{3\cdot4}=\frac{9}{12}-\frac{8}{12}=\frac{1}{12}$$

EXAMPLE 3.17. Perform the operation $7\frac{1}{5}-4\frac{5}{10}$.

The LCD of the fractions is 10. The desired difference can be expressed as $7\frac{2}{10}-4\frac{5}{10}$. But $\frac{5}{10}$ is larger than $\frac{2}{10}$. Therefore, borrowing 1 or $\frac{10}{10}$ from 7 and adding to $\frac{2}{10}$, and performing the operation, we obtain $6\frac{12}{10}-4\frac{5}{10}=2\frac{7}{10}$.

3.9 MULTIPLICATION OF FRACTIONS

To multiply two or more fractions multiply their numerators to obtain the numerator of the product. Multiply the denominators to obtain the denominator of the product.

EXAMPLE 3.18. Multiply $\frac{3}{7} \times \frac{2}{5}$.

The desired product is $\frac{3 \times 2}{7 \times 5} = \frac{6}{35}$.

The calculations are greatly simplified, and the product put in its lowest terms, by cancellation of common factors.

EXAMPLE 3.19. Determine the product $(2/5)(10/3)(6/7)$.

Express all the numerators and denominators in factored form, cancel the common factors, and carry out the multiplication of the remaining factors. The result is

$$\frac{2}{\cancel{5}} \times \frac{2 \cdot \cancel{5}}{\cancel{3}} \times \frac{2 \cdot \cancel{3}}{7} = \frac{8}{7} = 1\frac{1}{7}$$

Every integer can be written as a fraction with 1 as a denominator, since an integer divided by 1 is equal to the integer. Thus, to multiply an integer by a fraction, multiply the numerator of the fraction by the integer and retain the denominator of the given fraction.

EXAMPLE 3.20. Determine $\frac{3}{5} \times 6$.

Applying the rule, $\frac{3}{6} \times \frac{6}{1} = \frac{3 \times 6}{5} = \frac{18}{5} = 3\frac{3}{5}$.

The same answer is obtained if the position of the numbers is reversed:

$$6 \times \frac{3}{5} = \frac{6 \times 3}{5} = \frac{18}{5} = 3\frac{3}{5}$$

To multiply mixed numbers, convert them first into fractions and then apply the rule for multiplication of fractions.

EXAMPLE 3.21. Multiply $1\frac{3}{4}$ by $3\frac{1}{7}$.

Converting $1\frac{3}{4}$ and $3\frac{1}{7}$ to common fractions, we get $1\frac{3}{4} = \frac{4}{4} + \frac{3}{4} = \frac{7}{4}$ and $3\frac{1}{7} = \frac{21}{7} + \frac{1}{7} = \frac{22}{7}$. On multiplication we obtain $(7 \cdot 2 \cdot 11)/(2 \cdot 2 \cdot 7)$. Canceling 7 and 2, we get $11/2$ as the answer. This fraction can be converted to a mixed number by division: $11 \div 2$ is 5 with a remainder of 1, so that $5\frac{1}{2}$ is the final answer.

3.10 DIVISION OF FRACTIONS

To divide one fraction by another, invert the divisor and multiply the fractions.

EXAMPLE 3.22. Divide $\frac{3}{5}$ by $\frac{4}{7}$.

Inverting the divisor, we get $\frac{7}{4}$. Multiplying the dividend by $\frac{7}{4}$, the result is $\frac{3}{5} \times \frac{7}{4}$ or $\frac{21}{20}$.

EXAMPLE 3.23. Divide $\frac{3}{5}$ by 7.

The division operation can be indicated by $\frac{3}{5} \div \frac{7}{1}$.

Applying the rule, the result is $\frac{3}{5} \div \frac{7}{1} = \frac{3}{5} \times \frac{1}{7} = \frac{3}{35}$.

To divide mixed numbers, convert them first to fractions and apply the rule for division.

Division by Zero

It is important to emphasize the following reminder: Division by 0 is impossible. Therefore, a fraction cannot have 0 as a denominator.

3.11 RECIPROCALS

The reciprocal of a number is 1 divided by the number. It follows from the definition that the product of a number and its reciprocal is 1. The reciprocal of zero is not defined.

EXAMPLE 3.24. The reciprocal of 25 is $\frac{1}{25}$ or 0.04. Conversely, the reciprocal of 0.04 is 1/0.04, or 25.

EXAMPLE 3.25. The reciprocal of 0.0125 is 1/0.0125 or 80. Conversely, the reciprocal of 80 is 1/80, or 0.0125.

EXAMPLE 3.26. From the sum of the reciprocals of 5 and 20 subtract the reciprocal of 25.

Expressing the numbers as reciprocals and performing the operations,

$$\frac{1}{5}+\frac{1}{20}-\frac{1}{25}=\frac{4}{20}+\frac{1}{20}-\frac{1}{25}=\frac{5}{20}-\frac{1}{25}=\frac{25}{100}-\frac{4}{100}=\frac{21}{100}$$

EXAMPLE 3.27. If a resistance of 20 ohms is connected in parallel with a resistance of 60 ohms, the equivalent resistance R is given in reciprocal form by $\frac{1}{R}=\frac{1}{20}+\frac{1}{60}$. Determine the equivalent resistance.

Adding the reciprocals of 20 and 60,

$$\frac{1}{R}=\frac{1}{20}+\frac{1}{60}=\frac{3}{60}+\frac{1}{60}=\frac{4}{60}=\frac{1}{15}$$

Since R is the reciprocal of $1/R$, $R = 1/\frac{1}{15} = 15$. Therefore, the equivalent resistance is 15 ohms.

3.12 CONVERSION OF COMMON FRACTIONS INTO DECIMAL FRACTIONS

To convert a common fraction into a decimal equivalent form, divide the numerator by the denominator.

EXAMPLE 3.28. Convert 7/16 into a decimal fraction.

Since 16 is larger than 7, there will be no digits except 0 to the left of the decimal point in the quotient. Then, performing the division, 0 is adjoined to 7 to give 70, which is larger than 16. Proceeding as with integers, the decimal equivalent of 7/16 is 0.4375.

EXAMPLE 3.29. Convert 1/273 into a decimal fraction.

Since 1 is smaller than 273, begin with a decimal point in the quotient. Adjoining a 0 to 1 gives 10, which is also smaller than 273, so a 0 is entered after the decimal point in the quotient. Adjoining another 0 makes the dividend 100, which is also smaller than 273, and therefore a second zero is adjoined after the decimal place in the quotient. Adjunction of a third 0 makes the dividend 1000, which is larger than the divisor, and ordinary division rules are used to give the quotient $1/273 \approx 0.003663$.

The method of converting decimal fractions to common fractions has few applications in elementary science and will not be given here. In industry, it is common to use a conversion table.

Solved Problems

3.1. Read the following fractions: (*a*) 2/3,   (*b*) 7/8,   (*c*) 13/21,   (*d*) $5\frac{37}{64}$,   (*e*) 9/250.

(*a*) two thirds   (*c*) thirteen twenty-firsts   (*e*) nine two hundred fiftieths

(*b*) seven eighths   (*d*) five and thirty-seven sixty-fourths

3.2. Write the following fractions: (*a*) four fifths,   (*b*) three halves,   (*c*) eleven sixteenths,   (*d*) twenty-three and forty-five hundredths,   (*e*) one hundred thirty-seven five-hundreths.

(*a*) 4/5   (*b*) 3/2   (*c*) 11/16   (*d*) $23\frac{45}{100}$ (*e*) 137/500

3.3. Convert the following fractions to higher terms as indicated.

(*a*) 3/4 to twentieths   (*c*) 4/9 to sixty-thirds

(*b*) 2/5 to sixtieths   (*d*) 7/15 to one hundred twentieths

(*a*) Write $\frac{3}{4} = \frac{?}{20}$ because the denominator of the fraction after conversion is to be 20. The original denominator 4 has to be multiplied by 5 to give 20. Therefore, the original numerator 3 has to be multiplied by 5. The resulting fraction is $\frac{3 \cdot 5}{4 \cdot 5} = \frac{15}{20}$.

(*b*) $\frac{2}{5} = \frac{?}{60}$ or $\frac{2 \cdot 12}{5 \cdot 12} = \frac{24}{60}$

(*c*) $\frac{4}{9} = \frac{?}{63}$ or $\frac{4 \cdot 7}{9 \cdot 7} = \frac{28}{63}$

(*d*) $\frac{7}{15} = \frac{?}{120}$ or $\frac{7 \cdot 8}{15 \cdot 8} = \frac{56}{120}$

3.4. Reduce: (*a*) 12/16,   (*b*) 200/240,   (*c*) 108/135,   (*d*) 294/336 to lowest terms.

(*a*) Factor the numerator and denominator and divide out their common factors:

$$\frac{12}{16} = \frac{\cancel{2} \cdot \cancel{2} \cdot 3}{\cancel{2} \cdot \cancel{2} \cdot 2 \cdot 2} = \frac{3}{4}$$

(*b*) $\frac{200}{400} = \frac{\cancel{2} \cdot \cancel{2} \cdot \cancel{2} \cdot 5 \cdot \cancel{5}}{\cancel{2} \cdot \cancel{2} \cdot \cancel{2} \cdot 2 \cdot 3 \cdot \cancel{5}} = \frac{5}{6}$

(*c*) $\frac{108}{135} = \frac{2 \cdot 2 \cdot \cancel{3} \cdot \cancel{3} \cdot \cancel{3}}{\cancel{3} \cdot \cancel{3} \cdot \cancel{3} \cdot 5} = \frac{4}{5}$

(*d*) $\frac{294}{336} = \frac{\cancel{2} \cdot \cancel{3} \cdot 7 \cdot \cancel{7}}{\cancel{2} \cdot 2 \cdot 2 \cdot 2 \cdot \cancel{3} \cdot \cancel{7}} = \frac{7}{8}$

3.5. Express each of the following numbers as a product of prime factors:

(*a*) 48,   (*b*) 186,   (*c*) 735,   (*d*) 1260.

Divide the number by the lowest prime numbers, beginning with 2 if possible. Continue dividing the successive quotients until the remainder is a prime number. The divisors with the remainder will be the required factors.

(a) $2\overline{)48}$

$2\overline{)24}$

$2\overline{)12}$

$2\overline{)\ 6}$

$\quad 3$ The prime factors of 48 are $2 \cdot 2 \cdot 2 \cdot 2 \cdot 3$.

(b) $2\overline{)186}$

$3\overline{)\ 93}$

$\quad 31$ The prime factors of 186 are $2 \cdot 3 \cdot 31$.

(c) $3\overline{)735}$

$5\overline{)245}$

$7\overline{)\ 49}$

$\quad 7$ The prime factors of 735 are $3 \cdot 5 \cdot 7 \cdot 7$.

(d) $2\overline{)1360}$

$2\overline{)\ 630}$

$3\overline{)\ 315}$

$3\overline{)\ 105}$

$5\overline{)\ \ 35}$

$\quad 7$ The prime factors of 1260 are $2 \cdot 2 \cdot 3 \cdot 3 \cdot 5 \cdot 7$.

3.6. Without using longhand division determine whether the following numbers are divisible by 2, 3, 5, 6, 9, or 10.   (a) 72,   (b) 135,   (c) 576,   (d) 1250.

(a) Divisible by 2, 3, 6, and 9 since the last digit is even and the sum of the digits is divisible by 3 and by 9.

(b) Divisible by 3, 5, and 9 since the sum of the digits is divisible by 3 and by 9 and the last digit is 5.

(c) Divisible by 2, 3, 6, and 9, since the last digit is even and the sum of the digits is divisible by 3 and by 9.

(d) Divisible by 2, 5, and 10 since the last digit is zero.

3.7. Which of the two fractions is the larger?   (a) 4/5 or 5/6,   (b) 11/12 or 29/32.

(a) The least common denominator is $5 \cdot 6 = 30$.

$$\frac{4}{5} = \frac{4 \cdot 6}{5 \cdot 6} = \frac{24}{30} \qquad \text{and} \qquad \frac{5}{6} = \frac{5 \cdot 5}{6 \cdot 5} = \frac{25}{30}$$

The numerator of the second fraction is larger than the numerator of the first. Therefore $\frac{5}{6}$ is the larger of the two.

(b) Resolve 12 and 32 into products of prime factors:

$$12 = 2 \cdot 2 \cdot 3 \qquad 32 = 2 \cdot 2 \cdot 2 \cdot 2 \cdot 2 \qquad \text{LCD} = 3 \cdot 2 \cdot 2 \cdot 2 \cdot 2 \cdot 2 = 96$$

The two fractions become

$$\frac{11 \cdot 8}{12 \cdot 8} = \frac{88}{96} \qquad \frac{29 \cdot 3}{32 \cdot 3} = \frac{87}{96}$$

$\dfrac{11}{12}$ is larger than $\dfrac{29}{32}$.

3.8. Determine the least common multiple (LCM) of:

(*a*) 12, 16 and 30,   (*b*) 120 and 270,   (*c*) 108, 180, 252,   (*d*) 75, 630, 400, 225.

Determine the prime factors of each given number.   Take the product of the factors, using each factor the greatest numbers of times that it occurs in any number of the set.

(*a*) $12 = 2 \cdot 2 \cdot 3$ $16 = 2 \cdot 2 \cdot 2 \cdot 2$ $30 = 2 \cdot 3 \cdot 5$

  $\text{LCM} = 2 \cdot 2 \cdot 2 \cdot 2 \cdot 3 \cdot 5 = 240$

(*b*) $120 = 2 \cdot 2 \cdot 2 \cdot 3 \cdot 5$ $270 = 2 \cdot 3 \cdot 3 \cdot 3 \cdot 5$

  $\text{LCM} = 2 \cdot 2 \cdot 2 \cdot 3 \cdot 3 \cdot 3 \cdot 5 = 1080$

(*c*) $108 = 2 \cdot 2 \cdot 3 \cdot 3 \cdot 3$ $180 = 2 \cdot 2 \cdot 3 \cdot 3 \cdot 5$ $252 = 2 \cdot 2 \cdot 3 \cdot 3 \cdot 7$

  $\text{LCM} = 2 \cdot 2 \cdot 3 \cdot 3 \cdot 3 \cdot 5 \cdot 7 = 3780$

(*d*) $75 = 3 \cdot 5 \cdot 5$ $630 = 2 \cdot 3 \cdot 3 \cdot 5 \cdot 7$ $400 = 2 \cdot 2 \cdot 2 \cdot 2 \cdot 5 \cdot 5$ $225 = 5 \cdot 5 \cdot 3 \cdot 3$

  $\text{LCM} = 2 \cdot 2 \cdot 2 \cdot 2 \cdot 3 \cdot 3 \cdot 5 \cdot 5 \cdot 7 = 25{,}200$

3.9. Change the following mixed numbers to fractions: (*a*) $2\frac{1}{3}$,   (*b*) $4\frac{5}{8}$,   (*c*) $1\frac{33}{64}$,   (*d*) $20\frac{4}{5}$.

Multiply each integer by the denominator of the adjoining fraction.   Add the product to the numerator of the fraction and divide the sum by the denominator.

(*a*) $\dfrac{(2 \cdot 3) + 1}{3} = \dfrac{6 + 1}{3} = \dfrac{7}{3}$

(*b*) $\dfrac{(4 \cdot 8) + 5}{8} = \dfrac{32 + 5}{8} = \dfrac{37}{8}$

(*c*) $\dfrac{(1 \cdot 64) + 33}{64} = \dfrac{97}{64}$

(*d*) $\dfrac{(20 \cdot 5) + 4}{5} = \dfrac{104}{5}$

3.10. Add: (*a*) 3/8 and 7/16;   (*b*) 2/3, 3/4 and 1/5,   (*c*) $3\frac{5}{8}$ and $5\frac{1}{32}$,   (*d*) 8, $2\frac{2}{3}$, $1\frac{3}{4}$, and $3\frac{1}{2}$.

(*a*) $\text{LCD} = 16$

$$\frac{3 \cdot 2}{8 \cdot 2} + \frac{7}{16} = \frac{6}{16} + \frac{7}{16} = \frac{13}{16}$$

(*b*) $\text{LCD} = 3 \cdot 4 \cdot 5 = 60$

$$\frac{2 \cdot 20}{3 \cdot 20} = \frac{3 \cdot 15}{4 \cdot 15} + \frac{1 \cdot 12}{5 \cdot 12} = \frac{40}{60} + \frac{45}{60} + \frac{12}{60} = \frac{97}{60} = 1\frac{37}{60}$$

(c) First add the integers: $3 + 5 = 8$. LCD $= 32$. Add the fractions: $\dfrac{5 \cdot 4}{8 \cdot 4} + \dfrac{1}{32}$.

$$\frac{20}{32} + \frac{1}{32} = \frac{21}{32}$$

Add the two sums to get the result $8 + \dfrac{21}{32} = 8\frac{21}{32}$.

(d) $8 + 2 + 1 + 3 = 14$

$$\frac{2}{3} + \frac{3}{4} + \frac{1}{2} = \frac{2 \cdot 4}{3 \cdot 4} + \frac{3 \cdot 3}{4 \cdot 3} + \frac{1 \cdot 6}{2 \cdot 6} = \frac{8}{12} + \frac{9}{12} + \frac{6}{12} = \frac{23}{12} = 1\frac{11}{12}$$

Add the two sums to get the result $14 + 1\frac{11}{12} = 15\frac{11}{12}$.

3.11. Determine: (a) $5/8 - 7/12$, (b) $42\frac{3}{4} - 27\frac{2}{5}$.

(a) Determine the LCD, and express each fraction with the LCD as its denominator

$$8 = 2 \cdot 2 \cdot 2 \qquad 12 = 2 \cdot 2 \cdot 3 \qquad \text{LCD} = 2 \cdot 2 \cdot 2 \cdot 3 = 24$$

$$\frac{5 \cdot 3}{8 \cdot 3} - \frac{7 \cdot 2}{12 \cdot 2} = \frac{15 - 14}{24} = \frac{1}{24}$$

(b) Subtract separately the integers and then the fractions. Then add the two results.

$$42 - 27 = 15$$

$$\frac{3}{3} - \frac{2}{5} = \frac{3 \cdot 5}{4 \cdot 5} - \frac{2 \cdot 4}{5 \cdot 4} = \frac{15 - 8}{20} = \frac{7}{20}$$

The required result is $15 + \dfrac{7}{20} = 15\frac{7}{20}$.

3.12. From the sum of $7\frac{7}{8}$ and $3\frac{3}{4}$ subtract the sum of $2\frac{1}{5}$ and $4\frac{1}{2}$.

Add separately each pair of numbers. Then subtract the second sum from the first.

$$(7\tfrac{7}{8} + 3\tfrac{3}{4}) - (2\tfrac{1}{5} + 4\tfrac{1}{2}) = (7\tfrac{7}{8} + 3\tfrac{6}{8}) - (2\tfrac{2}{10} + 4\tfrac{5}{10})$$

$$= 10\tfrac{13}{8} - 6\tfrac{7}{10} = 11\tfrac{5}{8} - 6\tfrac{7}{10} = 11\tfrac{25}{40} - 6\tfrac{28}{40}$$

$$= 10\tfrac{65}{40} - 6\tfrac{28}{40} = 4\tfrac{37}{40}$$

3.13. Simplify by cancellation:

(a) $6 \times 2/3$ (b) $5.8 \times 4/15$ (c) $2\frac{1}{2} \times \dfrac{3}{15} \times 1\frac{1}{4}$ (d) $2\frac{3}{8} \times 16$

(a) $6 \times 2/3 = \dfrac{\overset{2}{\cancel{6}} \times 2}{\cancel{3}} = 4$

(b) $\dfrac{\overset{}{\cancel{5}} \times \cancel{4}}{\underset{2}{\cancel{2}} \times \underset{3}{\cancel{15}}} = \dfrac{1}{2 \times 3} = \dfrac{1}{6}$

(c) $2\frac{1}{2} = \dfrac{5}{2} \qquad 1\frac{1}{4} = \dfrac{5}{4}$

$\dfrac{5}{2} \times \dfrac{3}{15} \times \dfrac{5}{4} = \dfrac{5 \times \cancel{3} \times \cancel{5}}{2 \times \cancel{15} \times 4} = \dfrac{5}{8}$

(d) $2\frac{3}{8} = \frac{19}{8}$

$$\frac{19 \times \cancel{16}^{2}}{8} = 38$$

3.14. Multiply, and simplify if possible:

(a) $\frac{2}{3} \times \frac{3}{5} \times \frac{1}{4}$ (b) $\frac{3}{16} \times 9$ (c) $\frac{2}{3} \times 3\frac{9}{16}$ (d) $4\frac{2}{3} \times 12\frac{3}{5}$

(a) $\frac{\cancel{2} \times \cancel{3} \times 1}{\cancel{3} \times 5 \times \cancel{4}_{2}} = \frac{1}{10}$ (c) $\frac{\cancel{2} \times \cancel{57}^{19}}{\cancel{3} \times \cancel{16}_{8}} = \frac{19}{8} = 2\frac{3}{8}$

(b) $\frac{3 \times 9}{16} = \frac{27}{16} = 1\frac{11}{16}$ (d) $\frac{14}{\cancel{3}} \times \frac{\cancel{63}^{21}}{5} = \frac{294}{5} = 58\frac{4}{5}$

3.15. Divide and simplify:

(a) $\frac{5}{12}$ by $\frac{2}{3}$ (b) $\frac{9}{16}$ by 8 (c) $7 \div \frac{3}{4}$ (d) $\frac{4}{5} \div 2\frac{7}{15}$

Invert the divisor and multiply the resulting fractions.   Cancel common factors where possible.

(a) $\frac{5}{\cancel{12}_{4}} \times \frac{\cancel{3}}{2} = \frac{5}{8}$ (c) $\frac{7}{1} \times \frac{4}{3} = \frac{28}{3} = 9\frac{1}{3}$

(b) $\frac{9}{16} \times \frac{1}{8} = \frac{9}{128}$ (d) $\frac{4}{5} \div \frac{37}{15} = \frac{4}{\cancel{5}} \times \frac{\cancel{15}^{3}}{37} = \frac{12}{37}$

3.16. Simplify $2\frac{1}{2} \times \frac{7}{8} \div 3\frac{3}{4}$.

Convert the mixed numbers to fractions, invert the divisor fraction, and then multiply the three fractions.

$$\frac{5}{2} \times \frac{7}{8} \div \frac{15}{4} = \frac{\cancel{5}}{2} \times \frac{7}{\cancel{8}_{2}} \times \frac{\cancel{4}}{\cancel{15}_{3}} = \frac{7}{12}$$

3.17. Give the reciprocals of the following: (a) 12, (b) 1/6, (c) 5/4, (d) $3\frac{2}{5}$.

(a) $\frac{1}{12}$ (b) $\frac{1}{\frac{1}{6}} = 6$ (c) $\frac{1}{\frac{5}{4}} = \frac{4}{5}$ (d) $\frac{1}{3\frac{2}{5}} = \frac{1}{\frac{17}{5}} = \frac{5}{17}$

3.18. Determine the reciprocal of $\frac{3}{4} + \frac{2}{5}$.

Add the fractions and then invert.

$$\frac{3}{4} + \frac{2}{5} = \frac{3 \cdot 5}{4 \cdot 5} + \frac{2 \cdot 4}{5 \cdot 4} = \frac{15}{20} + \frac{8}{20} = \frac{23}{20}$$

The desired result is $\frac{1}{\frac{23}{20}} = \frac{20}{23}$.

3.19. Find the *difference* between $3\frac{1}{8}$ and its reciprocal.

$$3\frac{1}{8} - \frac{1}{3\frac{1}{8}} = \frac{25}{8} - \frac{1}{\frac{25}{8}} = \frac{25}{8} - \frac{8}{25} = \frac{25 \cdot 25}{8 \cdot 25} - \frac{8 \cdot 8}{25 \cdot 8}$$

$$= \frac{625}{200} - \frac{64}{200} = \frac{561}{200} = 2\frac{161}{200}$$

3.20. Simplify $(4\frac{1}{2} \div 2) + 6\frac{1}{4}$.

First do the division, then the addition.

$$4\frac{1}{2} \div 2 = \frac{9}{2} \div 2 = \frac{9}{2} \times \frac{1}{2} = \frac{9}{4}$$

$$\frac{9}{4} + 6\frac{1}{4} = \frac{9}{4} + \frac{25}{4} = \frac{\overset{17}{\cancel{34}}}{\underset{2}{\cancel{4}}} = \frac{17}{2} = 8\frac{1}{2}$$

Supplementary Problems

The answers to this set of problems are at the end of this chapter.

3.21. Read the following fractions: (*a*) 5/6, (*b*) 17/32, (*c*) 11/4, (*d*) $3\frac{9}{128}$.

3.22. Write the following fractions:

    (*a*) five ninths   (*c*) twelve and seven sixty-fourths

    (*b*) fifteen sixteenths   (*d*) two hundred ninety-three two hundred seventy-thirds

3.23. Change the following fractions to the indicated higher terms:

    (*a*) 9/10 to thirtieths   (*c*) 3/7 to twenty-eighths

    (*b*) 2/3 to twenty-fourths   (*d*) 5/6 to seventy-seconds

3.24. Reduce the following fractions to lowest terms:

    (*a*) $\dfrac{36}{27}$ (*b*) $\dfrac{60}{108}$ (*c*) $\dfrac{128}{224}$ (*d*) $\dfrac{336}{378}$

3.25. Find the prime factors (with multiplicities) of the following numbers:

    (*a*) 72   (*b*) 252   (*c*) 495   (*d*) 1040

3.26. By inspection, determine the divisibility of the given numbers by 2, 3, 4, 5, 6, 9, and 10.

    (*a*) 84   (*b*) 240   (*c*) 1512   (*d*) 2205

3.27. Find the least common multiple of the following:

    (*a*) 18, 27, 36   (*b*) 32, 48, 72   (*c*) 600 and 960   (*d*) 75, 225, 630

3.28. Change the following mixed numbers to fractions:

(a) $7\frac{5}{8}$,   (b) $2\frac{9}{32}$,   (c) $5\frac{3}{7}$,   (d) $16\frac{2}{3}$.

3.29. Change the following fractions to mixed numbers or integers:

(a) $\dfrac{25}{7}$ (b) $\dfrac{56}{8}$ (c) $\dfrac{86}{13}$ (d) $\dfrac{660}{25}$

3.30. Determine the sum of the following numbers and simplify the answers:

(a) 4/7 and 3/10   (b) 1/2, 5/6 and 3/15   (c) $5\frac{3}{4}$ and $3\frac{2}{5}$ (d) $12\frac{1}{3}$, $1\frac{1}{2}$, and $3\frac{5}{8}$

3.31. Subtract and simplify the answers to mixed numbers:

(a) 2/3 from 4/5   (b) $16\frac{3}{16} - 5\frac{7}{8}$ (c) $12\frac{1}{3} - 10\frac{2}{7}$ (d) $6\frac{7}{20} - 2\frac{3}{4}$

3.32. Multiply each of the following numbers by 3 and simplify the answer to a mixed number:

(a) 5/8   (b) 4/9   (c) 6/7   (d) $4\frac{1}{2}$

3.33. Divide each number by 5 and simplify the answer to a mixed number (where possible):

(a) 3/4   (b) 4/7   (c) 5/3   (d) $8\frac{1}{3}$

3.34. Multiply the following and simplify if possible:

(a) 3/5 by 20   (b) $5/8 \times 4/15$ (c) $8\frac{5}{8}$ by $\dfrac{2}{3}$ (d) $6\frac{2}{3} \div 2\frac{3}{4}$

3.35. Divide the following and simplify if possible:

(a) 9/32 by 3/16   (b) 5/3 by 5   (c) $46 \div 3\frac{2}{7}$ (d) $3\frac{2}{5} \div 2\frac{1}{8}$

3.36. Given the fractions 2/5, 1/4, 3/8:

(a) Determine their sum.

(b) Determine their product.

(c) Subtract the third from the sum of the first two.

(d) Find the sum of the first and third, and then divide this sum by the second.

3.37. (a) Simplify $1/20 + 1/40 + 1/50$.

(b) Invert the sum in (a) and simplify the answer.

3.38. Find the reciprocals of:

(a) 125   (b) $1\frac{5}{8}$ (c) $\dfrac{323}{373}$ (d) $\dfrac{15}{32}$

3.39. To the positive difference of the reciprocals of 4 and 12 add the reciprocal of 30.

3.40. Convert the following to decimals with three significant figures:

(a) 5/8   (b) 11/13   (c) 318/273   (d) 3937/3600

Answers to Supplementary Problems

3.21.   (*a*) five sixths   (*c*) eleven fourths

       (*b*) seventeen thirty-seconds   (*d*) three and nine one hundred twenty-eighths

3.22.   (*a*) 5/9   (*b*) 15/16   (*c*) $12\frac{7}{64}$ (*d*) $\dfrac{293}{273}$

3.23.   (*a*) 27/30   (*b*) 16/24   (*c*) 12/28   (*d*) 60/72

3.24.   (*a*) 4/3   (*b*) 5/9   (*c*) 4/7   (*d*) 8/9

3.25.   (*a*) 2, 2, 2, 3, 3   (*b*) 2, 2, 3, 3, 7   (*c*) 3, 3, 5, 11   (*d*) 2, 2, 2, 2, 5, 13

3.26.   (*a*) 2, 3, 4, 6   (*b*) 2, 3, 4, 5, 6, 10   (*c*) 2, 3, 4, 6, 9   (*d*) 3, 5, 9

3.27.   (*a*) 108   (*b*) 288   (*c*) 4800   (*d*) 3150

3.28.   (*a*) 61/8   (*b*) 73/32   (*c*) 38/7   (*d*) 50/3

3.29.   (*a*) $3\frac{4}{7}$ (*b*) 7   (*c*) $6\frac{8}{13}$ (*d*) $26\frac{2}{5}$

3.30.   (*a*) 61/70   (*b*) $1\frac{8}{15}$ (*c*) $9\frac{3}{20}$ (*d*) $17\frac{11}{24}$

3.31.   (*a*) 2/15   (*b*) $10\frac{5}{16}$ (*c*) $2\frac{1}{21}$ (*d*) $3\frac{3}{5}$

3.32.   (*a*) $1\frac{7}{8}$ (*b*) $1\frac{1}{3}$ (*c*) $2\frac{4}{7}$ (*d*) $13\frac{1}{2}$

3.33.   (*a*) 3/20   (*b*) 4/35   (*c*) 1/3   (*d*) $1\frac{2}{3}$

3.34.   (*a*) 12   (*b*) 1/6   (*c*) $5\frac{3}{4}$ (*d*) $18\frac{1}{3}$

3.35.   (*a*) $1\frac{1}{2}$ (*b*) 1/3   (*c*) 14   (*d*) $1\frac{3}{5}$

3.36.   (*a*) $1\frac{1}{40}$ (*b*) 3/80   (*c*) 11/40   (*d*) $3\frac{1}{10}$

3.37.   (*a*) 19/200   (*b*) $10\frac{10}{19}$

3.38.   (*a*) 1/125   (*b*) 8/13   (*c*) 373/323   (*d*) 32/15 or $2\frac{2}{15}$

3.39.   (*a*) 1/5

3.40.   (*a*) 0.625   (*b*) 0.846   (*c*) 1.16   (*d*) 1.09

Chapter 4

Percentage

Basic Concepts

4.1   TERMS

Percent means by the hundred; also, in or for every hundred, that is, *hundredths*.   If we consider that a quantity is subdivided into 100 equal parts, a certain percent of the quantity is the number of these hundredth parts involved.   The word *rate* is sometimes used for percent.

Percentage is the result of taking a specified percent of a quantity.   The process of using percent in calculations is also called percentage.

Base is the number of which a given percent is calculated.   It is the quantity which is divided into 100 parts, when percent is involved.

4.2   PERCENT AND FRACTIONS

Percent is a special type of a fraction with 100 as a denominator.   The symbol for percent is %.   Thus 20 percent (written 20%) means 20 hundredths or 20/100; 60 percent (written 60%) means 60 hundredths or 60/100.   A percent may always be reduced to a fraction and then this fraction may in turn be expressed in decimal form.   Thus 25 percent represents the same fractional measure as 25/100 or 1/4 or the same decimal measure as 0.25.   A fractional measure can always be expressed as a percent by writing an equivalent fraction with 100 as its denominator.   Thus $3/5 = (3 \cdot 20)/(5 \cdot 20) = 60/100$, and so 3/5 is the same fractional measure as 60 percent.   Since a decimal may be writen as a common fraction, it is also possible to express a decimal fraction as an equivalent percent.   Thus, $0.24 = 24/100 = 24$ percent; and $1.15 = 115/100 = 115$ percent.

The rules for converting a decimal fraction to a percent and vice versa can be summarized as follows:

Rule 1.   To convert a decimal fraction to equivalent percent, multiply the decimal fraction by 100.   This means that the decimal point has to be moved two places to the right.

Rule 2.   To convert a percent to a decimal fraction, divide the percent by 100.   This means that the decimal place has to be moved two places to the left.

EXAMPLE 4.1.   Express 2.15 as a percent.

Move the decimal point two places to the right:

$$2.15 = 215.\% = 215\%$$

Note that percent may be more than 100.   This means that more than one complete unit is involved.

EXAMPLE 4.2.   Express 12 percent as a decimal.

Move the decimal point two places to the left:

$$12\% = 0.12$$

To convert a common fraction to percent, convert the fraction so that its denominator is 100.   The numerator of the new fraction is the percent.   The common fraction may also be converted into its decimal equivalent and Rule 1 above applied.

51

EXAMPLE 4.3.   Express 6/25 as a percent.

Multiply the fraction by 4/4.

$$\frac{6}{25} \times \frac{4}{4} = \frac{24}{100} = 24\%$$

EXAMPLE 4.4.   Express 3/7 as a percent.

Converting 3/7 to its decimal equivalent,

$$3/7 = 0.42857\ldots \approx 42.86\%$$

Types of Percentage Problems

All percentage problems can be reduced to three types:

1. calculating the percent of a given quantity;

2. determining the rate or percent;

3. computing the base or the quantity.

4.3   PERCENT OF A QUANTITY

To calculate the percent of a given quantity, express the given percent as a common or decimal fraction and multiply it by the quantity (base).

EXAMPLE 4.5.   What is 12 percent of 70?

Since $12\% = 12/100 = 3/25$, it follows that 12% of 70 is $(3/25)70 = 42/5 = 8\frac{2}{5}$.

Alternate Solution

Since $12\% = 0.12$, it follows that 12% of 70 is $(0.12)(70) = 8.4$.

4.4   PERCENT FROM PERCENTAGE

To determine what percent one quantity is of another, divide the percentage by the base.

EXAMPLE 4.6.   72 is what percent of 120?

The base is 120 and the percentage is 72. 72 is $\frac{72}{120}$ of 120, that is $\frac{3}{5}$ of 120, and $\frac{3}{5} = \frac{3 \times 20}{5 \times 20} = 60/100 = 60\%$

Alternate Solution

72 is $\frac{72}{120}$ of 120; $\frac{72}{120} = 0.6$.   But $0.6 = 0.60 = \frac{60}{100} = 60\%$.

4.5   QUANTITY FROM PERCENTAGE AND PERCENT

To calculate the base, divide the percentage by the rate and multiply by 100.

EXAMPLE 4.7.   24 is 15% of what number?

Since 15% of the number is 24, then 1% is $24/15 = 1.6$.   Therefore 100% of the number is $(1.6)(100) = 160$.

Experimental Error or Uncertainty

It is usual to express the relative error in measurements (see Chapter 2) as a percent.   The accuracy or uncertainty of a measuring instrument is also stated by the manufacturer in percent.

4.6   PERCENT ERROR OR UNCERTAINTY

The percent error in an experiment is calculated from the following:

$$\% \text{ error} = \frac{\text{experimental value} - \text{accepted or ``true'' value}}{\text{accepted or ``true'' value}} \times 100$$

Percent error is a measure of experimental accuracy.

EXAMPLE 4.8.   The experimental value of the sound velocity determined by a student is 355 m/s.   The accepted value under the same conditions is 342 m/sec.   Calculate the percent error in the measurement.

$$\% \text{ error} = \frac{355 - 342}{342} \times 100$$

$$= \frac{13}{342} \times 100$$

$$\approx 3.8\%$$

4.7   PERCENT OF DIFFERENCE

Frequently one has to make experimental determinations of physical quantities for which there are no recorded accepted values.   A reasonable estimate of accuracy is obtained by finding the percent of difference between two determinations of the unknown quantity, using two independent methods.

EXAMPLE 4.9.   A resistor is measured by the voltmeter-ammeter method and by using a Wheatstone bridge: $R_{va} = 22.4$ ohms; $R_w = 22.2$ ohms.   Calculate the percentage difference between the two methods.

Since one can not tell from the data which of the values is more accurate, use the average as base.

$$R_{\text{average}} = (22.4 + 22.2)/2 = 22.3 \text{ ohms}$$

$$\% \text{ difference} = \frac{22.4 - 22.2}{22.3} \times 100 \approx 0.897\% \approx 0.9\%$$

4.8   UNCERTAINTIES IN SCALE READINGS

The fraction of the smallest scale subdivision estimated by the observer is recorded as the last significant figure in a measurement.   The observer's confidence in the limits of his estimate may be called the reading uncertainty of the instrument or the reading error or simply uncertainty.   For example, when an object's mass is recorded as 51.73 ± 0.01 g the observer is sure of the digits 51.7

but he estimates the last digit (3). His reading uncertainty of ± 0.01 g means that the slider on the beam balance is between 0.72 and 0.74, with the best estimate being 0.73.

The percent of uncertainty in reading an instrument is the ratio of the uncertainty to the measured value expressed in percent.

EXAMPLE 4.10. What is the percent reading uncertainty in the measurement, 1.67 ± 0.02 g?

The reading uncertainty is 0.02 g. Hence,

$$\% \text{ uncertainty} = \frac{0.02}{1.67} \times 100 \approx 1.2\%$$

EXAMPLE 4.11. A wattmeter has an accuracy of $\pm 2\%$ of the full scale reading. The scale range is 0 to 500 watts. If a measurement of 150 watts is made with this instrument, within what limits does the true value lie?

The measurement uncertainty of the wattmeter is 2% of 500 watts.

$$\text{wattmeter uncertainty} = \pm(0.02 \times 500) = \pm 10 \text{ watts}$$

The measured value will lie between $(150 - 10)$ watts and $(150 + 10)$ watts or between 140 and 160 watts and would be recorded as 150 ± 10 watts.

Solved Problems

4.1. Convert the following decimal fractions to percent forms:

(a) 0.125,   (b) 0.1,   (c) 3.67,   (d) 0.043,   (e) 0.0033

Multiply each decimal fraction by 100 (that is, move the decimal point two places to the right) and attach the % symbol.

(a) 12.5%   (b) 10%   (c) 367%   (d) 4.3%   (e) 0.33%

4.2. Convert each of the following percents to decimal fractions:

(a) 36%,   (b) 208%,   (c) 3%,   (d) 9.5%,   (e) 0.67%

Divide each percent by 100 (that is, move the decimal point two places to the left) and omit the % symbol.

(a) 0.36   (b) 2.08   (c) 0.03   (d) 0.095   (e) 0.0067

4.3. Change the following fractions to percent forms:

(a) 2/5,   (b) 3/8,   (c) 7/4,   (d) 23/10,000,   (e) 5/64

Convert the common fractions to their decimal equivalents. Then move their decimal points two places to the right and attach the % symbol.

(a) $2/5 = 0.4 = 40\%$ (d) $23/10,000 = 0.0023 = 0.23\%$

(b) $3/8 = 0.375 = 37.5\%$ (e) $5/64 \approx 0.0781 \approx 7.81\%$

(c) $7/4 = 1\frac{3}{4} = 1.75 = 175\%$

4.4. Determine:

    (*a*) 20% of 75   (*c*) 3/4% of 1600   (*e*) 350% of 0.018

    (*b*) 6% of 152   (*d*) 6.4% of 2.5

        Change the percents to decimal fractions.   Multiply the decimal fractions by the given quantites (bases).

    (*a*) 20% = 0.2   (*c*) 3/4% = 0.75% = 0.0075   (*e*) 350% = 3.5

         (0.2)(75) = 15   (0.0075)(1600) = 12   (3.5)(0.018) = 0.063

    (*b*) 6% = 0.06   (*d*) 6.4% = 0.064

         (0.06)(152) = 9.12   (0.064)(2.5) = 0.16

4.5. What percent of

    (*a*) 15 is 6?   (*c*) 480 is 1.6?   (*e*) 27.5 is 2.5?

    (*b*) 192 is 36?   (*d*) 400 is 900?

        The first of the two numbers is the base; the second number is the percentage. To find percent, divide percentage by the base. Multiply the quotient by 100 and attach the % symbol.

    (*a*) $6/15 = 0.4 = (0.4)(100)\% = 40\%$

    (*b*) $36/192 = 0.1875 = (0.1875)(100)\% = 18.75\%$

    (*c*) $1.6/480 \approx 0.00333 \approx (0.00333)(100)\% \approx 0.333\%$

    (*d*) $900/400 = 2.25 = (2.25)(100)\% = 225\%$

    (*e*) $2.5/27.5 \approx 0.0909 \approx (0.0909)(100)\% \approx 9.09\%$

4.6. Determine the number if:

    (*a*) 3% of the number is 18   (*d*) 0.35% of the number is 10.5

    (*b*) 60% of the number is 72   (*e*) 2.5% of the number is 0.012

    (*c*) 150% of the number is 390

        The number to be found is the base or the quantity which represents 100%. First determine 1% of the number, then multiply by 100 to obtain the number.

    (*a*) 1% is 18/3 = 6   100% = 6 × 100 = 600

    (*b*) 1% is 72/60 = 1.2   100% = 1.2 × 100 = 120

    (*c*) 1% is 390/150 = 2.6   100% = 2.6 × 100 = 260

    (*d*) 1% is 10.5/0.35 = 30   100% = 30 × 100 = 3000

    (*e*) 1% is 0.012/2.5 = 0.0048   100% = 0.0048 × 100 = 0.48

4.7. A copper compound contains 80 percent copper by weight. How much copper is there in a 1.95-g sample of the compound?

        Since 80% = 0.80, it follows that 80% of 1.95 = (0.80)(1.95) = 1.56. Therefore, the given sample contains 1.56 g of copper.

4.8.   An 8.04-g sample of iron sulfide contains 5.12 g of iron by weight.   What is the percent of iron in the iron sulfide?

5.12 is 5.12/8.04 of 8.04; 5.12/8.04 ≈ 0.637.   Since 0.637 = 0.637(100/100) = 63.7%, the given sample contains 63.7% iron.

4.9.   Common silver solder contains 63 percent silver by weight.   How much solder can be made with 16 oz of silver?

Since 63 percent of the solder must be 16 oz, it follows that 1 percent is 16/63 or 0.254 oz.   Hence 100 percent of the solder must be (0.254)(100) or 25.4 oz.   Therefore, 25.4 oz of the solder can be made with 16 oz of silver.

4.10.   The measured distance between two plates is 0.453 in.   The uncertainty in the measurement is ±0.0005 in.   Calculate the percent of measurement uncertainty or error.

$$\% \text{ uncertainty} = \frac{0.0005}{0.453} \times 100\% \approx 0.11\%$$

4.11.   A voltmeter reads 120 volts.   It is known that the voltmeter reading is 6% too high.   What should the corrected reading be?

The 120-volt reading corresponds to (100 + 6)% or 106%, 1% of the reading is $\frac{120}{106} = 1.13$. The corrected reading can be represented by 100%, or $1.13 \times 100 = 113$.   The corrected value for the voltmeter reading is 113 volts.

4.12.   If the present length of a brass rod is 1.4651 m, what will its length be after a 0.3 percent expansion caused by heating?

The decimal equivalent of 0.3% is 0.003. The added length or expansion is (1.4651)(0.003) = 0.0043953 m.

The final rod length will be 1.4651 + 0.0043953 = 1.4694953 m.   When rounded off to 5 significant figures, the expanded rod has a length of 1.4695 m.

4.13.   The volume of a certain gas is 46.0 ml.   If this is 8 percent lower than the volume under different conditions, what was this previous volume?

The given volume is (100 − 8)% or 92% of the previous volume.
1% of the previous volume is 46/92 or 0.5; 100% is (0.5)(100) or 50.   The previous volume was 50 ml.

4.14. The freezing point depression of a 1-molar solution has an accepted value of 1.86 °C.   A student determines this constant to be 1.90 °C.   Calculate the percent of error.

$$\% \text{ of error} = \frac{1.90 - 1.86}{1.86} \times 100\%$$

$$= \frac{0.04}{1.86} \times 100\%$$

$$\approx 2.15\% \approx 2\%$$

4.15. Two experimenters measure the volume of a liquid sample and obtain the results $V_1 = 50.2 \pm 0.05$ cm^3 and $V_2 = 50.0 \pm 0.05$ cm^3.   What is the percentage difference between the two values?

Since the measurement uncertainty is the same for both experimenters, the average of their measurements is used as a base:

$$\text{average volume} = \frac{50.2 + 50.0}{2} = 50.1 \text{ cm}^3$$

$$\% \text{ difference} = \frac{50.2 - 50.0}{50.1} \times 100\% = \frac{0.2}{50.1} \times 100\% \approx 0.4\%$$

4.16. A balance has a tolerance of $\pm 1.2\%$.

(a) If the mass of an object on this balance is 23.80 g, how large is the error or uncertainty?

(b) Between what limits does the mass lie?

(a) Convert 1/2% to a decimal fraction.

$1/2\% = 0.50\% = 0.005$ uncertainty $= (23.80)(0.005) = 0.119 \approx 0.12$ g

(b) The upper limit of the mass is $23.80 + 0.12 = 23.92$ g

The lower limit of the mass is $23.80 - 0.12 = 23.68$ g

Supplementary Problems

The answers to this set of problems are at the end of this chapter.

4.17. Convert each of the following fractions to decimal and percent forms:

(a) 3/5   (b) 7/3   (c) 15/8   (d) 4/9   (e) $\frac{13}{4}$

4.18. Convert each of the following decimal fractions to percent forms:

(a) 0.124   (b) 0.345   (c) 1.45   (d) 21.5   (e) 0.00345

4.19. Convert each of the following percents to decimal and common fraction forms:

(a) 4%   (b) 45%   (c) 4.7%   (d) 146%   (e) 0.015%   (f) 250%

4.20. Determine:

    (*a*) 5% of 140   (*c*) 0.64% of 15   (*e*) 250% of 23

    (*b*) 54% of 380   (*d*) 1.2% of 65   (*f*) $42\frac{2}{3}$% of 75

4.21. The number 15 is

    (*a*) 50% of what number?   (*c*) 1.5% of what number?   (*e*) $3\frac{4}{3}$% of what number?

    (*b*) 75% of what number?   (*d*) 150% of what number?   (*f*) 0.05% of what number?

4.22. Determine the number if

    (*a*) 5% of the number is 20   (*d*) 300% of the number is 35

    (*b*) 120% of the number is 50   (*e*) $3\frac{2}{3}$% of the number is 88

    (*c*) 0.15% of the number is 10

4.23. A sample of water is decomposed into 0.965 g of oxygen and 0.120 g of hydrogen. Calculate the mass percent of oxygen and hydrogen in the sample of water.

4.24. The length of a brass rod is 3.594 m. If the rod were to expand by 0.02 percent, what would be its increase in length?

4.25. To carry out a certain reaction a chemist needs 24.5 g of 100 percent acid. He has acid of 96 percent strength. How much of the weaker acid can be used for the reaction?

4.26. Decomposition of limestone yields 56 percent quicklime. How much limestone is needed to yield 2000 lb of quicklime?

4.27. If we assume that water increases in volume by 9 percent on freezing, determine how many cubic feet of water would be required to produce 545 ft^3 of ice.

4.28. An iron ore contains 58 percent iron oxide. Iron constitutes 70 percent of iron oxide. How much iron is there in 10,000 kg of this ore?

4.29. A gas contracts from a volume of 32 ml to a volume of 30.5 ml. What is the percentage change in volume?

4.30. The largest possible error in making a measurement of 45 ml is 0.5 ml. What is the largest possible percentage error?

4.31. Ammonia gas contains 82.4 percent nitrogen. Determine the mass of ammonia gas containing 300 kg of nitrogen.

4.32. An electrical resistance decreased 2 percent. If the present value of the resistance is 74.8 ohms, what was it before the change took place?

4.33. Silver solder contains 63% silver. How much of the solder can be made with 520 g of silver?

4.34. A 5.0-g sample of iron ore contains 3.1 g of iron. Express the iron content of the sample in percent form.

4.35. An ammeter reads 0.1 amp when disconnected. When the instrument is connected in circuit, the reading is 2.4 amp.

    (*a*) What is the corrected reading of the ammeter?

    (*b*) By what percent of error is the reading wrong if it is not corrected for zero error?

4.36. A micrometer has a zero error of -0.002 cm.   What percent of error will be introduced into a micrometer reading of 0.025 cm if a correction is not made for the zero error?

4.37. A flask is marked 250 ml.   Its volume measured with a standard container is 250.5 ml.   What is the percent error in the flask calibration?

4.38. A meterstick is compared with a standard meter.   The meterstick is 0.5 mm short.   What is the percent error? 1 meter $= 1000$ mm.

4.39. Determine to one significant figure the percent uncertainties in the following measurements:

   (a) 528 ± 1 g   (c) 0.80 ± 0.005 amp   (e) 21.3 ± 0.1 sec

   (b) 6.5 ± 1 cm   (d) 3.27 ± 0.03 ml

4.40. A speedometer reads 15% too low.   What is the actual speed to the nearest mile per hour if the indicated speed is 65 mph?

Answers to Supplementary Problems

4.17.   (a) 0.6, 60%   (c) 1.875, $187\frac{1}{2}\%$ (e) 3.25, 325%

   (b) $2.33\ldots, 233\frac{1}{3}\%$ (d) $0.44\ldots, 44\frac{4}{9}\%$

4.18.   (a) 12.4%   (b) 34.5%   (c) 145%   (d) 2150%   (e) 0.345%

4.19.   (a) 0.04, 1/25   (c) 0.047, 47/1000   (e) 0.00015, 3/20,000

   (b) 0.45, 9/20   (d) 1.46, 73/50   (f) 2.5, 5/2

4.20.   (a) 7   (b) 205.2   (c) 0.096   (d) 0.78   (e) 57.5   (f) 32

4.21.   (a) 30   (b) 20   (c) 1000   (d) 10   (e) $394\frac{14}{19}$ (f) 30,000

4.22.   (a) 400   (b) $41\frac{2}{3}$ (c) $6666\frac{2}{3}$ (d) $11\frac{2}{3}$ (e) 2400

4.23.   88.9%, 11.1% approximately

4.24.   0.000719 m approximately

4.25.   25.5 g approximately

4.26.   3570 lb approximately

4.27.   500 ft^3

4.28.   4060 kg

4.29. $4\frac{11}{16}\%$ or 4.69% approximately

4.30. $1\frac{1}{9}\%$ or 1.11% approximately

4.31. 364 kg approximately

4.32. 76.3 ohms approximately

4.33. 825 g approximately

4.34. 62%

4.35. (*a*) 2.3 amp   (*b*) 4.3% high

4.36. 7.4% low

4.37. 0.2% low approximately

4.38. 0.05% low

4.39. (*a*) 0.2%   (*b*) 20%   (*c*) 0.6%   (*d*) 0.9%   (*e*) 0.5%

4.40. 76 mph

Chapter 5

Essentials of Algebra

Basic Concepts

5.1 TERMINOLOGY

Algebra is a generalized and extended type of arithmetic. In addition to the numbers used in arithmetic, algebra includes one or more letters or symbols. Because relationships between quantities are expressed most efficiently by algebra, it may be called the short-hand of arithmetic.

Unknown. A symbol used in algebra to represent a quantity whose value is not known.

Literal symbol. A letter, such as x, y, g, v, t, that represents an unknown real number.

Algebraic operations with symbols are expressed in the same way as with the real numbers of arithmetic and include: addition, subtraction, multiplication, division, raising to a power, and extracting roots.

An *algebraic expression* is the result of a finite sequence of algebraic operations applied to a set of numbers and symbols. Typical algebraic expressions are

$$5x \qquad 2ab \qquad 3x + 4y \qquad (1/2)gt^2 \qquad (5/9)(F - 32) \qquad (W^2 + Z^2)/(W + Z)$$

Terms of an algebraic expression are any of its parts which are or may be connected by addition signs. The three terms of $3a^2 + 2b - 4xy$ are $3a^2, 2b$, and $-4xy$.

A *factor* is one of two or more quantities which may be multiplied together to form a *product*. In the product $2.73\,pq$, 2.73, p, and q are the displayed factors. The *literal factors* of $2.73\,pq$ are p and q.

Any factor in a product is called the *coefficient* of the product of the remaining factors. In $16at^2$, 16 is the coefficient of at^2; a of $16t^2$; t^2 of $16a$.

The *numerical coefficient* in a term is its numerical factor. The numerical coefficient in $9.80\,t$ is 9.80. The numerical coefficient of Fd is 1, implied though not written explicitly.

Like terms have the same literal factors. For example: $3x$ and $10x$, $2.5\,pv$ and $-7.4\,pv$.

Unlike terms have literal factors some of which are different. For example: $3x$ and $10x^2$, gt and $(1/2)gt^2$.

In an expression of the form t^n, n is an *exponent* and, *if n is a positive integer*, it denotes the number of times t is multiplied by itself to form the expression. The symbol t may be called the *base* for the exponential quantity t^n. For example: 2 is the exponent of t in t^2, 4 is the exponent of x in x^4; in the first expression t is the base, while x is the base in the second.

A *monomial* is an algebraic expression with only one term. For example: $2x$, $-3ax^2y$, $2\pi r$.

A *binomial* is an algebraic expression with two terms. For example: $2x - 3y$, $3a^2 - 6ab$, $v + at$.

A *polynomial* is an algebraic expression usually with three or more terms. For example: $3x^2y + 5xy - 6y^2$, $a^2 + 2ab + b^2$, $1 + b^2t + bt^2 + b^3t^3$. It is *usually* understood that the exponents associated with the variables in a polynomial are positive integers.

An *algebraic fraction* is the quotient of two algebraic expressions. For example: $1/f$, Q/r^2, $\sqrt{2s/g}$, $m/(r+x)^2$, $(500+C)/500C$, $M_0/\sqrt{1-v^2/c^2}$.

Signs of grouping are used to indicate certain terms or factors which are to be treated as a unit. The three commonly used types are: *parentheses* ( ), *brackets* [ ], *braces* { }, and the *vinculum* ——.

EXAMPLES: $(x+1)$; $[3+(a-2)]$; $\{7-6y\}$; $\overline{5+b}$

Signed Numbers

5.2   THE NUMBER LINE

In arithmetic, each number can be represented by a point on a horizontal line. If one point of the line is labeled 0, positive numbers are represented by points to the right and *negative numbers* by points to the left of 0. Such a representation is called the *number line* and it is shown in Fig. 5-1. The number line represents the set of all *real numbers*.

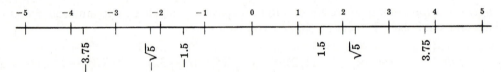

Fig. 5-1.   The number line

The use of negative numbers in elementary problems is usually associated with measurements in an "opposite" direction. We are familiar with such usage in temperature measurements, where positive and negative indicate measurements above and below zero; and in bookkeeping, where credits may be considered positive, and debits negative.

5.3   ABSOLUTE VALUE

The numerical value of absolute value of a number may be regarded intuitively as its magnitude without regard to sign. This absolute value is indicated by placing vertical bars around the number, and so $|a| = |-a| = a$, for a positive, and $|a| = |-a| = -a$, for a negative. The absolute value of a number is also its distance along the number line from 0.

EXAMPLE 5.1.   What is the absolute value of the number   (a) 6,   (b) -8,   (c) -12.5,   (d) $-gt$?

(a) $|6| = 6$ (b) $|-8| = 8$ (c) $|-12.5| = 12.5$ (d) $|-gt| = gt$, if $gt > 0$

5.4   OPERATIONS WITH SIGNED NUMBERS

If negative numbers are to be accepted as numbers, the basic laws must be obeyed, and this requires the following rules of operation with signed numbers:

Rule 1.   (a) To add two numbers of the same sign, add their absolute values and attach the common sign.

(b) To add two signed numbers of opposite sign, subtract the smaller absolute value from the larger and attach the sign of the larger.

(c) To subtract two signed numbers, change the sign of the subtrahend and add as in 1(a) or 1(b).

Rule 2.   To multiply two signed numbers, multiply their absolute values and attach a positive sign if the numbers are of like sign; perform the same multiplication and attach a negative sign if they are of unlike sign.

Rule 3.   The rule for multiplication has an analogous counterpart for division.

EXAMPLE 5.2.   What is the sum of   (a) -9 and -5,   (b) -9 and 5,   (c) 3.65 and -1.27?

(a) $|-9| + |-5| = 9 + 5 = 14$; attaching the common negative sign, we get -14.

(b) $|-9| - |5| = 9 - 5 = 4$; attaching the sign of the number with the larger absolute value, we get -4.

(c) $|3.65| - |-1.27| = 3.65 - 1.27 = 2.38$.

EXAMPLE 5.3.   What is the product of   (a) -9 and -5,   (b) -9 and 5,   (c) 2.80 and -1.25?

(a) $|-9| \cdot |-5| = 9 \cdot 5 = 45$.   Since the numbers are of like sign, the answer is 45.

(b) $|-9| \cdot |5| = 45$.   Since the numbers have unlike signs, we attach a minus sign and get -45 for the product.

(c) $|2.80| \cdot |-1.25| = 3.50$.   Since the numbers have unlike signs, we attach a minus sign and get -3.50.

If more than two numbers are involved in a computation, these rules may be easily generalized. For example, a product of several signed numbers is negative if an odd number of negative signs is involved; the product is positive if there is an even number of negative signs.

EXAMPLE 5.4.   Find the sum of the following numbers:

(a) 3, 12, 81, 10   (b) $-2, -5, -15, -30$ (c) 3, -6, 8, 10, -7, -9

(a) In this case the numbers are all positive, and addition is exactly as for natural numbers. Thus, $3 + 12 + 81 + 10 = 106$.

(b) In this case the numbers are all negative, and so we add their absolute values and attach a negative sign to the sum. Thus,

$$|-2| + |-5| + |-15| + |-30| = -(2 + 5 + 15 + 30) = -52$$

(c) In this case some signs are positive and some negative. We find the sum of the negatives and the sum of the positives and then add these two sums according to Rule 1(b) for addition of signed numbers, given above. Thus,

$$3 + -6 + 8 + 10 + -7 + -9 = (3 + 8 + 10) + [(-6) + (-7) + (-9)]$$
$$= 21 + (-22) = -1$$

EXAMPLE 5.5.   Determine the product of the following numbers:

(a) $-2, 4, 5, -10, -8$ (b) 4, $-2, -2, -6, -4$

(a) In this case there is an odd number of negative signs (three) and so the product is $-(2 \cdot 4 \cdot 5 \cdot 10 \cdot 8)$, or -3200.

(b) In this case there is an even number (four) of negative signs, and so the product is $+(4 \cdot 2 \cdot 2 \cdot 6 \cdot 4)$, or simply 384.

EXAMPLE 5.6.   Perform the indicated divisions: (a) $-24 \div 3$ (b) $-9 \div (-3/2)$ (c) $1.6 \div -0.4$

(a) $|-24| \div |3| = 8$.   Since the numbers have unlike signs, we attach the negative sign and obtain -8 for the quotient.

(b) $|-9| \div |-3/2| = 9 \times 2/3 = 6$.   Since the numbers have like signs, the quotient is $+6$ or 6.

(c) $|1.6| \div |-0.4| = 4$.   Since the numbers have unlike signs, we attach the negative sign and obtain -4 for the quotient.

Operations with Algebraic Expressions

5.5   OPERATIONS WITH MONOMIALS

(a) Addition and Subtraction

It is possible to use the "distributive law" to combine two or more monomials under addition or subtraction only if their symbolic or literal parts are identical; for example, $2ax$ and $3ax$. If this condition is met, the result is another monomial with the same symbolic part but whose numerical factor is the result of adding the numerical factors of the individual monomials. Monomials such as $2ax$ and $3by$ have unlike parts and cannot be combined in this way.

EXAMPLE 5.7.   Determine the sum of the following monomials: $3xy, -2xy, 5xy, xy$.

The result is $(3 - 2 + 5 + 1)xy = 7xy$.

EXAMPLE 5.8.   Determine the sum of the following monomials: $2xy, 3x^2, 5x^2, -6xy, -2x^2$.

In this case, there are two distinct symbolic parts in evidence, xy and x^2. The sum is $(2 - 6)xy + (3 + 5 - 2)x^2 = 6x^2 - 4xy$.

(b) Multiplication and Division

If monomials are to be multiplied or divided, the simplified notation of exponents may be used to advantage, along with the procedure of canceling. However, it is cautioned that simplification for multiplication and division may be applied only to exponential quantities with the same base.

EXAMPLE 5.9.   Determine the product of the following monomials: $3xy, 2x^2y^3, 5xz$.

The numerical factor of the product is observed to be $(3)(2)(5)$ or 30. We then obtain the product as $30xx^2xyy^3z$, which can be simplified to $30x^4y^4z$.

EXAMPLE 5.10.   Simplify the expression $(12xy^2z^3)/(4x^2yz^2)$.

The numerical factor of the quotient is $12/4$ or 3. By cancellation of like factors, we then find the final result to be $3yz/x$.

5.6   USE OF GROUPING SYMBOLS

Parentheses are used in arithmetical or algebraic expressions whenever a combination of numbers is to be treated as a unit. Thus, when we write $(m + n) + r$ we mean the sum of the number $m + n$ and the number r, while $m + (n + r)$ designates the sum of m and $n + r$. That these two sums are equal is the statement of the associative law of addition.

When several arithmetic operations are to be performed in succession, if care is not taken, it is quite possible that the expressed result will be completely ambiguous. The following rule may be helpful in avoiding this ambiguity:

Rule.   Whenever any combination of operations is involved, do all multiplications first, then all divisions in the order, and finally the additions and subtractions in the order given.

EXAMPLE 5.11.   Use the rule to determine the value of the expression $3 \cdot 5 - 8 \div 2$.

Do the multiplication first: $3 \cdot 5 = 15$; the division comes next: $8 \div 2 = 4$; finally, do the subtraction: $15 - 4 = 11$.

However, the best practice is to use grouping signs to make clear what is intended. For example in the expression $2 - (3 \div 4) + (2 \cdot 3)$, the parentheses show that division and multiplication are to be done before addition and subtraction.

EXAMPLE 5.12.   Simplify the following expression: $7 + 6(4 - 1) + (15 \div 3) - 2[20 - (2 \cdot 8) + 5]$.

We first perform the three operations indicated within the parentheses and obtain $7 + (6)(3) + 5 - 2[20 - 16 + 5]$. This can now be simplified into $7 + 18 + 5 - (2)(9) = 30 - 18 = 12$.

In order to "disentangle" or simplify expressions involving grouping symbols, we work "from the inside out" and apply the following rule: A pair of parentheses or equivalent grouping symbols may be simply removed if preceded by a plus sign; if preceded by a minus sign, the sign of each term inside should be changed after the grouping symbols have been removed.   If the grouping symbol is preceded by a multiplying factor, each inside term should be multiplied by that factor.

EXAMPLE 5.13.   Simplify $2x - [3 - (x + 1)]$.

We first remove the "inside" parentheses, which gives us $2x - [3 - x - 1]$ or $2x - [2 - x]$.   Next we remove the square brackets and obtain $2x - 2 + x$ or $3x - 2$.

EXAMPLE 5.14.   Simplify $t + \{2 - 3[t - (1 - t)] + 2\}$.

We perform the steps of the simplification process in the order suggested above.

$$t + \{2 - 3[t - 1 + t] + 2\}$$
$$t + \{2 - 3[2t - 1] + 2\}$$
$$t + \{2 - 6t + 3 + 2\}$$
$$t + \{7 - 6t\}$$
$$t + 7 - 6t$$

and finally $7 - 5t$

5.7   OPERATIONS WITH POLYNOMIALS

(a) Addition or Subtraction

To add (or subtract) two or more polynomials, arrange them vertically so that like terms and numerical terms are in vertical alignment.   Then add (or subtract).

EXAMPLE 5.15.   (*a*) Add $(2x^2 + x - 10)$, $(2x^2 + 6)$, and $(6x - 2)$.

(*b*) Subtract $6y - 2$ from $4y^2 + y - 8$.

Arrange the three polynomials in three lines as shown below. Perform the addition.

Arrange the polynomials as shown below.   Change the sign of the subtrahend and add.

$$\begin{array}{r} 2x^2 + x - 10 \\ 2x^2 + 6 \\ 6x - 2 \\ \hline 4x^2 + 7x - 6 \end{array}$$ is the required sum

$$\begin{array}{r} 4y^2 + y - 8 \\ -6y + 2 \\ \hline 4y^2 - 5y - 6 \end{array}$$ is the required difference

(b) Multiplication and Division

The rules for dealing with special cases which occur frequently are:

Rule 1.   To multiply (or divide) a polynomial by a monomial, multiply (or divide) each term of the polynomial by the monomial.

Rule 2.   To multiply two polynomials, multiply each term of one of the polynomials by each term of the other and combine the results according to the rules for monomials.

EXAMPLE 5.16.   Multiply $3x^2 + 4xy - y^2$ by $2a$.

An application of the above rule gives the result as $6ax^2 + 8axy - 2ay^2$.

EXAMPLE 5.17. Divide $6xy^2 + 2x^2y^3$ by $2xy^2$.

An application of the rule, in this case, gives the result $3 + xy$.

EXAMPLE 5.18. Divide $2ab - 6ab^2 + 3b$ by $2a$.

In this case, the divisor divides exactly only the first two terms of the polynomial, and we can write the result as $(b - 3b^2) + 3b/2a$. The quotient could also be expressed in the form $(2ab - 6ab^2 + 3b)/2a$, in which the division is merely indicated and not actually performed.

EXAMPLE 5.19. Perform the indicated multiplication: $(2x + 3)(4x - 2)$.

We first multiply $2x$ by each term of the second binomial and obtain $8x^2 - 4x$. The product of 3 by each term of the second is then found to be $12x - 6$. Combining $12x$ and $-4x$ we get $8x$. The product of the two binomials is $8x^2 + 8x - 6$.

If more than two polynomials are to be multiplied together, the complete product can be obtained by the successive multiplication of pairs.

EXAMPLE 5.20. Perform the indicated multiplication: $(r - t)(2r + t)(r + t)$.

We first multiply $(r - t)(2r + t)$, as in the preceding example, and obtain $2r^2 - rt - t^2$. The product $(2r^2 - rt - t^2)(r + t)$ is then found to be $2r^3 + 2r^2t - r^2t - rt^2 - t^2r - t^3$, and this can be simplified to $2r^3 + r^2t - 2rt^2 - t^3$.

There is a method which is sometimes useful for dividing polynomials by one another, the procedure being similar to the long division process of arithmetic. However, there are few elementary applications of this division process.

5.8   SIMPLE PRODUCTS

There are several simple products of such frequent occurrence that they need special consideration. These are the products or powers of simple binomials.

The rule for multiplication of polynomials can be restated in a slightly different form for binomials:

Obtain the sum of all possible products which can be formed by multiplying together only one term from each of the binomials. For example,

$$(a + b)^2 = (a + b)\,(a + b) = a^2 + ab + ab + b^2 = a^2 + 2ab + b^2$$

Similarly,
$$(a - b)^2 = (a - b)(a - b) = a^2 - 2ab + b^2$$

Thus the square of a binomial is the square of the first term, plus twice the cross-product, plus the square of the second term.

EXAMPLE 5.21. $(2x + y)^2 = (2x)^2 + 2(2x)y + y^2 = 4x^2 + 4xy + y^2$

EXAMPLE 5.22. $(3a - 2b)^2 = (3a)^2 + 2(3a)(-2b) + (-2b)^2 = 9a^2 - 12ab + 4b^2$

Applying the rule to binomials raised to the third power,

$$(a+b)^3 = (a+b)^2(a+b) = (a^2 + 2ab + b^2)(a+b) = a^3 + 3a^2b + 3ab^2 + b^3$$
$$(a-b)^3 = a^3 - 3a^2b + 3ab^2 - b^3$$

Thus, the cube of a binomial is the cube of the first term, plus three times the second term times the square of the first term, plus three times the first term times the square of the second term, plus the cube of the second term.

EXAMPLE 5.23. $(2a+b)^3 = (2a)^3 + 3(2a)^2b + 3(2a)b^2 + b^3$
$$= 8a^3 + 12a^2b + 6ab^2 + b^3$$

EXAMPLE 5.24. $(x-3y)^3 = x^3 + 3x^2(-3y) + 3x(-3y)^2 + (-3y)^3$
$$= x^3 - 9x^2y + 27xy^2 - 27y^3$$

Another common product is:

$$(a+b)(a-b) = a^2 + ab - ab + b^2 = a^2 - b^2$$

Thus, the product of the sum and difference of two quantities is the difference of their squares.

EXAMPLE 5.25. $(2x+y)(2x-y) = (2x)^2 - y^2 = 4x^2 - y^2$

EXAMPLE 5.26. $(3a+2b)(3a-2b) = (3a)^2 - (2b)^2 = 9a^2 - 4b^2$

It is sometimes convenient to think of the product of two binomials as formed by the multiplication of terms as indicated below.

$$(ax + by)(cx + dy)$$

EXAMPLE 5.27. $(2x+y)(3x-2y) = (2x)(3x) + y(3x) + 2x(-2y) + y(-2y)$
$$= 6x^2 + 3xy - 4xy - 2y^2$$
$$= 6x^2 - xy - 2y^2$$

EXAMPLE 5.28. $(a-2b)(3a+b) = a(3a) + (-2b)(3a) + ab + (-2b)b$
$$= 3a^2 - 5ab - 2b^2$$

EXAMPLE 5.29. $-(2x+y)^2 = -[(2x)^2 + 2(2x)(y) + (y)^2] = -[4x^2 + 4xy + y^2] = -4x^2 - 4xy - y^2$

Notice that the signs of the resulting terms were changed *after* squaring the binomial.

EXAMPLE 5.30. $(2x+y+z)^2 = ((2x+y)+z)^2 = (2x+y)^2 + 2(2x+y)(z) + (z)^2$
$$= (2x)^2 + 2(2x)(y) + (y)^2 + (4x+2y)(z) + z^2$$
$$= 4x^2 + 4xy + y^2 + 4xz + 2yz + z^2$$

Notice that, by grouping any two terms of the initial expression and treating them as a single term, the problem is reduced to that of squaring a binomial.

EXAMPLE 5.31. $(x+y+z)(x+y-z) = [(x+y)+z][(x+y)-z]$
$$= (x+y)^2 - z^2$$
$$= x^2 + 2xy + y^2 - z^2$$

Notice that by grouping conveniently the terms in both parentheses of the initial expression, the problem has been reduced to that of the product of the sum and difference of two quantities.

EXAMPLE 5.32. $(x + y + z)(x - y - z) = [x + (y + z)][x - (y + z)]$
$$= x^2 - (y + z)^2$$
$$= x^2 - (y^2 + 2xy + z^2)$$
$$= x^2 - y^2 - 2xy - z^2$$

Notice that by grouping conveniently the terms in both parentheses of the initial expression, the problem has been reduced to that of the product of the sum and difference of two quantities.

5.9 FACTORING

The process of factoring may be considered as the reverse of multiplication.   For we say that the product of 3 and 4 is 12, while 12 may be factored as the product (3)(4).   In the process of multiplication, we are combining the numbers, while the process of factoring breaks down the given expression into smaller multiplicative units or factors.

Removal of a Monomial Factor

The multiplication of a polynomial by a monomial is accomplished by multiplying each term of the polynomial by the monomial.   Hence, any monomial factor which is common to all terms of the polynomial can be factored out.

EXAMPLE 5.33.   Factor $4x^2 + 8x + 12$.

Since 4 is a common monomial factor of each term, we can factor the given trinomial into $4(x^2 + 2x + 3)$.

EXAMPLE 5.34.   Factor $2a^2b + 4ab^2 - 2ab^3$.

In this case, $2ab$ is a common monomial factor, and so the given polynomial can be factored as $2ab(a + 2b - b^2)$.

Difference of Two Squares

Since we have seen before that $a^2 - b^2 = (a + b)(a - b)$, we have the following rule for factoring: The difference of the squares of two quantities can be factored as the product of the sum and difference of the given quantities.

EXAMPLE 5.35.   Factor $9x^2 - 4y^2$.

Since this expression is $(3x)^2 - (2y)^2$, we can factor it as the product

$$(3x + 2y)(3x - 2y)$$

EXAMPLE 5.36.   Factor $4a^2b^2 - 16a^2c^2$.

In this case, we first remove the monomial factor $4a^2$ and obtain $4a^2(b^2 - 4c^2)$.   The factor $b^2 - 4c^2$ is a difference of two squares and so may be factored as $(b + 2c)(b - 2c)$.   The desired complete factorization is then $4a^2(b + 2c)(b - 2c)$.

Trinomials of the Type $x^2 + 2xy + y^2$

Since the result of squaring a binomial is a trinomial, one of the latter can often be factored into the product of two equal binomial factors.

EXAMPLE 5.37. Factor $m^2 + 6m + 9$.

Since $3^2 = 9$ and $2 \cdot m \cdot 3 = 6m$, the trinomial is the square of the binomial $m + 3$. Thus, $m^2 + 6m + 9 = (m + 3)(m + 3)$.

EXAMPLE 5.38. Factor $25r^2 + 10rs + s^2$

The given expression is a trinomial square since $(5r)^2 = 25r^2, (s)^2 = s^2, 2 \cdot 5r \cdot s = 10rs$. Therefore, $25r^2 + 10rs + s^2 = (5r + s)^2 = (5r + s)(5r + s)$.

Trinomials of the Type $x^2 + px + q$

Inasmuch as $(x + a)(x + b) = x^2 + (a + b)x + ab$, it appears that a trinomial of the given type can be factored if we can find two numbers a and b, such that $a + b = p$ and $ab = q$. If we can find such numbers, the desired factorization is then $(x + a)(x + b)$.

EXAMPLE 5.39. Factor $x^2 + 5x + 6$.

The above requirements are satisfied if $a = 2$ and $b = 3$; hence $x^2 + 5x + 6 = (x + 2)(x + 3)$.

EXAMPLE 5.40. Factor $x^2 - 8x + 12$.

We see that $(-2) + (-6) = -8$, and $(-2)(-6) = 12$, so that $x^2 - 8x + 12 = (x - 2)(x - 6)$.

General Trinomial of the Type $rx^2 + px + q$

In the case of a trinomial of this type, we can use the procedure shown below. This procedure does not guarantee that we can always factor the trinomial; however, it provides a more systematic and more effective procedure than any trial and error method.

EXAMPLE 5.41. Factor $5x^2 + 13x - 6$.

Multiply the entire trinomial by 5, the coefficient of the term x^2, and leave written explicitly the product of this coefficient and the coefficient of the term of x. That is, $25x^2 + 5(13x) - 30$.

Switch the coefficient of the term of x and the number by which the trinomial was multiplied. That is, write the expression as $25x^2 + 13(5x) - 30$. Since $25x^2 = (5x)^2$ rewrite the expression as $(5x)^2 + 13(5x) - 30$.

Try to find two numbers that add up to 13 and multiply to equal -30. Factor the expression using these two numbers and rewrite it as

$$(5x)^2 + 13(5x) - 30 = (5x + 15)(5x - 2) \leftarrow \text{Notice that } +15 - 2 = 13 \text{ and } (+15)(-2) = -30$$

Since the given trinomial was initially multiplied by 5, it is necessary to divide it by 5. Therefore,

$$\frac{(5x + 15)(5x - 2)}{5} = \frac{5(x + 3)(5x - 2)}{5} = (x + 3)(5x - 2)$$

The original expression can be rewritten as

$$5x^2 + 13x - 6 = (x + 3)(5x - 2)$$

EXAMPLE 5.42. Factor $6y^2 - 7y - 5$.

Multiply the trinomial by 6, the coefficient of the y^2 term, and leave indicated the product of this coefficient and the coefficient of y. That is,

$$6(6y^2 - 7y - 5) = 36y^2 - 6(7y) - 30$$

Switch the coefficient of the term of x and the number by which the trinomial was multiplied. That is, rewrite the expression as $36y^2 - 7(6y) - 30$. Since $36y^2 = (6y)^2$ rewrite the expression as

$$36y^2 - 6(7y) - 30 = (6y)^2 + 7(6y) - 30$$

Look for two numbers that add up to 7 and multiply to equal -30. Then, factor the expression using these two numbers and rewrite it as

$$(6y)^2 + 7(6y) - 30 = (6y - 10)(6y + 3) \leftarrow \text{Notice that } -10 + 3 = 7 \text{ and } (-10)(3) = -30.$$

Since the trinomial was initially multiplied by 6, it is necessary to divide it by 6. Therefore,

$$\frac{(6y - 10)(6y + 3)}{6} = \frac{\cancel{2}(3y - 5)\cancel{3}(2y + 1)}{\cancel{6}} = (3y - 5)(2y + 1)$$

In consequence, the original expression can be written as $6y^2 - 7y - 5 = (3y - 5)(2y + 1)$.

EXAMPLE 5.43. Factor $-10x^2 - 7x + 12$.

Multiply the trinomial by -10, the coefficient of the x^2 term, and leave indicated the product of this coefficient and the coefficient of x. That is,

$$(-10)(-10x^2 - 7x + 12) = 100x^2 + 10(7x) - 120.$$

Switch the coefficient of the term of x and the number by which the trinomial was multiplied. That is, rewrite the expression as $100x^2 + 7(10x) - 120$. Since $100x^2 = (10x)^2$ rewrite the expression as

$$100x^2 + 7(10x) - 120 = (10x)^2 + 7(10x) - 120.$$

Look for two numbers that add up to 7 and multiply to equal -120. Then, factor the expression using these two numbers and rewrite it as

$$(10x)^2 + 7(10x) - 120 = (10x + 15)(10x - 8) \leftarrow \text{Notice that } 15 - 8 = 7 \text{ and } (15)(-8) = -120.$$

Since the trinomial was initially multiplied by -10, it is necessary to divide it by -10. Therefore,

$$\frac{(10x + 15)(10x - 8)}{-10} = \frac{\cancel{5}(2x + 3)\cancel{2}(5x - 4)}{(-\cancel{5})(\cancel{2})} = -(2x + 3)(5x - 4) = (2x + 3)(4 - 5x).$$

In consequence, the original expression can be written as

$$-10x^2 - 7x + 12 = (2x + 3)(4 - 5x).$$

5.10 CANCELLATION

The process of simplifying a fraction or an equation by dividing out a common factor is called *cancellation*.

EXAMPLE 5.44. Simplify the following by cancellation.

(a) $12a^2b \div 3a$ (b) $7.5 \, drt \div 1.5 \, dt$ (c) $-10c^2v^2 \div 1.25 \, cv^4$

Write the quotients as fractions and divide the numerator and the denominator of each by the common factor or factors.

(a) $\dfrac{12a^2b}{3a} = \dfrac{\cancel{3} \cdot 4a^2b}{\cancel{3}\cancel{a}} = 4ab$

(b) $\dfrac{7.5 \, drt}{1.5 \, dt} = \dfrac{\cancel{5} \cdot \cancel{1.5}\, d\cancel{r}t}{\cancel{1.5}\, \cancel{d}\cancel{t}} = 5r$

(c) Divide the numerical coefficients in the usual way and cancel the common literal factors.

$$\frac{-10c^2v^2}{1.25 \, cv^4} = \frac{-10}{1.25} \cdot \frac{c^2v^2}{cv^4} = -8\frac{c}{v^2} = -\frac{8c}{v^2}$$

EXAMPLE 5.45.   Simplify $(4x^2 - y^2)/(2x + y)$.

The numerator is the difference of the two squares and can be factored into $(2x + y)(2x - y)$.   The common factor $2x + y$ is then canceled.

$$\frac{(4x^2 - y^2)}{(2x + y)} = \frac{(2x + y)(2x - y)}{(2x + y)} = 2x - y, \text{ provided } (2x + y) \neq 0$$

5.11   OPERATIONS WITH ALGEBRAIC FRACTIONS

Operations with algebraic fractions follow the same rules as for arithmetical fractions.   The operations and the final results are often simplified by reducing the fractions to their simplest terms by cancellation.   In a reduced fraction the numerator and the denominator have no common factors except 1.   The operations with algebraic fractions can best be illustrated by examples as given below.

EXAMPLE 5.46.   Reduce the fraction mgL/mL^2.

The factored form of the numerator is $m \cdot g \cdot L$; the factored form of the denominator is $m \cdot L \cdot L$.   The common factors are m and L, and these can be canceled.   The reduced fraction is then

$$\frac{m \cdot g \cdot L}{m \cdot L \cdot L} = g/L$$

EXAMPLE 5.47.   Reduce the fraction $(3s^2 - st - 2t^2)/(3s^2 + 5st + 2t^2)$.

The numerator can be factored into $(s - t)(3s + 2t)$ and the denominator into $(3s + 2t)(s + t)$.   Cancellation of the common factor $(3s + 2t)$ leaves the fraction in the reduced form $(s - t)/(s + t)$.

EXAMPLE 5.48.   Add the fractions $1/p$ and $1/q$.

The least common denominator is pq.   Therefore, multiplying the first fraction by q/q and the second by p/p, we obtain $q/pq + p/pq$.   The desired sum is $(q + p)/pq$.

EXAMPLE 5.49.   Subtract a/mn^2 from b/m^2n.

The least common denominator is m^2n^2.   Therefore, expressing the two fractions with their least common denominator, we have

$$bn/m^2n^2 - am/m^2n^2 = (bn - am)/m^2n^2$$

EXAMPLE 5.50.   Perform the indicated multiplication:

$$\frac{R}{V(K + L)} \cdot \frac{(K^2 - L^2)Q}{R^3}$$

Since $K^2 - L^2$ is a difference of two squares, this expression can be factored into $(K + L)(K - L)$.   Therefore, we get

$$\frac{R}{V(K + L)} \cdot \frac{(K + L)(K - L)Q}{R^{3\,2}}$$

The product of the two fractions, after canceling common factors, is

$$\frac{(K - L)Q}{VR^2}$$

EXAMPLE 5.51.   Perform the indicated divisions:

(a) $\dfrac{15a^2}{16b^3} \div \dfrac{3a}{4b^2}$ (b) $\dfrac{r}{t - 1} \div \dfrac{s^2}{t}$

As in arithmetic, invert the divisor; then multiply the numerators and multiply the denominators. Cancel any common factor(s).

(a) $\dfrac{15a^2}{16b^3} \div \dfrac{3a}{4b^2} = \dfrac{\overset{5}{\cancel{15}}a^2}{\underset{4}{\cancel{16}}b^3} \cdot \dfrac{\cancel{4}\cancel{b^2}}{\cancel{3}\cancel{a}} = \dfrac{5a}{4}$

(b) $\dfrac{r}{t-1} \div \dfrac{s^2}{t} = \dfrac{r}{t-1} \cdot \dfrac{t}{s^2} = \dfrac{rt}{(t-1)s^2}$

Note that t may not be canceled because t is *not* a factor of $t - 1$. This is a common error!

Equations

5.12 DEFINITIONS

An *equation* is a statement that two quantities are equal. For example, $1 + 1 = 2$ and $2 + 3 = 5$ are very simple examples of an arithmetic equation.

Members or *sides* of an equation are the expressions on the two sides of the equality sign.

Algebraic equations have one or more unknowns. The word "algebraic" is frequently omitted.

An *identity* is an equation which is true for all values of the unknown. For example, $2g + 6g = 8g$ is true for every value of g.

A *conditional equation* is true only for specific values of the unknown. For example, $2g = 64$ is correct only if $g = 32$.

A *linear equation* has the unknown(s) to the first power. For example, $3x + 7 = 16$ is a linear equation or of the *first degree equation* in x.

In a *quadratic equation* the highest exponent of any unknown is 2. For example, $y^2 - 8y + 15 = 0$ is a quadratic or *second-degree equation* in y.

There are equations of higher degree but they are seldom used in elementary applications.

5.13 SOLUTIONS OF LINEAR EQUATIONS

There is usually a question implied in connection with an algebraic equation: What actual number must each symbol represent if the equality is to hold arithmetically? Determining these numbers is known as *solving* the equation, while the numbers are its *solutions*.

To check the solution is to substitute the determined value of the unknown into the original equation. If an arithmetic identity results, then the solution is correct.

Each type of equation has its own peculiar methods of solution, but an important principle underlies all these methods. This principle is sometimes known as the "*golden rule of equations*":

What you do to one member of an equation, you must do to the other.

That is, if you add 2 to one member, you must add 2 to the other; if you multiply (or divide) the left member by 6, you must multiply (or divide) the right member by 6; if you square the right member, you must also square the left member, etc.

Consider the equation $a + b = c + d$. If we add $-b$ to both members, the equation becomes $a = c + d - b$; we note that b has disappeared from the left member and has made an appearance in the right member with its sign changed. Likewise, if we add $-c$ to both members of the original equation, the equation becomes $a + b - c = d$; c has disappeared from the right member and has appeared in the left member as $-c$. These results lead us to the following principle known as the *principle of transposition*:

> Any term of either member of an equation may be transposed to the other member, provided the sign of the term is changed.

We now apply this principle, along with the "golden rule," to the solution of any linear equation. The procedure for solving quadratic equations will be given in the next chapter.

EXAMPLE 5.52. Solve for the unknown: $3x - 4 = x + 10$.

Transposing (with changed signs) -4 to the right member and x to the left member of the equation, we get

$$3x - x = 4 + 10$$

Note that all the terms containing the unknown are on the left side. Collecting terms on each side, we obtain

$$2x = 14$$

Dividing each member by 2 gives the required solution, $x = 7$.

To check, substitute 7 for x in the original equation:

$$3(7) - 4 \overset{?}{=} 7 + 10$$
$$21 = 4 \overset{?}{=} 17$$
$$17 = 17$$

The resulting identity shows that the solution is correct.

EXAMPLE 5.53. Solve the following equation for v: $2v - 1 + 5v + 2 = v + 9$.

$$7v + 1 = v + 9 \qquad \text{collect terms on both sides}$$
$$7v - v = 9 - 1 \qquad \text{transpose the } v \text{ term to the left side, and the constant 1 to the right}$$
$$6v = 8 \qquad \text{collect terms}$$
$$v = \frac{8}{6} = \frac{4}{3} \qquad \text{divide both members by 6 and reduce}$$

The final test of a correct solution is that it satisfies the original equation. In the above example, we substitute ("plug in") the number $\frac{4}{3}$ for v in the given equation and obtain $2(\frac{4}{3}) = 1 + 5(\frac{4}{3}) + 2 \overset{?}{=} \frac{4}{3} + 9$. A bit of arithmetic reduces this expression to the equality $10\frac{1}{3} = 10\frac{1}{3}$, thus checking our solution.

EXAMPLE 5.54. Solve the following equation for t: $3.15t + 1.65 = 1.4t + 2$.

The first step is to eliminate the decimals by multiplying both members by 100.

$$315t + 165 = 140t + 200$$
$$315t - 140t = 200 - 165 \qquad \text{transpose the } 140t \text{ term to the left side, and the 165 term to the right side}$$
$$175t = 35 \qquad \text{collect terms}$$
$$t = \frac{35}{175} \qquad \text{divide both members by 175}$$
$$t = \frac{1}{5} = 0.2 \qquad \text{reduce by cancellation}$$

Check: $\quad 3.15(0.2) + 1.65 \overset{?}{=} 1.4(0.2) + 2$
$$0.63 + 1.65 \overset{?}{=} 0.28 + 2$$
$$2.28 = 2.28$$

EXAMPLE 5.55. Solve the following equation for F: $265 = \frac{5}{9}(F - 32)$.

$$265 \cdot \frac{9}{5} = \frac{\not{5}}{\not{9}}(F - 32)\frac{\not{9}}{\not{5}} \qquad \text{multiply both members by 9 and divide by 5}$$

$$265 \cdot \frac{9}{5} = F - 32 \qquad \text{remove parentheses}$$

$$265 \cdot \frac{9}{5} + 32 = F \qquad \text{transpose } -32 \text{ to the left side}$$

$$509 = F \qquad \text{carry out the indicated operations}$$

$$F = 509 \qquad \text{interchange the left and right members}$$

Check: $265 \stackrel{?}{=} \frac{5}{9}(509 - 32)$

$265 \stackrel{?}{=} \frac{5}{9}(477)$

$265 \stackrel{?}{=} 5(53)$

$265 = 265$

Solved Problems

SIGNED NUMBERS

5.1. Give the absolute values of the following:

(a) 7   (b) -15 (c) $+3\frac{1}{2}$ (d) -0.062 (e) $-v$

The absolute values of the given quantities are their magnitudes without the signs. Thus

(a) $|7| = 7$ (c) $|+3\frac{1}{2}| = 3\frac{1}{2}$ (e) $|-v| = v$, if $v \geq 0$

(b) $|-15| = 15$ (d) $|-0.062| = 0.062$ $= -v$, if $v < 0$

5.2. Find the sum of the following numbers:

(a) $-8, -6, -9, -4$ (c) $0.8, -1.6, 7.0, -0.21, 0.1$ (e) $1\frac{1}{2}, -2\frac{1}{3}, \frac{5}{6}$

(b) $12, -7, 5, -33$ (d) $-5, -5, 0, +5$

(a) Since all the numbers have the minus sign, add their absolute values and attach the common sign. $|-8| + |-6| + |-9| + |-4| = 27$. The sum is -27.

    In (b), (c), (d) and (e) both signs appear. Add absolute values of numbers with the same signs. Subtract the smaller sum from the larger and attach the sign of the larger absolute sum to the difference.

(b) $|12| + |5| = 17$ $|-7| + |-33| = 40$ $40 - 17 = 23$

The sum is -23.

(c) $|0.8| + |7.0| + |0.1| = 7.9$ $|-1.6| + |-0.21| = 1.81$ $7.9 - 1.81 = 6.09$

The sum is 6.09 since the numbers in the larger sum have a plus sign.

(d) The sum is -5.

(e) $|1\frac{1}{2}| + \left|\frac{5}{6}\right| = \left|\frac{3}{2}\right| + \left|\frac{5}{6}\right| = \left|\frac{9}{6}\right| + \left|\frac{5}{6}\right| = \frac{14}{6}$ $|-2\frac{1}{3}| = \left|-\frac{7}{3}\right| = \frac{14}{6}$

$\frac{14}{6} - \frac{14}{6} = 0$. The sum is 0.

5.3. In each of the following pairs of numbers subtract the second from the first:

(a) 12, 20   (b) 7, −7   (c) −7, 7   (d) −2.82, 0.98   (e) 0, −35

Change the sign of the second number and add the two by the addition rule for signed numbers.

(a) $12 - 20 = 12 + (-20) = -8$ (d) $-2.82 + (-0.98) = -3.80$

(b) $7 - (-7) = 7 + 7 = 14$ (e) $0 + (35) = 35$

(c) $-7 - 7 = -7 + (-7) = -14$

5.4. Multiply the following:

(a) $(-6)(+8)$ (c) $(-2)(-2)(-2)$ (e) $(-1.5)(-1)(-0.4)(-2)$

(b) $(-2)(-9)$ (d) $(6)(-5)(2)$

The number of negative signs is odd in parts (a), (c), (d). The minus sign is attached to each of these products, giving the answers: (a) -48, (c) -8, (d) -60.

The number of negative signs in parts (b) and (e) is even, and so the products have a plus sign. The answers are: (b) 18, (e) 1.2.

5.5. Simplify the following indicated divisions:

(a) $\dfrac{-24}{4}$ (b) $\dfrac{-24}{-2}$ (c) $7.5 \div -2.5$ (d) $\dfrac{+14}{-0.2}$ (e) $\dfrac{0}{-6}$

Applying the rule for division of signed numbers, the answers are:

(a) -6 (b) 12   (c) -3 (d) -70 (e) 0

OPERATIONS WITH MONOMIALS

5.6. Determine the sum of the following and simplify:

(a) $8a, -3a + 9a$ (c) $-qr, 9qr, -3qr, 7qr$ (e) $\dfrac{1}{2}x, -\dfrac{1}{3}y, \dfrac{3}{4}z$

(b) $\pi r^2, -3\pi r^2$ (d) $-5T, R, +7T^2 + 4T - 6T^2$

Add the coefficients of like terms and attach to the corresponding literal parts. Express the sum of the unlike terms.

(a) $(8 - 3 + 9)a = 14a$

(b) $(1 - 3)\pi r^2 = -2\pi r^2$

(c) $(-1 + 9 - 3 + 7)qr = 12qr$

(d) $(-5 + 4)T + (+7 - 6)T^2 + R = -T + T^2 + R = T^2 - T + R$

(e) There are no like terms. The sum is $\dfrac{1}{2}x - \dfrac{1}{3}y + \dfrac{3}{4}z$.

5.7. Subtract the second quantity from the first:

(a) $8a, -3a$ (b) $-13t, 5t$ (c) $-19m, -6m$

(d) $5.47\pi r, -2.12\pi r$ (e) $-5m, -8m^2$

Change the sign of the second number and add the numerical coefficients of like terms.

(a) $8a + 3a = (8 + 3)a = 11a$

(b) $-13t + (-5t) = (-13 - 5)t = -18t$

(c) $-19m + 6m = (-19 + 6)m = -13m$

(d) $5.47\pi r + 2.12\pi r = (5.47 + 2.12)\pi r = 7.59\pi r$

(e) $-5m + 8m^2$, or factoring out m and interchanging factors, the answer is $m(8m - 5)$.

5.8. Multiply the following pairs of monomials:

(a) $(3y)(-y)$ (c) $(-5mn)(2mn)$ (e) $(-3.25rt)(-0.2ts)$

(b) $(4ab)(7ab)$ (d) $(4a)^2$

Multiplying separately the numerical and literal coefficients of each pair, we obtain:

(a) $-3y^2$ (b) $28a^2b^2$ (c) $-10m^2n^2$ (d) $16a^2$ (e) $0.65rt^2s$

5.9. Divide the following:

(a) $-32t \div 4$ (c) $-7.5r^2s \div 1.5rs$ (e) $-32a^3b^4c \div -4ab^4c^2$

(b) $14mn$ by $-7n$ (d) $2(cD)/3$ by $c/2$

Set up each indicated division as a fraction.   Simplify by dividing the numerical coefficients and canceling common factors.

(a) $\dfrac{-32t}{4} = -8t$ (d) $\dfrac{2cD}{3} \Big/ \dfrac{c}{2} = \dfrac{2cD}{3} \cdot \dfrac{2}{c} = \dfrac{4}{3}D$

(b) $\dfrac{14mn}{-7n} = -2m$ (e) $\dfrac{-32a^3b^4c}{-4ab^4c^2} = 8a^2/c$

(c) $\dfrac{-7.5r^2s}{1.5rs} = -5r$

SIGNS OF GROUPING

5.10. Simplify:

(a) $3t - 5[(2t + 1) - (4 - t)]$ (d) $3 - (x - 2) - \{[3x - (2x + 1)] + 3\}$

(b) $5\{y + [2 - 3(2 - 3y) - 4y] - 2\}$ (e) $2 - \{x - 5y[2y + 2(x - 5)] + 4\}$

(c) $2 - 4a\{a - [5 + 2(3 - a) - a - b]\}$

Working from the "inside out", applying the rules for signed numbers at each step, and combining like terms, the results are:

(a) $3t - 5[2t + 1 - 4 + t] = 3t - 10t - 5 + 20 - 5t = -12t + 15$

(b) $5\{y + [2 - 6 + 9y - 4y] - 2\} = 5\{y + [-4 + 5y] - 2\} = 5\{y - 4 + 5y - 2\}$

$$= 5\{6y - 6\} = 30y - 30$$

(c) $2 - 4a\{a - [5 + 6 - 2a - a - b]\} = 2 - 4a\{a - [11 - 3a - b]\} = 2 - 4a\{a - 11 + 3a + b\}$

$$= 2 - 4a\{4a - 11 + b\} = 2 - 16a^2 + 44a - 4ab$$

(d) $3 - (x - 2) - \{[3x - (2x + 1)] + 3\} = 3 - x + 2 - [3x - 2x - 1 + 3]$

$$= 5 - x - [x + 2] = 5 - x - x - 2 = 3 - 2x$$

(e) $2 - \{x - 5y[2y + 2x - 10] + 4\} = 2 - \{x - 10y^2 - 10xy + 50y + 4\}$

$$= 2 - x + 10y^2 + 10xy - 50y - 4$$

$$= 10y^2 + 10xy - 50y - x - 2$$

5.11. Simplify the following expressions:

(a) $V = \dfrac{1}{2}[v_0 + (v_0 + at)]$ (c) $\left(\dfrac{2\pi r}{T}\right)^2 \Big/ r$

(b) $d = \left(v_0 + \dfrac{1}{2}at\right)t$ (d) $\left(\dfrac{mr^2}{2} + mr^2\right)\dfrac{a}{r}$

(a) $V = \dfrac{1}{2}[v_0 + v_0 + at] = \dfrac{1}{2}[2v_0 + at] = v_0 + at/2$

(b) $d = v_0 t + \dfrac{1}{2}at^2$

(c) $\dfrac{4\pi^2 r^2}{T^2} \div r = \dfrac{4\pi^2 r^2}{T^2 \cancel{r}} = \dfrac{4\pi^2 r}{T^2}$

(d) $\left(\dfrac{mr^2}{2} + mr^2\right)\dfrac{a}{r} = \dfrac{3mr^2}{2}\dfrac{a}{\cancel{r}} = \dfrac{3mra}{2}$

OPERATIONS WITH POLYNOMIALS

5.12. Add the following

(a) $2x^2 - 3x + 1$ and $5x^2 + 7x - 4$

(b) $a^3 - 3a^2 + 3a - 1$ and $-a^3 + 2a^2 + 2a$

(c) $26m + 10n + 14p,\ 12m + 15n - 20,\ -12n - 5p$

(d) $5an + 3bm + 4,\ -3an - 5bm - 1,\ 7an - 2bm - 2$

(e) $x^3 + y^3,\ x^3 + 6x^2y^2,\ -6x^2y^2 + y^3$

Arranging the polynomials vertically so that like terms are in vertical alignment and adding, we obtain

(a) $2x^2 - 3x + 1$
$\underline{5x^2 + 7x - 4}$
$7x^2 + 4x - 3$ *Ans.*

(c) $26m + 10n + 14p$
$12m + 15n \qquad - 20$
$\underline{\qquad -12n - \ 5p}$
$38m + 13n + \ 9p - 20$ *Ans.*

(e) $x^3 \qquad\qquad + \ y^3$
$x^3 + 6x^2y^2$
$\underline{\qquad -6x^2y^2 + \ y^3}$
$2x^3 \qquad\qquad + 2y^3$ *Ans.*

(b) $\quad a^3 - 3a^2 + 3a - 1$
$\underline{-a^3 + 2a^2 + 2a}$
$\qquad -a^2 + 5a - 1$ *Ans.*

(d) $\quad 5an + 3bm + 4$
$-3an - 5bm - 1$
$\underline{7an - 2bm - 2}$
$9an - 4bm + 1$ *Ans.*

5.13. Subtract the second expression from the first:

(a) $-6x + 4y, 7x + 5y$

(b) $3a^2 - 2b + 7, -6a^2 - 8$

(c) $4p^2 + 13p - 10, -3p^2 + p + 5$

(d) $6a - 7b - 10c, 12a - 7c$

(e) $12x^2 - 3x + 1, -2x^2 + 9x - 7$

Arrange the polynomials vertically so that like terms are in vertical alignment; change the signs of all the terms of the subtrahend, and add like terms.

(a) $-6x + 4y$
$\quad\,\underline{-7x - 5y}$
$\quad\,-13x - \,\,y$ Ans.

(b) $3a^2 - 2b + \,\,7$
$\quad\,\underline{6a^2 \qquad + \,\,8}$
$\quad\,9a^2 - 2b + 15$ Ans.

(c) $4p^2 + 13p - 10$
$\quad\,\underline{3p^2 - \,\,\,\,p - \,\,5}$
$\quad\,7p^2 + 12p - 15$ Ans.

(d) $\quad\,\,6a - 7b - 10c$
$\quad\,\underline{-12a \qquad + \,\,7c}$
$\quad\,-6a - 7b - \,\,3c$ Ans.

(e) $12x^2 - \,\,3x + 1$
$\quad\,\underline{2x^2 - \,\,9x + 7}$
$\quad\,14x^2 - 12x + 8$ Ans.

5.14. Multiply $2x^2 + 5xy - 3y^2$ by (a) -4 (b) $2xy$ (c) $-3y$ (d) x^2 (e) $5t$

Multiplying each term of the polynomial by the given monomial, we obtain:

(a) $-8x^2 - 20xy + 12y^2$

(b) $4x^3y + 10x^2y^2 - 6xy^3$

(c) $-6x^2y - 15xy^2 + 9y^3$

(d) $2x^4 + 5x^3y - 3x^2y^2$

(e) $10x^2t + 25xyt - 15y^2t$

5.15. Perform the indicated divisions:

(a) $(2x^2 - 4xy + 4y^2) \div 2$

(b) $(t^3 - 3t^2 + 2t) \div 4t$

(c) $(24m^2n^2 + 12mn^3 - 6mn) \div 6mn$

(d) $(10c^2 - 25c) \div (-5c)$

(e) $(x^3 - x^2y^3 + x^4y^4) \div x^2y^2$

Dividing each term of the polynomial by the given monomial, we obtain:

(a) $x^2 - 2xy + 2y^2$

(b) $\dfrac{t^2}{4} - \left(\dfrac{3}{4}\right)t + \dfrac{1}{2}$

(c) $4mn + 2n^2 - 1$

(d) $-2c + 5$

(e) $\dfrac{x}{y^2} - y + x^2y^2$

5.16. Perform the indicated multiplication:

(a) $(2x - 3y)(5x + y)$

(b) $(d + 2)(3d + f)$

(c) $(2x^2y - xy)(3xz + y)$

(d) $(p^2 + 3p - 4)(2p - 1)$

(e) $(2a^2 + ab - b^2)(3a - 2b)$

Multiply each term of the first expression by each term of the second and simplify where possible:

(a) $10x^2 + 2xy - 15xy - 3y^2 = 10x^2 - 13xy - 3y^2$

(b) $3d^2 + df + 6d + 2f$

(c) $6x^3yz + 2x^2y^2 - 3x^2yz - xy^2$

(d) $2p^3 - p^2 + 6p^2 - 3p - 8p + 4 = 2p^3 + 5p^2 - 11p + 4$

(e) $6a^3 - 4a^2b + 3a^2b - 2ab^2 - 3ab^2 + 2b^3 = 6a^3 - a^2b - 5ab^2 + 2b^3$

SIMPLE PRODUCTS

5.17. Expand the following expressions:

(a) $(3x - 2y)^2$ (c) $(m + 2n)^3$ (e) $(1 + 1.5t)^2$ (g) $(2r - p - m)(2r + p + m)$

(b) $(2a + 3b)^2$ (d) $(3p - q)^3$ (f) $-(2a + b + c)^2$ (h) $(2r - p - m)(2r - p + m)$

Using the rules for squaring and cubing binomials, we obtain:

(a) $(3x)^2 + 2(3x)(-2y) + (-2y)^2 = 9x^2 - 12xy + 4y^2$

(b) $(2a)^2 + 2(2a)(3b) + (3b)^2 = 4a^2 + 12ab + 9b^2$

(c) $(m)^3 + 3(m)^2(2n) + 3m(2n)^2 + (2n)^3 = m^3 + 6m^2n + 12mn^2 + 8n^3$

(d) $(3p)^3 + 3(3p)^2(-q) + 3(3p)(-q)^2 + (-q)^3 = 27p^3 - 27p^2q + 9pq^2 - q^3$

(e) $1^2 + 2(1)(1.5t) + (1.5t)^2 = 1 + 3t + 2.25t^2$

(f) $-(2a + b + c)^2 = -[(2a + b) + c]^2 = -[(2a + b)^2 + 2c(2a + b) + c^2]$

$$= -(4a^2 + 4ab + b^2 + 4ac + 2bc + c^2)$$

$$= -4a^2 - 4ab - b^2 - 4ac - 2bc - c^2$$

(g) $(2r - p - m)(2r + p + m) = [2r - (p + m)][2r + (p + m)]$

$$= (2r)^2 - (p + m)^2$$

$$= 4r^2 - (p^2 + 2mp + m^2)$$

$$= 4r^2 - p^2 - 2mp - m^2$$

(h) $(2r - p - m)(2r - p + m) = [(2r - p) - m][(2r - p) + m]$

$$= (2r - p)^2 - m^2$$

$$= 4r^2 - 4rp + p^2 - m^2$$

5.18. Expand the following by inspection:

(a) $(2x - y)(2x + y)$ (c) $\left(r + \dfrac{t}{2}\right)\left(r - \dfrac{t}{2}\right)$ (e) $(0.2 - g)(0.2 + g)$

(b) $(1 + p)(1 - p)$ (d) $(v_1 + v_2)(v_1 - v_2)$

Each expression is the product of the sum and difference of two quantities. This product is equal to the difference of the squares of the quantities.

(a) $4x^2 - y^2$ (b) $1 - p^2$ (c) $r^2 - t^2/4$ (d) $v_1^2 - v_2^2$ (e) $0.04 - g^2$

FACTORING

5.19. Factor the following:

(a) $d^2 - cd$ (c) $2\pi r^2 + 2\pi rh$ (e) $6m^3 + 3m^2n + 12mp$

(b) $ab^2 + a^2b + a^2b^2$ (d) $\dfrac{x^2}{9} - \dfrac{4y^2x^2}{9}$

There is a common monomial factor in each of the given expressions.   Removal of this factor yields the following answers:

(a) $d(d-c)$ (c) $2\pi r(r+h)$ (e) $3m(2m^2 + mn + 4p)$

(b) $ab(b + a + ab)$ (d) $\dfrac{x^2}{9}(1 - 4y^2) = \dfrac{x^2}{9}(1 + 2y)(1 - 2y)$

5.20. Factor completely the following:

(a) $9x^2 - 16y^2$ (c) $\dfrac{R^2}{9} - \dfrac{T^2}{16}$ (e) $8y^3 - 36y^2 t + 54yt^2 - 27t^3$

(b) $v^2 - 2vs + s^2$ (d) $n^2 + 10n + 25$

(a) The binomial is the difference of two squares.   Therefore, the factored form is $(3x + 4y)(3x - 4y)$.

(b) The given polynomial is the square of a binomial.   The factored form is $(v - s)(v - s)$.

(c) This difference of two squares factors into $\left(\dfrac{R}{3} + \dfrac{T}{4}\right)\left(\dfrac{R}{3} - \dfrac{T}{4}\right)$.

(d) The given polynomial is the square of a binomial.   The factored form is $(n + 5)(n + 5)$.

(e) The given polynomial is the cube of a binomial whose terms are $2y$ and $-3t$.   The factored form is $(2y - 3t)(2y - 3t)(2y - 3t)$ or $(2y - 3t)^3$.

5.21. Factor completely the following:

(a) $p^2 + 7p + 12$ (c) $j^2 - 4j - 45$ (e) $2x^2 - 11x + 15$.

(b) $b^2 - 3b - 4$ (d) $9x^2 + 10x + 1$

All the given trinomials can be factored into products of the type $(ax + b)(cx + d)$, with a and e factors of the leading coefficient, and b and d factors of the constant term.

(a) Since the coefficient of p^2 is 1, $a = 1$ and $c = 1$.   Therefore, the first step in factoring is $(p\ )(p\ )$.   In the cross-product $7p$, the coefficient 7 is the sum of the factors of 12, which may be (12, 1) or (6, 2) or (4, 3); only the last set of factors satisfies this condition.   Hence $p^2 + 7p + 12$ factors into $(p + 3)(p + 4)$.

(b) The factors are $(b + 1)(b - 4)$

(c) The factors are $(j + 5)(j - 9)$

(d) Multiply the trinomial by 9 and write it as $81x^2 + 9(10x) + 9$.   Switch the coefficients of x to obtain $81x^2 + 10(9x) + 9$.   Rewrite the expression as $(9x)^2 + 10(9x) + 9$.   Look for two numbers that add up to 10 and multiply equal to 9.   Factor the expression using these numbers and write it as $(9x)^2 + 10(9x) + 9 = (9x + 9)(9x + 1)$.   Divide the expression by 9.   The original trinomial can be written as $9x^2 + 10x + 1 = (x + 1)(9x + 1)$.

(e) Multiply the trinomial by 2 and write it as $4x^2 - 2(11x) + 30$.   Switch the coefficient of x to obtain $4x^2 - 11(2x) + 30$.   Rewrite the expression as $(2x)^2 - 11(2x) + 30$.   Look for two numbers that add up to -11 and multiply equal to $+30$.   Factor the expression using these two numbers and write it as $(2x)^2 - 11(2x) + 30 = (2x - 6)(2x - 5)$.   Divide the expression by 2.   The original trinomial can be written as $2x^2 - 11x + 15 = (x - 3)(2x - 5)$.

5.22. Simplify $(2x^2 - 3x - 2)(2x^2 + x - 1)/(x^2 - x - 2)$.

If the numerator and denominator of this algebraic fraction are factored, and the common factors are canceled, the result is

$$\frac{\cancel{(x-2)}(2x+1)\cancel{(x+1)}(2x-1)}{\cancel{(x-2)}\cancel{(x+1)}} = (2x+1)(2x-1) = 4x^2 - 1$$

5.23. Perform the indicated operations and simplify:

(a) $\dfrac{16a^2b^3}{8a^2b} \cdot \dfrac{4ab}{2b}$ (c) $\dfrac{2ab(c^4 - 1)}{c - 1}$ (e) $\dfrac{2a^2 + a - 1}{a^2 - 1} \div \dfrac{2a^2 - a}{a(a + 1)}$

(b) $\dfrac{4x^2 - 9y^2}{2x + y} \cdot \dfrac{2x^2 - xy - y^2}{2x + 3y}$ (d) $\dfrac{p^2 - 4q^2}{pq + 2q^2} \cdot \dfrac{q}{p - 2q}$

(a) Canceling the common factors, the result is $4ab^2$.

(b) Factoring,

$$4x^2 - 9y^2 = (2x + 3y)(2x - 3y)$$
$$2x^2 - xy - y^2 = (2x + y)(x - y)$$

The given expressions can be written as

$$\frac{\cancel{(2x+3y)}(2x - 3y) \cdot \cancel{(2x+y)}(x - y)}{\cancel{(2x+y)}\cancel{(2x+3y)}}$$

After cancellation the result is

$$(2x - 3y)(x - y)$$

(c) Factor $c^4 - 1$ into $(c^2 + 1)(c^2 - 1)$ and $(c^2 - 1)$ into $(c + 1)(c - 1)$. Then the given expression can be written as

$$\frac{2ab(c^2 + 1)(c + 1)\cancel{(c - 1)}}{\cancel{(c - 1)}}$$

The result is $2ab(c^2 + 1)(c + 1)$.

(d) The factors of $p^2 - 4q^2$ are $(p + 2q)$ and $(p - 2q)$. Rewriting the given expression and canceling common factors, the result is

$$\frac{\cancel{(p+2q)}\cancel{(p-2q)} \cdot q}{q\cancel{(p+2q)}\cancel{(p-2q)}} = 1$$

(e) First invert the second fraction and multiply. Then factor and cancel the common terms

$$\frac{(2a - 1)(a + 1)}{(a + 1)(a - 1)} \cdot \frac{\cancel{a}\cancel{(a + 1)}}{\cancel{a}\cancel{(2a - 1)}}$$

The result is $\dfrac{a + 1}{a - 1}$.

FRACTIONS

5.24. Express each of the following in the form of a single fraction:

(a) $\dfrac{5}{T} - \dfrac{1}{3T}$ (b) $1 + \dfrac{Y}{1 - Y}$ (c) $\dfrac{3}{x - 1} + \dfrac{2}{1 - x}$

(d) $\dfrac{-B}{2R^2} - \dfrac{A}{R}$ (e) $\dfrac{4}{m^2 + 2m + 1} - \dfrac{2}{m + 1}$

The procedure is the same as for the addition of arithmetical fractions. After determining the least common denominator, add the numerators

(a) LCD $= 3T$

$$\frac{3 \cdot 5}{3T} - \frac{1}{3T} = \frac{15 - 1}{3T} = 14/(3T)$$

(b) LCD $= 1 - Y$

$$\frac{1 - Y}{1 - Y} \cdot 1 + \frac{Y}{1 - Y} = \frac{1 - Y + Y}{1 - Y} = 1/(1 - Y)$$

(c) LCD $= x - 1$

If the numerator and the denominator of the second fraction are multiplied by -1, the expression becomes

$$\frac{3}{x - 1} + \frac{(-1)2}{(-1)(1 - x)} = \frac{3}{x - 1} - \frac{2}{x - 1} = \frac{3 - 2}{x - 1} = 1/(x - 1)$$

(d) LCD $= 2R^2$

$$\frac{-B}{2R^2} - \frac{2R}{2R} \cdot \frac{A}{R} = \frac{-B - 2RA}{2R^2} = \frac{-(B + 2RA)}{2R^2}$$

(e) Factor $m^2 + 2m + 1$ into $(m + 1)(m + 1)$; this is the LCD, since the denominator of the second fraction is $m + 1$

$$\frac{4}{(m + 1)(m + 1)} - \frac{m + 1}{m + 1} \cdot \frac{2}{m + 1} = \frac{4 - 2(m + 1)}{(m + 1)(m + 1)} = \frac{4 - 2m - 2}{(m + 1)(m + 1)} = \frac{2(1 - m)}{(m + 1)^2}$$

5.25. Perform the indicated operations and simplify:

(a) $\left(\dfrac{5m^2n^3}{8m^3n}\right)\left(\dfrac{2mn}{15m^2n^2}\right)$ (c) $\left(\dfrac{2s + 1}{s - 1}\right)\left(\dfrac{s + 1}{2s - 1}\right)$ (e) $\dfrac{2d^2 + 5d - 3}{(d - 6)^2} \div \dfrac{2d - 1}{d^2 - 36}$

(b) $\dfrac{3ab}{2c} \div \dfrac{3c}{2ab}$ (d) $\dfrac{x}{80} \div \dfrac{(0.05 - x)}{100}$

(a) Cancel common factors first. Then multiply the numerators and multiply the denominators

$$\frac{5m^2n^3}{8m^3n} \cdot \frac{2mn}{15m^2n^2} = n/(12m^2)$$

(b) Invert the second fraction; cancel the common factors, then multiply the numerators and multiply the denominators

$$\frac{3ab}{2c} \cdot \frac{2ab}{3c} = a^2b^2/c^2$$

(c) There are no common factors and hence no cancellation is possible.   Multiplying the numerators and the denominators gives the result

$$\frac{2s^2 + 3s + 1}{2s^2 - 3s + 1}$$

(d) Invert the second fraction, multiply by the first, and simplify

$$\frac{x}{80} \frac{100}{(0.05 - x)} = \frac{100x}{80(0.05 - x)} = \frac{100x}{4 - 80x} = \frac{25x}{1 - 20x}$$

(e) Invert the second fraction and multiply by the first:

$$\frac{2d^2 + 5d - 3}{(d - 6)^2} \frac{d^2 - 36}{(2d - 1)}$$

Factor and cancel like terms; multiply the remaining factors in the numerator and denominator, respectively.

$$\frac{(2d - 1)(d + 3)}{(d - 6)(d - 6)} \frac{(d + 6)(d - 6)}{(2d - 1)} = \frac{(d + 3)(d + 6)}{d - 6} = \frac{d^2 + 9d + 18}{d - 6}$$

EQUATIONS

5.26. Solve the following equations:

(a) $3h = 12$ (b) $40 = 8a$ (c) $0.1p = 5$ (d) $3y = 13\frac{1}{2}$

(e) $0.96 = 0.8r$

   To solve an equation of the first degree it is necessary to have the unknown on one side of the equation with coefficient 1.   To accomplish this, divide each side of the equation by the original coefficient of the unknown

(a) $\dfrac{3h}{3} = \dfrac{12}{3}$ $h = \dfrac{12}{3} = 4$

(b) $\dfrac{40}{8} = \dfrac{8a}{8}$ $a = \dfrac{40}{8} = 5$

(c) $\dfrac{0.1p}{0.1} = \dfrac{5}{0.1}$ $p = \dfrac{5}{0.1} = 50$

(d) $\dfrac{3y}{3} = \dfrac{13\frac{1}{2}}{3}$ $y = \dfrac{13\frac{1}{2}}{3} = \dfrac{\frac{27}{2}}{3} = 9/2$

(e) $\dfrac{0.096}{0.8} = \dfrac{0.8r}{0.8}$ $r = \dfrac{0.96}{0.8} = 1.2$

5.27. Solve the following equations:

(a) $y/3 = 12$ (b) $5 = (1/2)n$ (c) $(1/8)d = 2.5$ (d) $\dfrac{3n}{5} = 15$

(e) $\dfrac{8x}{18} = 26$

   If the given fractional coefficients of the unknown are multiplied by their reciprocals, the coefficients become 1.   The other member of the equation must also be multiplied by the same reciprocal.

(a) $\dfrac{y}{\not3} \cdot \not3 = 12 \cdot 3$ $y = 36$

(b) $5 \cdot 2 = \dfrac{1}{\not2} \cdot \not2 n$ $n = 10$

(c) $\dfrac{1}{\not8} \cdot \not8 d = 2.5 \cdot 8$ $d = 20$

(d) $\dfrac{\not3 n}{\not3} \cdot \dfrac{\not3}{\not3} = \overset{5}{\not{15}} \cdot \dfrac{5}{\not3}$ $n = 25$

(e) $\dfrac{\not3 x}{18} \cdot \dfrac{18}{\not8} = \overset{13}{\not{26}} \cdot \dfrac{\overset{9}{\not{18}}}{\underset{\underset{2}{4}}{\not8}}$ $x = 117/2 = 58\tfrac{1}{2}$

5.28. Solve and check the following equations:

    (a) $a + 6 = 8$ (c) $b - 3 = 7$ (e) $S + 2.25 = 1.75$

    (b) $9 = 5 + d$ (d) $12 = c - 2$

        Transpose the numerical terms so that the unknown is alone on one side of the equation.

    (a) $a = 8 - 6$ $a = 2$ *Check:* $2 + 6 \overset{?}{=} 8$ $8 = 8$

    (b) $9 - 5 = d$ $d = 4$ *Check:* $9 \overset{?}{=} 5 + 4$ $9 = 9$

    (c) $b = 7 + 3$ $b = 10$ *Check:* $10 - 3 \overset{?}{=} 7$ $7 = 7$

    (d) $12 + 2 = c$ $c = 14$ *Check:* $12 \overset{?}{=} 14 - 2$ $12 = 12$

    (e) $S = 1.75 - 2.25$ $S = -0.50$ *Check:* $-0.50 + 2.25 \overset{?}{=} 1.75$ $1.75 = 1.75$

5.29. Solve for the unknown and check the following equations:

    (a) $16 = 2(t + 3)$

              $16 = 2t + 6$ remove parentheses

              $16 - 6 = 2t$ transpose 6

                 $10 = 2t$ collect terms on left side

                  $5 = t$ divide both members by 2

           or $t = 5$

         Check: $16 \overset{?}{=} 2(5) + 6$ $16 \overset{?}{=} 10 + 6$ $16 = 16$

    (b) $2x - 4 = 2(x - 6) - (x - 10)$

            $2x - 4 = 2x - 12 - x + 10$ remove parentheses

            $2x - 2x + x = -12 + 10 + 4$ transpose the unknown terms to the
                                                left side: transpose -4 to the right side

        Check: $2(2) - 4 \overset{?}{=} 2(2 - 6) - (2 - 10)$

                 $4 - 4 \overset{?}{=} 2(-4) - (-8)$

                     $0 \overset{?}{=} -8 + 8$

(c) $4 + \dfrac{3b}{10} = 7$

$\dfrac{3b}{10} = 7 - 4$ transpose 4

$\dfrac{3b}{10} = 3$ collect terms on right side

$\dfrac{\cancel{10}}{\cancel{3}} \cdot \dfrac{\cancel{3}b}{\cancel{10}} = \dfrac{10}{\cancel{3}} \cdot \cancel{3}$ multiply both sides by 10/3, the reciprocal of 3/10

$b = 10$

Check: $4 + \dfrac{3(10)}{10} \overset{?}{=} 7$ $4 + \dfrac{30}{10} \overset{?}{=} 7$ $4 + 3 \overset{?}{=} 7$ $7 = 7$

(d) $\dfrac{2a - 3}{6} = \dfrac{2a + 3}{10}$

$\overset{5}{\cancel{30}} \dfrac{(2a - 3)}{\cancel{6}} = \overset{3}{\cancel{30}} \dfrac{(2a + 3)}{\cancel{10}}$ multiply both members by 30, LCD of 6 and 10; cancel common factors

$10a - 15 = 6a + 9$ remove parentheses

$10a - 6a = 9 + 15$ transpose

$4a = 24$ collect terms

$a = 6$ divide both sides by 4

Check: $\dfrac{2(6) - 3}{6} \overset{?}{=} \dfrac{2(6) + 3}{10}$ $\dfrac{12 - 3}{6} \overset{?}{=} \dfrac{12 + 3}{10}$ $\dfrac{9}{6} \overset{?}{=} \dfrac{15}{10}$ $\dfrac{3}{2} = \dfrac{3}{2}$

(e) $0.75i - 0.41 = 0.25i + 2.09$

$75i - 41 = 25i + 209$ multiply both sides by 100

$75i - 25i = 209 + 41$ transpose terms

$50i = 250$ collect terms

$i = 5$ divide both sides by 50

Check: $0.75(5) - 0.41 \overset{?}{=} 0.25(5) + 2.09$

$3.75 - 0.41 \overset{?}{=} 1.25 + 2.09$

$3.34 = 3.34$

5.30. Solve the following equation for S: $300S(99 - 15) = 320(15 - 8)$.

$300 \cdot S \cdot 84 = 320 \cdot 7$ remove parentheses on both sides

$S = \dfrac{320 \cdot 7}{300 \cdot 84}$ divide both members by $300 \cdot 84$

$= \dfrac{4}{45}$ cancel like factors

≈ 0.0889 carry out indicated division

Check: $300(0.0889)(84) \overset{?}{\approx} 320 \cdot 7$

$2240 = 2240$

5.31. Solve the following equation for E: $12 = \dfrac{115 - E}{0.25}$.

$$(12)(0.25) = 115 - E \qquad \text{multiply both members of the equation by } 0.25$$

$$3 = 115 - E \qquad \text{simplify the left member}$$

$$E = 115 - 3 \qquad \text{transpose } -E \text{ to the left side and 3 to the right side}$$

$$E = 112 \qquad \text{simplify the right member}$$

$$Check: \quad 12 \overset{?}{=} \frac{115 - 112}{0.25}$$

$$12 \overset{?}{=} \frac{3}{0.25}$$

$$12 = 12$$

5.32. Solve the following equation for P: $\dfrac{21}{50} = \dfrac{16}{P - 15}$.

$$21(P - 15) = 50 \cdot 16 \qquad \text{since the given equation may be regarded as a proportion, the product of the means equals the product of the extremes}$$

$$21 \cdot P - 21 \cdot 15 = 50 \cdot 16 \qquad \text{remove parentheses}$$

$$21P - 315 = 800 \qquad \text{simplify}$$

$$21P = 800 + 315 = 1115 \qquad \text{transpose 315 to the right side}$$

$$P = \frac{1115}{21} \approx 53.1 \qquad \text{divide both members by 21}$$

$$Check: \quad 21(53.1 - 15) \overset{?}{=} 50 \cdot 16$$

$$21(38.1) \overset{?}{=} 800$$

$$800.1 \approx 800$$

5.33. Solve the following equation for V:

$$\frac{1}{60} + \frac{1}{V} = (1.5 - 1)\left(\frac{1}{30} - \frac{1}{10}\right)$$

$$\frac{1}{60} + \frac{1}{V} = (0.5)\left(\frac{-2}{30}\right) \qquad \text{perform operations in parentheses}$$

$$\frac{1}{60} + \frac{1}{V} = \frac{-1}{30} \qquad \text{simplify the right member}$$

$$\frac{1}{V} = -\frac{1}{60} - \frac{1}{30} \qquad \text{transpose } \frac{1}{60} \text{ to the right side}$$

$$\frac{1}{V} = -\frac{3}{60} \qquad \text{simplify right member}$$

$$-3V = 60 \qquad \text{multiply both members by } 60V$$

$$V = \frac{60}{-3} = -20 \qquad \text{divide both members by } -3$$

$$\textit{Check:} \quad \frac{1}{60} + \frac{1}{-20} \overset{?}{=} (1.5 - 1)\left(\frac{1}{30} - \frac{1}{10}\right)$$

$$-\frac{1}{30} = -\frac{1}{30}$$

FORMULAS

Many of the formulas used in science and technology are first-degree equations.

5.34. Solve the following for the indicated symbol:

(a) $C = K - 273$, for K

$$C + 273 = K \qquad\qquad \text{transpose } -273$$

or $\quad K = C + 273$

(b) $F = 1.8C + 32$, for C

$$F - 32 = 1.8C \qquad\qquad \text{transpose } 32$$

$$\frac{F - 32}{1.8} = C \qquad\qquad \text{divide both members by } 1.8$$

or $\quad C = \dfrac{F}{1.8} - \dfrac{32}{1.8} \equiv 0.56F - 17.8$

(c) $v = v_0 + 32t$, for v_0; for t

$$v - 32t = v_0 \qquad\qquad \text{transpose } -32t$$

or $\quad v_0 = v - 32t$

Also, $\quad v - v_0 = 32t \qquad\qquad \text{transpose } v_0 \text{ and } -32t$

$$\frac{v - v_0}{32} = t \qquad\qquad \text{divide both members by } 32$$

or $\quad t = (v - v_0)/32$

(d) $P = \pi d + 2w$, for d; for w

$$P - 2w = \pi d \qquad\qquad \text{transpose } 2w$$

$$\frac{P - 2w}{\pi} = d \qquad\qquad \text{divide both members by } \pi$$

or $\quad d = (P - 2w)/\pi$

Also, $\quad P - \pi d = 2w \qquad\qquad \text{transpose } \pi d$

$$\frac{P - \pi d}{2} = w \qquad\qquad \text{divide both members by } 2$$

or $\quad w = (P - \pi d)/2$

(e) $V = V_0(1 + bt)$, for V_0; for t

$$\frac{V}{1 + bt} = V_0 \qquad\qquad \text{divide both sides by } (1 + bt)$$

or $\quad V_0 = V/(1 + bt)$

Also, $V = V_0 + V_0bt$ remove parentheses

$V - V_0 = V_0bt$ transpose V_0

$\dfrac{V - V_0}{V_0b} = t$ divide both members by V_0b

or $t = (V - V_0)/(V_0b)$

5.35. Evaluate the unknown with the given values for the other symbols:

(a) $C = \pi d$; $\pi = 3.14$, $C = 12.56\,\text{cm}$

$12.56 = 3.14d$ substitute the given values in the formula

$\dfrac{12.56}{3.14} = \dfrac{\cancel{3.14}}{\cancel{3.14}}d$ divide both sides by 3.14

$4 = d$ or $d = 4\,\text{cm}$

(b) $V = V_0 + at$; $V = 20\,\text{m/sec}$, $V_0 = 80\,\text{m/sec}$, $t = 5\,\text{sec}$

$20 = 80 + a \cdot 5$ substitute the given values

$20 - 80 = 5a$ transpose 80

$-60 = 5a$ collect terms on the left side

$\dfrac{-60}{5} = \dfrac{\cancel{5}a}{\cancel{5}}$ divide both members by 5

$-12 = a$ or $a = -12\,\dfrac{\text{m/sec}}{s} = -12\,\text{m/sec}^2$

(c) $C = (5/9)(F - 32)$; $F = 50\,°\text{F}$

$C = \dfrac{5}{9}(50 - 32)$ substitute the given values in the formula

$C = \dfrac{5}{\cancel{9}}\overset{2}{\cancel{(18)}}$ perform the operation inside the parentheses and cancel

$C = 5(2)$ or $C = 10\,°\text{F}$

(d) $R = R_0(1 + at)$; $R = 101\,\text{ohm}$, $R_0 = 100\,\text{ohm}$, $t = 200\,°\text{C}$

$101 = 100(1 + a \cdot 200)$ substitute the given values

$101 = 100 + 20{,}000a$ remove parentheses

$101 - 100 = 20{,}000a$ transpose 100

$1 = 20{,}000a$ collect terms on left side

$\dfrac{1}{20{,}000} = \dfrac{\cancel{20{,}000}}{\cancel{20{,}000}}a$ divide both members by 20,000

$a = \dfrac{1}{20{,}000} = 0.00005 = 5 \times 10^{-5}/°\text{C}$

(e) $V = \dfrac{bhH}{3}$; $V = 100\,\text{ft}^3$, $b = 10\,\text{ft}$, $h = 5\,\text{ft}$

$100 = \dfrac{10 \cdot 5 \cdot H}{3}$ substitute the given values

$$100 \cdot 3 = \frac{50H \cdot \cancel{3}}{\cancel{3}}$$ multiply both sides by 3 and cancel

$$300 = 50H$$

$$\frac{\overset{6}{\cancel{300}}}{\cancel{50}} = \frac{\cancel{50}}{\cancel{50}}H$$ divide both members by 50

$$6 = H \quad \text{or} \quad H = 6\,\text{ft}$$

Supplementary Problems

The answers to this set of problems are at the end of this chapter.

5.36. Find the absolute value of each of the following:

(a) -5 (b) 10 (c) -6 (d) -1 (e) $-g$

5.37. In each of the following pairs of numbers, subtract the first number from the first:

(a) $3, -6$ (b) $5, 10$ (c) $-3, -6$ (d) $0, -5$ (e) $-9, 7$

5.38. In each of the following pairs of numbers, subtract the first number from the second:

(a) $0, -5$ (b) $-4, -4$ (c) $-8, 6$ (d) $3, -9$ (e) $-1, 0$

5.39. Find the sum of the following numbers:

(a) $-3, 6, -5, -2$ (c) $1/2, -3/4, -5/8$ (e) $3, -1, -4, 5, 10, -2, -12$

(b) $-2, -5, 6, 8, 10$ (d) $3, 4, 7, -4, -20, -30$

5.40. Find the products of the numbers in the sets of Problem 5.39.

5.41. Determine the sum of the following expressions:

(a) $3x, -5x, x$ (c) $x^2, -xy, 3x^2$ (e) $5, -3xy, x^2y^2, -6xy, -5x^2y^2, -12$

(b) $-3t, 6t, -5t, 2t$ (d) $pq, -5pq, p^3q, p^2q^2$

5.42. Simplify each of the following products:

(a) $(xy)(x^2y)$ (c) $(3x)(5xy)(-2x^2y^3)$ (e) $(0.2x)(-1.2xy)(-0.5xyz)$

(b) $(3ab)(-2ab^2)$ (d) $(-2abc)(3a^2bc^2)$

5.43. Divide the first expression by the second:

(a) $-10mn, 2m^2$ (c) $6p^3q^2, -3pq^3$ (e) $24ab^2c^3, 12a$

(b) $8a^2bc^3, 4ab$ (d) $0.56xy^3, 0.7xy^4$

5.44. Multiply as indicated:

(a) $(2x - y)^2$ (c) $(4a - 2b)^2$ (e) $(r + 0.1)^2$ (g) $(2x + 3y + 5)^2$

(b) $(3x + 2z)^2$ (d) $(1 + at)^2$ (f) $-(4x + 7)^2$ (h) $(x + y + 1)(x - y - 1)$

5.45. Multiply as indicated:

    (a) $(2x + y)^3$ (c) $(4x + 3y)^3$ (e) $(a + 0.1)^3$

    (b) $(5p - q)^3$ (d) $(1 + bt)^3$

5.46. Simplify:

    (a) $2x - 3[(2x - 3y) + x - 2]$ (c) $4 - 5a\{2a + 6[-4(a - 2)]\}$ (e) $2y - 3x[2x + (1 - (4x + y) + 2]$

    (b) $2t - [t^2 + 1 - 3(t - 1)]$ (d) $t[3t^2 - (4t + 4t - 2t^2) + 1 + t]$

5.47. Multiply $3x^3 - 2x^2 + x - 2$ by:

    (a) -2 (b) $-3x$ (c) $x - y$ (d) $2x - 2$ (e) $5xy$

5.48. Perform the indicated multiplications:

    (a) $(3x - 2y)(x + 5y)$ (c) $(pq - 2)(p^2 + 1)$ (e) $(2u + 3v)(4u - 2v)$

    (b) $(3x - 2y)(6x + y)$ (d) $(2t^2 - 1)(t + 2)$

5.49. Perform the indicated divisions:

    (a) $(10c^2 - 25c) \div (-5c)$ (c) $(24m^3n^2 + 16m^2n^3) \div (-8m^3n^3)$ (e) $(5pq - 3pq^2 + 2p) \div (pq)$

    (b) $(12a^3 + 6a^2) \div (3a^2)$ (d) $2(y - 3)(3y + 1) \div 4(y - 3)$

5.50. Factor a monomial from each of the following polynomials:

    (a) $5x^2 - 10xy + 15xy^2$ (c) $2a + 4abc - 2ac + 2a^3$ (e) $\dfrac{Ba}{2} + \dfrac{ba}{2}$

    (b) $3abc - 3ac + 6acd$ (d) $2r^3 - r^2$

5.51. Factor each of the following polynomials:

    (a) $16y^2 - x^2$ (c) $4a^2b^2 - 25c^2$ (e) $s(t_1^4 - t_2^4)$

    (b) $9x^2y^2 - 4z^2$ (d) $\dfrac{1}{4x^2} - 4y^2$

5.52. Factor each of the following polynomials:

    (a) $v^2 - 2vs + s^2$ (c) $a^2b^2 - 2ab + 1$ (e) $8a^3 - 36a^2b + 54ab^2 - 27b^3$

    (b) $9n^2 + 30n + 25$ (d) $rs^2 + 8rs + 16r$

5.53. Factor each of the following:

    (a) $2x^3 - 2xy^2$ (c) $2m^2 + 2m - 4$ (e) $x^2 - 8x + 7$

    (b) $12ax^2 - 3ay^2$ (d) $t^2 + 2t - 15$ (f) $p^2 + 3p - 10$

5.54. Factor the following:

    (a) $2x^2 + 3x - 2$ (d) $4a^2 - 2ab - 2b^2$ (g) $4x^2 + x - 33$

    (b) $3p^2 + 7pq - 6q^2$ (e) $6s^2 - 7st - 5t^2$ (h) $20x^2 + x - 1$

    (c) $2m^2 + 13m - 7$ (f) $21x^2 + 11x - 2$ (i) $-15x^2 + 16x - 4$

5.55. Simplify each of the following expressions:

    (a) $(2p^2 - 128) \div (p - 8)$ (b) $\dfrac{d^2 - 4n^2}{dn + 2n^2} \cdot \dfrac{2n^2}{d - 2n}$

(c) $ab(a^2 - b^2)/(a^2 + ab)$ (d) $(3x - y)(2x^2 - xy - y^2)/(2x + y)$

5.56. Which of the following expressions can be simplified? Perform simplifications, when possible.

(a) $\dfrac{n^2 - 1}{n^2 + 2}$ (b) $\dfrac{(m - 1)^2}{(m + 1)^2}$ (c) $\dfrac{a^2 + b^2}{a + b}$ (d) $\dfrac{a^3 + y^3}{a^2 + y^2}$ (e) $\dfrac{(p - z)^3}{p^2 - z^2}$

5.57. Simplify the following sums:

(a) $\dfrac{1}{u} + \dfrac{1}{v}$ (c) $\dfrac{q_1}{r_1} + \dfrac{q_2}{r_2}$ (e) $\left(\dfrac{mR^2}{2} + mR^2\right)\dfrac{a}{R}$

(b) $\dfrac{1}{d} + \dfrac{bt}{d}$ (d) $\dfrac{1}{12}ML^2 + M\dfrac{L^2}{2}$

5.58. Simplify the indicated differences:

(a) $\dfrac{1}{p} - \dfrac{1}{q}$ (c) $\dfrac{(h_1/h_2) - 1}{t}$ (e) $m\left(d + \dfrac{h}{2}\right)^2 - m\left(d - \dfrac{h}{2}\right)^2$

(b) $\dfrac{Q}{5}\left(\dfrac{1}{r_1} - \dfrac{1}{r^2}\right)$ (d) $\dfrac{m}{(r - x)^2} - \dfrac{m}{(r + x)^2}$

5.59. Simplify the following products:

(a) $\dfrac{21rs^3}{(2p + q)^2} \cdot \dfrac{4p^2 - q^2}{(15r^2s)}$ (c) $\dfrac{GMm}{(R + h)^2} \cdot \dfrac{R + h}{m}$ (e) $\dfrac{GME}{r^2}\left(\dfrac{1}{Mr}\right)$

(b) $f\left(\dfrac{V}{V - s}\right)\left(1 + \dfrac{E}{V}\right)$ (d) $\left(\dfrac{2\pi MR^4}{4}\right)\left(\dfrac{1}{\pi R^2}\right)$

5.60. Simplify:

(a) $\dfrac{1}{8} \div \dfrac{2d}{c}$ (c) $\dfrac{3}{2}Mr^2 \div Mgr$ (e) $\dfrac{2\pi ML(R_2^4 - R_1^4)}{4\pi(R_2^2 - R_1^2)L}$

(b) $q - \dfrac{A(t_2 - t_1)}{L}$ (d) $\dfrac{2}{(1/R) + (1/1100)}$

5.61. Solve the following equations for the unknown and check the solutions:

(a) $c + 3 = 8$ (b) $h - 3 = 6$ (c) $12 = d - 2$ (d) $0.6U = 1.44$ (e) $(2a)/3 = 4$

5.62. Solve if possible the following equations and check the solutions:

(a) $3x - 4 + x = 5x - 6$

(b) $2y + 6 - 3y = 3 - 2y + 12$

(c) $4t - 6 + 2t = t + 2t + 3t - 10$

(d) $5(h + 3) - 2(4h - 10) = 14$

(e) $19 - 5E(4E + 1) = 40 - 10E(2E - 1)$

5.63. Solve the following equations and check the solutions:

(a) $18(10 - E) + 8.5E = 130$

(b) $(0.05)(9.8)(10) = (0.05)(0.64)F + 0.77F$

(c) $120.7S(98.7 - 30.3) = 94.5(98.7 - 26.3)$

(d) $0.75i - 0.41 = 0.25i + 2.09$

(e) $0.6r - 0.56 + 10(1 - 0.4r) = 0$

5.64. Solve the following equations and check the solutions:

(a) $32a = 640 - 20a$ (c) $258 = 250\left(\dfrac{330 + y}{330}\right)$ (e) $212 = \dfrac{9}{4}R + 32$

(b) $4 = \dfrac{90 - 12}{d}$ (d) $294 = 250\left(\dfrac{V + 88}{V - 88}\right)$

5.65. Solve for the indicated symbol:

(a) $3ax = 12$, for x (c) $R = \dfrac{E}{I}$, for E (e) $5R - 12S = 18$, for S

(b) $v = v_0 + gt$, for t (d) $PV = 12T$, for V

5.66. Solve the following equations and check the solutions:

(a) $\dfrac{1}{10} = \dfrac{1}{60} + \dfrac{1}{R}$ (c) $\dfrac{2}{45} = \dfrac{1}{p} - \dfrac{1}{15}$ (e) $\dfrac{1}{F} = (1.6 - 1)\left(\dfrac{1}{-60} + \dfrac{1}{40}\right)$

(b) $\dfrac{1}{-10} = \dfrac{1}{30} + \dfrac{1}{q}$ (d) $3 = \dfrac{25}{f} + 1$

5.67. Substitute the given values for the symbols and determine the value of the unknown:

(a) $V = E - 5I$; $E = 122$, $I = 4$

(b) $P = 2(L + h)$; $P = 15$, $L = 4.5$

(c) $F = \dfrac{9}{5}C + 32$; $F = 68$

(d) $\dfrac{1}{R} = \dfrac{1}{R_1} + \dfrac{1}{R_2}$; $R_1 = 150$, $R_2 = 300$

(e) $\dfrac{\lambda - \lambda_0}{\lambda_0} = \dfrac{v}{e}$; $\lambda = 5005$, $\lambda_0 = 5000$, $c = 3 \times 10^{10}$

Answers to Supplementary Problems

5.36. (a) 5   (b) 10   (c) 6   (d) 1   (e) g, if $g \geq 0$

5.37. (a) 9   (b) -5 (c) 3   (d) 5   (e) -16

5.38. (a) -5 (b) 0   (c) 14   (d) -12 (e) 1

5.39. (a) -4 (b) 17   (c) $-\frac{7}{8}(d)$ (d) -40 (e) -1

5.40. (a) -180 (b) 4800   (c) $\dfrac{15}{64}$ (d) $-201,600$ (e) 14,400

5.41. (a) $-x$ (b) 0   (c) $4x^2 - xy$ (d) $p^3q + p^2q^2 - 4pq$ (e) $-4x^2y^2 - 9xy - 7$

5.42. (a) x^3y^2 (b) $-6a^2b^3$ (c) $-30x^4y^4$ (d) $-6a^3b^2c^3$ (e) $0.12x^3y^2z$

5.43.   (a) $-5n/m$ (b) $2ac^3$ (c) $-2p^2/q$ (d) $0.8/y$ (e) $2b^2c^3$

5.44.   (a) $4x^2 - 4xy + y^2$ (d) $1 + 2at + a^2t^2$ (g) $4x^2 + 12xy + 20x + 9y^2 + 30y + 25$

        (b) $9x^2 + 12xz + 4z^2$ (e) $r^2 + 0.2r + 0.01$ (h) $x^2 - y^2 - 2y - 1$

        (c) $16a^2 - 16ab + 4b^2$ (f) $-16x^2 - 56x - 49$

5.45.   (a) $8x^3 + 12x^2y + 6xy^2 + y^3$ (d) $1 + 3bt + 3b^2t^2 + b^3t^3$

        (b) $125p^3 - 75p^2q + 15pq^2 - q^3$ (e) $a^3 + 0.3a^2 + 0.03a + 0.001$

        (c) $64x^3 + 144x^2y + 108xy^2 + 27y^3$

5.46.   (a) $9y - 7x + 6$ (c) $110a^2 - 240a + 4$ (e) $6x^2 + 3xy - 9x + 2y$

        (b) $-t^2 + 5t - 4$ (d) $5t^3 - 7t^2 + t$

5.47.   (a) $-6x^3 + 4x^2 - 2x + 4$ (d) $6x^4 - 10x^3 + 6x^2 - 6x + 4$

        (b) $-9x^4 + 6x^3 - 3x^2 + 6x$ (e) $15x^4y - 10x^3y + 5x^2y - 10xy$

        (c) $3x^4 - 2x^3 + x^2 - 2x - 3x^3y + 2x^2y - xy + 2y$

5.48.   (a) $3x^2 + 13xy - 10y^2$ (c) $p^3q - 2p^2 + pq - 2$ (e) $8u^2 + 8uv - 6v^2$

        (b) $18x^2 - 9xy - 2y^2$ (d) $2t^3 + 4t^2 - t - 2$

5.49.   (a) $-2c + 5$ (b) $4a + 2$ (c) $-3/n - 2/m$ (d) $(3y + 1)/2$ (e) $5 - 3q + 2/q$

5.50.   (a) $5x(x - 2y + 3y^2)$ (c) $2a(1 + 2bc - c + a^2)$ (e) $\dfrac{a}{2}(B + b)$

        (b) $3ac(b - 1 + 2d)$ (d) $r^2(2r - 1)$

5.51.   (a) $(4y - x)(4y + x)$ (c) $(2ab - 5c)(2ab + 5c)$ (e) $s(t_1^2 + t_2^2)(t_1 - t_2)(t_1 + t_2)$

        (b) $(3xy - 2z)(3xy + 2z)$ (d) $\left(\dfrac{1}{2x} - 2y\right)\left(\dfrac{1}{2x} + 2y\right)$

5.52.   (a) $(v - s)(v - s)$ (c) $(ab - 1)(ab - 1)$ (e) $(2a - 3b)(2a - 3b)(2a - 3b)$

        (b) $(3n + 5)^2$ (d) $r(s + 4)(s + 4)$

5.53.   (a) $2x(x - y)(x + y)$ (c) $2(m - 1)(m + 2)$ (e) $(x - 7)(x - 1)$

        (b) $3a(2x - y)(2x + y)$ (d) $(t - 3)(t + 5)$ (f) $(p + 5)(p - 2)$

5.54.   (a) $(2x - 1)(x + 2)$ (d) $2(2a + b)(a - b)$ (g) $(4x - 11)(x + 3)$

        (b) $(3p - 2q)(p + 3q)$ (e) $(3s - 5t)(2s + t)$ (h) $(4x + 1)(5x - 1)$

        (c) $(2m - 1)(m + 7)$ (f) $(3x + 2)(7x - 1)$ (i) $(3x - 2)(2 - 5x)$

5.55.   (a) $2p + 16$ (b) $2n$ (c) $b(a - b)$ (d) $(3x - y)(x - y)$

5.56.   (a) no   (b) no   (c) no   (d) no   (e) $(p - z)^2/(p + z)$

5.57.   (a) $(u + v)/uv$ (c) $(r_2q_1 + q_2r_1)/(r_1r_2)$ (e) $3amR/2$

        (b) $(1 + bt)/d$ (d) $7ML^2/12$

5.58. (a) $(q - p)/pq$ (c) $(h_1 - h_2)/h_2 t$ (e) $2mdh$

       (b) $Q(r_2 - r_1)/5r_1 r_2$ (d) $4mrx/(r^2 - x^2)^2$

5.59. (a) $7s^2(2p - q)/5r(2p + q)$ (c) $\dfrac{GM}{R + h}$ (e) GE/r^3

       (b) $f(V + E)/(V - s)$ (d) $MR^2/2$

5.60. (a) $c/16d$ (c) $3r/2g$ (c) $M(R_2^2 + R_1^2)/2$

       (b) $\dfrac{qL - A(t_2 - t_1)}{L}$ (d) $1100R/(1100 + R)$

5.61. (a) $c = 5$ (b) $h = 9$ (c) $d = 14$ (d) $u = 2.4$ (e) $a = 6$

5.62. (a) $x = 2$ (b) $y = 9$ (c) no solution   (d) $h = 7$ (e) $E = -7/5$

5.63. (a) $E \approx 5.26$ (b) $F \approx 6.11$ (c) $S \approx 0.829$ (d) $i = 5$ (e) $r \approx 2.78$

5.64. (a) $a \approx 12.31$ (b) $d = 19.5$ (c) $y = 10.56$ (d) $V = 1088$ (e) $R = 80$

5.65. (a) $x = 4/a$ (c) $E = IR$ (e) $S = (5R - 18)/12$

       (b) $t = (v - v_0)/g$ (d) $v = 12T/P$

5.66. (a) $R = 12$ (b) $q = -7.5$ (c) $p = 9$ (d) $f = 12.5$ (e) $F = 200$

5.67. (a) $V = 102$ (b) $h = 3$ (c) $C = 20$ (d) $R = 100$ (e) $v = 3 \times 10^7$

Chapter 6

Ratio and Proportion

Ratio

6.1 BASIC CONCEPTS

A *ratio* is the comparison by division of two quantities expressed in the same units. For example, the heights of two trees may be in the ratio of 2 to 5, meaning that if the height of the first tree is considered to be 2 meters, the height of the second will be 5 meters; i.e., the first tree is 2/5 as high as the second. The idea of ratio is also used in connection with maps or drawings. If a map indicates a scale of 1 in. = 10 miles, the ratio of distances on the map to actual ground distances is 1 to 633,600, since 10 miles = 633,600 inches.

The symbolism for the ratio of a number a to a number b is $a:b$, which is equivalent to $a \div b$. Thus, a ratio is often written as a fraction, a/b. The two quantities in a ratio are called its *terms*.

A ratio is unchanged in value if both terms are multiplied or divided by the same number. Thus, any ratio of rational numbers can be expressed so that both terms are integers without a common factor. For example, the ratio $12:15$ can be simplified to $4:5$ by dividing both terms by 3; and the ratio $1\frac{1}{2}:3\frac{1}{4}$ can be reduced to $6:13$ by multiplying both terms by 4.

The magnitudes of two physical quantities must be expressed in the same units if their ratio is to be meaningful. Thus, the ratio of the length of a 1-ft rule to the length of a 1-yd stick is not $1:1$ but $1:3$, since 1 yard is equal to 3 feet.

EXAMPLE 6.1. Express in simplest form the ratio of two time intervals measured as 12 seconds and 3 minutes.

Since 3 minutes is equivalent to 180 seconds, the indicated ratio is $12:180$, which can be simplified to $1:15$.

A *scale* on a map or a drawing is the ratio between any represented dimension to the actual dimension. Scales are usually shown by one of the following methods:

1. Scale $\overset{0}{\underset{\rule{3cm}{0.4pt}}{\qquad\qquad}}\,^{10}\text{m}$

2. Scale 1 inch equals approximately 13.7 miles.

3. Scale 1/500,000. In this case, 1 inch on the map corresponds to 500,000 inches, or about 8 miles, on the ground.

The reciprocal of a given ratio is called the *inverse ratio*. Thus, $6:8$ is the ratio inverse to $8:6$.

Proportion

6.2 BASIC CONCEPTS

A statement that two ratios are equal is called a *proportion*. For example, the ratios $2:3$ and $4:6$ are equal and form a proportion, which may be written as $2:3=4:6$, or $2:3::4:6$.

The proportion $a:b=c:d$ is read: "a is to b as c is to d". The inside members (b and c) are called the *means*, and the outside members (a and d) are called the *extremes* of the proportion.

The above proportion can be written in fractional form as $a/b = c/d$.

Rule. In any proportion, the product of the means is equal to the product of the extremes.

For instance, $3:4=6:8$ is a proportion and the product of the means is $(4)(6)$ or 24 and the product of the extremes is $(3)(8)$ or 24.

If we are given any three members of a proportion, we can use this rule to determine the fourth member.   We illustrate with an example.

EXAMPLE 6.2.   Determine b if $3:5=b:15$.

The means of this proportion are 5 and b while the extremes are 3 and 15.   An application of the above rule then gives $5b=45$, from which we obtain $b=9$.

In any proportion the mean terms can be interchanged; this is also true of the extreme terms.   In Example 2, the proportion could be written as $3:b=5:15$ or $15:5=b:3$.

EXAMPLE 6.3.   Is $2:13::5:33$ a valid proportion?

The product of 13 and 5 is 65; the product of 2 and 33 is 66.   Since the two products are not equal, the ratios $2:13$ and $5:33$ should not be equated as a proportion.

6.3   DIRECT PROPORTION

When a given quantity is changed in some ratio and another quantity is always changed in the same ratio or by the same factor, then the two quantities are said to be in *direct proportion* to each other.

EXAMPLE 6.4.   If the diameter of a circle is doubled, the circumference will also double.   If the diameter is divided by 5, the circumference is also divided by 5, etc.   Hence the circumference is in direct proportion to the diameter.   The relationship between the circumference and diameter can also be represented by two other statements:

"The circumference is *directly proportional* to the diameter" and the "The circumference *varies directly* as the diameter."

The direct proportion between the circumference and corresponding radii of any two circles can be symbolized as:

$$C_1 : C_2 :: R_1 : R_2 \quad \text{or} \quad \frac{C_1}{C_2} = \frac{R_1}{R_2}$$

which can also be written in the form

$$\frac{C_1}{R_1} = \frac{C_2}{R_2} = \text{constant}$$

The constant in the final proportion is $2\pi \approx 6.28$.   However, its value need not be known to determine the value of one of the quantities if the other three are known.

EXAMPLE 6.5.   A man who is 6 feet tall casts a shadow 5 feet long.   If a building at the same time has a shadow 225 feet long, how high is the building?

The lengths of shadows at a given time are in direct proportion to the heights of the objects casting the shadows.   Therefore

$$\text{man's height}:\text{man's shadow}=\text{building's height}:\text{building's shadow}$$

Substituting the given values, we obtain

$$6:5=H:225 \quad \text{or} \quad \frac{6}{5}=\frac{H}{225}$$

Cross-multiplying,

$$(6)(225) = 5H$$
$$1350 = 5H$$
$$\text{or} \quad H = \frac{1350}{5} = 270$$

The height of the building is 270 feet.

EXAMPLE 6.6.   If $36.00\,\text{in.} = 91.44\,\text{cm}$, how many centimeters are there in $12.00\,\text{in.}$?

If the number of inches decreases in the ratio of $36.00:12.00$, then the number of centimeters must decrease in the same ratio.   The following proportion can then be set up

$$36 : 12 :: 91.44 : N \quad \text{or} \quad \frac{36}{12} = \frac{91.44}{N}$$
$$36N = (91.44)(12)$$
$$N = 30.48 \text{ cm}$$

6.4 INVERSE PROPORTION

If a given quantity is changed in some ratio and another quantity is changed in the inverse ratio, then the two quantities are said to be *inverse proportion* to each other.

EXAMPLE 6.7.   If the speed of a car is doubled, the time needed to travel a given distance will be halved.   Speed is in inverse proportion to the time of travel.   This relationship can also be represented by the statements: "Speed is inversely proportional to the time of travel" and "Speed varies inversely as the time of travel."

If the car speed is S and the time of travel is t, then the inverse proportion can be expressed as $S_1 : S_2 :: t_2 : t_1$, or

$$\frac{S_1}{S_2} = \frac{t_2}{t_1}$$

EXAMPLE 6.8.   The rotational speeds of two connected pulleys are inversely proportional to their diameters.   A pulley with a diameter of 4 in. and a speed of 1800 rpm is connected to a pulley with a 6-in. diameter.   What is the rotational speed of the second pulley?

The pulley diameter increases in the ratio $6:4$; therefore, the speeds will be related as the inverse ratio $4:6$ or $2:3$.   Let the speed of the second pulley be S.   Then, expressing the relationship as a proportion, we get

$$S : 1800 :: 2 : 3$$
$$3600 = 3S$$
$$S = 1200$$

Thus, the speed of the second pulley is 1200 rpm.

If $S_1 =$ rotational speed of larger pulley
$\quad S_2 =$ rotational speed of smaller pulley
$\quad d_2 =$ diameter of larger pulley
$\quad d_2 =$ diameter of smaller pulley

then the inverse proportion could also be represented by the equation

$$\frac{S_2}{S_1} = \frac{d_1}{d_2} \quad \text{or} \quad \frac{S_1}{S_2} = \frac{d_2}{d_1}$$

The subject of direct and inverse proportion will be discussed in greater detail in Chapter 8.

Solved Problems

6.1. Simplify each of the following ratios:

(a) $3:24$ (b) $135:15$ (c) $0.5:2.5$ (d) $0.42 \div 0.12$ (e) $1\frac{3}{4} \div \frac{5}{8}$

(a) Divide both terms by the common factor 3.   The simplified ratio is $1:8$.

(b) Divide out the common factor 15 to obtain the ratio $9:1$.

(c) Divide both terms by 0.5.   The simplified ratio is $1:5$.

(d) Multiply both terms by 100.   The ratio becomes $42:12$.

Dividing both terms by 6, the ratio becomes $7:2$.

(e) Changing $1\frac{3}{4}$ to a fraction, the ratio becomes $\frac{7}{4} : \frac{5}{8}$.

Multiplying both terms by 8, the simplified ratio is $14:5$.

6.2. Represent in simplest forms the ratio of each of the following measurements:

(a) 3 nickels to a quarter   (d) 2 hours to 25 minutes

(b) 2 inches to 3 feet   (e) 0.130 and 0.325 cm

(c) 3/4 pint to $1\frac{1}{2}$ gallons

(a) 3 nickels $= 15¢$ and 1 quarter $= 25¢$.   Therefore, the ratio is $15¢:25¢$, or $3:5$.

(b) 3 feet $= 3 \times 12 = 36$ inches.   The ratio is $2:36$ or $1:18$.

(c) 1 gallon $= 4$ quarts $= 8$ pints. $1\frac{1}{2}$ gallons $= \left(\frac{3}{2}\right)(8) = 12$ pints.

The ratio is $3/4 : 12$.   Multiplying both terms by 4, the simplified ratio is $3:48$, or $1:16$.

(d) 2 hours $= (2)(60) = 120$ minutes.   The ratio is $120:25$ or $24:5$.

(e) Multiply both terms by 1000 to give $130:325$.   Dividing by 65, the simplified ratio is $2:5$.

6.3. Which of the following denote correct proportions?

(a) $5/9 = 4/7$ (c) $1\frac{1}{4} : 3 = 7/8 : 2$ (e) $15:24 = 40:64$

(b) $5:75 :: 6:9$ (d) $2/3.6 = 5/8.6$

Multiply the means and the extremes.

(a) $(5)(7) \overset{?}{=} (4)(9)$, $35 \neq 36$: not a proportion

(b) $(5)(9) \overset{?}{=} (7.5)(6)$, $45 = 45$: a proportion

(c) $(1\frac{1}{4})(2) \overset{?}{=} (7/8)(3)$, $10/4 \neq 21/8$: not a proportion

(d) $(2)(8.6) \overset{?}{=} (5)(3.6)$, $17.2 \neq 18$: not a proportion

(e) $(15)(64) \overset{?}{=} (24)(40)$, $960 = 960$: a proportion

6.4. Solve each of the following proportions for the unknown member:

(a) $x : 5 :: 6 : 15$ (c) $\dfrac{7.62}{2.54} = \dfrac{d}{9}$ (e) $4 : 90 = t : 130$

(b) $5/7 = 2.5/S$ (d) $\dfrac{5}{m} = \dfrac{1}{72}$

Equate the product of the means to the product of the extremes.   Solve for the unknown by dividing both sides by the coefficient of the unknown.

(a) $15x = 30$ $\qquad$ $x = 30/15 = 2$

(b) $5S = 17.5$ $\qquad$ $S = 17.5/5 = 3.5$

(c) $(7.62)(9) = 2.54d$ $\qquad$ $d = \dfrac{\overset{3}{\cancel{(7.62)}}(9)}{\cancel{2.54}} = 27$

(d) $m = (5)(72) = 360$

(e) $90t = 520$ $\qquad$ $t = \dfrac{520}{90} = \dfrac{52}{9} = 5\frac{7}{9} \approx 5.778$

6.5. A U.S. Forest Service map has a scale $1/4$ inch $= 1$ mile.

(a) What distance is represented by 3 inches on this map?

(b) To how many inches on the map does a distance of $3\frac{1}{2}$ miles correspond?

(a) Set up a proportion and solve $1/4$ inch : 3 inch :: 1 mile : D mile

$$3 = D/4 \qquad D = 12 \quad \text{or the distance is 12 miles}$$

(b) $1/4 : x :: 1 : 3\frac{1}{2}$

$$x = \left(\frac{1}{4}\right)\left(3\frac{1}{2}\right) = \left(\frac{1}{4}\right)\left(\frac{7}{2}\right) = 7/8$$

The given distance is represented on the map by 7/8 inch.

6.6. The map scale shows 5 cm representing 150 km.

(a) What actual distance is represented by a map distance of 3.2 cm?

(b) Determine the map scale if 1 km $= 1000$ m; 1 m $= 100$ cm.

(a) The proportion is $\dfrac{5 \text{ cm}}{3.2 \text{ cm}} = \dfrac{150 \text{ km}}{d \text{ km}}$, where D is the unknown distance.   Cross-multiplying and solving for D:

$$5D = (150)(3.2) \qquad D = \frac{(150)(3.2)}{5} = 96$$

The distance is 96 km.

(b) $150 \text{ km} = (250)(1000)(100) \text{ cm} = 15{,}000{,}000 \text{ cm}$

$$\text{The scale is} \quad \frac{5 \text{ cm}}{15{,}000{,}000 \text{ cm}} = \frac{1}{3{,}000{,}000}.$$

6.7. If 25 feet of wire weighs $2\frac{3}{4}$ pounds, what is the weight of 6 feet 4 inches of this wire?

Convert 6 feet 4 inches to $6\frac{4}{12}$ or $6\frac{1}{3}$ ft.   Set up the proportions with W as the unknown weight.

$$\frac{25 \text{ feet}}{6\frac{1}{3} \text{ feet}} = \frac{2\frac{3}{4} \text{ pounds}}{W \text{ pounds}}$$

Cross-multiplying:

$$25W = \left(2\tfrac{3}{4}\right)\left(6\tfrac{1}{3}\right) = \frac{11}{4}(19/3) = \frac{209}{12}$$

$$W = \frac{209}{12} \div 25 = \frac{209}{(12)(25)} = \frac{209}{300} \approx 0.697$$

The weight of 6'4'' of the wire is $0.697 \approx 0.7$ pound.

6.8. 7 grams of iron combines with 4 grams of sulfur to form iron sulfide. How much sulfur will combine with 56 grams of iron?

The ratio of iron to sulfur is $7:4$, and this ratio must remain constant for this reaction. Therefore, if the unknown amount of sulfur is S grams, the proportion $7:4=56:S$ must hold. Equating the product of the means to the product of the extremes, $7S = 4(56)$, from which

$$S = \frac{4(56)}{7} = 32$$

Thus, 32 g of sulfur will combine with 56 g of iron.

6.9. Analysis of carbon dioxide shows that for every 12 grams of carbon there are present 32 grams of oxygen.

(a) How many grams of carbon are combined with 100 grams of oxygen in a sample of carbon dioxide?

(b) If 25 grams of carbon dioxide is decomposed into carbon and oxygen, how much of each is present?

(a) The ratio of carbon to oxygen is $12:32$ or $3:8$. Set up the proportion with C representing the unknown mass of carbon:

$$3:8 :: C:100$$

Equating the product of the means to the product of the extremes, $300 = 8C$, from which $C = \dfrac{300}{8} = 37.5$, or 37.5 gram of carbon is present in the sample.

(b) Let C represent the amount of carbon. Then $(25 - C)$ represents the amount of oxygen. Set up the proportion $3:8 :: C:(25 - C)$. Equating the product of the means to the product of the extremes, $3(25 - C) = 8C$. Removing the parentheses, $75 - 3C = 8C$. Transposing, $11C = 75$, or $C = \dfrac{75}{11} \approx 6.82$ gram. The mass of oxygen is $25 - 6.82 \approx 18.18$ gram.

6.10. A stake 10 feet high casts a shadow 8 feet long at the same time that a tree casts a shadow 60 feet long. What is the height of the tree?

If H is the height of the tree, the proportion for the four quantities is $10:H :: 8:60$.

$$8H = 600 \qquad \text{or} \qquad H = \frac{600}{8} = 75$$

The tree is 75 feet high.

6.11. The ratio of a propeller rotation rate to that of an engine is $2:3$.   If the engine is turning at 4200 rpm, what is the propeller rate?

Let the propeller rate of turning be R.   The direct proportion is $2:3::R:4200$.   Then

$$(2)(4200) = 3R \qquad 8400 = 3R \qquad R = \frac{8400}{3} = 2800$$

The propeller is turning at 2800 rpm.

6.12. The tax on a property assessed at \$12,000 is \$800.   What is the assessed value of a property taxed at \$1100.

Let the unknown value be V; then the direct proportion is

$$\frac{\$12,000}{\$800} = \frac{\$V}{\$1100}$$

Then, $(12,000)(1100) = 800V$,

$$V = \frac{(12,000)(1100)}{800} = 16,500$$

The assessed value is \$16,500.

6.13. A photographic negative is $2\frac{1}{2}$ in. by $1\frac{1}{4}$ in. If the width of an enlarged print is $3\frac{3}{4}$ in., what should its length be?

Let the required length be L; then the proportion is

$$2\frac{1}{2} : 1\frac{1}{4} :: L : 3\frac{3}{4}$$

from which

$$\left(2\frac{1}{2}\right)\left(3\frac{3}{4}\right) = \left(1\frac{1}{4}\right)L \qquad L = \frac{\left(2\frac{1}{2}\right)\left(3\frac{3}{4}\right)}{1\frac{1}{4}} = \frac{(5/2)(15/4)}{5/4} = \frac{75/8}{5/4} = 15/2 = 7\frac{1}{2}$$

The length of the enlargement should be $7\frac{1}{2}$ in.

6.14. An airplane travels 550 miles in 1 hour 50 minutes.   At the same speed, what distance would the plane cover in 2 hours 10 minutes?

Let the required distance be D; then the proportion is

$$\frac{D \text{ miles}}{550 \text{ miles}} = \frac{2^h 10^m}{1^h 50^m} = \frac{130^m}{110^m}$$

$$110D = (550)(130) \qquad D = \frac{\overset{5}{\cancel{(550)}}(130)}{\cancel{110}} = 650$$

The plane will cover 650 miles.

6.15. The volume of a gas at constant pressure is directly proportional to its absolute temperature.   The volume is 23.5 ml at 300 K.   What is the temperature of this gas when the volume is 19.5 ml?

Let T be the unknown temperature.   Then

$$23.5 : 19.5 :: 300 : T$$

from which

$$23.5\,T = (19.5)(300) \qquad T = \frac{(19.5)(300)}{23.5} \approx 249$$

The sought temperature is about 249 K.

6.16. An airplane flying at 450 km/hr covers a distance in 3 hours 15 minutes.   At what speed would it have to fly to cover the same distance in 2 hours 30 minutes?

For a given distance, the speed is inversely proportional to the time of travel.   If the required speed is v, then the inverse proportion is

$$\frac{450}{v} = \frac{2^h 30^m}{3^h 15^m} = \frac{150 \text{ min}}{195 \text{ min}}$$

Solving the proportion,

$$(450)(195) = 150v \qquad v = \frac{\overset{3}{\cancel{(450)}}(195)}{\cancel{150}} = 585$$

The plane would have to fly at 585 km/hr.

6.17. The rate of rotation of meshed gears is inversely proportional to the number of teeth.   A gear with 24 teeth turns at 1200 rpm.   What is the rotation rate of a meshed 18-tooth gear?

Let the unknown rotation rate be R; then the inverse proportion is

$$R : 1200 :: 24 : 18$$

from which

$$18R = (1200)(24) \qquad R = \frac{\overset{400}{\cancel{(1200)}}\overset{4}{\cancel{(24)}}}{\underset{3}{\cancel{18}}} = 1600$$

The 18-tooth gear turns at 1600 rpm.

6.18. If 10 men can complete a project in 12 days, how many days will it take 15 men to complete it?

Assuming a constant performance for each man, the time for a job is inversely proportional to the number of men.   The inverse proportion is $10 : 15 :: N : 12$, where N is the unknown number of days.

$$15N = 120 \qquad N = \frac{120}{15} = 8$$

The 15 men can complete the project in 8 days.

6.19. The current in an electrical circuit is in inverse ratio to the resistance.   If the current is 2.40 amperes when the resistance is 45 ohms, what is the current when the resistance is 25 ohms?

From the given data the related inverse proportion is $\dfrac{2.4}{C} = \dfrac{25}{45}$, where C is the unknown current. Cross-multiplication gives

$$(2.4)(45) = 25C \qquad C = \dfrac{(2.4)\overset{9}{\cancel{(45)}}}{\underset{5}{\cancel{25}}} = \dfrac{21.6}{5} = 4.32$$

The current is 4.32 amperes.

6.20. Two pulleys connected by a belt rotate at speeds in inverse ratio to their diameters. If a 10-inch driver pulley rotates at 1800 rpm, what is the rotation rate of an 8-inch driven pulley?

Let the unknown rate be R; then the inverse proportion is

$$\dfrac{R}{1800} = \dfrac{10}{8}$$

from which

$$8R = 18,000 \qquad R = \dfrac{18,000}{8} = 2250$$

The driven pulley rotates at 2250 rpm.

Supplementary Problems

The answers to this set of problems are at the end of this chapter.

6.21. Simplify each of the following ratios:

  (a) $3:6$ (b) $4:12$ (c) $15:25$ (d) $2:6$ (e) $360:48$

6.22. Indicate in simplest form the ratio of each of the following measurements:

  (a) 3 in. and 2 ft   (c) $4\frac{1}{2}$ in. and $3\frac{1}{4}$ in.   (e) $2\frac{1}{2}$ min and $6\frac{1}{4}$ min

  (b) $2\frac{1}{4}$ ft and 3 yd   (d) 4 min and 34 sec

6.23. Which of the following designated equalities denote correct proportions?

  (a) $5/6 \overset{?}{=} 15/24$ (c) $2.5/10 \overset{?}{=} 3/6$ (e) $0.5/3.5 \overset{?}{=} 1.5/10.5$

  (b) $12/5 \overset{?}{=} 6/2.5$ (d) $0.005/0.15 \overset{?}{=} 3/6$

6.24. Solve each of the following proportions for the unknown member:

  (a) $2:5 = 4:x$ (b) $3:x = 5:3$ (c) $2:6 = x:18$ (d) $x:3 = 4:9$

6.25. Find the value of the unknown number in the following:

  (a) $\dfrac{40}{5} = \dfrac{24}{S}$ (c) $\dfrac{7}{U} = \dfrac{21}{5}$ (e) $\dfrac{0.84}{R} = \dfrac{6}{120}$

  (b) $\dfrac{1.5}{P} = \dfrac{2.5}{0.2}$ (d) $\dfrac{5}{4} = \dfrac{125}{M}$

6.26. A map has the scale of 1.5 in. = 2.0 miles.   What distance is represented by 18 in. on this map? Set up a proportion and solve for the unknown distance.

6.27. If the scale of a map is 1 in. = 15 miles, determine the distance on the map corresponding to an actual distance of (*a*) 45 miles; (*b*) 100 miles; (*c*) 250 miles.

6.28. Calcium and chlorine combine in the weight ratio of 36 : 64.   How much chlorine will combine with 5 g of calcium?

6.29. Analysis of carbon disulfide shows that for every 3 g of carbon 16 g of sulfur is present.   How much carbon disulfide contains 96 g of sulfur?

6.30. If a jet plane can travel 4500 km in 5.0 hr, how many kilometers will it travel in 25 min?

6.31. A certain car used 8 gal of gasoline to travel 150 miles.   Assume the same rate of consumption of gasoline to determine how much gasoline will be required to go a distance of 400 miles.

6.32. If 2 g of hydrogen are obtained by a certain process from 18 g of water, how much water would be required to produce 0.52 g of hydrogen by the same process?

6.33. If the volume of a gas is kept constant, the ratio of absolute temperature is equal to the ratio of corresponding pressures.   A certain volume of a gas at 300 K is subjected to a pressure of 80,000 dyne/cm^2.   If the temperature is increased to 373 K, find the new pressure.

6.34. In Problem 6.33, if the original pressure is reduced to 50,000 dyne/cm^2, what will be the resulting temperature?

6.35. Silver combines with sulfur in the ratio of 27 : 4.   How much sulfur will combine with 135 g of silver?

6.36. In a certain reaction, 11.2 l of chlorine are liberated from 58.5 g of sodium chloride.

(*a*) How many grams of sodium chloride are needed to liberate 19.2 l of chlorine under the same conditions?

(*b*) How many liters of chlorine can be liberated from 200 g of sodium chloride under the same conditions?

6.37. The distance of two children from the center of a teeter-totter are in inverse ratio to their weights.   A 60-lb boy sits 5 ft from the center.   How far should a 48-lb boy sit so as to counterbalance the first boy?

6.38. The fundamental frequency of a string is inversely proportional to its length.   If the fundamental frequency of a 50-cm string is 250 cycles/sec, what is the fundamental frequency of a 25-cm string?

6.39. The rotational speeds of two meshed gears vary inversely as their numbers of teeth.   If a gear with 24 teeth rotates at 420 rpm, how many teeth should a meshing gear have to rotate at 360 rpm?

6.40. The volume of a gas at a constant temperature is inversely proportional to the pressure. The volume is 4.63 liter when the pressure is 752 mm Hg. What is the volume of this gas at a pressure of 1380 mm Hg?

Answers to Supplementary Problems

6.21. (*a*) 1 : 2   (*b*) 1 : 3   (*c*) 3 : 5   (*d*) 1 : 3   (*e*) 15 : 2

6.22. (*a*) 1 : 8   (*b*) 1 : 4   (*c*) 18 : 13   (*d*) 120 : 17   (*e*) 2 : 5

6.23. (*b*); (*e*)

6.24. (*a*) 10   (*b*) 9/5   (*c*) 6   (*d*) 4/3

6.25. (*a*) 3   (*b*) 3/25, or 0.12   (*c*) 5/3   (*d*) 100   (*e*) 16.8

6.26. 24 mi

6.27. (*a*) 3 in   (*b*) $6\frac{2}{3}$ in.   (*c*) $16\frac{2}{3}$ in.

6.28. $8\frac{8}{9}$ or 8.89 g, approximately

6.29. 114 g

6.30. 375 km

6.31. $21\frac{1}{3}$ gal

6.32. 4.68 g

6.33. 99,500 dyne/cm^2 approximately

6.34. 188 K approximately

6.35. 20 g

6.36. (*a*) 100.3 g approximately   (*b*) 38.3 liter approximately

6.37. 6 ft 3 in. or $6\frac{1}{4}$ ft

6.38. 500 cycles/sec

6.39. 28

6.40. 2.52 liter

Chapter 7

Linear Equations

The Linear Function

7.1 FUNCTION

If x and y represent real numbers, then y is a *function* of x if y is uniquely determined by the value of x.

EXAMPLE 7.1. The equation $y = 3x$ states that y is a function of x because for every real number substituted for x there can be only one real number for the value of y. Thus, if $x = 1$, $y = 3$; if $x = -2$, $y = -6$; etc.

7.2 VARIABLES

If y is a function of x, the symbol x is called the *independent variable* and y the *dependent variable*. However, the symbols themselves do not vary, but the selection of real numbers which they represent does.

7.3 LINEAR FUNCTION

An equation of the form $y = mx + b$, for numbers m and b, defines a *linear function*. The name "linear" is derived from the fact that the graph of any linear function, as just defined, is a straight line.

(NOTE):   The definition that we have just given is the one usually found in elementary books, and we are going along with this common practice in order to avoid confusion in the mind of the student. We feel obligated to point out, however, that more advanced mathematics accepts a function as "linear" only if $b = 0$ in the elementary definition.

7.4 FUNCTIONAL REPRESENTATION

The linear and other functions can be represented by (1) a table of number pairs, (2) a graph, (3) an equation, (4) a verbal statement.

EXAMPLE 7.2.   The ratio of the y-values to the corresponding x-values is 2.   We can represent this function in four different ways:

(1)   Table of Pairs.   Choose a few whole numbers for the values of x; doubling each value of x gives the corresponding y-value.   Each number pair has the ratio $y/x = 2$.

x	-3	-2	-1	1	2	3
y	-6	-4	-2	2	4	6

(2)   Graph.   The two basic components of an ordinary *rectangular coordinate* graph are a pair of mutually perpendicular number lines marked off as scales, the two lines intersecting at their common 0 point, *the origin*.   It is customary to designate the horizontal scale the x-axis and the vertical scale the y-axis (see Fig. 7-1).   Every point of the plane has associated with it a pair (x, y) of coordinates, which designate the distances of the point from the two axes.   The number x (the *abscissa*) will give the distance from the y-axis in the units of the x-axis, while the number y (the *ordinate*) will give the distance from the x-axis in the units of the y-axis.   For example, the pair

(1, 2) is the coordinate pair of the point which is one unit in the positive direction from the y-axis and two units in the positive direction from the x-axis. Other points, along with their coordinates, are illustrated in Fig. 7-1. All the points lie on a straight line, thus satisfying the graphical test for the linearity of a function. The placement of coordinate pairs is called plotting or graphing. This procedure is greatly simplified by the use of graph paper.

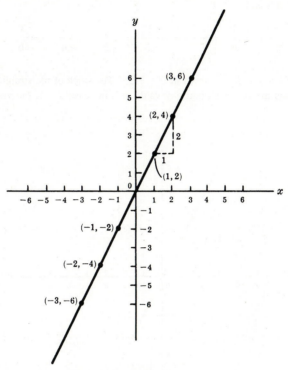

Fig. 7-1. Graph of $y = 2x$

(3) Equation. The given relationship between x- and y-values is $y/x = 2$. Multiplying both sides of the equation by x gives $y = 2x$, which is of the form $y = mx$, and this equation represents the function.

(4) Verbal Statement. The above three representations can be expressed verbally: y is a linear function of x through the origin with a slope of 2.

7.5 SLOPE

When any value of x in the graph of Fig. 7-1 is increased by 1, the value of y is increased by 2; when any value of x is increased by 2, the value of y is increased by 4, and so on. In each case the ratio of the increase in the y-value to the increase in the x-value is 2. The line has a *slope* of $+2$.

Definition. The ratio of the change in the y-value to the change in the corresponding x-value for any two points of a straight line is called the slope of the line.

Let the change in the y-value be denoted by Δy and the change in the x-value by Δx. That is, let $\Delta y = y_2 - y_1$ and $\Delta x = x_2 - x_1$; then, the slope, m, can be defined as $m = \Delta y/\Delta x = (y_2 - y_1)/(x_2 - x_1)$. The slope of every straight line is a constant. For equations of the form $y = mx$, m is the slope of its graph. The slope tells us the direction (uphill, downhill) of the line. Lines with a positive slope rise uphill to the right; lines with negative slope fall downhill to the right. The absolute value of the slope, its steepness, tells us how fast the line rises or falls. The greater the absolute value of the slope, the faster the rise or fall.

7.6   NEGATIVE SLOPE

In the case of some linear functions the value of m is negative.

EXAMPLE 7.3.   The ratio of nonzero y-values to x-values is -2.   The four representations of this function are:

(1)   A partial table of pairs of x and y.

x	-3	-2	-1	1	2	3
y	6	4	2	-2	-4	-6

(2)   Graph.   A plot of the pairs x and y is shown in Fig. 7-2.   The slope of the straight line (which goes through the origin) is *negative*.   This means than when any value of x increases by 1, the value of y decreases by 2.   The slope m is -2.

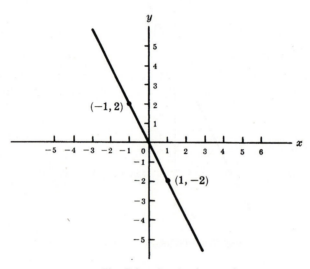

Fig. 7-2.   Graph of $y = -2x$

(3)   Equation.   The function may be represented by the equation $y = -2x$.

(4)   Verbal Statement. y is a linear function of x through the origin with a slope of -2.   That is, y is a *decreasing* linear function of x.

7.7   DIRECT PROPORTION

From the table of pairs in Example 7.2, it is seen that when the values of x are doubled the values of y are also doubled.   Thus, the relationship between the variables can also be stated: *y is directly proportional to x*, or in symbols $y \propto x$, where $\propto$ is the symbol for "proportional to." The statement for Example 7.3 is: y is directly proportional to $-x$, or $y \propto (-x)$.   The direct proportionality can also be represented by the equation:

$$\frac{y_1}{x_1} = \frac{y_2}{x_2} \qquad \text{or} \qquad \frac{y_1}{y_2} = \frac{x_1}{x_2}$$

where x_1, x_2 and y_1, y_2 are respective corresponding values of x and y.

In a linear function $y = mx$, or direct proportion, when one variable is multiplied or divided by a number, the other variable is multiplied or divided by the same number.

7.8 APPLICATIONS

There are many applications to the linear function through the origin in pure science, industry, and elsewhere.   To avoid confusion, letters other than x and y are used to represent the variables.

EXAMPLE 7.4.   The interest I earned in a savings account is a constant multiple of the money on deposit M, since the amount of interest is directly proportional to the known deposited sum.   The independent variable is M.   The equation that represents the function is $I = RM$, where R is the constant ratio that represents the interest rate.   If R is 5% per annum, then $R = I/M = 0.05$.   The graph of the interest function is a straight line through the origin with a slope of 0.05.

EXAMPLE 7.5.   The weight W of water is a constant multiple of its volume V because weight is directly proportional to volume.   The dependent variable is W.   The equation that represents the function is $W = kV$, where k denotes the weight of one unit of volume, and the function is seen to be linear through the origin.

The Linear Equation

7.9 LINEAR EQUATION

An equation of the form $Ax + By + C = 0$ is called *linear* if A, B, and C are constants and A and B are not simultaneously zero.   This equation is a *first-degree* equation in x and y and defines y as a function of x if $B \neq 0$.

7.10 REPRESENTATION

The linear equation can be represented by (1) a table of number pairs, (2) a graph, (3) an equation, (4) a verbal statement.

EXAMPLE 7.6.   The function $5y - 10x - 5 = 0$ can be represented in four ways.

(1)   Table.   Assigning some values to x and calculating the values of y gives a partial table of number pairs.

x	0	1	2	3
y	1	3	5	7

(2)   Graph.   The graph of this function is the straight line A shown in Fig. 7-3.   When the value of x in graph A is increased by 1, the value of y is increased by 2.   The line has a slope of $+2$ and crosses the y-axis 1 unit above the origin.   The *y-intercept* of the straight line is thus $+1$.

(3)   Equation.   When the equation of a straight line is expressed as $y = mx + b$, for numbers m and b, it is said to be in *slope intercept form*.   Also, the equation defines y as an *explicit* function of x.   In this equation, m is the slope and b is the y-intercept of its graph.   From the graph, it is seen that $m = +2$, $b = +1$, and the equation that represents the given function is $y = 2x + 1$.   This form can also be found by solving the given equation for y.

(4)   Verbal Statement.   The above three representations are equivalent to the statement: y is a linear function of x through the point $(0, 1)$ with a slope of 2.

Caution.   The two variables in a linear relationship are *not necessarily* directly proportional to each other.   Doubling the value of x in the above linear equation will *not* double the value of y.   Direct proportionality applies only to the linear functions with zero y-intercept, i.e. to those of the form $y = mx + b$ with $b = 0$.

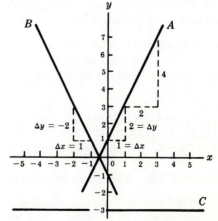

Fig. 7-3.   Graphs   of $y = 2x + 1$ (A), $y = 2x - 1$ (B), $y = -3$ (C)

7.11   NEGATIVE SLOPE

The slope of the graph of a first-degree equation can be negative.

EXAMPLE 7.7.   The equation $5y + 10x + 5 = 0$, when solved for y, is in the slope–intercept form

$$y = -2x - 1$$

The slope is -2 and the y-intercept is -1. Since the equation is of the first degree, we know that it defines a linear function and only two coordinate points are needed for a graph of the function. One of the points is the y-intercept point $(0, -1)$; an additional point is readily obtained by substituting for x some small integer (for $x = -2, y = 3$). The graph is shown in Fig. 7-3 as line B.

7.12   ZERO SLOPE

When the slope of a straight line is zero, the value of y is independent of x; that is, the value of y is constant.

EXAMPLE 7.8.   The graph of the constant function defined by $y = -3$ is a straight line parallel to the x-axis and 3 units below it; it is represented by line C in Fig. 7-3.

7.13   APPLICATIONS

Many relationships in the physical sciences have linear graphs, as for example: $v = v_0 + at$; $F = (9/5)C + 32$; $E = V + Ir$; $L_t = L_0(1 + \beta t)$; $T = 273 + t$.

7.14   INTERCEPTS OF A LINE

An intercept is the point where a line crosses an axis. Intercepts are useful because they can help us to sketch quickly the graph of a linear equation. Since every straight line is completely defined by any two of its points, if we know these points we can sketch its graph.

A straight line may have up to two intercepts. The x-intercept represents the point where the line crosses the horizontal axis (x-axis). Since any point a on the x-axis is of the form $(a, 0)$ we can calculate this intercept by letting y be equal to zero in the equation $Ax + By + C = 0$ and solving for x.

To find the y-intercept, notice that any point on the vertical axis (y-axis) is of the form $(0, b)$. Therefore, we can find this intercept by letting x be equal to 0 in the equation $Ax + By + C = 0$ and solving the resulting equation for y.

EXAMPLE 7.9.   What are the intercepts of the straight line defined by $2x + y - 6 = 0$?

To find the x-intercept, set $y = 0$ and solve for x. In this case, if we set $y = 0$, then $x = 3$. To find the y-intercept, set $x = 0$ and solve for y. In this case, if we set $x = 0$, then $y = 6$. Therefore, the line crosses the axis at $(3, 0)$ and $(0, 6)$. Since a straight line is completely defined by two of its points, we can plot the graph by $2x + y - 6 = 0$ by drawing the line that goes through these two points.

Application to Empirical Data-Graphing

7.15 EMPIRICAL EQUATIONS

An equation which can be obtained from experimental data is called *empirical*. The use of graph paper provides a convenient method for determining empirical equations for the representation of functions. It is recommended that the data be graphed on full-size sheets of graph paper.

7.16 GOOD GRAPHING PRACTICES

It is customary to plot the independent variable along the horizontal axis. In experimental work, the independent variable usually refers to the physical quantity which is readily controlled or which can be measured with the highest precision.

One of the most important requirements of a good graph is the choice of scales for the two coordinate axes. The "mathematical" functions like $y = 5x^2$ are usually graphed so that both variables are plotted to the same scale; this is seldom possible with experimental data. Also, in the case of mathematical graphs, letters such as y and x identify the axes; the "pure" numbers mark the scale subdivisions. However, to be meaningful, a graph of laboratory data must have each of its axes labeled to indicate the quantity being measured and the measurement units.

The following are suggestions for plotting a graph.

1. Use a sharp pencil.

2. Choose scales which make both plotting and reading of points easy. Some suggested scales of this kind are:

    One large scale division on the graph paper to represent

    (*a*) $0.1, 0.01$, or $0.001\ldots$; $1, 10$, or $100\ldots$ units ⎫
    (*b*) $0.2, 0.02$, or $0.002\ldots$; $2, 20$, or $200\ldots$ units ⎬ of the plotted quantity
    (*c*) $0.5, 0.05$, or $0.005\ldots$; $5, 50$, or $500\ldots$ units ⎭

3. Choose scales so that the graph occupies most of the graph sheet.

4. Leave wide margins (1/2 to 1 in.) unless the graph paper already has wide margins.

5. It is not necessary to have both quantities plotted to the same scale.

6. It is conventional to plot the quantity which you choose or vary (independent variable) along the horizontal axis and the quantity which results or which you observe (dependent variable) along the vertical axis.

7. Number the major scale divisions on the graph paper from left to right below the horizontal axis, and from the base line upward at the left of the vertical axis.

8. Name the quantity plotted and the units in which it is expressed along each axis, below the horizontal axis (extending to the right) and to the left of the vertical axis (extending upward).

9. Print all symbols and words.

10. Plot all the observed data. Mark experimental points clearly; a point surrounded by a small circle, $\odot$, a small triangle, $\triangle$ or some similar symbol is suggested. Connect the points by a smooth curve if warranted by the nature of the data.

11. Place a title in the upper part of the graph paper. The title should not cross the curve. Examples of suitable working titles are: "Relationship between Distance and Time,"

"Distance versus Velocity," "Velocity as a Function of Distance." The first quantity stated is usually the dependent variable.

12. Place your name or initials and the date in the lower right-hand corner of the graph.

13. In drawing a curve, do not try to make it go through every plotted point.   The graphing of experimental data is an averaging process.   If the plotted points do not fall on a smooth curve, the best or most probable smooth curve should be drawn in such a way that there are, if possible, as many randomly plotted points above the curve as below it.   If the graph is not a straight line, then a special device called a French curve may be used for the construction of the graph.

14. A graph is a picture of a relationship between variables.   It ought to be neat and legible.   It should tell a complete story.

7.17   GRAPHING THE SPECIAL LINEAR FUNCTION $y = mx$

If the graph of data on an ordinary sheet of graph paper is a straight line through the origin then its equation is of the form $y = mx$.   The slope m is obtained directly from the graph.

EXAMPLE 7.10.   The length L and weight W of a number of wooden cylinders with identical cross sections were measured as indicated:

L, cm	0	2.5	5.1	7.5	10	12.5	15.0
W, g-wt	0	7.0	13.7	20.6	27.7	34.7	42.5

To determine the empirical equation connecting L and W from the data in the table above, the data is graphed on cartesian (ordinary) graph paper.   The graph is shown in Fig. 7-4.   The length L was chosen as the independent variable.

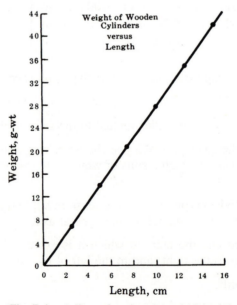

Fig. 7-4.   A linear function through the origin

Since the graph goes through the origin, the W-intercept is zero.   Therefore the equation is of the form $W = mL$.   To find the slope of the line, divide any ordinate by the corresponding abscissa.   In this case, the 10-cm abscissa is used because it simplifies division.   The ordinate as read from the graph is 28 g-wt.   Therefore, the slope

m is $28/10 = 2.8$ g-wt/cm.   This is a *physical slope* because units are involved and its value was calculated by using the scales of measured physical quantities.

The empirical equation giving the relationship between the weights and lengths is $W = 2.8\,L$.   This equation tells us that the weight of a wooden cylinder of given cross section is directly proportional to its length.   The proportionality constant is 2.8; it means that the weight of a cylinder 1 cm long is 2.8 g-wt.   The proportionality constant will depend on the material of the cylinder and on the units used.   For example, if pounds and inches were used, the constant would be 0.016 lb/in.   Had aluminum cylinders been used, the constant would have been 7.5 g-wt/cm.

If the scales were greater and/or the divisions were more evident, the values could be determined more accurately with approximately values of $m = 2.78$ g-wt/cm and $W = 2.78\,L$.

7.18   GRAPHING THE GENERAL LINEAR FUNCTION $y = mx + b$

If the graph of the experimental data is a straight line *not* through the origin, then the associated empirical equation is of the form $y = mx + b$ with $b \neq 0$.   The slope m and the y-intercept b are obtained directly from the graph.

EXAMPLE 7.11.   Data for the relationship between the electrical resistance R of a coil of wire and its temperature t are shown in the table below.   Temperature t is the independent variable since it can be varied easily and continuously.

The graph of the data is shown in Fig. 7-5.   Since it is a straight line, its equation is of the form $R = mt + b$.   The vertical scale need not begin at zero.   However, in order to obtain the empirical equation directly from the graph, the temperature scale must begin at zero.   The physical slope of the line is $1.7/100 = 0.017$ milliohm/degree.   The y-intercept is 4.95 milliohm.   Therefore, the empirical equation is $R = 0.017t + 4.95$.   Thus, the resistance is related to the temperature in a linear way.   However, these two physical equantities are *not* directly proportional to each other.

t, °C	27.5	40.5	56.0	62.0	69.5	83.0
R, milliohms	5.38	5.64	5.88	6.00	6.12	6.34

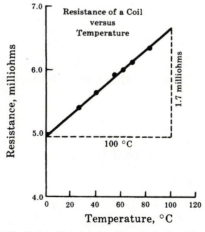

Fig. 7-5.   Graph of a linear relationship

7.19   GRAPHING LINEAR FUNCTIONS WITH A CALCULATOR

To sketch a linear equation using a graphing calculator, you can use the following general guidelines:

1. Solve the equation for y in terms of x.

2. Set the calculator mode to rectangular (this step does not apply to all graphing calculators).

3. Set the size of the viewing window, the range, by defining the maximum and minimum x-values of the abscissa and the maximum and minimum y-values of the ordinate.

4. Always use parentheses if you are unsure of the order in which the calculator processes mathematical operations.

EXAMPLE 7.12.   Sketch the graph of $y = 0.017x + 4.95$ using the graphing calculators shown below.

(NOTE):   Function keys are enclosed in curly brackets in CAPITALS, for example, {ENTER}.   Options on screen are shown underlined, for example, **EDIT**.   Comments about the function of a key or option are enclosed in quotes.

Using an HP-38G

(1) Define the function.
     {LIB}
     Function {ENTER}
     EDIT

     0.017{*}{X,T,θ}{+}4.95{ENTER}

(2) Set the range.

     {SHIFT}{SETUP-PLOT}   "Press the shift key (the turquoise key) first and then press the PLOT key in the
                            SETUP Menu."

     XRNG: 0 {ENTER} "Minimum horizontal value"   120 {ENTER} "Maximum horizontal value"
     YRNG: 0 {ENTER} "Minimum vertical value"   7 {ENTER} "Maximum vertical value"
     XTICK: 20 {ENTER}"Horizontal tick spacing"   YTICK: 1 {ENTER} "Vertical tick spacing"

(3) Graph the function.

     {PLOT}.

Using a TI-8

(1) Define the function.
     {Y = }

     0.017{x}{X,T,θ}{+}4.95{ENTER}   "Use {X,T,θ} to write the independent variable X"

(2) Set the range.

     {WINDOW}{ENTER}

     Xmin = 0 {ENTER}   Ymin = 4 {ENTER}
     Xmax = 120 {ENTER}   Ymax = 7 {ENTER}
     Xscl = 20 {ENTER}   Yscl = 1 {ENTER}

(3) Graph the function.

     {GRAPH}

     Figure 7-6 shows the graph of this function.

EXAMPLE 7.13.   In adult humans, the height of a person can be estimated as a linear function of the length of some of his bones.   The following table shows some of the values relating the height of a person and the length of his tibia (connects the knee and the ankle).   What is the linear relationship between these two quantities? What is the height of a man whose tibia is about 48 cm long?

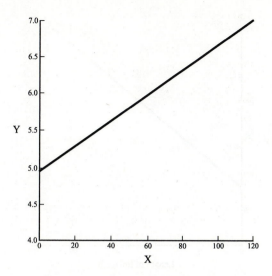

Fig. 7-6   Graph of $y = 0.017x + 4.95$

Height of a person, cm	173	188
Length of the tibia bone, cm	38	44

Let us assume that $(x_1, y_1) = (38, 173)$ and $(x_2, y_2) = (44, 188)$. The slope can be calculated as follows: $m = (y_2 - y_1)/(x_2 - x_1)$. That is, $m = (188 - 173)/(44 - 38) = 2.5$.

To calculate the equation of this straight line we can use the point–slope equation:

$$y = y_1 + m(x - x_1)$$

Replacing the value of the slope and any of the given points we obtain:

$$y = y_1 + m(x - x_1)$$
$$y = 173 + 2.5(x - 38) \qquad \leftarrow \text{Notice that we have replaced } x_1 \text{ by 38 and } y_1 \text{ by 173}$$
$$y = 173 + 2.5x - 95$$
$$y = 2.5x + 78$$

Figure 7-7 shows the graph of this line. From the figure, we can observe that the height of a person with a tibia of 48 cm is approximately 1.90 m.

To sketch this graph using a graphing calculator follow the steps indicated below.

Using an HP-38G

(1)   Define the function.

{LIB}
Function {ENTER}
EDIT

2.5{*}{X,T,θ}{+}78{ENTER}

(2)   Set the range.

{SHIFT}{SETUP-PLOT}   "Press the shift key (the turquoise key) first and then press the PLOT key in the SETUP Menu"

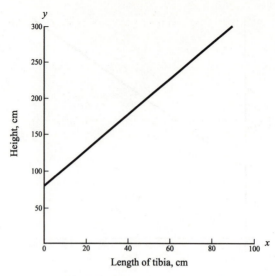

Fig. 7-7   Graph of $y = 2.5x \times 78$

XRNG: 0 {ENTER} "Minimum horizontal value"   100 {ENTER} "Maximum horizontal value"
YRNG: 0 {ENTER} "Minimum vertical value"   300 {ENTER} "Maximum vertical value"
XTICK: 20 {ENTER} "Horizontal tick spacing"   YTICK: 50 {ENTER} "Vertical tick spacing"

(3) Plot the graph.

  {PLOT}

Using a TI-82

(1) Define the function.

  {Y=}

  2.5{x}{X,T,θ}{+}78{ENTER}   "Use {X,T,θ} to write the variable X"

(2) Set the range.

  {WINDOW}{ENTER}

  Xmin = 0 {ENTER}   Ymin = 0 {ENTER}
  Xmax = 100 {ENTER}   Ymax = 300 {ENTER}
  Xscl = 20 {ENTER}   Yscl = 50 {ENTER}

(3) Graph the function.

  {GRAPH}

EXAMPLE 7.14.   The Celsius and Fahrenheit scales are widely used to measure temperature.   The following table shows the relationship between these two scales.   What is the linear equation that allows us to calculate a temperature in Celsius degrees as a function of a given measure expressed in Fahrenheit degrees? What is the physical meaning of the slope?

Fahrenheit	32	212
Celsius	0	100

Let $(x_1, y_1) = (32, 0)$ and $(x_2, y_2) = (212, 100)$.   Then, the slope m can be calculated as follows:

$$m = (y_2 - y_1)/(x_2 - x_1) = (100 - 0)/(212 - 32) = 100/180 = 5/9$$

Replacing the value of the slope and any of the given points in the point–slope equation we obtain:

$$y = y_1 + m(x - x_1)$$
$$y = 32 + (5/9)(x - 0) \leftarrow \text{Replacing } (x_1, y_1) \text{ with } (0, 32)$$
$$y = (5/9)x + 32$$

The slope 5/9 may be interpreted to mean that for every 9-degree increase in the Fahrenheit temperature, there is an increase of 5 degrees Celsius.   Figure 7-8 shows the graph of this function.

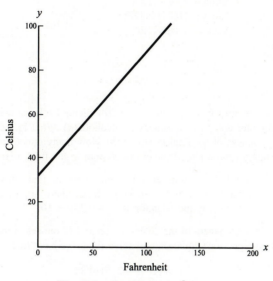

Fig. 7-8   Graph of $y = (\frac{5}{9})x + 32$

To sketch this graph using a graphing calculator follow the steps indicated below.

Using an HP-38G

(1) Define the function.

{LIB}
<u>Function</u> {ENTER}
<u>EDIT</u>

{(}5{/}9{)}{*}{X,T,θ}{+}32{ENTER}

(2) Set the range.

{SHIFT}{SETUP-PLOT}   "Press the shift key (the turquoise key) first and then press the PLOT key in the SETUP Menu"

XRNG: 0 {ENTER} "Minimum horizontal value"   200 {ENTER} "Maximum horizontal value"
YRNG: 0 {ENTER} "Minimum vertical value"   100 {ENTER} "Maximum vertical value"
XTICK: 50 {ENTER} "Horizontal tick spacing"   YTICK: 20 {ENTER} "Vertical tick spacing"

(3) Graph the function.

{PLOT}

Using a TI-82

(1)   Define the function.

   {Y = }

   {()5{÷}9()}{x}{X,T,θ}{+}32{ENTER}   "Use {X,T,θ} to write the variable X"

(2)   Set the range.

   {WINDOW}{ENTER}

   Xmin = 0 {ENTER}   Ymin = 0 {ENTER}
   Xmax = 200 {ENTER}   Ymax = 100 {ENTER}
   Xscl = 50 {ENTER}   Yscl = 20 {ENTER}

(3)   Graph the function.

   {GRAPH}

EXAMPLE 7.15.   Assume that a chemist mixes x ounces of a 20% alcohol solution with y ounces of another alcohol solution at 35%.   If the final mixture contains 10 ounces of alcohol: (*a*) What is the linear equation relating both solutions and the total number of ounces of the final mixture? (*b*) How many ounces of the 35% solution need to be added to 5 ounces of the 20% solution to obtain 12 ounces of alcohol in the final mixture?

(*a*) If we have x ounces of an alcohol solution at 20%, this implies that only 20% of the x ounces are alcohol.   Likewise, only 35% of the y ounces are alcohol.   Since there are 10 ounces of alcohol in the final mixture, the relationship can be defined by the equation $0.2x + 0.35y = 10$.

(*b*) In this case, we know that there are 5 ounces of the 20% solution and 12 ounces of the final solution.   Therefore, if we substitute these values in the previous equation and solve for y, we have that

$$y = (10 - 0.2*5)/0.35$$
$$y = 25.7 \text{ ounces}$$

   Therefore, add 25.7 ounces (of the solution at 35%) to the 5 ounces (of the solution at 20%) to obtain 10 ounces of alcohol in the final mixture.   Figure 7-9 shows the graph of this function.

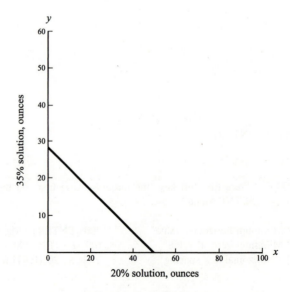

Fig. 7-9   Graph of $y = \dfrac{1}{0.35}(10 - 0.2x)$

To sketch this graph using a graphing calculator follow the steps indicated below.

Using an HP-38G

(1) Define the function.

{LIB}
<u>Function</u> {ENTER}
<u>EDIT</u>

{(}10{−}0.2{*}{X,T,θ}{)}{/}0.35{ENTER}

(2) Set the range.

{SHIFT}{SETUP-PLOT}   "Press the shift key (the turquoise key) first and then press the PLOT key in the
                              SETUP Menu"

XRNG: 0 {ENTER} "Minimum horizontal value"   100 {ENTER} "Maximum horizontal value"
YRNG: 0 {ENTER} "Minimum vertical value"   100 {ENTER} "Maximum vertical value"
XTICK: 4 {ENTER} "Horizontal tick spacing"   YTICK: 4 {ENTER} "Vertical tick spacing"

(3) Graph the function.

{PLOT}

Using a TI-82

(1) Define the function.

{Y=}

{(}10{−}0.2{x}{X,T,θ}{)}{÷}0.35{ENTER}   "Use {X,T,θ} to write the variable X"

(2) Set the range.

{WINDOW}{ENTER}

Xmin = 0 {ENTER}   Ymin = 0 {ENTER}
Xmax = 100 {ENTER}   Ymax = 100 {ENTER}
Xscl = 4 {ENTER}   Yscl = 4 {ENTER}

(3) Graph the function.

{GRAPH}

Solved Problems

7.1. (a) Which of the following equations represent linear functions of the form $y = mx$?
$2y - 10x + 4 = 0$, $3y - 12 = 0$, $10y - 15x = 0$.

(b) Sketch the graphs of the three equations

(a) Solve each equation explicitly for y. This will transform the equations into the slope–intercept form,
$y = mx + b$.

$$
\begin{array}{ccc}
2y - 10x + 4 = 0 & 3y - 12 = 0 & 10y - 15x = 0 \\
y - 5x + 2 = 0 & y - 4 = 0 & y - 1.5x = 0 \\
y = 5x - 2 & y = 4 & y = 1.5x
\end{array}
$$

$10y - 15x = 0$ represents a linear function of the desired type, since $y = 1.5x$ is of the form $y = mx$. The other equations yield general linear functions.

(b) Each of the equations will plot as a straight line. Substitute 0 and 1 for x in $2y - 10x + 4 = 0$ to get the coordinate points $(0, -2)$ and $(1, 3)$. A line through these two points is the graph of the equation. The graph of $3y - 12 = 0$ is a straight line parallel to the x-axis and 4 units above it, since y has a constant value of 4. The graph of $10y - 15x = 0$ is a straight line through the origin $(0, 0)$. A second coordinate point $(1, 1.5)$ is obtained by substituting $x = 1$ in the equation. The three graphs are sketched in Fig. 7-10.

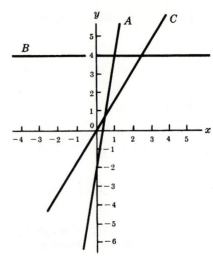

Fig. 7-10. $2y - 10x + 4 = 0$ (A)
$3y - 12 = 0$ (B)
$10y - 15x = 0$ (C)

7.2.   Determine the slope and the y-intercept for the graph of $4x + 3y - 6 = 0$.

Solve the equation explicitly for y.

$$4x + 3y - 6 = 0$$

$$3y = -4x + 6$$

$$y = -\frac{4}{3}x + 2$$

which is of the form $y = mx + b$. The slope is $-\frac{4}{3}$ and the y-intercept is $+2$.

7.3.   What is the equation of the graph shown in Fig. 7-11?

The slope is *negative*, since y decreases when x increases. When x increases 2 units, y decreases 1 unit. Therfore the slope is $-\frac{1}{2}$ or -0.5. The y-intercept is $+1$. The equation of the graph is $y = -\frac{1}{2}x + 1$, or $y = -0.5x + 1$.

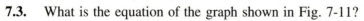

Fig. 7-11.

7.4.   In the equation $2p - 5q + 8 = 0$, q is the independent variable. (a) Write an equation which defines p as an explicit function of q. (b) Determine the slope and the p-intercept of the graph of the function. (c) Sketch the graph.

(a) Solve the given equation for p.

$$2p - 5q + 8 = 0$$

$$2p = 5q - 8$$

$$p = \frac{5q}{2} - \frac{8}{2}$$

$$p = 2.5q - 4$$

(b) The slope of the graph is 2.5. The p-intercept is -4.

(c) If $q = 1$, $p = -1.5$; $q = 2$, $p = 1$. See Fig. 7-12, and note the different scales for p and q.

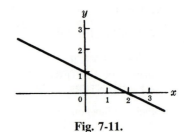

Fig. 7-12.

7.5. The distances covered by a car and the corresponding elapsed times are shown in the table below.

Time, hours	0	0.5	1.5	2.5	3.5
Distance, kilometers	0	50	150	250	350

(a) Plot on graph paper the distance as a function of time.

(b) Determine the slope and the equation of the graph.

(c) What is the physical meaning of the slope?

(d) Write symbolically in two forms "distance is directly proportional to time."

(a) Let the time t be the independent variable and the distance d be the dependent variable. Let the side of one large square of the graph paper represent 1 hour on the horizontal axis and 100 kilometers on the vertical axis. Plot the points and join them as shown in Fig. 7-13. Label the axis with quantities plotted and the measurement units.

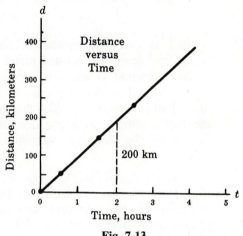

Fig. 7-13.

(b) Since the graph is a straight line, the function is linear. The equation is of the form $d = mt$, where m is the physical slope. Construct a triangle as shown. Since m is the change in the dependent variable divided by the corresponding change in the independent variable,

$$m = \frac{200}{2} = 100$$

The equation for the data is $d = 100t$.

(c) The slope represents a constant speed of 100 km/hr.

(d) $d \propto t$; $d_1/d_2 = t_1/t_2$ where d_1, d_2 and t_1, t_2 are respective corresponding values of d and t.

7.6. A spring is stretched by means of a weight. The associated data are reproduced in the table below.

Weight, g-wt	0	100	200	300	400
Elongation, cm	0	4.2	8.5	12.6	16.6

(a) Show that the empirical equation for the data is linear and determine the slope of the graph.

(b) Interpret the physical meaning of the slope.

(c) Is the elongation directly proportional to the weight? Why?

(a) Let the weight W be the independent variable, with the elongation denoted by the variable E. A convenient scale for the horizontal axis is 100 g-wt to a side of one large square; for the vertical, 5 cm to a side of one large square. Plot the points as shown in Fig. 7-14. Draw the best straight line through the points. Construct a right triangle with 500 g-wt as a base; read off the scale the number of cm in the triangle altitude (21 cm). The slope is

$$m = \frac{21}{500} = 0.042$$

The empirical equation is $E = 0.042\ W$.

(b) The physical slope of 0.042 cm/g-wt means that the spring will stretch 0.042 cm for each additional gram of weight.

(c) Yes, because the equation is of the form $y = mx$; the graph goes through the origin.

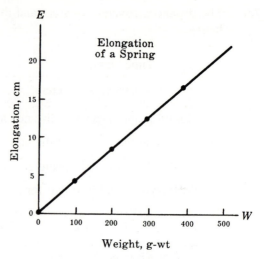

Fig. 7-14.

7.7. A ball is given an initial velocity of 30 cm/sec down an inclined plane. The speed of the ball increases uniformly with time. After 2 sec has elapsed, the velocity of the ball is 70 cm/sec. Sketch a graph of the ball's velocity as a function of time. Determine the defining equation of the function.

Since the velocity v is to be a function of time t, plot v vertically and t horizontally. The point $(0, 30)$ defines the y-intercept, since at $t = 0$, $y = 30$ cm/sec. The slope is $(70 - 30)/2 = 40/2 = 20$; this means that for every additional second the velocity increases by 20 cm/sec as shown in Fig. 7-15 (solid curve). The defining equation is of the form $y = mx + b$ or (in the variables of the problem) $v = 20t + 30$, and so the velocity of the ball is related to the time in a graphically linear way. It is important to note that the velocity is *not* directly proportional to the time.

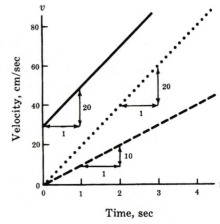

Fig. 7-15.

7.8. Rework Problem 7.7 with the ball starting from rest. The ball reaches a velocity of 40 cm/sec at the end of 2 seconds.

Since the initial velocity is zero, the y-intercept is 0, and so the slope is $40/2 = 20$, which is to say that the acceleration is 20 cm/cm^2. Since the y-intercept is zero, $b = 0$, and the equation of the graph is of the form $y = mx$. Thus, the graph would have the same slope as before but would go through the origin as shown in Fig. 7-15 (dotted). The equation of this graph is $v = 20t$. In this case the velocity is directly proportional to time. Thus, we see that the two graphs define linear functions whose corresponding values differ by a constant 30. The identical slope in both graphs means that the velocity of the ball increases at the same rate for both cases.

7.9. Rework Problem 7.7 with the angle of the plane changed. The ball is released from rest, and its velocity at the end of 2 sec is 20 cm/sec.

The y-intercept b is 0, the intercept point being $(0, 0)$, and so the graph passes through the origin. The slope is $(20 - 0)/2 = 10$, and the equation of this graph is $v = 10t$. The velocity is again directly proportional to time, but the rate of change of the velocity is 10/20 or 1/2 of that in the two previous examples.   See Fig. 7-15 (dashed).

7.10. Transform the equation $12 = V + 0.05I$ into the slope–intercept form for the equation of a straight line, with V defined as a function of I.   What are the values of the slope and the y-intercept? What is the interpretation of the slope and the y-intercept if V is measured in volts and I is measured in amperes? Sketch the graph of the function.

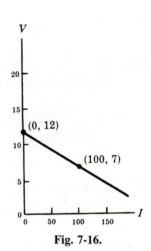

Transposing, $-V = 0.05I - 12$

Changing signs, $V = -0.05I + 12$

Therefore, the slope of the line is -0.05, and its y-intercept is $+12$.   The interpretation of -0.05 is that if I is increased by 1 amp, V will decrease by 0.05 volt.   The interpretation of $+12$ is that when the current is zero, V will be 12 volts.   The graph is shown in Fig. 7-16 with V along the vertical axis.   One point of the graph is the intercept point $(0, 12)$; an additional point can be obtained from the value of the slope.   Thus, for a 100-amp increase in I, V will drop 5 volts, so that $(100, 7)$ is a point on the graph.

Fig. 7-16.

7.11. The temperature of air in a closed container is varied and the resulting pressure is read on a gauge.   The table below shows the data obtained.

$t,\,^\circ C$	-80	0	23.5	33	99
p, psi	14.4	20.1	21.9	22.6	27.2

(*a*) Determine the empirical equation for the data.

(*b*) What is the physical meaning of the slope?

(*c*) Determine algebraically the temperature at which the pressure becomes zero?

(*d*) Is the pressure directly proportional to the temperature?

(*a*) Plot the temperature horizontally, since it is controlled by the observer.   Convenient scales are shown in Fig. 7-17.   Note that the temperature has both negative and positive values.   The slope is

$$m = \frac{12.8}{179} = 0.0715$$

as found by direct measurement. The equation of the graph is $p = 20.1 + 0.0715t$.

(*b*) A slope of 0.0715 psi/°C means that if the temperature increases by 1 °C, the pressure increases by 0.0715 psi.

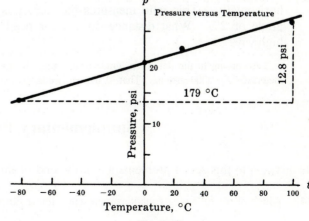

Fig. 7-17.

(c) Substitute $p=0$ in the empirical equation and solve for t.

$$0 = 20.1 + 0.0715t$$

$$t = \frac{-20.1}{0.0715} = -281 \ ^\circ C$$

(d) No. The equation is not of the form $y = mx$. The graph does not go through $(0, 0)$.

7.12. The solubility of calcium chromate in 100 ml of water is a function of the solution temperature t, as shown in the table below.

Temperature, $^\circ$C	0	20	40	60	80	100
Solubility, g/100 ml	13.0	10.4	8.5	6.1	4.4	3.2

(a) Determine the equation to fit the data.

(b) Interpret the physical slope.

(c) Is solubility directly proportional to the temperature?

(a) The plot of the data is shown in Fig. 7-18. The slope is negative.

$$m = -\frac{13 - 3.2}{100 - 0} = \frac{-9.8}{100} = -0.098$$

The s-intercept is 13.0. The empirical equation is $s = 13.0 - 0.098t$.

(b) The physical slope -0.098 means that increasing the temperature of the solution 1 $^\circ$C will decrease the solubility of calcium chromate by 0.098 g in 100 ml of water.

(c) No. The equation is not of the form $y = mx$. The graph does not go through the origin. However, the solubility of calcium chromate is a linear function of the temperature.

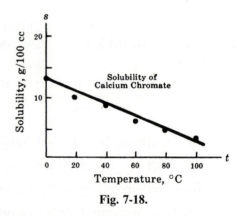

Fig. 7-18.

7.13. Civil engineers call the grade of a roadbed the ratio obtained by dividing the distance the road rises or falls to the distance it runs horizontally. This ratio measures the steepness of the road. Highways have an average grade of 2%. What distance does a car need to travel to rise 180 meters on a typical highway?

According to the definition, grade $=$ rise/run. In this case, the rise is 180 meters, therefore, run $= 180$ meters/0.02 $= 9000$ meters. That is, the car needs to travel a distance of 9 km.

Supplementary Problems

The answers to this set of problems are at the end of this chapter.

7.14. Which of the following equations represent a linear function of the form $y = mx$?

$$4x - 3y = 16 \qquad 3x + y = 1 \qquad y^2 - 5x = 1 \qquad 2x + y = 0 \qquad 5y - 20 = 0$$

7.15. (*a*) Which of the following equations have a straight line graph passing through the origin?

$$v + 10 = -32t \qquad P = 0.433h \qquad C = -T - 273 \qquad S = 5t - 16t^2 \qquad C = 6.28R \qquad F + 64 = 8a$$

(*b*) Which of the equations graph as straight lines with negative slopes?

7.16. Transform the following equations to define y as an explicit function of x.

(*a*) $3x + 2y - 5 = 0$ (*b*) $5y - 20x = 15$ (*c*) $4y - 5 = 0$

7.17. (*a*) Determine the slopes of the graphs of the functions in Problem 7.15.

(*b*) What are the y-intercepts?

(*c*) Using the y-intercept and one other coordinate point, draw the graphs for the functions in Problem 7.15.

7.18. (*a*) Write the empirical equation $P - 14 = 0.05t$ in the slope–intercept form, with P as the dependent variable.

(*b*) Determine the P-intercept.

(*c*) Determine the slope.

7.19. (*a*) Which is the independent variable in Fig. 7-19?

(*b*) Determine the slope of the graph in Fig. 7-19.

(*c*) Write the equation represented by the graph.

(*d*) Write *two* verbal statements which describe the relationship between the variables.

(*e*) What would be the equation of the graph, if it were drawn parallel to the given line through the point (0, 200)?

7.20. (*a*) In the equation $V = 1.2 - 0.25R$, what is the value of the V-intercept?

(*b*) What is the value of the slope?

(*c*) Interpret the meaning of the slope.

(*d*) Sketch a graph of this equation.

7.21. (*a*) Transform the following equation to define C as an explicit function of F: $5F - 9C + 160 = 0$.

(*b*) What is the slope of the line graph?

(*c*) What is the C-intercept?

(*d*) Sketch a graph of this function.

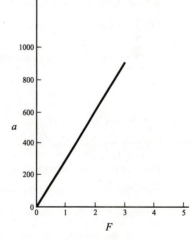

Fig. 7-19.

7.22. (*a*) What is the slope in Fig. 7-20? What does it mean physically?

(*b*) What is the v-intercept? What does it mean physically?

(*c*) What is the physical meaning of the point (0, 15)?

7.23. (*a*) Express the following equation in slope–intercept form to define R as a function of t: $R = 500(1 - 0.004t)$.

(*b*) What is the slope of the line graph?

(*c*) What is the R-intercept of the graph?

(d) If t is measured in degrees Celsius and R is ohms, what is the interpretation of the slope? Of the y-intercept?

7.24. (a) Transform $\dfrac{L - L_0}{L_0 t} = a$ into the slope–intercept form of the equation of a straight line with L as the dependent and t the independent variable. Regard L_0 and a as constants.

   (b) What is the expression for the slope of the line graph of the defined function? For the L-intercept?

7.25. (a) Express the equation $F - mg = ma$ in the slope–intercept form with a regarded as a function of F.

   (b) If $g = 9.8$ and $m = 2$, what is the slope?

   (c) What is the F-intercept of the graph of part (a) of this question?

   (d) Under what conditions would a be directly proportional to F of part (a) of this question?

   (e) In what respect would the graphs in (a) and (d) be alike? In what respect different?

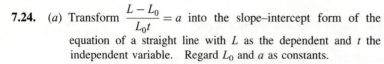

Fig. 7-20.

7.26. (a) From the data below, graph the spring elongation S versus the applied force F.

F, g-wt	0	100	200	300	400	500
S, cm	0	10.5	20.6	31.1	41.4	51.8

   (b) What conclusion can you draw about the relationship between S and F?

   (c) Determine the empirical equation of the graph.

   (d) A force of 160 g-wt is applied to the spring. From the graph determine the elongation.

   (e) An unknown force causes an elongation of 49.8 cm. From the graph determine the unknown force.

   (f) Find the unknowns in (d) and (e) above by substitution in the empirical equation.

7.27. The solubility S (in 100 g of water) of magnesium sulfate as a function of temperature t is shown in the table below.

t, °C	0	10	20	30	40	50	60	70	80	90	100	110
S, g	26.9	31.5	36.2	40.9	45.6	50.3	55.0	59.6	64.2	68.9	73.8	77.9

   (a) Plot a graph of the data.

   (b) Determine the empirical equation of the graph.

   (c) What are the units of the physical slope?

   (d) Read from the graph the solubility at 83 °C.

   (e) Use the empirical equation to calculate the temperature at which the solubility is 33.8 g. Check your answer on the graph.

7.28. The thermal conductivity C of a silica block varies with temperature t as shown by the following table.

t, °F	0	200	400	600	800	1000	1200	1400	1600	1800	2000
C, BEU	6.0	6.8	7.5	8.3	9.0	9.8	10.5	11.3	12.0	12.8	13.5

   (a) Plot a graph of the above data.

   (b) Determine the empirical equation.

7.29. The surface tension T of benzene depends on its temperature t.   The related data obtained by a student in the laboratory are shown below.

$t, {}^\circ C$	25.7	38.7	48.7	57.8	64.0
T, dyne/cm	28.4	26.9	25.7	24.7	24.0

(*a*) Plot a graph of the data on cartesian graph paper.

(*b*) What conclusions can be drawn from (*a*)?

(*c*) Determine the empirical equation for the data.

7.30. The relationship between the approximate height of a person and the length of his radius bone (connects the elbow and the wrist) is shown below.   What is the linear relationship between the height of the person (*y*) and the length of his radius bone (*x*)?

Height, cm	153	178
Length radius, cm	20	27

7.31. Suppose that x ounces of a 45% alcohol solution are combined with y ounces of a 75% alcohol solution.

(*a*) What is the linear equation that relates the different solutions and the total number of ounces of alcohol of the final mixture?

(*b*) How many ounces of the 75% solution must be added to 10 ounces of the 45% solution to obtain 15 ounces of alcohol in the final mixture?

7.32. What is the grade of a road that rises 68 feet over a horizontal distance of 850 feet? The United States Department of Transportation requires that roads longer than 750 feet with a grade of 8% or higher have a sign to warn drivers about the steepness of the road.   Does this road require a sign?

Answers to Supplementary Problems

7.14. $2x + y = 0$

7.15. (*a*) $P = 0.433h$, $C = 6.28R$ (*b*) $v + 10 = -32t$, $C = -T - 273$

7.16. (*a*) $y = -\dfrac{3}{2}x + \dfrac{5}{2}$ (*b*) $y = 4x + 3$ (*c*) $y = \dfrac{5}{4}$

7.17. (*a*) $-3/2$; 4; 0   (*b*) 5/2; 3; 5/4   (*c*) See Fig. 7-21.

7.18. (*a*) $P = 0.05t + 14$ (*b*) $+14$ (*c*) 0.05

7.19. (*a*) F

(*b*) 300

(*c*) $a = 300F$

(*d*) (1) a is a linear function of F through the origin;
    (2) a is directly proportional to F.

(*e*) $a = 300F + 200$

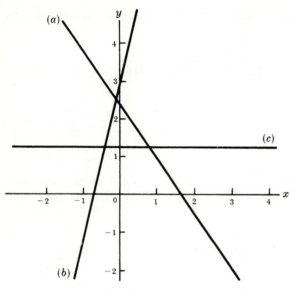

Fig. 7-21.

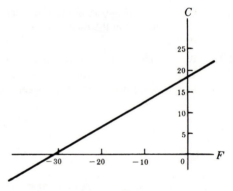

Fig. 7-22.

7.20. (a) 1.2

(b) −0.25

(c) When R increases by one unit, V decreases by 0.25 units.

(d) See Fig. 7-22.

7.21. (a) $C = (5/9)F + 160/9$

(b) 5/9

(c) 160/9 or $17\frac{7}{9}$ or 17.8

(d) See Fig. 7-23.

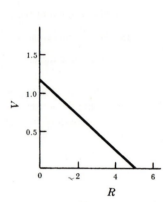

Fig. 7-23.

7.22. (a) −100/15 or −20/3 or −6.67 negative acceleration (deceleration).

(b) 100. Initial speed.

(c) After a lapse of 15 sec, the speed is zero.

7.23. (a) $R = -2t + 500$

(b) −2

(c) 500

(d) Resistance change for each degree rise in temperature; resistance at 0 °C.

7.24. (a) $L = aL_0 t + L_0$ (b) $aL_0; L_0$

7.25. (a) $a = F/m - g$ (c) $-g$

(b) 1/2   (d) if $g = 0$

(e) Both are straight lines; graph of (d) is through the origin.

7.26. (*a*) See Fig. 7-24.

(*b*) S is a linear function of F, or S is directly proportional to F.

(*c*) $S = 0.10F$

(*d*) $S \approx 15$ cm

(*e*) $F \approx 500$ g-wt

(*f*) 16 cm; 498 g-wt

7.27. (*a*) See Fig. 7-25.

(*b*) $S = 0.46t + 27$

(*c*) g/°C

(*d*) 67 g (approx.)

(*e*) 14.8 °C $\approx$ 15 °C

7.28. (*a*) See Fig. 7-26.

(*b*) $C \approx 0.0038t + 6.0$

7.29. (*a*) See Fig. 7-27.

(*b*) Surface tension is a decreasing linear function of temperature.

(*c*) $T \approx -0.11t + 31$

7.30. $y = 3.6x + 81$

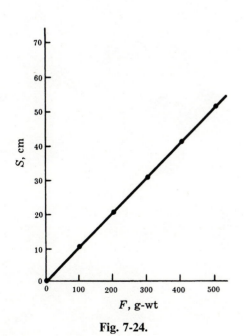

Fig. 7-24.

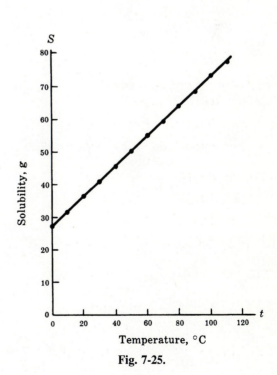

Fig. 7-25.

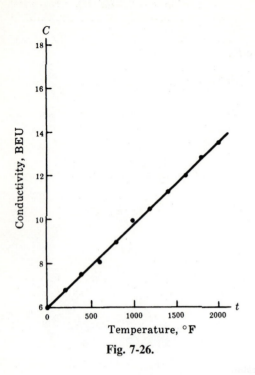

Fig. 7-26.

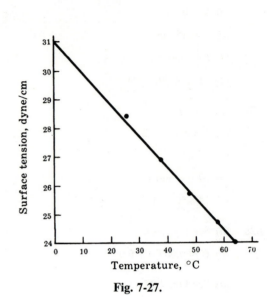

Fig. 7-27.

7.31. (*a*) $0.45x + 0.75 = T$ where T is the total number of ounces of alcohol.

   (*b*) 14 ounces.

7.32. The road has a grade of 8%.   Yes! it requires a sign since it is also longer than 750 feet.

Chapter 8

Exponents and Radicals

Definitions and Symbols

8.1 EXPONENTIAL FORM

Exponential notation may be used to show that the same number is repeated several times in a product. Thus, $5 \cdot 5$ is simplified to 5^2 and $r \cdot r \cdot r$ to r^3. The numbers 5^2 and r^3 are in *exponential form*. The repeated factor is called the *base*. In 5^2 the base is 5 and in r^3 it is r.

8.2 POSITIVE INTEGRAL EXPONENT

The positive integer which shows the number of times the base is used as a factor is called the positive integral *exponent*. In 5^2 the exponent is 2; in r^3 it is 3. In the product a^n, the factor a occurs n times.

8.3 READING EXPONENTIAL NOTATION

The notation 5^2 is read "the second power of 5" or "5 raised to the second power" or "5 squared." The number r^3 is read "the third power of r" or "r raised to the third power" or "r cubed." The product a^n is read "a raised to the nth power" or the "nth power of a."

8.4 ZERO EXPONENT

By definition, $a^0 = 1$, if a is not zero.

EXAMPLES: $5^0 = 1$ $\qquad r^0 = 1$ if $r \neq 0$ $\qquad 10^0 = 1$ $\qquad (-2)^0 = 1$ $\qquad \left(\dfrac{a}{b}\right)^0 = 1$ if $\dfrac{a}{b} \neq 0$
$(0.05)^0 = 1$ $\qquad (2/3)^0 = 1$

8.5 NEGATIVE INTEGRAL EXPONENT

By definition, $a^{-n} = \dfrac{1}{a^n}$ if n is a positive integer and a is not zero.

EXAMPLES: $3^{-2} = \dfrac{1}{3^2} = \dfrac{1}{9}$ $\qquad 2^{-5} = \dfrac{1}{2^5} = \dfrac{1}{32}$ $\qquad 10^{-6} = \dfrac{1}{10^6}$ $\qquad t^{-1} = \dfrac{1}{t}$ $\qquad -5h^{-3} = \dfrac{-5}{h^3}$
$(1 + bT)^{-1} = \dfrac{1}{1 + bT}$

8.6 ROOTS

If two numbers a and b satisfy the equation $a^n = b$ with n a positive integer, then a is defined as an *n*th *root* of b.

EXAMPLE 8.1. Since $3^2 = (-3)^2 = 9$, the numbers 3 and -3 are the *second* or *square* roots of 9.

EXAMPLE 8.2. Since $(-5)^3 = (-5)(-5)(-5) = -125$, a *third* or *cube* root of -125 is -5.

No even root of a negative number can be expressed as a real number.

EXAMPLE 8.3. The square roots of -2 and the fourth roots of -16 are not real numbers.

8.7   PRINCIPAL *n*th ROOT

If such a number exists, the unique positive or the unique negative (if no positive exists) *n*th root of the number b is called its *principal n*th root.   If $a^n = b$, then $\sqrt[n]{b}$ is the principal *n*th root of b.

EXAMPLE 8.4. $\sqrt[2]{9} = 3$ is the principal second or square root of 9, usually written $\sqrt{9}$.

EXAMPLE 8.5. $\sqrt[3]{-27} = -3$ is the principal third or cube root of -27.

8.8   RADICAL, RADICAND, AND INDEX

The principal *n*th root $\sqrt[n]{b}$ of a number is also called a *radical*; b is the *radicand*; the positive integer n is called the *index*.

8.9   FRACTIONAL EXPONENTS

Definition.   If m and n are positive integers, $a^{m/n} = \sqrt[n]{a^m}$.

EXAMPLE 8.6. $2^{3/2} = \sqrt{2^3} = \sqrt{8}$

EXAMPLE 8.7. $5^{2/3} = \sqrt[3]{5^2} = \sqrt[3]{25}$

EXAMPLE 8.8. $4^{-3/2} = \dfrac{1}{4^{3/2}} = \dfrac{1}{\sqrt{4^3}} = \dfrac{1}{\sqrt{64}} = \dfrac{1}{8}$

EXAMPLE 8.9. $p^{-2/3} = \dfrac{1}{p^{2/3}} = \dfrac{1}{\sqrt[3]{p^2}}$

Laws of Exponents

The so-called *laws of exponents* follow from the definition of an exponent.

Law 1. $(a^n)(a^m) = a^{n+m}$

EXAMPLE 8.10. $2^3 \cdot 2^4 = 2^{3+4} = 2^7 = 128$

EXAMPLE 8.11. $T^2 \cdot T^3 = T^{2+3} = T^5$

Law 2. $(a^n)/(a^m) = a^{n-m}$

EXAMPLE 8.12. $3^5/3^2 = 3^{5-2} = 3^3 = 27$

EXAMPLE 8.13. $R^3/R = R^{3-1} = R^2$

Law 3. $(a^n)^m = a^{nm}$

EXAMPLE 8.14. $(2^3)^2 = 2^6 = 64$

EXAMPLE 8.15. $(3p^3)^2 = 3^2(p^3)^2 = 9p^6$

Law 4. $(a^n)(b^n) = (ab)^n$

EXAMPLE 8.16. $(3^2)(5^2) = 15^2 = 225$

EXAMPLE 8.17. $2^2 d^3 \cdot 4^2 v^3 = 8^2 d^3 v^3 = 64(dv)^3$

In the statements of the above laws, it is assumed that a and b are real numbers, while m and n are integers.

Power Functions

Many of the relationships between variables in the physical world can be expressed by power functions.

8.10 POWER FUNCTION

A *power function* is defined by the equation $y = kx^n$. The variables are x and y; k and n are constants.

EXAMPLES: $C = 3.14D$ $d = 4.9t^2$ $p = 10^5 V^{-1}$ $V = \frac{4}{3}\pi R^3$ $v = 8h^{1/2}$

In most elementary physical problems only the positive values of the functions are calculated and have meaning. Only the simplest power functions will be considered here—those with $n = 1, 2, 3, 1/2, -1, -2$.

8.11 VARIATION

One important application of the power functions is in variation problems. Variation predicts the change in one variable as the result of a specified change in the other variable.

If both variables increase or both decrease in the same ratio, the variation is *direct*. For example, the weekly pay of a factory worker increases with the number of hours worked. If one of the variables increases while the other decreases in the same ratio, the variation is *inverse*. For example, the time needed to make a trip between two cities decreases as the travel speed increases.

The mathematical symbol for direct variation is $\propto$.

8.12 DIRECT VARIATION

The connection between direct variation and a power function is made by the following definition. Let x and y be numerical measures of two dependent quantities. Then y is said to vary directly as the nth power of x if y/x^n is a constant for some positive number n.

If we let the constant be k, then $y/x^n = k$ or $y = kx^n$. This relationship between y and x can also be expressed by the statement: y is directly proportional to the nth power of x. When $n = 1$, it is customary to omit the phrase "the 1st power of." Thus, $y = kx$ would be stated: y varies directly as x, or y is directly proportional to x.

When $n = 2$, the variation expressed by $y = kx^2$ would be stated: y varies directly as x squared, or y is directly proportional to the square of x. When $n = 1/2$, the variation expressed by $y = kx^{1/2} = k\sqrt{x}$ would be stated: y varies directly as x to the one half power, or y is directly proportional to the square root of x.

8.13   INVERSE VARIATION

The connection between inverse variation and a power function can be made as follows.

Let x and y be numerical measures of two dependent quantities. Then y is said to vary inversely as the nth power of x if $y = k/x^n$, for some positive number n and some real number k. The power function $y = k/x^n$ may be written also as $y = kx^{-n}$ or $yx^n = k$. An equivalent statement which expresses the inverse variation is: y is inversely proportional to the nth power of x. For a given x_1 there is a corresponding y_1, and so $y_1 x_1^n = k$; similarly $y_2 x_2^n = k$ for corresponding x_2 and y_2. Therefore

$$y_1 x_1^n = y_2 x_2^n \qquad \text{or} \qquad y_1/y_2 = x_2^n/x_1^n$$

Either of these two equations may be used to represent an inverse variation.

When $n = 1$, the inverse variation expressed by $y = k/x = kx^{-1}$, or $yx = k$, would be stated: y varies inversely as x, or y is inversely proportional to x.

When $n = 2$, the inverse variation expressed by $y = k/x^2 = kx^{-2}$, or $yx^2 = k$, would be stated: y varies inversely as x squared, or y is inversely proportional to the square of x.

8.14   GRAPHS OF POWER FUNCTIONS

The graphs of power functions are called *parabolic* if n is positive and *hyperbolic* if n is negative. The graphs for $y = x^n$, $n = 1, 2$, and $1/2$ are shown in Fig. 8-1; Fig. 8-2 shows the graphs for $n = -1$ and $n = -2$.

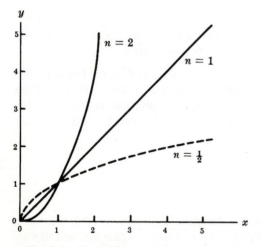

Fig. 8-1.   Parabolic power function $y = x^n$

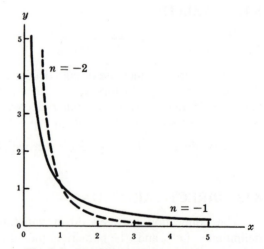

Fig. 8-2.   Hyperbolic power function $y = x^n$

8.15   GRAPHING POWER FUNCTIONS WITH A CALCULATOR

A graphing calculator is a handy tool for visualizing power functions. However, it is necessary to keep in mind that the viewing window may not accommodate the entire graph. In some occasions, it may be necessary to modify the range to find additional information about the nature of the graph. The ZOOM option of menu of PLOT is useful for viewing different sections of the graphs.

EXAMPLE 8.18. Graph Figs. 8-1 and 8-2 with a graphing calculator.

Convention: Function keys are enclosed in curly brackets in CAPITALS, for example, {ENTER}
Options on screen are shown underlined, for example, EDIT
Comments about the function of a key or option are enclosed in quotes

HP-38G (Functions of Fig. 8-1)

As a previous step it may be necessary to clear all functions, if any, previously defined. To do this, press the sequence {SHIFT}{CLEAR} and answer YES to the question "Clear all expressions?"

(1)   Define the function.

"First function, $f(x) = x^2$"

{LIB} Function {ENTER}
EDIT
{X,T,θ}   "This key writes an X"
{x^y} 2   "This raises the X to the 2nd power"
{ENTER}   "F1(x) = x^2

"Second function, $f(x) = x^{1/2}$"

EDIT
{X,T,θ}   "This key writes an X"
{x^y}{(}1{/}2{)}   "This allows to calculate the square root of X"
{ENTER}   F2(x) = x$^{1/2}$

"Third function, $f(x) = x$"

EDIT
{X,T,θ}   "This key writes an X"
{ENTER}   "F3(x) = x"

(2)   Set the range.

{SHIFT}{SETUP-PLOT}

XRNG: 0 {ENTER}   5 {ENTER}
YRNG: 0 {ENTER}   5 {ENTER}
XTICK: 1 {ENTER}   1 {ENTER}

(3)   Plot the functions.

{PLOT}

To move from one graph to another use the directional keys ($\blacktriangleleft$, $\blacktriangleright$, $\blacktriangle$, $\blacktriangledown$). You will visit the graphs in the same order in which they were defined.

TI-82 (Functions of Fig. 8-1)

(1)   Define the functions.

"First function, $f(x) = x^2$"

{Y=}
{X,T,θ}   "This key writes an X"
{^}   "This key is the 'raise to the power' function"
2{ENTER}

"Second function, $f(x) = x^{1/2}$"

{X,T,θ}
{^}
{()1{÷}2()}
{ENTER}

"Third function, $f(x) = x$"

{X,T,θ}   "This key writes an X"
{ENTER}

(2)   Set the range.

{WINDOW}{ENTER}
Xmin = 0 {ENTER}
Xmax = 5 {ENTER}
Xscl = 1 {ENTER}
Ymin = 0 {ENTER}
Ymax = 5 {ENTER}
Yscl = 1 {ENTER}

(3)   Plot the functions.

{GRAPH}

To move from one graph to another use the directional keys (◄, ►, ▲, ▼).   You will visit the graphs in the same order in which they were defined.

HP-38G (*Function of Fig. 8-2*)

(1)   Define the functions.

"First function, $f(x) = x^{-2}$"
{LIB} <u>Function</u> {ENTER}
<u>EDIT</u>
{X,T,θ}   "This key writes an X"
{x^y}{−x}2   "This raises the X to the negative 2nd power"

{ENTER}   "Second function, $f(x) = x^{-1}$"

{X,T,θ}   "This key writes an X"
{x^y}{−x}1   "This raises the X to the negative 1st power"
{ENTER}

If the previous range settings, defined for the functions of Fig. 8-1, are still in effect, press PLOT to sketch the graphs.   If necessary, reset the standard range by pressing {SHIFT}{SETUP-PLOT} followed by {SHIFT}{CLEAR}. You may have to use the ZOOM option of the PLOT menu to view different parts of the graph.

8.16   DETERMINING EMPIRICAL EQUATIONS OF POWER FUNCTIONS

It was shown in Chapter 7 how the equation of a straight line can be obtained from empirical data.   Another example is given here to emphasize direct variation.

EXAMPLE 8.19. $n = 1$.   The diameters D and circumferences C of several disks were measured by a student.   The data are shown in the table below.

D, cm	5.10	7.90	10.20	12.10
C, cm	16.45	25.20	31.95	38.20
C/D	3.23	3.19	3.13	3.16

The graph of the circumference as a function of the diameter is shown in Fig. 8-3.

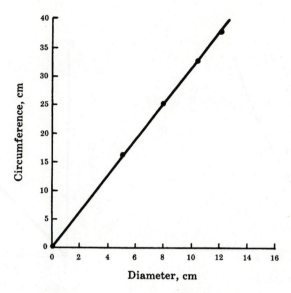

Fig. 8-3. Circumference versus diameter

Since the graph is a straight line and goes through the origin, the circumference varies directly as the diameter, or the circumference is directly proportional to the diameter. In mathematical symbols, $C \propto D$.

The equation is $C = k/D$, where k is a constant whose value is the physical slope of the graph, and $k = 31.7/10 = 3.17$. Thus, the equation is $C = 3.17D$. As a check, the ratio C/D for each measurement was calculated; the ratios are shown in the above table. The average ratio is 3.18, which agrees quite well with the value 3.17 from the graph. From the well-known formula we know that $C = \pi D$, where $\pi \approx 3.14$. The 1% difference between the experimental and theoretical values of π is caused by measurement uncertainties.

EXAMPLE 8.20. $n = 2$. The weights W of plywood squares of sides S are shown in the following table.

S, ft	0	1.0	1.5	2.0	2.5
W, lb	0	0.50	1.1	2.0	3.2
W/S		0.50	0.73	1.00	1.28

The variation is direct: when S increases, W also increases. However, W is *not* directly proportional to S, as can be seen from the graph in Fig. 8-4. Also, the ratio W/S is not constant. Choosing $n = 2$ as the next simplest integer for the exponent of S, a table of number pairs is prepared, as shown below.

S^2, ft^2	0	1.00	2.25	4.00	6.25
W, lb	0	0.50	1.10	2.00	3.20
W/S^2		0.50	0.49	0.50	0.51

The graph of W versus S^2 is shown in Fig. 8-5. Since the plot is a straight line through the origin, the weight varies directly as the *square* of the side length, or the weight is directly proportional to the square of the side length.

The ratio W/S^2 is constant within limits of measurement uncertainty. The physical slope of the graph is $2.5/5.0 = 0.50$ lb/ft^2. The equation of the power function is $W = 0.50S^2$.

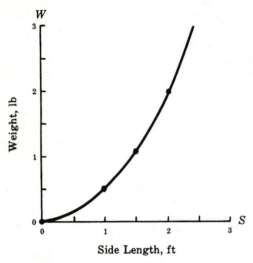

Fig. 8-4. Weight versus side length

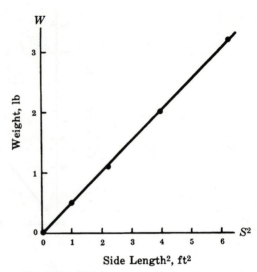

Fig. 8-5. Weight versus side length squared

EXAMPLE 8.21. $n = 1/2$. Several rods were suspended from a nail through a hole at one end and allowed to swing as pendulums. The period of oscillation P and the length of each rod L was measured. The data are shown below.

L, cm	0	12.6	29.4	45.1	91.5
P, sec	0	0.60	0.90	1.08	1.52
P/L		0.048	0.031	0.024	0.017

The data show that P and L increase or decrease together, but since the ratio is not constant the variation is not a direct proportion. The plot of the data is shown in Fig. 8-6. The trend indicates a possible fractional exponent.

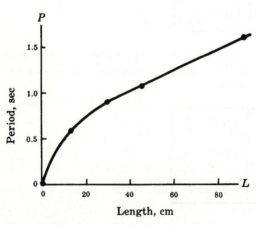

Fig. 8-6. Period versus length

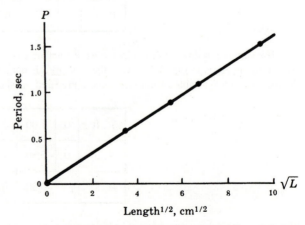

Fig. 8-7. Period versus length$^{1/2}$

Trying $n = 1/2$, we make a table of pairs for $L^{1/2}$ and P.

$L^{1/2}$, cm$^{1/2}$	3.55	5.42	6.72	9.57
P, sec	0.60	0.90	1.08	1.52
$P/L^{1/2}$	0.17	0.17	0.16	0.16

The graph of P versus $L^{1/2}$ is shown in Fig. 8-7. Since the graph is a straight line through the origin, P varies directly as $L^{1/2}$, or P is proportional to $L^{1/2}$. The equation of the power function is $P = kL^{1/2}$, or $P = k\sqrt{L}$. The constant k is the physical slope in Fig. 8-7, and $k = 1.6/10 = 0.16$ sec/cm$^{1/2}$. The equation then becomes $P = 0.16L^{1/2}$ or $P = 0.16\sqrt{L}$.

EXAMPLE 8.22. $n = -1$. The fundamental frequency F of a vibrating string changes with its length L. It is seen from the data in the table below that the frequency increases as the string length decreases.

L, meters	2.02	1.02	0.66	0.33
F, cycles/sec	9.93	19.61	30.72	60.45

The plot of F versus L is shown in Fig. 8-8. There is an inverse relationship between the two variables. To test the hypothesis that the frequency is a power function of string length with an exponent of -1, one plots F versus $1/L$ (reciprocal of L). Frequencies and calculated values of $1/L$ are paired in the following table.

$1/L$, m^{-1}	0.50	0.98	1.52	3.03
F, cps	9.93	19.61	30.72	60.45

The recently adopted unit of frequency is the *hertz* (Hz); it is equivalent to cycles per second (cps).

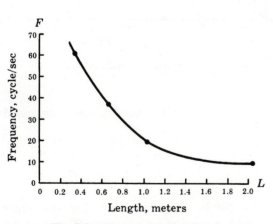

Fig. 8-8. Frequency versus length

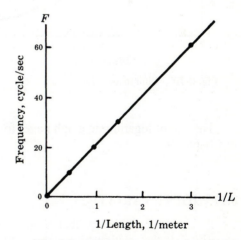

Fig. 8-9. Frequency versus 1/length

The graph of Fig. 8-9 is a straight line through the origin. This means that the frequency F varies as $1/L$, or the frequency is inversely proportional to the length. The equation is of the form $FL = k$, or $F = k/L$, where $k = \dfrac{40}{2}$ cps-m $= 20$ cps-m, is the value of the slope in Fig. 8-9. Another test of the relationship is to multiply together each pair of F–L values: each product is approximately 20.

EXAMPLE 8.23. $n = -2$. The dependence of illumination E on the distance d from a light source is shown in the table below.

d, ft	4	6	8	10	12
E, ft-candle	87	41	23	14	10

The graph of the data is shown in Fig. 8-10. There is an inverse relationship between the two variables. However, if E is a power function of d, the exponent cannot be -1, since doubling d does *not* halve E. The next simplest negative integer is -2. Values of $1/d^2$ are calculated and reproduced below.

$1/d^2$, ft^{-2}	0.063	0.028	0.016	0.010	0.007
E, ft-candle	87	41	23	14	10

The plot of E versus $1/d^2$ in Fig. 8-11 is a straight line through the origin. Therefore, the illumination varies as $1/\text{distance}^2$, or the illumination varies inversely as the square of the distance. An equivalent statement is: illumination is inversely proportional to the square of the distance. The equation is of the form $Ed^2 = k$ or $E = k/d^2$, where $k = 70/0.05 = 1400$ ft-candle-ft^2, the value of the slope in Fig. 8-11. The constancy of the $E \cdot d^2$ products verifies the conclusion drawn from the graph.

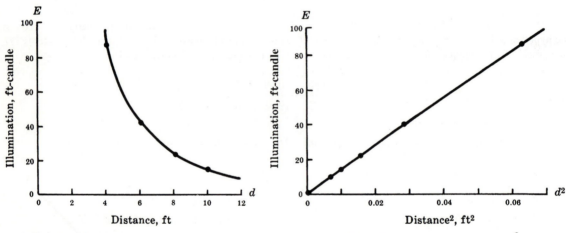

Fig. 8-10. Illumination versus distance

Fig. 8-11. Illumination versus distance^2

The use of logarithmic graph paper for determining the equation of a power function is discussed in Chapter 9.

8.17 JOINT VARIATION

It sometimes happens that a quantity is dependent for its value on more than one other quantity. If the relationship is either direct or inverse with respect to some power of each of the individual independent quantities, we have a problem in joint variation. The connecting equation does not define a simple power function but rather a power function with respect to each individual independent quantity.

EXAMPLE 8.24. The area A of a triangle varies jointly as its altitude h and its base b.

This statement in mathematical notation is $A \propto bh$, or $A = kbh$. The constant $k = 1/2$. It is also correct to say: The area of a triangle varies as the product of its altitude and base.

EXAMPLE 8.25.   The power P in an electrical current varies jointly as the resistance R and the square of the current I.

The mathematical notation for the statement is $P \propto RI^2$, or $P = KRI^2$.   If the resistance is doubled, but the current is constant, the power is doubled.   If the resistance is constant and the current is doubled, the power is quadrupled (2^2).   If both the resistance and the current are doubled, the power is increased 8 times.

8.18   FORMULAS

An equation which expresses in symbols the relationship between two or more physical quantities is called a formula.   Many of the important rules and principles of pure and applied science are expressed as power function formulas.   In general a formula expresses one quantity in terms of the others; i.e., the symbol for the expressed quantity is on the left side with its coefficient and exponent equal to unity.   Examples of formulas are $C = \pi D$, $E = (1/2)mv^2$; $v = \sqrt{2gh}$.

It is frequently desirable to solve a formula for some other symbol.   This is called *changing the subject of a formula*.

EXAMPLE 8.26.   Solve the formula $v = 980t$ for t.

Dividing both sides of the equation by 980, we get $v/980 = t$ or $t = v/980$.

EXAMPLE 8.27.   Solve the formula $H = P/550$ for P.

Multiplying both sides of the equation by 550, we get $550H = P$ or $P = 550H$.

EXAMPLE 8.28.   Solve the formula $A = (1/2)bh$ for h.

Multiplying both sides by 2, we get $2A = 2(1/2)bh$ or $2A = bh$.

Dividing both sides by b, we obtain $2A/b = h$ or $h = 2A/b$.

EXAMPLE 8.29.   Solve $S = (1/2)gt^2$ for t.

Multiplying both sides by 2, we get $2S = 2(1/2)gt^2$

Dividing both sides by g, we obtain $\dfrac{2S}{g} = \dfrac{\cancel{g}t^2}{\cancel{g}} = t^2$

Taking the square root of each side we get $\sqrt{2S/g} = \sqrt{t^2}$
$$t = \sqrt{2S/g} \quad \text{or} \quad t = (2S/g)^{1/2}$$

Solved Problems

EXPONENTS AND RADICALS

8.1.   Write each of the following in exponential form:

   (a) $2 \cdot 2 \cdot 2 \cdot 2 \cdot 2$ (c) $\dfrac{1}{3 \cdot 3 \cdot 3 \cdot 3}$ (e) $\sqrt[3]{a \cdot a \cdot a \cdot a \cdot a}$

   (b) $m \cdot m \cdot m$ (d) $(-5)(-5)(-5)$ (f) $\sqrt{\dfrac{a}{b} \cdot \dfrac{a}{b} \cdot \dfrac{a}{b}}$

Applying the definition of exponents, we get:

(a) 2^5 (b) m^3 (c) 3^{-4} (d) $(-5)^3$ (e) $\sqrt[3]{a^3} = a^{5/3}$ (f) $\sqrt{(a/b)^3} = (a/b)^{3/2}$

8.2. Write each of the following radicals in exponential form:

(a) $\sqrt[3]{4}$ (c) $\sqrt[5]{4^{-2}}$ (e) $(\sqrt[3]{t})^2$

(b) $\sqrt{3^4}$ (d) $(\sqrt{S})^3$ (f) $\sqrt[4]{p^{-2}}$

Applying the definition of radical and exponent, we obtain:

(a) $4^{1/3}$ (b) $3^{4/2} = 3^2$ (c) $4^{-2/5}$ (d) $S^{3/2}$ (e) $t^{2/3}$ (f) $p^{-2/4} = p^{-1/2}$

8.3. Write each of the following numbers in fractional or radical form:

(a) 3^{-2} (b) $3^{1/2}$ (c) $2^{3/4}$ (d) $5^{1/3}E^{1/2}$ (e) $T^{2/3}$ (f) $R^{-1/2}$

From definitions of exponent and radical, we obtain:

(a) $\dfrac{1}{3^2} = \dfrac{1}{9}$ (d) $5^{2/6}E^{3/6} = (5^2E^3)^{1/6} = \sqrt[6]{5^2E^3} = \sqrt[6]{25E^3}$

(b) $\sqrt{3}$ (e) $\sqrt[3]{T^2}$

(c) $\sqrt[4]{2^3}$ (f) $\dfrac{1}{R^{1/2}} = \dfrac{1}{\sqrt{R}}$

8.4. Simplify the following expressions:

(a) $\dfrac{3^3 \cdot a^0}{3}$ (c) $\dfrac{(3^{-4})(3^{+2})(3^{+5})}{(3^{+2})(3^{+3})}$ (e) $\sqrt{16ab^4c^8d}$

(b) $\dfrac{5^2}{5^3 \cdot t^0}$ (d) $\dfrac{2^3 \cdot 2^{-6} \cdot 2^6 \cdot 2^{-7}}{2^4 \cdot 2^{-5} \cdot 2^0}$ (f) $\sqrt{(36x)(9xy^4)}$

Using the laws of exponents and the definition $x^0 = 1$, we obtain:

(a) $3^3 \cdot 3^{-1} \cdot 1 = 3^2 = 9$

(b) $5^2 \cdot 5^{-3} \cdot 1 = 5^{-1} = \dfrac{1}{5}$

(c) $3^{-4} \cdot 3^{+2} \cdot 3^{+5} \cdot 3^{-2} \cdot 3^{-3} = 3^{-9} \cdot 3^{+7} = 3^{-2} = \dfrac{1}{3^2} = \dfrac{1}{9}$

(d) $2^3 \cdot 2^{-6} \cdot 2^6 \cdot 2^{-7} \cdot 2^{-4} \cdot 2^{+5} \cdot 1 = 2^{14} \cdot 2^{-17} = 2^{-3} = \dfrac{1}{2^3} = \dfrac{1}{8}$

(e) $4b^2c^4\sqrt{ad}$

(f) $\sqrt{(36)(9)x^2y^4} = (6^2 \cdot 3^2 \cdot x^2 \cdot y^4)^{1/2} = 6 \cdot 3 \cdot xy^2 = 18xy^2$

DIRECT VARIATION, $n = 1$

8.5. The amount of gasoline consumed by a car traveling at a given speed varies directly with the distance traveled.   In covering 225 miles, 18 gal were used.

(*a*) Find the equation relating gasoline consumption and distance.

(*b*) What is the meaning of the equation constant, and what are the units?

(*c*) Plot a graph of the function defined by (*a*).

(*d*) Determine the number of gallons needed to travel 350 miles.

(*e*) Find the distance that can be covered with 50 gal.

(*a*) Let G represent the amount of gasoline in gallons and d be the distance traveled in miles. The statement in the problem implies that $G/d = k$, or $G = kd$, where k is a constant. Substituting the values $G = 18$ and $d = 225$ in the equation results in $18 = k(225)$. Solving for k, we get $k = 18/225 = 0.080$. The functional equation is then $G = 0.080d$.

(*b*) The constant 0.080 is interpreted as the ratio of gasoline consumption to the distance traveled. The units associated with the constant are gallons/mile or gallons per mile. The car will consume 0.080 gal for each mile of travel.

(*c*) To graph the function defined by the equation, substitute a few simple numbers for d and find the corresponding values of G as shown in the table below. Since d is the independent variable, it is graphed along the horizontal axis; the dependent variable G is plotted vertically as shown in Fig. 8-12.

d, miles	0	100	200	300	400	500
G, gal	0	8	16	24	32	40

(*d*) Substituting 350 for d in the equation, we get

$$G = (0.080)(350) = 28 \text{ gal}$$

(*e*) Substituting 50 for G in the equation, we get $50 = (0.080)d$, and solving for d

$$d = \frac{50}{0.080} = 625 \text{ miles}$$

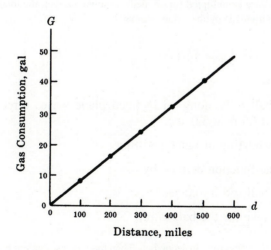

Fig. 8-12. Gas consumption versus distance

The direct variation expressed by $y = kx$ is often given a different form. To a given value x_1 there corresponds a value y_1 such that $y_1/x_1 = k$. Similarly, corresponding to x_2 there is a value y_2 such that $y_2/x_2 = k$. Since k is the same in a given variation, we can write

$$\frac{y_1}{x_1} = \frac{y_2}{x_2} \qquad \text{or} \qquad \frac{y_1}{y_2} = \frac{x_1}{x_2}$$

This expression is in the form of a proportion, so that if any three of the quantities are known, the fourth one can be obtained. For instance, part (e) of the above example could have been obtained by setting up the proportion

$$\frac{G_1}{G_2} = \frac{d_1}{d_2}$$

Substituting the given values,

$$\frac{18}{50} = \frac{225}{d_2} \qquad \text{or} \qquad 18d_2 = (225)(50) \qquad \text{or} \qquad d_2 = \frac{(225)(50)}{18} = 625 \text{ miles}$$

Thus, the three expressions $\frac{G}{d} = k$, $G = kd$, $\frac{G_1}{G_2} = \frac{d_1}{d_2}$ are equivalent; each can be translated into the same statement of variation or proportionality.

8.6. At constant pressure and when well above the liquefaction temperature, the volume of a gas varies directly with its absolute temperature.

(a) Express the above relationship between the two quantities by an equivalent verbal statement.

(b) Write three mathematical expressions equivalent to the given statement.

(c) Explain in simple words the statement requested in (a).

(a) The volume of a gas is directly proportional to its absolute temperature.

(b) Let V = measure of volume of the gas, T = absolute temperature of the gas, k = constant. Then

$$\frac{V}{T} = k \qquad V = kT \qquad \frac{V_1}{V_2} = \frac{T_1}{T_2}$$

(c) If the absolute temperature is multipled (or divided) by some number, the measure of the volume of the gas will be multiplied (or divided) by the same number.

DIRECT VARIATION, $n = 2$

8.7. The distance that a ball rolls along an inclined plane varies directly as the square of the time. The ball rolls 13.5 ft in 3.0 sec.

(a) Find the equation relating distance and time.

(b) Plot a graph of the function defined by (a).

(c) How far will the ball roll in 6.0 sec?

(d) How long will it take for the ball to roll 24 ft?

(a) Let d = distance in feet and t = time in seconds. Then the given statement means that $d/t^2 = k$ or $d = kt^2$, where k is a constant. Substituting the given values in the equation,

$$13.5 = k(3.0)^2 \qquad \text{or} \qquad \frac{13.5}{3^2} = \frac{13.5}{9} = 1.5$$

The exact functional equation is then $d = 1.5t^2$.

(b) To graph the function, substitute a few simple numbers for t in the equation and find the corresponding values of d. Tabulate the results as shown in the table below. Plot t horizontally, as it is the independent variable. Connect the points which we have plotted with a smooth curve. The resulting graph shown in Fig. 8-13 is a portion of a curve called a *parabola*.

t, sec	0	1	2	3	4	500
d, ft	0	1.5	6	13.5	24	37.5

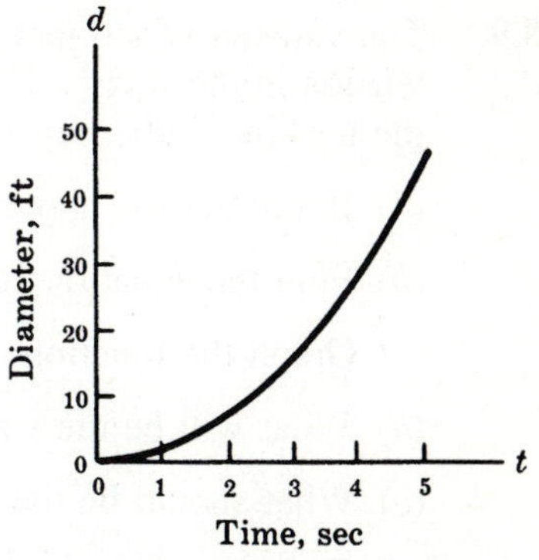

(c) Substitute 6.0 sec for t and solve for d:

$$d = 1.5(6^2) = (1.5)(36) = 54 \text{ ft}$$

Fig. 8-13. Distance–time graph

(d) Substitute 24 ft for d and solve for t:

$$24 = (1.5)t^2 \qquad \text{or} \qquad t^2 = \frac{24}{1.5} = 16$$

$$t = \pm\sqrt{16} = \pm 4 \text{ sec}$$

Although there are two possible solutions, we reject $t = -4$ because negative time has no physical meaning in this problem, and conclude that $t = 4$.

8.8. Given $P = 3I^2$, where $P =$ rate of heat generation in an electrical conductor and $I =$ current:

(a) Write two equivalent forms of the given equation.

(b) Translate the equation into statements of variation and proportionality.

(c) Explain the meaning of the equation in simple words.

(d) What happens to the value of P when I is increased 5 times?

(e) What happens to the value of I if P is divided by 16?

(a) $\dfrac{P}{I^2} = 3 \qquad \dfrac{P_1}{P_2} = \dfrac{I_1^2}{I_2^2}$

(b) The rate of heat generation in an electrical conductor varies directly as the square of the current; or, the rate of heat generation in an electrical conductor is directly proportional to the square of the current.

(c) When the current is doubled, the rate of heat generation is increased four times. If the current is reduced to one-third, the rate of heat generation is one-ninth of its former value. This can be worked out by substituting simple numbers for I in the equation as shown in the table below.

I	0	1	2	3
P	0	$3 \cdot 1^2 = 3$	$3 \cdot 2^2 = 12$	$3 \cdot 3^2 = 27$

(d) P is increased 25 times.

(e) I is divided by 4 or reduced to one-fourth of its previous value.

DIRECT VARIATION, $n = 1/2$

The power function $y = kx^{1/2}$ may be written also as $y = k\sqrt{x}$.

8.9. The velocity of a transverse wave in a stretched wire varies directly as the square root of the tension in the wire. The velocity of a transverse wave in a certain wire is 300 m/sec when the tension is 900 newtons.

(*a*) Represent the above variation by four equivalent equations.

(*b*) Find the equation constant and write the functional equation.

(*c*) Graph the function defined in (*b*).

(*d*) What will be the wave velocity when the tension is 3600 newtons?

(*e*) What should be the tension in the wire for a wave velocity of 250 m/sec?

(*f*) Translate the variation into a statement of proportionality.

(*g*) Explain the variation using small integers.

Let v = velocity of the transverse wave, t = tension in the wire, and k = constant.

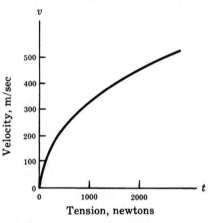

Fig. 8-14. Graph of $v = 10\sqrt{t}$

(*a*) $\dfrac{v}{\sqrt{t}} = k$ $v = k\sqrt{t}$ $v = kt^{1/2}$ $\dfrac{v_1}{v_2} = \dfrac{\sqrt{t_1}}{\sqrt{t_2}}$

(*b*) Substitute the given values of v and t into the first equation in (*a*) to obtain

$$\frac{300}{\sqrt{900}} = k \qquad \text{or} \qquad k = \frac{300}{30} = 10$$

The exact functional equation connecting v and t is $v = 10\sqrt{t}$, or $v = 10t^{1/2}$.

(*c*) Make a table of values by substituting a few integral squares for t and calculating v. Plot the values, with t along the horizontal axis, as shown in Fig. 8-14. The graph is of the parabolic type.

t, newtons	0	100	400	900	1600	2500
v, m/sec	0	100	200	300	400	500

(*d*) Substitute $t = 3600$ newton into the equation in (*b*):

$$v = 10\sqrt{3600} = 10 \cdot 60 = 600 \text{ m/sec}$$

(*e*) Substitute $v = 250$ m/sec into the equation in (*b*):

$$250 = 10\sqrt{t} \qquad \frac{250}{10} = \sqrt{t} \qquad 25 = \sqrt{t}$$

Squaring both sides

$$25^2 = t \qquad \text{or} \qquad 625 = t$$

Therefore, the required tension is 625 newtons.

(*f*) The velocity of the wave is directly proportional to the square root of the tension.

(*g*) If the tension is multiplied by 4, the velocity is doubled; if the tension is divided by 9, the velocity is divided by 3.

8.10. The period of a loaded oscillating coiled spring is directly proportional to the square root of the vibrating mass. The period is 1.33 second when the mass is 0.350 kilogram.

(*a*) Write the variation in mathematical symbols.

(b) Write three equivalent equations for the above variation.

(c) Determine the equation constant and write the functional equation.

(d) Graph the function defined in (c).

(e) What would be the value of the period for a mass of 0.150 kilogram?

(f) What should be the value of the mass so that the period is 1.15 second?

Let P = period of the spring; M = suspended mass; C = constant.

(a) $P \propto \sqrt{M}$ or $P \propto M^{1/2}$

(b)
$$\frac{P}{\sqrt{M}} = C \qquad P = C\sqrt{M} \qquad \frac{P_1}{P_2} = \frac{\sqrt{M_1}}{\sqrt{M_2}}$$

(c) Substituting the given values for P and M into the second equation of (b) we obtain:

$$1.33 = C\sqrt{0.350}$$

Solving for C,

$$C = \frac{1.33}{\sqrt{0.350}} \approx \frac{1.33}{0.592} \approx 2.25$$

Therefore the functional equation is $P = 2.25\sqrt{M}$ or $P = 2.25M^{1/2}$.

(d) The results of substituting a few integral squares for M and calculating P are shown in the table below.

M, kilograms	0	1	4	9	16
P, seconds	0	2.25	4.50	6.75	9.00

The graph of P versus M is shown in Fig. 8-15.

(e) Substitute 0.150 kilogram into the functional equation:

$$P = 2.25\sqrt{0.150} \approx (2.25)(0.387) \approx 0.871 \text{ sec}$$

(f) Substitute 1.15 second for P in the functional equation:

$$1.15 = 2.25\sqrt{M}$$

$$\frac{1.15}{2.25} = \sqrt{M} \qquad 0.511 = \sqrt{M}$$

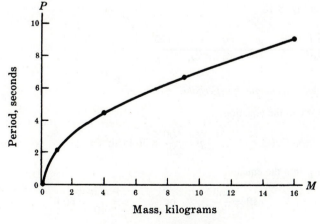

Fig. 8-15.   Period versus mass

Square both sides to obtain

$$M = (0.511)^2 \qquad M \approx 0.261 \text{ kilogram}$$

INVERSE VARIATION, $n = 1$

8.11. The pressure of a gas at constant temperature varies inversely as the volume.   The pressure
of 50 ft^3 of the gas is 12 lb/in^2.

(a) Write the statement in terms of proportionality.

(b) Write four equivalent forms of the equation representing the statement.

(c) Determine the equation constant and write the exact functional equation.

(d) Graph the function defined in (c).

(e) Determine the pressure of the gas if the volume is reduced to 20 ft^3.

(f) Determine the volume of the gas if the pressure is reduced to 10 lb.

(g) Without making calculations on paper, predict what will happen to the volume if the
pressure is (1) doubled, (2) tripled, (3) halved, (4) quartered.

Let p = pressure of the gas, v = volume of the gas, and k = constant.

(a) The pressure of a gas at constant temperature is
inversely proportional to its volume.

(b) $pv = k \qquad p = \dfrac{k}{v} \qquad p_1 v_1 = p_2 v_2 \qquad p = kv^{-1}$

(c) Substitute the given values of p and v into the first
equation in (b):

$$(12)(50) = k \qquad \text{or} \qquad k = 600$$

The exact functional equation connecting p and v is

$$pv = 600 \quad \text{or} \quad p = \dfrac{600}{v} \qquad \text{or} \qquad p = 600v^{-1}$$

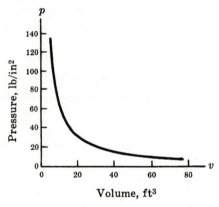

(d) Make a table of values by substituting a few simple
values for v and calculating p.   Plot the values with v
along the horizontal axis and connect with a smooth
curve as shown in Fig. 8-16.

Fig. 8-16.   Pressure–volume graph

v	6	10	20	30	60	100	120
p	100	60	30	20	10	6	5

The graph is called a *rectangular hyperbola*.

(e) Substitute 20 for v into the equation:

$$p = \frac{600}{20} = 30 \text{ lb/in}^2$$

(f) Substitute 10 for p into the equation:

$$10 = \frac{600}{v} \qquad \text{or} \qquad v = \frac{600}{10} = 60 \text{ ft}^3$$

(g) (1) Volume will be reduced to one-half of its former value.

(2) Volume will be reduced to one-third of its former value.

(3) Volume will be doubled.

(4) Volume will be quadrupled.

8.12. When drilling brass at high speed, the drill diameter in inches is inversely proportional to the drill's speed in revolutions per minute.   When the drill diameter is 0.50 inch, the drill speed is 2300 rpm.

(a) Write the statement of proportionality in mathematical symbols.

(b) Express the relationship by means of four equivalent equations.

(c) Determine the equation constant and write the functional equation.

(d) Graph the equation in (c).

(e) Calculate the drill speed for a diameter of 0.75 inch.

(f) Making only metal calculations, by what factor should the drill diameter be changed if the speed is quadrupled?

Let s = the drill speed, d = drill diameter, c = constant.

(a) $d \propto \dfrac{1}{s}$ or $d \propto s^{-1}$

(b) $ds = c$ $\qquad d = \dfrac{c}{s}$ $\qquad d = cs^{-1}$ $\qquad d_1 s_1 = d_2 s_2$

(c) Substitute the given values of d and s in the first equation of (b):

$$(0.50)(2300) = c$$
$$1150 = c \qquad \text{or} \qquad c = 1150$$

The functional equation is

$$ds = 1150 \qquad \text{or} \qquad d = \frac{1150}{s}$$

(d) Substitute a few values for s and calculate corresponding values of d as shown in the table below.

s, rpm	575	1150	2300	4600
d, inches	2	1	0.5	0.25

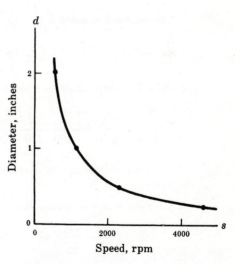

Fig. 8-17.   Diameter versus speed

The graph of d versus s is shown in Fig. 8-17.

(e) Substitute 0.75 for d into the equation:

$$0.75 = \frac{1150}{s} \qquad \text{or} \qquad s = \frac{1150}{0.75} \approx 1530 \text{ rpm}$$

(f) The drill diameter should be quartered or reduced to one-fourth of its former value.

INVERSE VARIATION, $n = 2$

8.13. The dose rate from a radioactive point source in air is inversely proportional to the square of the distance from the source. For a distance of 2 meters the dose rate is 80 milliroentgens per hour.

(a) State in words the relationship between the variables as a power variation.

(b) Write five equivalent forms of the equations representing the relationship.

(c) Determine the proportionality constant and write the exact functional equation.

(d) Graph the equation in (c).

(e) Determine the dose rate at a distance of 5 meters.

(f) Determine the distance at which the dose rate is 0.5 milliroentgen/hour.

(g) Without paper calculations predict the effect on dose rate if the distance is (a) halved, (b) multiplied by 4.

(h) Without paper calculations predict the required change in distance if the dose rate is to (1) increase by a factor of 9, (2) be reduced by a factor of 25.

Let I = dose rate, r = distance, c = constant.

(a) The dose rate from a radioactive source varies inversely as the second power of the distance from the source (or inversely as the square of the distance).

(b)
$$I = \frac{c}{r^2} \qquad Ir^2 = c \qquad I = cr^{-2} \qquad I_1 r_1^2 = I_2 r_2^2 \qquad \frac{I_1}{I_2} = \frac{r_2^2}{r_1^2}$$

(c) Substitute the given values of I and r in the second equation of (b)

$$(80)(4) = c \qquad c = 320$$

The functional equation is

$$Ir^2 = 320 \qquad \text{or} \qquad I = \frac{320}{r^2}$$

(d) Substitute a few values for r and calculate the corresponding values of I as shown in the table below.

r, meters	1	2	4
I, mr/hr	320	80	20

The graph of dose rate versus distance is shown in Fig. 8-18.

(e) Substitute 5 for r in the equation.

$$I = \frac{320}{5^2} = \frac{320}{25}$$

$$I = 12.8 \text{ milliroentgen/hour}$$

(f) Substitute 0.5 for I in the equation.

$$0.5 = \frac{320}{r^2} \qquad 0.5r^2 = 320 \qquad r^2 = \frac{320}{0.5} = 640$$

$$r = \sqrt{640} \approx 25.3 \text{ meter}$$

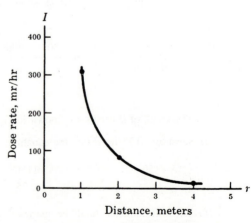

Fig. 8-18. Dose rate versus distance

(g) (1) The dose rate will be multiplied by 4.

    (2) The dose rate will be reduced by a factor of 16.

(h) (1) The distance should be reduced by a factor of 3.

    (2) The distance should be increased by a factor of 5.

8.14. The exposure time for photographing an object varies inversely as the square of the lens diameter.   The exposure time is 1/400 sec when the lens diameter is 2 cm.

(a) Translate the above into a statement of proportionality.

(b) Write five equivalent forms of the equation representing the statement.

(c) Determine the equation constant and write the exact functional equation.

(d) Graph the function defined in (c).

(e) Find the exposure time for a lens diameter of 4 cm.

(f) Find the lens diameter needed to give an exposure time of 1/25 sec.

(g) Without making calculations on paper, predict what will happen to the exposure time if the lens diameter is: (1) halved; (2) tripled; (3) divided by 4.

(h) Without making calculations on paper, predict what should be the lens diameter if the exposure time is to be: (1) quadrupled; (2) divided by 4; (3) multiplied by 9.

    Let t = exposure time, d = lens diameter, and k = constant.

(a) The exposure time for photographing an object is inversely proportional to the square of the lens diameter.

(b) $$t = \frac{k}{d^2} \qquad td^2 = k \qquad t = kd^{-2}$$

$$t_1 d_1^2 = t_2 d_2^2 \qquad \frac{t_1}{t_2} = \frac{d_2^2}{d_1^2}$$

(c) Substitute the given values of t and d into the second equation in (b):

$$\frac{1}{400}(2^2) = k \qquad k = \frac{4}{400} = 0.01$$

The exact functional equation is

$$td^2 = 0.01 \qquad \text{or} \qquad t = \frac{0.01}{d^2}$$

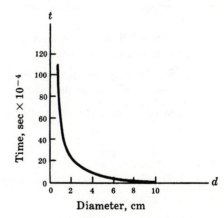

Fig. 8-19. Exposure time versus lens diameter

(d) Make a table of values by substituting a few simple values for d and calculating t.   Plot the values with d along the horizontal axis and connect with a smooth curve as shown in Fig. 8-19.

d, cm	1	2	3	4	5	10
t, sec	0.010	0.0025	0.0011	0.00063	0.0040	0.0001

The graph is of the hyperbolic type.

(e) Substitute 4 for d in the equation:

$$t = \frac{0.01}{4^2} = \frac{0.01}{16} = 0.00062 \approx 0.006 \text{ sec}$$

(*f*) Substitute $1/25$ for t in the equation:

$$\frac{1}{25} = \frac{0.01}{d^2}$$
$$d^2 = (0.01)(25) = 0.25$$
$$d = \pm\sqrt{0.25} = 0.5 \text{ cm}$$

The negative value of d is discarded because it has no physical meaning here.

(*g*) (1) multiplied by 4; (2) divided by 9; (3) multiplied by 16.

(*h*) (1) halved; (2) multiplied by 2; (3) divided by 3.

JOINT VARIATION

8.15. The centripetal force on a rotating object varies jointly as its mass and the square of its speed and inversely with the radius of rotation. For a mass of 90 g rotating at 140 cm/sec in a circle with a 60-cm radius the force is 29,400 dynes.

(*a*) Write equivalent forms of the equation representing the statement.

(*b*) Find the equation constant and write the exact functional equation.

(*c*) Determine the centripetal force when the mass is 200 g, the speed is 98 cm/sec, and the radius of the circular path is 40 cm.

(*d*) Without making calculations on paper, predict what the effect will be on the centripetal force if (1) the mass alone is doubled; (2) the mass alone is reduced to one-fourth of its former value; (3) the speed alone is tripled; (4) the speed alone is divided by 2; (5) the radius alone is multiplied by 6; (6) the radius alone is divided by 3; (7) the mass, the speed, and the radius are each doubled.

Use the symbols F, m, v, and r to represent measures of the centripetal force, the mass, the speed, and radius of rotation, respectively.

(*a*) Since F varies directly as the product of m and v^2 and inversely with r, we must have $Fr/mv^2 = k$, for some constant k. Solving the equation for F, we have $F = kmv^2/r$ as the equivalent form.

(*b*) Substitute the given values of m, v, and r in the first equation in (*a*):

$$k = \frac{(29,400)(60)}{90(140)^2} = 1$$

Therefore, the exact functional equation is

$$F = \frac{mv^2}{r}$$

(*c*)
$$F = \frac{200(98)^2}{40} \approx 48,000 \text{ dynes}$$

(*d*) Centripetal forces will be: (1) doubled; (2) divided by 4; (3) increased 9 times; (4) divided by 4; (5) divided by 6; (6) multiplied by 3; (7) multiplied by 4.

TRANSLATING EQUATIONS

Translate the equations in Problems 8.16 through 8.23 into statements of variation and proportionality.

8.16. $P = 4S$, where P = perimeter of a square, S = length of a side.

Both P and S are to the first power. Therefore, the variation is direct with the exponent of the power function $n = 1$. The statements are:

The perimeter of a square varies directly as the length of the side.

The perimeter of a square is directly proportional to the length of the side.

8.17. $V = ct$, where V = instantaneous speed of a freely falling body, t = time of fall, and c = constant.

The equation is of the power function form with the exponent $n = 1$. The required statements are:

The instantaneous speed of a freely falling body varies directly as the time of fall.

The instantaneous speed of a freely falling body is directly proportional to the time of fall.

8.18. $P = CE^2$, where P = power dissipated in a circuit, E = voltage, and C = constant.

The equation is of the power function type with the exponent $n = 2$. Therefore the appropriate statements are:

The power dissipated in a circuit varies directly as the square of the voltage.

The power dissipated in a circuit is directly proportional to the square of the voltage.

8.19. $H = k/n$, where H = horsepower transmitted by a shaft, n = number of revolutions per minute, and k = constant.

The equation is of the form $yx = k$ or $y = kx^{-1}$, which is an inverse variation to the first power. The equation can be expressed by the statements:

The horsepower transmitted by a shaft varies inversely with the number of revolutions per minute.

The horsepower transmitted by a shaft is inversely proportional to the number of revolutions per minute.

8.20. $f = C\sqrt{T}$, where f = frequency of a vibrating string, T = tension and C = constant.

The equation is of the power function form with the exponent $n = 1/2$. The relevant statements are:

The frequency of a vibrating string varies directly as the square root of the tension.

The frequency of a vibrating string is directly proportional to the square root of the tension.

8.21. $t = k/d^2$, where t = the time needed to fill a tank with water, d = the diameter of the pipe delivering the water.

The equation is of the form $y = kx^{-2}$ or $yx^2 = k$. This is an inverse variation to the second power. The required statements are:

The time needed to fill a tank with water varies inversely as the square of the diameter of the pipe delivering the water.

The time needed to fill a tank with water is inversely proportional to the square of the diameter of the pipe delivering the water.

8.22. $M = kit$, where $M =$ mass of electroplated metal, $i =$ current, $t =$ time, and $k =$ constant.

This is a joint variation since the mass depends on two variables. The appropriate statements are:

The mass of electroplated metal varies jointly with the current and the time.

The mass of electroplated metal varies directly as the product of the current and the time.

8.23. $\dfrac{a_2}{a_1} = \dfrac{F_2}{F_1}$, where a_2 and a_1 are the accelerations produced by the forces F_2 and F_1 applied to a body.

The equation can be rewritten as $\dfrac{a_1}{F_1} = \dfrac{a_2}{F_2}$. The equality of the two ratios means that $\dfrac{a}{F}$ is constant, or $\dfrac{a}{F} = k$, or $a = kF$. Therefore, the equation can be translated into statements:

The acceleration of a body varies directly as the applied force.

The acceleration of a body is directly proportional to the applied force.

EMPIRICAL EQUATIONS OF POWER FUNCTIONS

8.24. The area A and weight W of a number of metal plates with identical thickness were measured.

A, ft^2	0	2.5	5.1	7.5	10	12.5	15.0
W, lb	0	14.0	27	42	56	70	85

Determine the empirical equation from the data in the table above.

Graph the data on cartesian (ordinary) graph paper. The graph is shown in Fig. 8-20.

Since the graph goes through the origin, the intercept is zero. Therefore the equation is of the form $W = mA$. To find the slope of the line, divide any ordinate by the corresponding abscissa. In this case, the 10-ft^2 abscissa is used because it simplifies division. The ordinate as read from the graph is 56 lb. Therefore the slope m is $56/10 = 5.6$ lb. This is a *physical slope* because it has units and its value was calculated by using the scales of measured physical quantities.

The empirical equation giving the relationship between the weights of plates and their areas is $W = 5.6A$. This equation tells us that the weight of a metal plate of given thickness is directly proportional to its area. The proportionality constant is 5.6; it means that the weight of a plate 1 ft^2 in area is 5.6 lb.

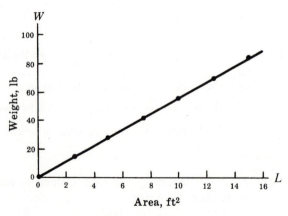

Fig. 8-20. Weight versus area

8.25. A ball is allowed to roll along an inclined plane, and the distance (S) are recorded for several elapsed times (t).

t, sec	0	1	2	3	4	5	6	7
S, m	0	0.15	0.30	0.65	1.20	1.90	2.70	3.6

Find the empirical relationship between the distance and time.

The graph of the data is shown in Fig. 8-21. Since the graph is nonlinear, we plot the distance versus the squares of the respective times. The result is seen in Fig. 8-22. Thus, one concludes that the distance is proportional to the square of the time, and the empirical equation is of the form $S = kt^2$. The value of k is obtained by finding the physical slope of the graph in Fig. 8-22. The value of k is 0.076 m/sec^2. Therefore the equation is $S = 0.075t^2$. The value of k depends on the angle of the inclined plane and on the units of measurement; had the plot of distance versus time squared been nonlinear, we would have tried a plot of distance versus time cubed.

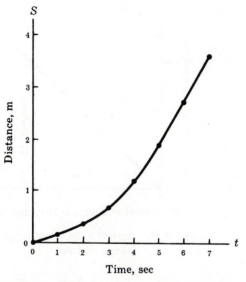

Fig. 8-21. Distance traveled along incline versus time

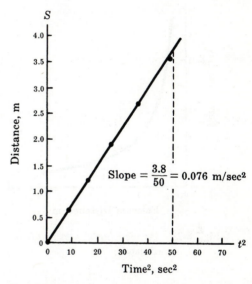

Fig. 8-22. Distance along incline versus time squared

8.26. The force needed to counterbalance a weight by means of a lever depends on the distance of the force from the lever pivot (fulcrum). Determine the empirical equation between the force F and its distance d from the fulcrum by using the data below.

d, m	0.30	0.38	0.50	0.6	1.1	1.9	2.9
F, newtons	25	20	15	13	7	4.0	2.6

The graph of the data is shown in Fig. 8-23. The relationship is nonlinear. The curve indicates some sort of inverse relationship. Therefore, one might try plotting the force versus the reciprocal of distance, $1/d$. Calculating the value of $1/d$ and retabulating, one gets:

$1/d$, m^{-1}	3.3	2.6	2.0	1.7	0.91	0.53	0.34
F, newtons	25	20	15	13	7.0	4.0	2.6

The graph of F versus $1/d$ shown in Fig. 8-24 is a straight line.   Therefore, the force is proportional to the reciprocal of distance, or $F = k(1/d) = k/d$.   This means that the force varies inversely with distance; i.e., the two variables are inversely proportional to each other.   The value of k is determined from the physical slope in Fig. 8-24:

$$k = \frac{19}{2.5} \approx 7.6 \text{ newton/meter}$$

Therefore, the empirical equation is $F = 7.6/d$, or $dF = 7.6$.

The inverse relationship in this example could also be detected quickly by multiplying each of the distances by the corresponding force.   A glance at the data shows that these products are nearly constant; therefore, the two variables are inversely proportional to each other.

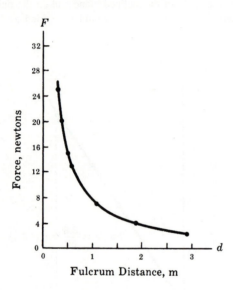

Fig. 8-23.   Applied force versus fulcrum distance

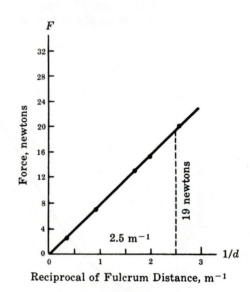

Fig. 8-24.   Applied force versus reciprocal of fulcrum distance

FORMULAS

8.27. Solve the formula $V = kq/r$ for q.

Multiplying both sides of the formula by r, we get

$$Vr = \frac{kq\cancel{r}}{\cancel{r}}$$

Dividing both sides by k, we obtain

$$\frac{Vr}{k} = \frac{\cancel{k}q}{\cancel{k}} \qquad \text{or} \qquad q = \frac{Vr}{k}$$

8.28. Solve the formula of Problem 8.27 for r.

Again multiplying both members of the formula by r, we get $Vr = kq$.   Dividing both members by V, we have

$$\frac{\cancel{V} \cdot r}{\cancel{V}} = \frac{kq}{V} \qquad \text{or} \qquad r = \frac{kq}{V}$$

8.29. Solve $P = E^2/R$ for E.

Multiplying both members by R, we have

$$PR = \frac{E^2}{R} R$$

Taking square roots of both sides, we obtain $\pm\sqrt{PR} = E$, or since the negative value of the square root has no meaning in this physical situation, $E = \sqrt{PR}$.

8.30. $v = \sqrt{2gd}$ for d.

Squaring both members, we obtain $v^2 = 2gd$.

Dividing both members by $2g$, we obtain $d = v^2/2g$.

8.31. Solve $F = G\dfrac{mM}{d^2}$ (a) for m, (b) for d.

(a) Dividing both sides by GM, we get

$$\frac{F}{GM} = \frac{G}{GM}\frac{mM}{d^2}$$

Multiplying both sides by d^2, we obtain

$$\frac{Fd^2}{GM} = \frac{md^2}{d^2} \qquad \text{or} \qquad m = \frac{Fd^2}{GM}$$

(b) Multiplying both sides by d^2, we get

$$Fd^2 = \frac{GmMd^2}{d^2}$$

Dividing both sides by F, we get

$$\frac{Fd^2}{F} = \frac{GmM}{F}$$

Taking the square root of both sides, we obtain

$$d = \sqrt{\frac{GmM}{F}}.$$

Supplementary Problems

The answers to this set of problems are at the end of this chapter.

8.32. Write each of the following in exponential form:

(a) $3 \cdot 3 \cdot 3 \cdot 3 \cdot 3 \cdot 3 \cdot 3$ (b) $\dfrac{1}{5 \cdot 5 \cdot 5 \cdot 5 \cdot 5 \cdot 5}$ (c) $\dfrac{1}{16 \cdot 16 \cdot 16}$ (d) $\sqrt[3]{\dfrac{f}{G} \cdot \dfrac{f}{G} \cdot \dfrac{f}{G} \cdot \dfrac{f}{G}}$

8.33. Write the results of the indicated operations in simplified exponential form:

(a) $2^5 \cdot 2^{-3} \cdot 2^6$ (b) $3^{-5} \cdot 3^6 \cdot 3^{-2}$ (c) $5^2 \cdot 5^6 \cdot 5^{-6}$ (d) $(4p)(4p)^3(4p)^{-4}$

8.34. Write the results of the indicated operations in simplified exponential form:

(a) $\dfrac{4^3}{4^{-2}}$ (c) $\dfrac{5^3}{5^3 \cdot 5^5}$ (e) $\dfrac{Mr^4}{Mr^2}$ (g) $\dfrac{H^{5/2}}{\sqrt{H}}$

(b) $\dfrac{3^4}{3^2}$ (d) $\dfrac{2^4 \cdot 2^6}{2^3 \cdot 2^{-5}}$ (f) $h^{1/2} \cdot h^{3/2}$

8.35. Write each of the following as an ordinary integer or fraction:

(a) $4^{3/2}$ (b) $27^{4/3}$ (c) 8^{-2} (d) 5^0 (e) $16^{3/2}$

8.36. Determine the principal: (a) square root of 9; (b) cube root of 27; (c) cube root of -64; (d) fourth root of 81.

8.37. Express each of the following in radical or fractional form:

(a) 2^{-1} (c) $5^{1/2}$ (e) 3^{-3} (g) $V^{3/2}$

(b) $3^{2/5}$ (d) $4^{1/3}$ (f) 2^{-2} (h) $3^{1/2}p^{1/2}d^{-1/2}$

8.38. Express each of the following in exponential form:

(a) $\sqrt{35}$ (c) $\sqrt[3]{3^2}$ (e) $\sqrt[4]{125}$ (g) $\sqrt[5]{b^2}$

(b) $\sqrt[3]{15}$ (d) $\sqrt[5]{5^3}$ (f) $\sqrt[3]{18}$ (h) $I^{1/2}L^{-1/2}$ (i) $\dfrac{p^{1.67}}{p}$

8.39. Multiply the following monomials:

(a) $3x^{1/2}y, 2xy^{1/4}$ (b) $2x^{2/3}, 5xy^{1/2}$ (c) $2xy^{3/4}, 4x^{1/3}y, 2xy$

8.40. The velocity v of a body falling freely from a height of 16 ft is $v = \sqrt{2 \cdot 32 \cdot 16}$ ft/sec.

(a) Express the velocity in simple exponential form with the lowest possible base.

(b) Calculate this velocity.

8.41. By Kepler's third law the distance d from a planet to the sun in astronomical units is equal to the period p of the planet's revolution in years raised to the 2/3 power, or $d = p^{2/3}$. Find the distance of a planet to the sun if the planet's period is 8 years.

8.42. Translate the following equations into statements of variation and/or proportionality:

(a) $W = RH$, where W = weekly wage, H = number of hours worked, R = constant.

(b) $m = dr^2$, where m = mass of a disk, r = radius of disk, d = constant.

(c) $I = cV$, where I = current in a conductor, V = potential difference across conductor, and c = constant.

(d) $FL = V$, where F = frequency of radio waves, L = wavelength of radio waves, and V = constant.

(e) $E = kv^2$, where E = kinetic energy of a body, v = its speed, k = constant.

(f) $AS = 6.3$, where A = atomic weight of an element and S = specific heat of the element.

(g) $V = k\sqrt{T}$, where V = velocity of transverse waves in a cord, T = tension in cord, and k = constant.

(h) $E = c/r^2$, where E = electric field intensity due to a charge, r = distance from the charge, and c = constant.

(i) $m = eit$, where m = mass of element liberated by electrolysis, i = current, t = time, and e = constant.

(j) $d_1/d_2 = M_1/M_2$, where d_1 and d_2 are the densities of two gases under the same conditions and M_1 and M_2 are the respective molecular weights of the gases.

(k) $R = kL/A$, where R = electrical resistance of a wire, L = its length, A = its cross-sectional area, and k = constant.

(l) $P_1/P_2 = r_2/r_1$, where P_1 and P_2 are the pressures due to surface tension on two drops of a liquid with radii r_1 and r_2, respectively.

8.43. In the equation $I = E/R$, what happens to the value of I if: (a) E alone is tripled;   (b) E alone is halved;   (c) R alone is divided by 3;   (d) R alone is multiplied by 5;   (e) E is doubled and R is divided by 4?

8.44. In the equation $F = KQ_1Q_2/d^2$, what happens to the value of F if: (a) Q_1 alone is quadrupled;   (b) Q_2 alone is divided by 5;   (c) Q_1 is doubled and Q_2 is tripled;   (d) d alone is tripled;   (e) Q_1, Q_2, and d are each doubled?

8.45. The pressure P of a given volume of gas varies directly with the absolute temperature T.

(a) If the pressure of a certain volume of gas is 200 lb-f/in^2 at 1500 °C, find the exact functional equation.

(b) Using the functional equation determine the pressure if the temperature is raised 500 °C.

8.46. The altitude of a triangle of constant area varies inversely as the length of the base.

(a) If the altitude of a triangle is 20 in. when the base is 5 in. in length, find the exact functional equation.

(b) Using the functional equation, determine the altitude when the base is 12 in.

8.47. Neglecting air resistance, the distance traveled by a freely falling body varies directly with the square of the time of fall.

(a) If a body falls 1600 ft in 10 sec, find the exact functional equation.

(b) Using the equation in (a), determine how far the body will fall in 18 sec.

8.48. The period T of a simple pendulum of length L varies directly with the square root of its length.

(a) If the period is 2.0 sec for a pendulum 100 cm long, find the exact functional equation.

(b) Using the equation in (a) find the period of a pendulum 25 cm long.

(c) How long should the pendulum be to give a period of 4 sec?

8.49. The volume V of a cone varies jointly as the altitude A and the square of the radius R of the base.

(a) If the volume of a cone is 30 in^3 when the altitude and radius are 10 in. and 3 in. respectively, find the exact functional equation.

(b) Using the equation in (a), find the volume of a cone whose altitude is 15 in. and the radius of whose base is 10 in.

8.50. The volume V of a sphere varies directly as the cube of its radius.   How does the volume of a sphere change if its radius is   (a) doubled;   (b) tripled;   (c) halved;   (d) quadrupled?

8.51. The average velocity V of a gas molecule is proportional to the square root of the Kelvin temperature K.

(a) Express the above statement by three equivalent equations.

(b) The average velocity of a molecule is 1500 m/sec at 400 K? Determine the equation constant from this data.   Write the exact functional equation.

(c) What will be the average velocity if the temperature is raised to 6400 K?

(d) At what temperature will the average velocity become 15,000 m/sec?

(e) Plot a graph of the relationship between molecular velocity and the Kelvin temperature. (*Hint*: Molecular velocity is the dependent variable.)

8.52. Neglecting air resistance, the speed v of a freely falling body varies directly with the time of fall.

(a) Express the above statement by three equivalent equations.

(b) In 5 sec a body attains a speed of 49 m/sec. Determine the equation constant from the given data and write the exact functional equation.

(c) Translate the equations in (a) and (b) into a statement of proportionality.

(d) What speed will the body attain in 8 sec?

(e) In what time will the body acquire a speed of 28 m/sec?

(f) Plot a graph of the relationship between speed and time. (*Hint*: Time is the independent variable.) What is the shape of the graph?

8.53. The time t is needed to travel a given distance varies inversely with the speed S of travel.

(a) Express the above statement by three equivalent equations.

(b) To travel the distance at 50 mph, $3\frac{1}{2}$ hr is required. Find the equation constant from the given data and write the exact functional equation.

(c) Translate the equations in (a) and (b) into a statement of proportionality.

(d) What time would be required to cover the given distance at a speed of 70 mph?

(e) What speed would be required to travel the given distance in 5 hr?

(f) Plot a graph of the relationship between speed and time. What type of curve is it?

(g) Without making calculations on paper, predict what would happen to the time if the speed is   (1) divided by 3;   (2) multiplied by 4?

(h) What would the speed have to be to   (1) double the time;   (2) divide the time by 3?

8.54. The luminous flux L passing through a circular opening varies directly as the square of the opening diameter D.

(a) Express the above statement by three equivalent equations.

(b) A circular diaphragm 5 mm in diameter placed in front of a light source admits 50 lumens of luminous flux. Find the equation constant from the given data and write the exact functional equation.

(c) Translate the equations in (a) and (b) into a statement of proportionality.

(d) How much light will be admitted by a diaphragm 10 mm in diameter when placed in front of the given source?

(e) What should be the diaphragm diameter to admit 0.5 lumen of light?

(f) Plot a graph of the relationship between the amount of light and the diameter of the opening. What type of curve is it?

(g) How does the amount of light change when the opening diameter is   (1) doubled;   (2) tripled;   (3) halved; (4) quadrupled?

8.55. The force F between two given electrical charges is inversely proportional to the square of the distance d between the charges.

(a) Translate the above statement into a statement of variation.

(b) Express the above statement by three equivalent equations.

(c) Two charges exert a force of 180 dynes on each other when they are 5 cm apart. Find the equation constant from the given data and write the exact functional equation.

(d) What force will be exerted by the charges when they are 3 cm apart?

(e) What should be the distance between the charges to give a force of 45 dynes?

(f) Plot a graph of the relationship between force and distance.   What type of curve is it? How does it differ from the graph of Problem 8.53?

(g) How does the force change if the distance is   (1) halved;   (2) tripled;   (3) quartered;   (4) multiplied by 5?

8.56. The energy E emitted each second by a black body varies jointly with the area of the body and the fourth power of its Kelvin temperature T.

(a) Express the above statement by three equivalent equations.

(b) When the temperature is 100 K, the rate of energy emission per unit area is 5.7 joule/(m^2)(sec). Determine the equation constant and write the exact functional equation.

(c) What happens to the value of E when the area alone is increased 10 times?

(d) What happens to the value of E when the temperature alone is increased 10 times?

(e) How should the temperature of a black body change so as to reduce the radiation to 1/256 of its former value?

8.57. For a constant load the effort E in pounds is inversely proportional to its distance D from the fulcrum.   When the lever arm is 2 ft the effort is 50 lb.

(a) Determine the equation expressing the relationship between the effort and the distance from the fulcrum.

(b) If the lever arm is 0.5 ft, how large is the effort?

(c) If the effort is 20 lb, what is the distance to the fulcrum?

8.58. The formula for centripetal acceleration is $a = v^2/r$, where $a =$ centripetal acceleration, $v =$ linear speed, and $r =$ radius of rotation.   What happens to the value of a if   (1) v alone is doubled;   (2) v alone is divided by 3;   (3) r alone is tripled?

8.59. The voltmeter reading V changes with the current I in a circuit as shown in the following table.

I, amp	0	0.100	0.195	0.290	0.390	0.485
V, volt	0	20	40	60	80	100

(a) Plot a graph of the given data.

(b) Determine the empirical equation of the graph.

(c) What are the units of the slope?

8.60. The density d of methane at 0 °C was determined for several pressures P.   The data are given in the table below.

P, cm Hg	19	38	57	76
d, g/liter	0.18	0.36	0.54	0.72

(a) Plot a graph of the data.

(b) What conclusion can you draw about the relationship between the variables?

(c) Determine the empirical equation of the graph.

(d) Use the equation to calculate the density of methane at a pressure of 45 cm Hg.   Check the calculated value against the value read from the graph.

(e) Use the equation to compute the value of the pressure for a density of 0.63 g/liter.   Check the computation against the value read from the graph.

8.61.   Oil is forced into a closed tube containing air.   The height H of the air column varies with pressure P as shown in the table below.

P, lb-f/in^2	6.7	8.1	10.2	12.9	14.4	17.6	21.4	25.1	29.4
H, in	13.3	10.7	9.2	7.1	6.3	5.1	4.1	3.5	3.0

(a) Plot a graph of the above data.

(b) What conclusions can be drawn from (a)?

(c) Plot a graph of H versus $1/P$.

(d) What conclusions can be drawn from (c)?

(e) Determine the empirical equation for the given data.

8.62.   The safe working load S of a wrought iron crane chain depends on the chain diameter d as shown in the following table.

d, in.	0.25	0.50	0.75	1.0	1.25	1.5	1.75	2.0
S, lb-f	1100	4200	9500	16,000	25,000	36,000	48,000	63,000

(a) Plot the preceding data.

(b) Does the graph in (a) represent a direct or an inverse variation? A direct or an inverse proportion?

(c) Plot S as a function of d^2.

(d) Plot S as a function of d^3.

(e) What conclusions can be drawn from (c) and (d)?

(f) Determine the equation relating S to d.   Check the equation by substituting one of the given points.

8.63.   The period P of an oscillating spring varies with the mass M attached to one end of the spring.   The experimental data are reproduced below.

M, kg	0	0.10	0.20	0.30	0.40	0.50
P, sec	0	0.65	0.90	1.12	1.30	1.45

(a) Plot the data.

(b) Does the graph show a direct or inverse variation?

(c) Plot P as a function of $\sqrt{M}$.

(d) Plot P as a function of $\sqrt[3]{M}$.

(e) What conclusions can be drawn from (c) and (d)?

(f) Determine the functional relationship between P and M.   Check the equation by substituting one of the given points.

(g) Find graphically the mass which will cause a period of 1.0 sec.

(h) Find graphically the period for a mass of 0.56 kg.

8.64. The power P generated by a wind-driven 30-cm two-bladed propeller depends on the wind speed v. The data are shown in the table below.

v, km/hr	13	16	19	26	32	40	48
P, kw	0.09	0.17	0.29	0.70	1.4	2.7	4.6

(a) Plot the data.

(b) What conclusions can be drawn from (a)?

(c) Plot P as a function of v^2.

(d) Plot P as a function of v^3.

(e) What conclusions can be drawn from (c) and (d)?

(f) Determine the equation connecting P and v. Check by substituting one of the given points.

8.65. (a) Given the equation $T = 630,000/N$, what quantities should be plotted along the axes to give a straight-line graph?

(b) What would be the numerical value of the slope of the line in (a)?

(c) What relationship exists between T and N?

8.66. (a) Given the equation $P = 3.2(10^7)t^3$. What quantities should be plotted along the axes to give a straight-line graph?

(b) What would be the numerical value of the slope of the line in (a)?

(c) What is the functional relationship between P and t?

8.67. The acceleration a varies with time t as shown in the following table.

t, sec	29.9	20.0	16.0	13.7	12.3	11.1	10.1
a, cm/sec^2	0.38	0.85	1.32	1.81	2.25	2.86	3.34

(a) Plot a graph of the data.

(b) What conclusion can be drawn from (a)?

(c) Plot a versus $1/t$.

(d) Plot a versus $1/t^2$.

(e) What conclusions can be drawn from (c) and (d)?

(f) Determine the equation relating a and t. Check it by substituting one of the given points.

Solve the following formulas for the indicated symbols.

8.68. $W = Fd$ for d; for F.

8.69. $A = (1/2)ah$ for a; for h.

8.70. $R = \dfrac{V}{I}$ for V; for I.

8.71. $m = eit$ for t; for i.

8.72. $E = (1/2)mv^2$ for m; for v.

8.73. $F = \dfrac{q_1 q_2}{d^2}$ for q_2; for d.

8.74. $\dfrac{P_1}{P_2} = \dfrac{T_1}{T_2}$ for T_1; for P_2.

8.75. $\dfrac{P_1 \bar{V}_1}{T_1} = \dfrac{P_2 \bar{V}_2}{T_2}$ for $\bar{V}_1$; for T_2.

8.76. $C = \dfrac{KA}{4\pi d}$ for A; for d.

8.77. $f = \dfrac{1}{2L}\sqrt{\dfrac{T}{m}}$ for L; for T; for m.

8.78. $H = \dfrac{I^2 Rt}{4.18}$ for R; for I.

8.79. $\dfrac{E_1}{E_2} = \dfrac{r_2^2}{r_1^2}$ for E_2; for r_1.

8.80. $p = \dfrac{1}{3}dv^2$ for v.

8.81. $T = 2\pi\sqrt{\dfrac{L}{g}}$ for L; for g.

8.82. $pV = \dfrac{m}{M}RT$ for m; for T.

8.83. $\dfrac{r_1}{r_2} = \dfrac{\sqrt{d_2}}{\sqrt{d_1}}$ for r_2; for d_1.

8.84. $F = 4\pi^2 mf^2 r$ for r; for f.

8.85. $v^2 = \dfrac{GM}{r}$ for M; for v.

8.86. $a = \dfrac{4\pi^2 R}{T^2}$ for R; for T.

Answers to Supplementary Problems

8.32. (a) 3^7 (b) 5^{-6} (c) 16^{-3} or 2^{-12} (d) $(f/G)^{4/3}$

8.33. (a) 2^8 (b) 3^{-1} (c) 5^2 (d) $(4p)^0$ or 1

8.34. (a) 4^5 (b) 3^2 (c) 5^{-5} (d) 2^{12} (e) r^2 (f) h^2 (g) H^2

8.35. (a) 8   (b) 81   (c) 1/64   (d) 1   (e) 64

8.36. (a) 3   (b) 3   (c) -4 (d) 3

8.37. (a) 1/2   (b) $\sqrt[5]{9}$ (c) $\sqrt{5}$ (d) $\sqrt[3]{4}$ (e) 1/27   (f) 1/4
       (g) $\sqrt{V^3}$ or $V\sqrt{V}$ (h) $\sqrt{3p/d}$

8.38. (a) $35^{1/2}$ (b) $15^{1/3}$ (c) $3^{2/3}$ (d) $5^{3/5}$ (e) $5^{3/4}$ (f) $18^{1/3}$
       (g) $b^{2/5}$ (h) $(I/L)^{1/2}$ (i) $p^{0.67}$

8.39. (a) $6x^{3/2}y^{5/4}$ (b) $10x^{5/3}y^{1/2}$ (c) $16x^{7/3}y^{11/4}$

8.40. (a) 2^5 ft/sec   (b) 32 ft/sec

8.41. 4 astronomical units.

8.42. (a) The weekly wages varies directly with the number of hours worked, or
       The weekly wages is directly proportional to the number of hours worked.

      (b) The mass of a disk varies as the square of its radius, or
       The mass of a disk is directly proportional to the square of its radius.

      (c) The current in a conductor varies directly with the potential difference across it, or
       The current in a conductor is directly proportional to the potential difference across it.

(d) The frequency of radio waves varies inversely with the wavelength, or
The frequency of radio waves is inversely proportional to the wavelength.

(e) The kinetic energy of a body varies as the square of its speed, or
The kinetic energy of a body is directly proportional to the square of its speed.

(f) The atomic weight of an element varies inversely with its specific heat, or
The atomic weight of an element is inversely proportional to its specific heat.

(g) The velocity of transverse waves in a cord varies as the square root of its tension, or
The velocity of transverse waves in a cord is directly proportional to the square root of its tension.

(h) The electric field intensity due to a charge varies inversely as the square of the distance from it, or
The electric field intensity due to a charge is inversely proportional to the square of the distance from it.

(i) The mass of an element liberated by electrolysis varies jointly with the current and the time, or
The mass of an element liberated by electrolysis is proportional to the product of the current and the time.

(j) The densities of gases vary directly with their molecular weights, or
The densities of gases are directly proportional to their molecular weights.

(k) The electrical resistance of a wire varies directly with its length and inversely proportional with its cross-sectional area, or
The electrical resistance of a wire is directly proportional to its length and inversely proportional to its cross-sectional area.

(l) The pressures due to surface tension on liquid drops vary inversely with their radii, or
The pressures due to surface tension on liquid drops are inversely proportional to their radii.

8.43. (a) tripled   (b) halved   (c) tripled   (d) one-fifth as large   (e) increased eightfold

8.44. (a) quadrupled   (b) one-fifth as large   (c) increased sixfold   (d) one-ninth as large

(e) unchanged

8.45. (a) $P = (2/15)T \approx 0.133T$ (b) 266 lb-f/in$^2 \approx 270$ lb-f/in^2

8.46. (a) $ab = 100$ (b) $8\frac{1}{3}$ in. ≈ 8.3 in.

8.47. (a) $d = 16t^2$ (b) 5184 ft ≈ 5200 ft

8.48. (a) $T = \frac{1}{5}\sqrt{L}$ or $T = 0.2\sqrt{L}$ (b) 1 sec   (c) 400 cm

8.49. (a) $V = AR^2/3$ or $V = 0.33AR^2$ (b) 500 in^3

8.50. (a) increased eightfold   (b) increased 27-fold   (c) one-eighth as great   (d) increased 64-fold

8.51. (a) $V = c\sqrt{K}$; $V/\sqrt{K} = c$ (c) 6000 K

$V_1/V_2 = \sqrt{K_1}/\sqrt{K_2}$ (d) $40{,}000$ K

(b) $c = 75$; $V = 75\sqrt{K}$ (e) See Fig. 8-25.

8.52. (a) $v = kt$; $v/t = k$; $v_1/v_2 = t_1/t_2$ (d) $78\frac{2}{5}$ or 78.4 m/sec

(b) $k = 49/5$; $v = (49/5)t = 9.8t$ (e) 2.86 sec

(c) The speed of a falling body is directly proportional to its time of fall.   (f) See Fig. 8-26.

8.53. (a) $tS = k$; $t = k/S$; $t_1S_1 = t_2S_2$; $t_1/t_2 = S_2/S_1$

(b) $tS = 175$

(c) The time to travel a given distance is inversely proportional to the speed of travel.

(d) $2\frac{1}{2}$ hr

(e) 35 mph

(f) hyperbola (see Fig. 8-27).

(g) (1) tripled; (2) one-fourth as long.

(h) (1) halved; (2) tripled.

8.54. (a) $L = kD^2$; $L/D^2 = k$; $L_1/L_2 = D_1^2/D_2^2$

(b) $k = 2$; $L = 2D^2$

(c) The luminous flux through a circular opening is directly proportional to the square of its diameter.

(d) 200 lumens

(e) 1/2 mm or 0.5 mm

(f) parabola (see Fig. 8-28).

(g) (1) quadrupled; (2) increased ninefold; (3) one-quarter as great; (4) increased 16-fold.

8.55. (a) The force F between two given electrical charges varies inversely as the square of the distance d between the charges.

(b) $F = k/d^2$; $Fd^2 = k$; $F_1/F_2 = d_2^2/d_1^2$

(c) $k = 4500$; $F = 4500/d^2$ or $Fd^2 = 4500$

(d) 500 dynes

(e) 10 cm

(f) hyperbola (see Fig. 8-29); graph approaches horizontal axis much faster than in Fig. 8-23.

(g) (1) quadrupled; (2) one-ninth as large; (3) increased 16-fold; (4) 1/25 as large.

8.56. (a) $E = kAT^4$; $E/AT^4 = k$; $E_1/A_1T_1^4 = E_2/A_2T_2^4$

(b) $k = 5.7 \times 10^{-8}$; $E = 5.7 \times 10^{-8}AT^4$

(c) increased 10-fold.

(d) increased 10,000-fold.

(e) Reduce temperature to one-fourth.

8.57. (a) $ED = 100$ (b) 200 lb   (c) 5 ft

8.58. (1) quadrupled   (2) one-ninth as great   (3) one-third as great.

8.59. (a) See Fig. 8-30   (b) $V \approx 210I$ (c) volt/ampere

8.60. (a) See Fig. 8-31   (b) direct variation   (c) $d = 0.0095P$ (d) $d = 0.43$ g/liter
(e) $P \approx 66$ cm Hg

8.61. (a) See Fig. 8-32.

(b) H is not a linear function of P; there may be an inverse power relationship between H and P.

(c) See Fig. 8-33.

(d) H is inversely proportional to P.

(e) $H = 90/P$ or $HP = 90$.

8.62. (a) See Fig. 8-34   (c) See Fig. 8-35   (e) S varies as d^2

(b) direct; neither   (d) See Fig. 8-36   (f) $S = 16{,}000d^2$

8.63. (a) See Fig. 8-37   (c) See Fig. 8-38   (e) P varies as $\sqrt{M}$ (g) 0.24^+ kg

(b) direct   (d) See Fig. 8-39   (f) $P = 2.1\sqrt{M}$ or $P = 2.1M^{1/2}$ (h) 1.5^+ sec

8.64. (*a*) See Fig. 8-40   (*c*) See Fig. 8-41 (*e*) P varies as v^3

   (*b*) P varies with v, but not directly (*d*) See Fig. 8-42 (*f*) $P = 0.000042v^3$

8.65. (*a*) T versus $1/N$ (*b*) 630,000 (*c*) inverse proportionality

8.66. (*a*) P and t^3 (*b*) 3.2×10^7 (*c*) P varies as t^3.

8.67. (*a*) See Fig. 8-43.   (*c*) See Fig. 8-44.

   (*b*) Acceleration is not a linear function of time; (*d*) See Fig. 8-45.
     there may be an inverse power relationship   (*e*) a varies as $1/t^2$.
     between a and t.   (*f*) $a = 340/t^2$

8.68. $d = W/F$; $F = W/d$ **8.78.** $R = 4.18H/I^2t$; $I = \pm\sqrt{4.18H/Rt}$

8.69. $a = 2A/h$; $h = 2A/a$ **8.79.** $E_2 = E_1 r_1^2/r_2^2$; $r_1 = \pm r_2\sqrt{E_2/E_1}$

8.70. $V = RI$; $I = V/R$ **8.80.** $v = \pm\sqrt{3p/d}$

8.71. $t = m/ei$; $i = m/et$ **8.81.** $L = gT^2/4\pi^2$; $g = 4\pi^2 L/T^2$

8.72. $m = 2E/v^2$; $v = \pm\sqrt{2E/m}$ **8.82.** $m = MpV/RT$; $T = MpV/mR$

8.73. $q_2 = Fd^2/q_1$; $d = \pm\sqrt{q_1 q_2/F}$ **8.83.** $r_2 = r_1\sqrt{d_1/d_2}$; $d_1 = r_2^2 d_2/r_1^2$

8.74. $T_1 = P_1 T_2/P_2$; $P_2 = P_1 T_2/T_1$ **8.84.** $r = F/4\pi^2 mf^2$; $f = \pm\dfrac{1}{2\pi}\sqrt{F/mr}$

8.75. $V_1 = P_2 V_2 T_1/P_1 T_2$; $T_2 = T_1 P_2 V_2/P_1 V_1$ **8.85.** $M = rv^2/G$; $v = \pm\sqrt{GM/r}$

8.76. $A = 4\pi Cd/K$; $d = KA/4\pi C$ **8.86.** $R = aT^2/4\pi^2$; $T = \pm 2\pi\sqrt{R/a}$

8.77. $L = \dfrac{1}{2f}\sqrt{T/m}$; $m = T/4f^2 L^2$

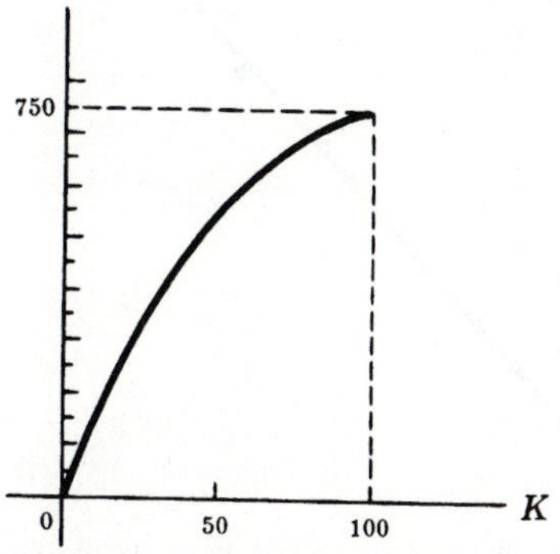

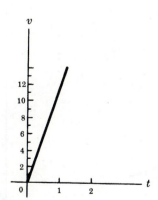

Fig. 8-25. Relationship between
the molecular velocity
of a gas molecule and its
Kelvin temperature

Fig. 8-26. Speed of a freely
falling body
versus the time of
fall

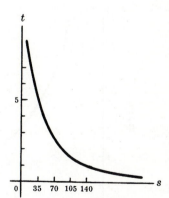

Fig. 8-27. Time needed to travel
a given distance
versus the speed of
travel

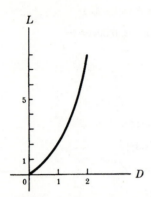

Fig. 8-28. Graph of the luminous flux as a function of the diameter of the opening through which it passes

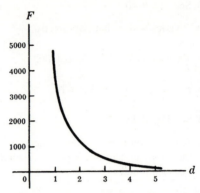

Fig. 8-29. Graph of the force between two given electric charges and the square of the distance that separate them

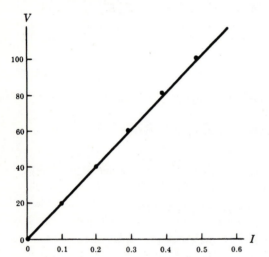

Fig. 8-30. Graph of a voltmeter reading as a function of the current in a circuit

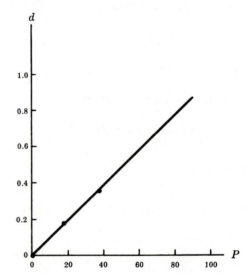

Fig. 8-31. Graph of the density of methane at 0°C at different pressures

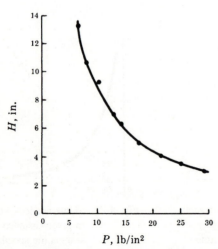

Fig. 8-32. Graph of the height of a column of air as a function of the pressure applied

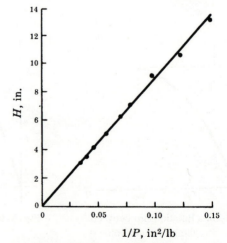

Fig. 8-33. Graph of the height of a column of air as a function of the increase of the pressure applied

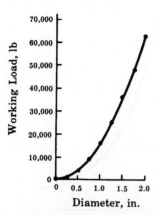

Fig. 8-34. Graph of the working load of a wrought iron crane as a function of the chain diameter

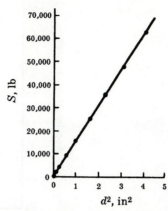

Fig. 8-35. Graph of the working load of a wrought iron crane as a function of the square of the chain diameter

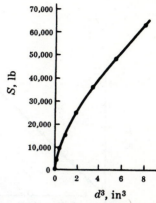

Fig. 8-36. Graph of the working load of a wrought iron crane as a function of the cube of the chain diameter

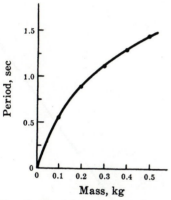

Fig. 8-37. Graph of the period P of an oscillating spring as a function of the mass attached to its end

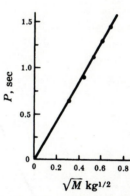

Fig. 8-38. Graph of the period P of an oscillating spring as a function of the square root of the mass attached to its end

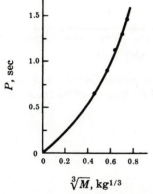

Fig. 8-39. Graph of the period P of an oscillating spring as a function of the cubic root of the mass attached to its end

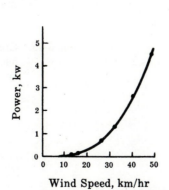

Fig. 8-40. Graph of the power P generated by a propeller as a function of the wind speed

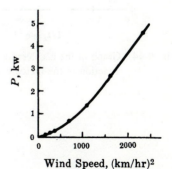

Fig. 8-41. Graph of the power P generated by a propeller as a function of the square of the wind speed

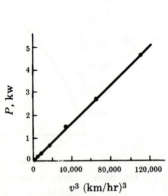

Fig. 8-42. Graph of the power generated by a propeller as a function of the cube of the wind speed

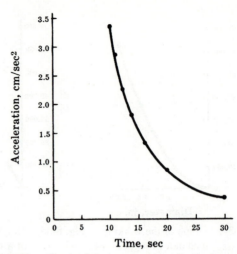

Fig. 8-43. Graph of the acceleration of a body as a function of time

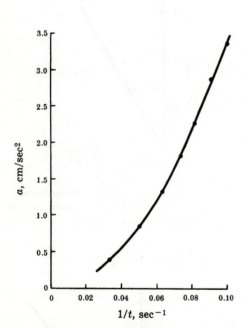

Fig. 8-44 Graph of the acceleration of a body as a function of the increase of time

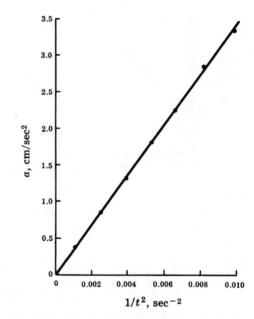

Fig. 8-45. Graph of the acceleration of a body as a function of the increase of the square of time

Chapter 9

Logarithms

Basic Concepts

9.1 NUMBERS AS POWERS

It can be proved that every positive real number N can be expressed as a power of a given positive real number b, if b is not equal to one. This fact makes it possible to simplify computations by the use of *logarithms*.

EXAMPLE 9.1. Calculate $(27)(81)(27^4)/(243)(9)(243)$.

Each number is an integral power of 3; the above can be written as

$$(3^3 3^4 3^{12})/(3^5 3^2 3^5)$$

Applying the laws of exponents, the expression equals:

$$3^{3+4+12-5-2-5} = 3^7$$

The answer is 3^7, or 2187.

This power can be calculated with calculator as follows:

HP-38G	TI-82
3{x^y}7{ENTER}	3{^}7{ENTER}

9.2 DEFINITION OF A LOGARITHM

The logarithm of a number N to a given base b is the power p to which the base must be raised to equal the number.

In mathematical symbols, $p = \log_b N$, which is exactly equivalent to $N = b^p$.

Any positive number except 1 may be used as a base.

EXAMPLE 9.2. The exponential equation $8 = 2^3$ can be written as the logarithmic equation $3 = \log_2 8$.

EXAMPLE 9.3. Conversely, the logarithmic equation $\log_3 81 = 4$ can be written as the exponential equation $81 = 3^4$.

9.3 COMMON LOGARITHMS

Logarithms to base 10 are called *common* or *Briggsian*. It is common practice to write $\log N$ instead of $\log_{10} N$.

EXAMPLE 9.4. Find the common logarithms of 10, 1000, 0.1, and 0.001.

Since $10 = 10^1$,	$\log 10 = 1$
Since $1000 = 10^3$,	$\log 1000 = 3$
Since $0.1 = 10^{-1}$,	$\log 0.1 = -1$
Since $0.001 = 10^{-3}$,	$\log 0.001 = -3$

171

To find the logarithm of 1000 with a calculator you may proceed as follows

HP-38G   *TI-82*

{SHIFT}{LOG}1000{)}{ENTER}   {LOG}1000{ENTER}

9.4 NEGATIVE LOGARITHMS

A logarithm will have a negative value if the number is smaller than 1 and the base is greater than 1, as illustrated in Example 9.4.

9.5 PARTS OF A LOGARITHM

To find the common logarithm of a number, express it first in scientific notation, $N = a(10^x)$ with $0 < a < 10$.   For example, the number 7340 is written $(7.34)(10^3)$.   The logarithm of every number has two parts: the *characteristic* and the *mantissa*.   The characteristic is the exponent of 10; the mantissa is the logarithm of a.

Since $\log 1 = 0$ and $\log 10 = 1$, $\log a$ will be a decimal between 0 and 1, which may be obtained from the table of logarithms on page 465.   Note that the decimal points are omitted in the table.

9.6 LOGARITHMIC TABLES

The table of common logarithms on page 465 is a four-place table.   It is precise to four significant figures and is adequate for most elementary applications.   For greater precision, tables with more places are available.   The tables contain positive mantissas only.

The mantissas of three-digit numbers can be read directly from the table in this book. Proportional parts are used with four-digit numbers.

EXAMPLE 9.5.   Determine the logarithm of 981.5

Express the number in scientific notation:

$$981.5 = (9.815)(10^2)$$

The characteristic is the exponent of 10, or 2.   Turn to page 465.   Locate the number 98 in the first column and follow it along the same line to the number 9917 under the column labeled 1.   Continue along the same line to the proportional parts column labeled 5 and find the digit 2.   Adding $9917 + 2 = 9919$.   Then mantissa is 0.9919; the characteristic is 2. $\log 981.5 = 2.9919$.   It may also be written as $0.9919 + 2$.

EXAMPLE 9.6.   Determine the logarithm of 0.0000589.

Express the number in scientific notation:

$$0.0000589 = (5.89)(10^{-5})$$

The characteristic is -5.   From the tables the mantissa is $+0.7701$.   The complete logarithm is written as $\bar{5}.7701$ or $0.7701 - 5$.

Caution: Writing -5.7701 would mistakenly indicate that both parts of the logarithm were negative.

9.7 ANTILOGARITHMS

The number whose logarithm is given is called the *antilogarithm* or *antilog* of the logarithm.

The process of finding the antilogarithm is just the reverse of finding a logarithm. The logarithm is written in the form $\log a + x$. The logarithm table is used to find a and the antilog is of the form $a(10^x)$.

EXAMPLE 9.7. Determine the antilogarithm of 2.4857.

Write $2.4857 = 0.4857 + 2$. From the table on page 465, $\log 3.06 = 0.4857$. Therefore, antilog $2.4857 = 3.06(10^2)$ or 306.

The antilogarithm of 2.4857 can be calculated with a calculator as follows:

HP-38G	TI-82
{SHIFT}{10^x}2.4857{ENTER}	{2^{nd}}{10^x}2.4857{ENTER}
The display shows 305.984903466.	The result is given as 305.9849035.

Both results can be rounded to 306.

EXAMPLE 9.8. Use a table of logarithms to determine the antilogarithm of $0.8537 - 3$.

From the table on page 465, $\log 7.14 = 0.8537$. Therefore, antilog $(0.8537 - 3) = 7.14(10^{-3}) = 0.00714$.

Laws of Logarithms

9.8 THE FIRST LAW OF LOGARITHMS

The logarithm of a product of two numbers A and B is equal to the sum of their logarithms: $\log(AB) = \log A + \log B$. The computation of AB is simplified because multiplication is replaced by addition.

EXAMPLE 9.9.
$$\log(25)(0.137) = \log 25 + \log 0.137$$
$$= 1.3979 + 0.1367 - 1 = 0.5346$$

To find this logarithm with a calculator you may proceed as follows:

HP-38G	TI-82
{SHIFT}{LOG}25{*}0.137{)}{ENTER}	{LOG}{(}25{*}0.137{)}{ENTER}
	Expression needs to be enclosed in parentheses.

If you are not sure of the order in which the calculator carries out its operations, use extra parentheses. Notice that, depending on the model of the calculator, operations like {LOG}25{*}0.137 may produce different results. Some calculators may interpret it as log(25)*0.137; others may interpret it as log(25*0.137).

9.9 THE SECOND LAW OF LOGARITHMS

The logarithm of a quotient of two numbers A and B is equal to the difference of their logarithms:

$$\log\frac{A}{B} = \log A - \log B.$$

EXAMPLE 9.10.
$$\log\frac{452.9}{0.00668} = \log 452.9 - \log 0.00668$$
$$= 2.6560 - (0.8248 - 3) = 4.8312$$

With a calculator, the logarithm of this quotient can be calculated as indicated below. Always use extra parentheses when calculating this type of operation to ensure that the calculations are carried out correctly.

HP-38G	TI-82
{SHIFT}{LOG}452.9{/}0.00668{)}{ENTER}	{LOG}{(}452.9{÷}0.00668{)}{ENTER}
The display shows 4.83122585821	The display shows 4.831225858

Both results can be rounded to 4.8312.

9.10 THE THIRD LAW OF LOGARITHMS

The logarithm of a number N raised to a power p is equal to the product of the power and the logarithm of the number: $\log N^p = p \log N$.

EXAMPLE 9.11. $\log(37.5)^2 = 2 \log 37.5 = 2(1.5740) = 3.1480$

With a calculator this logarithm can be calculated as follows:

HP-38G	TI-82
{SHIFT}{LOG}37.5{x^y}2{)}{ENTER}	{LOG}{(}37.5{^}2{)}{ENTER}
The display shows 3.14806253546	The display shows 3.148062535

Both results can be rounded to 3.1480.

EXAMPLE 9.12. $\log \sqrt[3]{163.2} = \log(163.2)^{1/3} = \dfrac{1}{3} \log 163.2$

$$= \dfrac{1}{3}(2.2127) \approx 0.7376$$

With a calculator this logarithm can be calculated as follows:

HP-38G

{SHIFT}{LOG}3{SHIFT}{$\sqrt[x]{x}$}163.2{)}{ENTER}

or

TI-82

{LOG}{(}163.2{^}{(}1{÷}3{)}{ENTER}

{SHIFT}{LOG}163.2{x^y}{(}1{/}3{)}{)}{ENTER}

The display shows 0.7375733848

The display shows 0.737573384806

Both results can be rounded to 0.7376.

Natural Logarithms

The laws of logarithms are applicable to any base. Many theoretical derivations in science are best accomplished when numbers are expressed to a base whose value is $e = 2.71828\ldots$. Thus, any positive real number N may be written as $N = e^x$. From the definition of a logarithm, $\log_e N = x$. The logarithm of a number N to base e is called a *natural* or *Naperian* logarithm and is written $\ln N$.

9.11 CONVERSION TO A DIFFERENT BASE

Although there are special tables of natural logarithms, it is frequently more practical to use tables of common logarithms. The following rule is used for changing the logarithm of a number to a different base:

$$\log_a N = (\log_b N)(\log_a b)$$

In changing from base 10 to e and vice versa it is necessary to use the following values:

$$\log e \approx 0.4343$$

$$\ln 10 \approx 2.303$$

EXAMPLE 9.13.   Find the value of $\ln 12$.

Applying the above rule,

$$\log_{10} 12 = \log_e 12 \cdot \log_{10} e$$

From tables of common logs, $\log_{10} 12 = 1.0792$.   Therefore,

$$1.0792 = (\ln 12)(0.4343) \qquad \text{or} \qquad \ln 12 = \frac{1.0792}{0.4343} \approx 2.485$$

Generally, natural logarithms can be calculated with a calculator using the **Ln** key.   The laws of logarithms, explained in sections 9.8, 9.9 and 9.10, are also valid for the natural logarithms.

With a calculator $\ln 12$ can be calculated as follows:

HP-38G	*TI-82*
{SHIFT}{Ln}12{()}{ENTER}	{Ln}12{ENTER}
The display shows 2.48490664979	The display shows 2.48490665

Both results can be rounded to 2.485.

Example 9.13 illustrates the fact that the logarithm to base e of a number exceeds in absolute value its logarithm to base 10.   In particular, $\log N = 0.4343 \ln N$; or conversely, $\ln N = 2.303 \log N$.

Exponential Functions

9.12   EXPONENTIAL FUNCTIONS

The function $f(x) = a^x$ is defined as an exponential function if a is a positive real number.

If $a = e$ (the base for natural logarithms), e^x is often written as $\exp x$.

In Fig. 9-1, we have shown the graph of an exponential function where $a > 1$ and also one where $0 < a < 1$.   The graphs of the functions defined by $y = 10^x$, $y = e^x$, and $y = 2^x$ are shown in Fig. 9-2.

There are many exponential functions in science which are defined by equations of the form $y = ce^{kx}$, where c and k are constants.

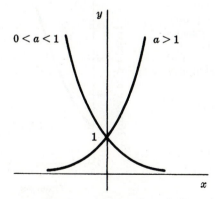

Fig. 9-1.   Graphs of $y = a^x$

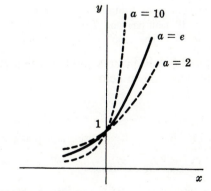

Fig. 9-2.   Graphs of $y = 2^x$, $y = e^x$, and $y = 10^x$

When the value of the exponent is positive, the function increases at a rate proportional to itself. The compound interest law is derived from a function of this kind.

To plot the graph of Fig. 9-1 with an HP-38G graphing calculator follow the steps indicated below. As a preparatory step clear any function previously defined. Do this by pressing {LIB}<u>FUNCTION</u>{ENTER} followed by {SHIFT}{CLEAR}. Answer "Yes" to the question "Clear all expressions." Reset also the viewing window settings by pressing {SHIFT} {SETUP PLOT} followed by {SHIFT}{CLEAR}.

(1) Define the functions.

   {LIB}<u>FUNCTION</u>{ENTER}EDIT   "$y = a^x$ for $0 < a < 1$"

   0.5{x^y}{X,T,θ}{ENTER}   "value of a set to 0.5"

                                                 "$y = a^x$ for $1 < a$"

   1.5{x^y}{X,T,θ}{ENTER}   "value of a set to 1.5"

(2) Plot the graphs.

   Press {PLOT}   "Using the standard viewing window settings"

With a TI-82 graphing calculator the graph of Fig. 9-2 can be plotted as follows:

(1) Define the functions.

   {Y=}

   2{^}{X,T,θ}{ENTER} $y = 2^x$

   {2^{nd}}{e^x}{X,T,θ}{ENTER} $y = e^x$

   {2^{nd}}{10^x}{X,T,θ}{{ENTER} $y = 10^x$

(2) Plot the graphs   "Using the standard viewing window settings"

   {GRAPH}

EXAMPLE 9.14. The number of neutrons in a nuclear reactor can be predicted from the equation $n = n_0 e^{t/T}$, where n = the number of neutrons at time t, n_0 = the number of neutrons at zero time, and T = reactor period. Thus, when $n_0 = 10$ neutrons and $T = 20$ seconds, the equation becomes $n = 10e^{t/20} = 10e^{0.05t}$. The graph of this equation is shown in Fig. 9-3.

The graph of Fig. 9-3 can be plotted with a graphing calculator following the steps shown below.

EXAMPLE 9.15. Plot $y = 10e^{0.05x}$ using a graphing calculator.

Using an HP-38G graphing calculator

(1) Define the function ($y = 10e^{0.05x}$) ← Notice that the function needs to be defined in terms of x.

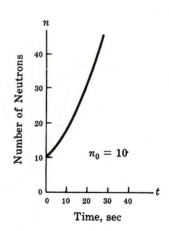

Fig. 9-3. Graph of $n = 10e^{0.05t}$

{LIB}<u>FUNCTION</u>{ENTER}<u>EDIT</u>

10{*}{SHIFT}{e^x}{()0.05{*}{X,T,θ}()}{ENTER}

(2)   Set the range.

{SHIFT}{SETUP-PLOT}

XRNG: 0 {ENTER}   50 {ENTER}

YRNG: 0 {ENTER}   50 {ENTER}

XTICK: 10 {ENTER}   YTICK: 10 {ENTER}

(3)   Plot the graph.

{PLOT}

Using a TI-82

(1)   Define the function ($y = 10e^{0.05x}$) ← Notice that the function needs to be defined in terms of *x*.

{Y =}10{x}{2nd}{e^x}{()0.05{x}{X,T,θ}()}ENTER}

(2) Set the range.

{WINDOW}{ENTER}

Xmin = 0 {ENTER}

Xmax = 50 {ENTER}

Xscl = 10 {ENTER}

Ymin = 0 {ENTER}

Ymax = 50 {ENTER}

Yscl = 10 {ENTER}

(3) Plot the graph using the standard viewing window settings.

{GRAPH}

9.13 EXPONENTIAL EQUATIONS

In an exponential equation the unknown is in the exponent. Simple exponential equations are solved by applying the definitions and laws of exponents and logarithms. However, there is one rule of quite general applicability.

Rule.   If only products are involved in the equation, and the unknown appears as an exponent, take the logarithm of both members of the equation and solve the resulting equation.

EXAMPLE 9.16.   Solve for *x*: $2^{x-1} = 16$.

Since $2^4 = 16$, it is clear that $x - 1 = 4$, so that $x = 5$.

EXAMPLE 9.17.   Solve for *n*: $3(2^n) = 12$.

This problem fits into the category of problems to which the above rule is applicable, so that $\log 3 + n \log 2 = \log 12$. A reference to the log table gives $\log 3 = 0.4771$, $\log 2 = 0.3010$, $\log 12 = 1.0792$. Hence, the equation becomes

$$0.4771 + (0.3010)n = 1.0792 \qquad \text{or} \qquad 0.3010n = 0.6021$$

Thus, $n = 0.6021/0.3010 \approx 2$.

This particular problem could also be solved more easily by noting that $2^x = 4$, where $x = 2$.

Graphing calculators can also be used to solve exponential equations. To solve equations of this type transform them such that one of their sides is zero. Consider the nonzero side to be a function on the unknown variable. Plot the graph and find its intercepts. The intercepts represent the solution of the exponential equation.

EXAMPLE 9.18. Solve $1.5^x = 2^{x+3}$.

Rewrite the equation such that one of its sides is zero.

$$1.5^x - 2^{x+3} = 0$$

Consider the function $y = 1.5^x - 2^{x+3}$. Plot this function and find its intercepts.

Using an HP-38G

(1) Define the function.

{LIB}<u>FUNCTION</u>{ENTER}

<u>EDIT</u>

$1.5\{x^y\}\{X,T,\theta\}\{-\}2\{x^y\}\{(\}\{X,T,\theta\}\{+\}3\{)\}\{ENTER\}$

(2) Plot the graph.

{PLOT} ← "Using the standard window settings".

(3) Find the x-intercept.

<u>MENU</u> <u>FNC</u> <u>Root</u>{ENTER}

The root value is -7.228.

Another method to solve exponential equations of this type requires that you consider each side of the equation to define a function of the unknown variable. That is, consider the two new functions $y = 1.5^x$ and $y = 2^{x+3}$. Plot both functions and find their intercepts. These intercepts represent the solution of the exponential equation. This process is illustrated below using a TI-82 graphing calculator.

(1) Define the functions.

{Y = }

$1.5\{^\wedge\}\{X,T,\theta\}\{ENTER\}$ ← First function $y = 1.5^x$

$2\{^\wedge\}\{(\}\{X,T,\theta\}\{+\}3\{)\}\{ENTER\}$ ← Second function $y = 2^{(x+3)}$

(2) Plot the graphs.

{GRAPH}

(3)   Find the intercepts.

$\{2^{nd}\}\{CALC\}\underline{intersect}\{ENTER\}$ ← "Use the directional keys to highlight the intersect option and then press ENTER"

First curve? {ENTER}

Second curve? {ENTER}

Guess? $-5.74\{ENTER\}$ ← This is an "educated" guess to help the calculator. Use the directional keys to decrease or increase the value of the independent variable x

The result is $x = -7.228$.

Due to the approximation of the intersect valve and accuracy of the calculator the value of the function $y = 1.5^x - 2^{x+3}$ for $x = -7.228$ may not be exactly zero.

9.14   LOGARITHMIC FUNCTIONS

The inverse of an exponential function is called a *logarithmic function*, according to the following definition.   The function $f(x) = \log_a x$ is called logarithmic if a is a positive number not equal to one.

The graphs of the functions defined by $y = \log_2 x$ and $y = \log_{10} x$ are shown in Fig. 9.4; the graphs for $y = \log_{1/2} x$ and $\log_{1/10} x$ are reproduced in Fig. 9.5.

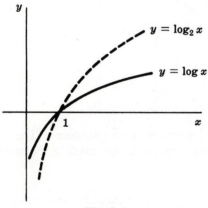

Fig. 9-4

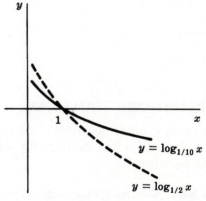

Fig. 9-5

To plot the function $y = \log_2 x$ with a graphing calculator express $\log_2 x$ in terms of $\log x$ using the rules of section 9.11.   The steps shown below illustrate this process.

Writing $\log_2 x$ in terms of $\log x$ results in $\log_2 x = \log x / \log 2$.

Using an HP-38G

(1)   Define the function.

$\{LIB\}\underline{FUNCTION}\{ENTER\}\underline{EDIT}$

$\{SHIFT\}\{LOG\}\{X,T,\theta\}\{/\}\{SHIFT\}\{LOG\}2\{)\}\{)\}\{ENTER\}$

(2)   Plot the graph.

   {PLOT} ← "Using the standard window settings".

Using a TI-82

(1)   Define the function.

   {Y=}

   {LOG}{(}{X,T,θ}{)}{÷}{LOG}{(}2{)}{ENTER}

(2)   Set the range

   {WINDOW}{ENTER}

   Xmin = −10 {ENTER}

   Xmax = 10 {ENTER}

   Xscl = 1 {ENTER}

   Ymin = −10 {ENTER}

   Ymax = 10 {ENTER}

   Yscl = 1 {ENTER}

(3)   Plot the graph.

   {GRAPH}

9.15   LOGARITHMIC EQUATIONS

An equation is called *logarithmic* if it contains the logarithm of an unknown. The definition and laws of logarithms and the rules for solving ordinary equations are used in the solutions of logarithmic equations.

EXAMPLE 9.19.   Solve for x: $2\log(x - 2) = 6.6064$.

Dividing both sides of the equation by 2, first obtain $\log(x - 2) = 3.3032$, and a glance at the log table shows that $\log 2010 = 3.3032$. Hence, $x - 2 = 2010$, so that $x = 2012$.

EXAMPLE 9.20.   Solve for t: $2\log t + \log t = 6$.

Consider $\log t$ as the unknown of the equation. Then adding the two terms on the left side, we have

$$3\log t = 6$$

Dividing both sides by 3, $\log t = 2$ and

$$t = \text{antilog } 2 = 100$$

To solve logarithmic equations with a graphing calculator follow a similar process to the one used to solve exponential equations. You can rewrite the equation so that one of its sides is zero and consider the nonzero side to define a function of the unknown variable. Plot the graph and find its intercepts. Alternatively, you can consider each side of the equation to define a function of the unknown variable. Plot their graphs and find their intercepts. These intercepts represent the solution of the original equation. Both methods are illustrated below.

EXAMPLE 9.21. Solve $\log(x + 3) = \log(7 - 3x)$ using a graphing calculator.

Using an HP-38G (method 1)

(1) Define the function.

{LIB}<u>FUNCTION</u>{ENTER}<u>EDIT</u>

{SHIFT}{LOG}{X,T,θ}{+}3{)}{−}{SHIFT}{LOG}7{−}3{*}{X,T,θ}{)}{ENTER}

(2) Plot the graph.

{PLOT}

(3) Find the *x*-intercept.

<u>MENU</u> <u>FCN</u> <u>Root</u> {ENTER}

The root value is 1.

Using a TI-82

(1) Define the functions.

{Y =}

{LOG}{(}{X,T,θ}{+}3{)}{ENTER}   "First function. $y = \log(x + 3)$"

{LOG}{(}7{−}3{×}{X,T,θ}{)}{ENTER}   "Second function. $y = \log(7 - 3x)$"

(2) Plot the graph.

{GRAPH}   "Set the range for the *x* and *y* values between −5 and 5 with scale of 1 for both axis"

(3) Find the intercept.

{2nd}{CALC}<u>intersect</u> {ENTER}

First curve? {ENTER}

Second curve? {ENTER}

Guess? {ENTER}

The intersection occurs at $x = 1$.

9.16 GROWTH AND DECAY

Exponential functions can be used to describe exponential growth and decay. Growth allows us to calculate how an amount will increase as a function of time given an initial amount or population. Decay allows us to calculate how an amount will decrease as a function of time from a given initial amount or population. Both, growth and decay, can be modeled by a function of the form $f(x) = ab^t$, however, they are generally written in some other form such as those listed below.

1. $A(t) = A_0(1 + r)^t$ where A_0 is the existing amount at time $t = 0$ and r is growth rate. If $r > 0$, then the initial amount grows exponentially. If $-1 < r < 0$, then the initial amount decays exponentially. The quantity $1 + r$ is called the growth factor.

2. $A(t) = A_0 b^{(t/k)}$ where k is the amount of time needed to multiply A_0 by b.

EXAMPLE 9.22.   In 1996, the population of the town of New Virginia was 35,000.   It has been determined that due to the new industries coming into the area, the population is expected to grow at the rate of 4%.   (*a*) What will be the population of the town 5 years from now?   (*b*) Draw the corresponding graph and determine when the population will reach 50,000?

Call $A(5)$ the population in 5 years.   Then, $A(5) = 35,000*(1 + 0.04)^5 \approx 42,583$ individuals.   Figure 9-6 shows the graph of this function.

From the graph we can observe that the population will reach the 50,000 mark in approximately 9 years.   To graph this function using a graphing calculator follow the steps indicated below.

Using an HP-38G

(1)   Define the function.

      {LIB}<u>FUNCTION</u>{ENTER}<u>EDIT</u>

      35000{*}{({)}1{+}0.04{)}}{x^y}{X,T,θ}{ENTER}

(2)   Set the range.

      {SHIFT}{SETUP-PLOT}

      XRNG: 0 {ENTER}   14 {ENTER}

      YRNG: 30000 {ENTER}   50000 {ENTER}

      XTICK: 2 {ENTER}   YTICK: 5000 {ENTER}

(3)   Plot the graph.

      {PLOT}

Using a TI-82

(1)   Define the function.

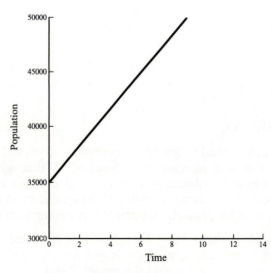

Fig. 9-6

{Y =}

$35000\{x\}\{(\}1\{+\}0.04\{)\}\{^\}\{X,T,\theta\}$

(2) Set the range.

{WINDOW}{ENTER}

Xmin = 0 {ENTER}

Xmax = 14 {ENTER}

Xscl = 2 {ENTER}

Ymin = 30000 {ENTER}

Ymax = 50000 {ENTER}

Yscl = 5000 {ENTER}

(3) Plot the graph.

{GRAPH}

EXAMPLE 9.23.   The atoms of a radioactive element are said to be unstable because they go spontaneously through a process called radioactive decay.   In this process each atom emits some of its mass as radiation, the remainder of the atom reforms to make an atom of some new element.   For example, radioactive carbon-14 decays into nitrogen; radium decays into lead.   One method of expressing the stability of a radioactive element is known as the half-life of the element.   That is, the time required for half of a given sample of a radioactive element to change to another element.   Assume that a radioactive isotope has a half-life of 5 days, At what rate does the substance decay each day?

We will use the formula $A(t) = A_0 b^{(t/k)}$ where k is the amount of time needed to multiply A_0 by b.   Since the half-life of this radioactive isotope is 5 days, this implies that after 5 days only half of the original sample remains.   In consequence, $b = 1/2$ and $t = 5$.   Replacing these values in the given formula, we have

$$A(5) = A_0(1/2)^{t/5} = A_0(0.5)^{t/5}$$

To find the decay rate, we need to rewrite the formula given above so that it looks like the formula $A(t) = A_0(1 + r)^t$.

Notice that we can rewrite $A(5) = A_0(0.5^{1/5})^t = A_0(0.87)^t = A_0(1 - 0.13)^t$.   Therefore, the decay rate is approximately 13%.

EXAMPLE 9.24.   The half-life of carbon-14 is 5750 years.   Samples taken at an excavation site show that 1 pound of a charcoal sample has 65% of carbon-14 compared to a standard sample.   What is the age of the sample?

$A_0 = 1$, $k = 5750$ and $b = 1/2$.   Therefore,

$$A(t) = A_0 b^{t/k}$$
$$0.65 = A_0(1/2)^{t/5750}$$

To solve for t proceed as follows:

$$\log(0.65) = \log[(0.5)^{t/5750}]$$
$$\log(0.65) = (t/5750)\log(0.5)$$
$$t = \log(0.65)*5750/\log(0.5)$$
$$t \approx 3574 \text{ years.}$$

Logarithmic and Semilogarithmic Graph Paper

9.17 LOGARITHMIC GRAPH PAPER

Logarithmic or log–log paper is a graph paper on which the rulings are not spaced uniformly. The lines are laid off at distances which are proportional to the logarithms of numbers from 1 to 10. Logarithmic paper can have one or more logarithmic cycles along each axis; if there is just one it is called single logarithmic paper. Thus, logarithmic paper makes it possible to plot logarithms of numbers without the use of logarithmic tables.

Logarithmic paper is used:

(1)   to graph a power function from a known equation.

(2)   to determine the equation of a power function which satisfies empirical data.

Taking the logarithm of the power function $y = Cx^m$, we have

$$\log y = \log C + m \log x$$

Now, let $\log y = Y$; $\log C = b$; $\log x = X$. Then, by substitution $Y = b + mX$, which is an equation of a straight line, with slope m and Y-intercept b. Thus, if one plots the logarithm of one variable against the logarithm of the second variable and the graph is a straight line, then the equation connecting the two variables is of the form $y = Cx^m$. The geometrical slope m is measured as on a uniform scale; the coefficient C is read as the intercept on the vertical axis through $x = 1$.

It is important to note that since the logarithm of zero is not defined, there is no zero on either of the two scales of logarithmic paper.

EXAMPLE 9.25.   Plot a graph of $T = 2\sqrt{m}$ on logarithmic paper.

Assume m is the independent variable. Assign a few values to m and calculate the corresponding values of T as shown in the table below.

m	1	4	9
T	2	4	6

Plot m along the horizontal scale of single logarithmic paper and T along the vertical scale. The graph is shown in Fig. 9-7.

The graph is a straight line. The measured geometrical slope is $m = \dfrac{h}{a} = 0.5$; the T-intercept is 2. These values agree with the constants of the equation:

$$T = 2\sqrt{m} = 2m^{1/2} = 2m^{0.5}$$

EXAMPLE 9.26.   The breaking strength S of three-strand manila rope is a function of its diameter D. Determine the empirical equation from the data in the following table.

D, in.	1.0	1.25	1.50	1.75	2.0	2.25
S, lb-f	1700	2700	3700	5300	6200	8200

Use single logarithmic paper. Plot diameter as the independent variable along the horizontal axis. The vertical scale will have a range of 1000 to 10,000 lb. The graph of the data is shown in Fig. 9-8. Since the graph is a straight line, the relationship between the two variables can be expressed as a power function of the form $y = Cx^m$. The geometrical slope is 1.9; the S-intercept is 1700. Therefore, the equation is $S = 1700D^{1.9}$.

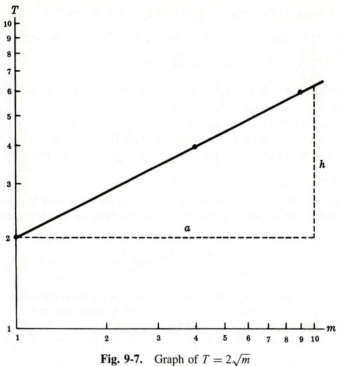

Fig. 9-7.　Graph of $T = 2\sqrt{m}$

9.18　SEMILOGARITHMIC PAPER

Semilogarithmic paper is a graph paper on which one of the axes has uniform subdivisions, while those on the second axis are proportional to the logarithms of numbers.

Semilogarithmic paper is used:

(1)　to graph an exponential function,

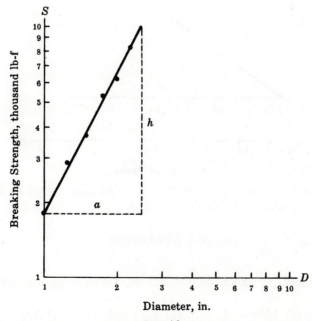

Fig. 9-8.　The power function $S \approx 1700D^{1.9}$ (Slope $= h/a \approx 1.9$; S-intercept ≈ 1700)

(2)   to determine the equation of an exponential function which satisfies empirical data.

Taking the logarithm of both terms of the equation $Y = C(10)^{mx}$, one gets

$$\log Y = \log C + mx$$

Now, let $\log Y = y$ and $\log C = b$.   Then, by substitution $y = b + mc$, which is an equation of a straight line, with slope m and y-intercept b.   Thus, if one plots the logarithm of one variable versus the values of the second variable and the graph is a straight line, then the equation is of the form $Y = C(10)^{mx}$.   The geometric slope m is measured; the coefficient C is read as the intercept on the Y-axis.   Semilogarithmic paper enables one to plot the logarithms of numbers without having to look them up in tables.

EXAMPLE 9.27.   If a rope is wound around a wooden pole, then the frictional force between the pole and the rope depends on the number of turns.   From the data in the table below determine the equation relating the force of friction F to the number of turns N.

N	0.25	0.5	0.75	1.0
F, lb-f	21	33	47	70

Use one-cycle semilogarithmic paper.   Plot the number of turns N along the horizontal axis and the force F along the vertical axis as shown in Fig. 9-9.   Since the graph is a straight line, its equation is of the form $F = C \cdot 10^{mN}$.   The slope of the line is 0.70; the y-intercept is 14.   Therefore, the equation is $F = 14 \cdot 10^{0.70N}$.   The geometric slope is measured on uniform scales by means of a rule.   Do *not* use the scales on the graph to determine geometric slope.

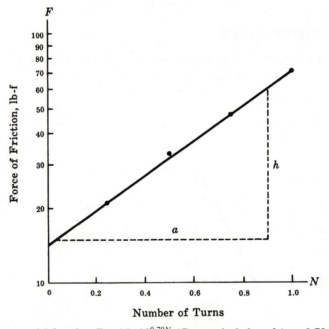

Fig. 9-9.   The exponential function $F \approx 14 \cdot 10^{0.70N}$ (Geometrical slope $h/a \approx 0.70$; F-intercept ≈ 14)

Solved Problems

9.1.   Write each of the following exponential equations in equivalent logarithmic form:

(a) $5^3 = 125$ (b) $16^{1/2} = 4$ (c) $\left(\dfrac{1}{3}\right)^r = k$ (d) $10^{-4} = 0.0001$

A direct application of the definition leads us to the following results:

(a) $\log_5 125 = 3$ (b) $\log_{16} 4 = 1/2$ (c) $\log_{1/3} k = r$ (d) $\log_{10} 0.0001 = -4$

9.2. Write each of the following logarithmic equations in equivalent exponential form:

(a) $\log_{10} 1000 = 3$ (b) $\log_4 2 = 1/2$ (c) $\log_b 1 = 0$ (d) $-2 = \log_3 1/9$

Again the definition leads us to the solutions:

(a) $10^3 = 1000$ (b) $4^{1/2} = 2$ (c) $b^0 = 1$ (d) $1/9 = 3^{-2}$

9.3. Determine the common logarithm of

(a) 10,000   (b) 100,000,000   (c) 0.001   (d) 0.0000001

Express the numbers in scientific notation. All of them will be in the form 10^p. The exponent p is the common logarithm.

(a) $\log 10,000 = \log 10^4 = 4$ (c) $\log 0.001 = \log 10^{-3} = -3$

(b) $\log 100,000,000 = \log 10^8 = 8$ (d) $\log 0.0000001 = \log 10^{-7} = -7$

With a calculator these logarithms can be calculated as follows.

Using an HP-38G:

(a) {SHIFT}{LOG}10000{ENTER}   (c) {SHIFT}{LOG}0.001{ENTER}

Using a TI-82:

(b) {LOG}100000000{ENTER}   (d) {LOG}0.0000001{ENTER}

9.4. Determine N if we know that $\log N$ is: (a) 5,   (b) -2,   (c) 1/3,   (d) 1.5.

The common logarithm is the exponent to base 10. Therefore, the values of N are:

(a) $N = 10^5 = 100,000$ (c) $N = 10^{1/3} = \sqrt[3]{10} \approx 2.154$

(b) $N = 10^{-2} = 0.01$ (d) $N = 10^{1.5} = 10^{3/2} = (10^3)^{1/2} = \sqrt{1000} = 10\sqrt{10} \approx 31.62$

The values of N can be calculated with the help of a calculator as follows:

Using an HP-38G:

(a) {SHIFT}{10^x}5{ENTER}   (c) {SHIFT}{10^x}{(}1/3{)}{ENTER}

Using a TI-82:

(b) {2^{nd}}{10^x}{(−)}2{ENTER}   (d) {2^{nd}}{10^x}1.5{ENTER}

9.5. Determine the logarithm of: (a) 463.8,   (b) 0.00396,   (c) $(5.893)(10^{-5})$.

(a) We first express 463.8 in scientific notation: $463.8 = (4.638)(10^2)$. The exponent of $10 =$ characteristic $= 2$. To find the mantissa, we turn to the table on page 465. Locate the number 46 in the first column and the number 6656 along the same line under the column labeled 3. Continue along the same line to proportional parts column labeled 8 and find the digit 7. Adding, $6656 + 7 = 6663$. Thus, the mantissa of $\log 4.638 = 0.6663$. Therefore, $\log 463.8 = 0.6663 + 2$.

Another way of writing this logarithm is $\log 463.8 = 2.6663$.

(b) Similarly, $0.003967 = 3.967(10^{-3})$. The characteristic of the logarithm of this number is -3 and the mantissa is 0.5985. Therefore, $\log 0.003967 = 0.5985 - 3$.

There are three additional ways in which this negative logarithm can be written: $\bar{3}.5985$, $7.5985 - 10$, and -2.4015.

(c) Since the number is already in scientific notation, the characteristic $= -5$. The mantissa from the table is 0.7703. Therefore, $\log(5.893)(10^{-5}) = 0.7703 - 5$. The three other ways of writing this logarithm are $\bar{5}.7703$, $5.7703 - 10$, and -4.2297.

9.6. Determine the antilogarithm of:

(a) 4.5378   (b) 0.6484 − 3   (c) −4.3842   (d) 8.0170 − 10

(a) We write $4.5378 = 0.5378 + 4$ and note in the table that $\log 3.45 = 0.5378$. Hence the required antilogarithm is $3.45(10^4)$ or 34,500.

(b) Similarly, $0.6484 - 3$ is seen to be the logarithm of $4.45(10^{-3})$ and so the desired antilogarithm is 0.00445.

(c) We emphasize again that the mantissa of a logarithm is always positive, and so a logarithm must be expressed with a positive mantissa if its antilogarithm is to be found. Adding and subtracting 5 to -4.3842 will give a positive mantissa. Thus, $5 + (-4.3842) - 5 = 0.6158 - 5$. A glance at the table reveals that $\log 4.129 = 0.6158$, and so the desired antilogarithm is $4.129(10^{-5})$ or 0.00004129.

(d) From the table, $\log 1.04 = 0.0170$. The characteristic is $8 - 10 = -2$. The desired antilogarithm is $1.04(10^{-2}) = 0.0104$.

The antilogarithms of the numbers 4.5378 and -4.3842 can be found with a calculator as indicated below.

Using an HP-38G:

(a) {SHIFT}{10ˣ}4.5378{ENTER}   (b) {SHIFT}{10ˣ}{−x}4.3842{ENTER}

Due to internal approximations the results obtained with the calculator may differ slightly of those obtained by table. In this case for (a) the calculator shows 34498.4831363. For (b) the result is $4.1257330378\mathrm{E} - 5$. In calculators the letter E is used to represent powers of 10. This expression is equivalent to $4.1257330378 \times 10^{-5}$.

Using a TI-82:

(a) {2ⁿᵈ}{10ˣ}4.5378{ENTER}   (b) {2ⁿᵈ}{10ˣ}{(−)}4.3842{ENTER}

In this case, internal errors and approximations also affect the results. These results are for (a) 34498.48314; for (b) $4.128573304\mathrm{E} - 5$. This latter value is equivalent to $4.128573304 \times 10^{-5}$.

9.7. Use logarithms to compute 243.2×12.54.

$$\log 243.2 = \log 2.432(10^2) = 0.3860 + 2$$
$$\log 12.54 = \log 1.254(10^1) = \underline{0.0983 + 1}$$
$$\text{sum of logarithms} = 0.4843 + 3 = \text{logarithm of product}$$

The desired product $=$ antilogarithm of $0.4843 + 3 = 3.050(10^3) = 3050$.

To calculate the logarithm of this product using a calculator follow the steps indicated below. Remember that due to internal errors and approximations the results may differ from those obtained when using a table of logarithms.

Using an HP-38G:

{SHIFT}{LOG}243.2{*}12.54{)}{ENTER}

The result is 3.4842611071.

STO▶{A . . . Z}{A}{ENTER}   ← This sequence stores the result into the variable A.

{SHIFT}{10^x}{A . . . Z}{A}{ENTER} ← This sequence calculates the antilogarithm.

The result is 3049.72800003.   This result can be rounded to 3050.

Using a TI-82:

{LOG}{(}243.2{x}12.54{)}{ENTER}

The result is 3.484261107.

{STO▶}{ALPHA}{A}{ENTER} ← This sequence stores the result into the variable A.

{2nd}{10^x}{2nd}{RCL}{ALPHA}{A}{ENTER}{ENTER} ← This sequence calculates the antilogarithm.

The result is 3049.727999.   Again, this result can be rounded to 3050.

9.8. Use logarithms to compute (73.29)(0.000428).

$$\log 73.29 = \log 7.329(10^1) = 0.8650 + 1$$
$$\log 0.000428 = \log 4.28(10^{-4}) = \underline{0.6314 - 4}$$
$$\text{sum of logarithms} = 1.4964 - 3 = 0.4964 - 2 = \text{logarithm of product}$$

The desired product = antilogarithm of $0.4964 - 2 = 3.136(10^{-2}) = 0.03136$.

9.9. Use logarithms to compute 754.2/12.5.

$$\log 754.2 = \log 7.542(10^2) = 0.8875 + 2$$
$$\log 12.5 = \log 1.25(10^1) = \underline{0.0969 + 1}$$
$$\text{difference of logarithms} = 0.7806 + 1 = \text{logarithm of quotient}$$

The desired quotient = antilogarithm of $0.7806 + 1 = 6.034(10^1) = 60.34$.

This logarithm can be calculated with a calculator as follows.

Using an HP-38G:

{SHIFT}{LOG}{754.2{/}12.5{)}{ENTER}

The result is 1.78057651506.

STO▶{A . . . Z}{A}{ENTER} ← store result in variable A

{SHIFT}{10^x}{A . . . Z}{A}{ENTER} ← find antilogarithm.

The result is 60.3359999998.   This result can be rounded to 60.34.

9.10. Use logarithms to compute 343.2/0.0429.

$$\log 343.2 = \log 3.432(10^2) = 0.5356 + 2 = 1.5356 + 1$$
$$\log 0.0429 = \log 4.29(10^{-2}) = \underline{0.6325 - 2} = 0.6325 - 2$$
$$\text{difference of logarithms} = 0.9031 + 3 = \text{logarithm of quotient}$$

The desired quotient = antilogarithm of $0.9031 + 3 = 8.000(10^3) = 8000$.

Since the logarithm must be expressed with a positive mantissa before the antilogarithm can be determined, the minuend in the subtraction process must be arranged so the subtraction contains a positive decimal portion.   This was the reason for rearranging the characteristic of $\log 343.2$ as shown.

9.11. Use logarithms to compute $\sqrt[3]{752}$.

$$\sqrt[3]{752} = 752^{1/3} \qquad \frac{1}{3}\log 752 = \frac{1}{3}(2.8762) \approx 0.9587$$

The desired number = antilog $0.9857 \approx 9.092$.

This logarithm can be calculated with a TI-82 graphing calculator as follows:

{LOG}{(}752{^}{(}1{÷}3{)}{)}{ENTER}

The result is 0.9587392802.

{STO▶}{ALPHA}{A}{ENTER} ← "store result in variable A"

{2^{nd}}{10^x}{2^{nd}}{RCL}{ALPHA}{A}{ENTER}{ENTER} ← "find antilogarithm"

The result is 9.093671888.

9.12. Use logarithms to calculate $(0.387)^{3/2}$.

$$\frac{3}{2}\log 0.387 = 1.5\log(0.387) = 1.5(0.5877 - 1) = 0.8816 - 1.5$$

Since the exponent of 10 in scientific notation should be an integer, we add -0.5 to 0.8816 and subtract -0.5 from -1.5. The value of the logarithm is unchanged, but it takes on the form $0.3816 - 1$.

The desired number = antilog $0.3816 - 1 \approx 2.408(10^{-1}) \approx 0.2408$.

9.13. Use logarithms to calculate the volume V of a cylinder from the formula $V = \dfrac{\pi D^2}{4}h$, if $\pi = 3.142$, $D = 1.25$ cm and $h = 23.7$ cm.

Taking logarithms,

$$\log V = \log \pi + 2\log D + \log h - \log 4$$

Before looking up logarithms, arrange the quantities in a column as shown below. This is called *blocking out*.

$$
\begin{aligned}
\log \pi = \log 3.142 &= 0.4972 \\
+2\log D = 2\log 1.25 &= 0.1938 \\
+ \log h = \log 23.7 &= \underline{1.3747} \\
\text{sum} &= 2.0657 \\
- \log 4 &= \underline{0.6021} \\
\log V &= 1.4636
\end{aligned}
\qquad
\begin{aligned}
&\textit{Work} \\
\log 1.25 &= 0.0969 \\
2\log 1.25 &= 0.1938
\end{aligned}
$$

$V =$ antilog $1.4636 = 29.8$; the volume of the cylinder is 29.08 cm^3.

9.14. Use logarithms to compute $\dfrac{(43.68)(12.5^4)(0.00127)}{(3.142)(954)}$.

Logarithm of answer = log of numerator − log of denominator.

Numerator

$$\log 43.68 = \log 4.368(10^1) = 0.6403 + 1$$
$$\log 12.5^4 = 4\log 1.25(10^1) = 0.3876 + 4$$
$$\log 0.00127 = \log 1.27(10^{-3}) = \underline{0.1038 - 3}$$
$$1.1317 + 2$$
$$0.1317 + 3$$

Work

$$\log 1.25 = 0.0969$$
$$4\log(1.25)(10^1) = 4(0.0969 + 1) = 0.3876 + 4$$

Denominator

$$\log 3.142 = \log 3.142(10^0) = 0.4972 + 0$$
$$\log 954 = \log 9.54(10^2) = \underline{0.9795 + 2}$$
$$\text{logarithm of denominator} = 1.4767 + 2$$
$$= 0.4677 + 3$$

Quotient

$$0.1317 + 3 = 1.1317 + 2$$
$$\underline{0.4767 + 3}$$
$$\log \text{answer} = 0.6550 - 1$$

The desired answer = antilog $(0.6550 - 1) = 0.4519$.

9.15. Determine the value of $\ln 8.42$.

Use the formula for conversion from one base to another.

$$\log_e N = (\log_{10} N)(\log_e 10)$$
$$(\log 8.42)(\ln 10)$$
$$= (\log 8.42)(2.303)$$
$$= (0.9253)(2.303) \approx 2.131$$

9.16. Find $\log_2 6$.

Use the base-conversion formula.

$$\log_{10} 6 = \log_2 6 \cdot \log_{10} 2$$
$$\log_2 6 = \frac{\log 6}{\log 2} = \frac{0.7782}{0.3010} \approx 2.585$$

9.17. Evaluate $\ln\left(\dfrac{343}{293}\right)$.

$$\ln\left(\frac{343}{293}\right) = \log\left(\frac{343}{293}\right)\ln 10 = (\log 343 - \log 293)(2.303)$$
$$= (2.5353 - 2.4669)(2.303)$$
$$= (0.0684)(2.303) \approx 0.158$$

9.18. Evaluate $5e^{-0.3}$.

Take logarithms to base 10.

$$\log 5e^{-0.3} = \log 5 - 0.3 \log e$$

Using tables of common logarithms,

$$\log 5e^{-0.3} = 0.6990 - 0.3(0.4343) = 0.6990 - 0.1303 = 0.5687$$

Since 0.5687 is the log of 3.704, $5e^{-0.3} = 3.704$.

9.19. Solve for x: $3(2^x) = 15$.

Take logs of both sides and obtain $\log 3 + x \log 2 = \log 15$, which becomes

$$0.4771 + 0.3010x = 1.1761 \qquad \text{or} \qquad 0.3010x = 0.6990$$

Hence, $x = 0.6990/0.3010 \approx 2.32$.

9.20. Solve $e^{-kt} = 10^{-6}$ for k, where $t = 2.5$.

Substituting 2.5 for t and taking logarithms of each side, we obtain $(-2.5k) \log e = -6$. Solving for k,

$$k = \frac{-6}{-2.5 \log e} = \frac{-6}{(-2.5)(0.4343)} \approx 5.53$$

This equation can be solved with an HP-38G graphing calculator as follows:

(1) Define the function.

{LIB}<u>FUNCTION</u>{ENTER}

<u>EDIT</u>

{SHIFT}{e^x}{$-$x}{(}{X,T,θ}{*}2.5{)}{$-$}{SHIFT}{10^x}{$-$x}6{ENTER}

(2) Plot the graph.

{PLOT} $\leftarrow$ "Using the standard window settings."

(3) Find the intercept.

<u>MENU</u> <u>FCN</u> <u>Root</u> {ENTER}

The result is 5.52620422318. This value can be rounded to 5.53.

Notice that regardless of the variable name in the problem the calculator can only use x as a variable name.

9.21. Find the height h in kilometers above sea level at which the barometric pressure p is 200 mm of mercury. The functional equation is $p = 760e^{-0.128h}$.

Substituting 200 for p, $200 = 760e^{-1.28h}$. Taking logarithms of each member, we obtain

$$\log 200 = -0.128h \log e + \log 760$$

Using log tables,

$$2.3010 = -0.128h(0.4343) + 2.8808 \qquad \text{or} \qquad -0.5798 = -0.128h(0.4343)$$

Therefore, $h = \dfrac{0.5798}{(0.128)(0.4343)} \approx 10.4$. The height h is 10.4 kilometers.

9.22. Find I from the equation $50 = 10 \log \left(\dfrac{I}{10^{-12}} \right)$.

Dividing both members by 10, we get $5 = \log\left(\dfrac{I}{10^{-12}}\right)$. But 5 is the common logarithm of 100,000 or 10^5; therefore, $10^5 = \dfrac{I}{10^{-12}}$.

Solving for I, $I = (10^5)(10^{-12}) = 10^{-7}$.

9.23. Given $S = S_0 e^{-kd}$. If $k = 2$, what should be the value of d to make $S = S_0/2$?

Divide both sides of the equation by S_0:

$$S/S_0 = e^{-kd}$$

Substitute $1/2$ for S/S_0 and 2 for k:

$$1/2 = e^{-2d}$$

Take logs of both sides of the equation: $\log 1/2 = -2d \log e$.   But

$$\log 1/2 = \log 1 - \log 2 = 0 - \log 2 = -\log 2$$

Therefore, $-\log 2 = -2d \log e$.   Solving for d,

$$d = \frac{-\log 2}{-2 \log e} = \frac{0.3010}{2(0.4343)} \approx 0.346$$

Note: $\dfrac{\log 2}{2}$ is *not* log 1! This is a common error.

9.24. From the equation $M = m - 5 \log\left(\dfrac{d}{10}\right)$ calculate d if $M = 15$ and $m = 5$.

Substituting the given values in the equation:

$$15 - 5 = 5 \log\left(\frac{d}{10}\right)$$

$$10 = 5 \log\left(\frac{d}{10}\right)$$

Dividing both sides by 5, $2 = \log\left(\dfrac{d}{10}\right)$.

$$\text{antilog}\left[\log\left(\frac{d}{10}\right)\right] = \text{antilog } 2 = 100$$

$$\frac{d}{10} = 100$$

Therefore, $d = (10)(100) = 1000$.

9.25. Solve the equation $\log p + \log 3p = 5$ for p.

By the first law of logarithms:

$$\log 3p = \log 3 + \log p$$

Therefore,

$$\log p + \log 3 + \log p = 5$$

$$2 \log p = 5 - \log 3$$

$$\log p = \frac{5 - \log 3}{2} = \frac{5 - 0.4771}{2} \approx 2.2614$$

$$p = \text{antilog } 2.2614 \approx 182.6$$

This equation can be solved using a graphing calculator as follows:

Using an HP-38G:

(1)   Define the function.

   {LIB}FUNCTION{ENTER}EDIT

   {SHIFT}{LOG}{X,T,θ}{)}{+}{SHIFT}{LOG}3{*}{X,T,θ}{)}{−}5{ENTER}

(2)   Plot the graph.

   {PLOT} ← "Using the standard window settings."

(3)   Find the intercept.

   MENU FCN Root

The result is 182.574185835.   This result can be rounded to 182.6.

9.26. If $\log(3x + 21) - \log(2x + 1) = \log 8$, find x.

   If we combine the terms on the left side of the equation according to the rules of logarithms, we obtain

$$\log \frac{3x + 21}{2x + 1} = \log 8$$

or,

$$\frac{3x + 21}{2x + 1} = 8$$

This equation becomes $3x + 21 = 16x + 8$ or $13x = 13$, and $x = 1$.

A check reveals that $x = 1$ satisfies the original equation.

This equation can be solved with a graphing calculator as follows:

Using an HP-38G:

(1)   Define the function.

   {LIB}FUNCTION{ENTER}

   EDIT

   {SHIFT}{LOG}3{*}{X,T,θ}{+}21{)}{−}{SHIFT}{LOG}2{*}{X,T,θ}{+}1{)}

   {−}{SHIFT}{LOG}8{)}{ENTER}

(2)   Plot the graph.

   {PLOT}

(3)   Calculate the intercept.

   MENU FCN Root {ENTER}

The solution is 0.999999999994.   This result can be rounded to 1.

9.27. Find the slope and the vertical intercept on logarithmic paper of:

   (a) $P = 3.5V^2$ (b) $v = 8\sqrt{h}$ (c) $a = \dfrac{4.56}{D}$

   Assume that the independent variable is on the right side of each equation.   Then,

   (a) slope = 2; intercept = 3.5

(b) slope $= 1/2$, since $\sqrt{h} = h^{1/2}$; intercept $= 8$

(c) slope $= -1$, since $\dfrac{1}{D} = D^{-1}$; intercept $= 4.56$

9.28. A car advertisement gives the time t of reaching certain speeds v with the car starting from rest, as shown in the following table.

t, sec	3.7	5.3	7.2	9.6
v, mph	30	40	50	60

Does the data obey a power function law? If it does, determine the equation satisfied by the data.

Plot the data on logarithmic paper, with t along the horizontal axis as shown in Fig. 9-10. Label the vertical axis with a scale running from 10 to 100.

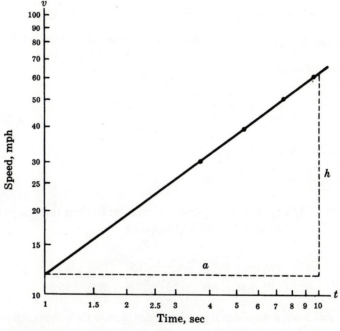

Fig. 9-10. The power function $v = 12t^{0.72}$ (Slope $h/a = 0.72$)

Since the graph is a straight line, the data may be represented by an equation of the form $v = ct^m$, where c is the intercept on the vertical axis at $t = 1$ and m is the geometrical slope. The value of the slope from the measurements of h and a on the graph is $m \approx 0.72$. The intercept has a value of $11.7 \approx 12$. Therefore the equation which fits the data is $v \approx 12t^{0.72}$.

9.29. The length L of the effort arm of a lever is related to the effort E as shown by the data in the following table. Determine the equation connecting the data.

E, g-wt	100	150	200	250	300	350	400
L, cm	74	51	38	30	25	22	19

Single logarithmic paper will contain the given data, as shown in Fig. 9-11. The effort arm is plotted vertically, with the scale range from 10 to 100 cm. The effort is plotted horizontally with the scale range from 100 to 1000. The graph is a straight line. The slope of the line is negative. Its value is $-0.99 \approx -1$. Therefore, the power function has the exponent -1. Since the scale of the independent variable begins with 100, the y-intercept cannot be obtained by reading the graph. Substituting the coordinates of one of the points (250, 30) on the line into the equation,

$$30 = C(250^{-1}) \quad \text{or} \quad 30 = \frac{C}{250}$$

Solving for C:

$$C = (30)(250) = 7500$$

Therefore, the empirical equation which satisfies the data is

$$L \approx 7500E^{-1} \quad \text{or} \quad L \approx \frac{7500}{E} \quad \text{or} \quad LE \approx 7500$$

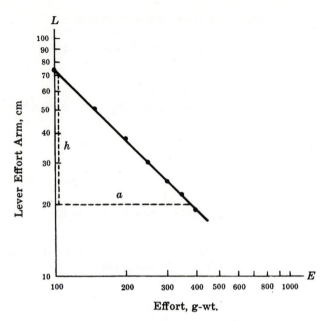

Fig. 9-11. The power function $L \approx 7500E^{-1}$ (Slope $= h/a \approx 1$)

9.30. The weights (W) of 3/4-inch plywood disks in relation to their diameters (D) are given in the following table. Determine the empirical equation.

D, cm	2.5	4.9	7.5	10.0	13.8
W, g-wt	3.0	12.3	28.9	50.0	95.6

To graph the data, *multiple* logarithmic paper is needed because the weight ranges over 2 cycles: 1–10 and 10–100. The graph on 2 × 2-cycle logarithmic paper is shown in Fig. 9-12. Since the graph is a straight line, its equation is of the form $y = Cx^m$. The geometrical slope of the line is 2.0. However, the y-intercept cannot be read from the graph. The y-intercept is the value read on the vertical axis when the value on the horizontal axis is 1. To find C, take any one of the given points, preferably one which is on the line, viz., the point (7.5, 28.9). One can write the equation $28.9 = C(7.5)^2$ or

$$C = \frac{28.9}{(7.5)^2} = \frac{28.9}{56.25} \approx 0.51$$

Therefore, the empirical equation between the weight and diameter is

$$W \approx 0.51D^2$$

It is well to remember that this particular equation will hold only for those disks which have the same thickness and density as the disks for which the data were obtained. Although the weight of disks can be expressed by the formula $W = CD^2$, the value of C must always be obtained by experiment. Thus, the value of C will depend not only on the physical properties of the disk material, but also on the units of measurement.

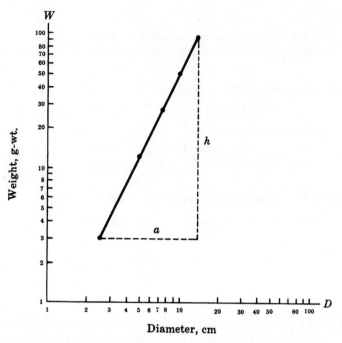

Fig. 9-12. The power function $W \approx 0.51D^2$ (Slope $= h/a = 2.0$)

9.31. The amount A of a bank deposit increases with the number of years n as shown below.

n, years	0	5	10	15	20	25
A, dollars	200	268	358	480	642	858

Can the data be expressed as an exponential function? If so, determine the exact functional equation.

Plot the number of years along the horizontal axis of semilogarithmic paper and the amount along the vertical axis as shown in Fig. 9-13. Label the vertical axis from 100 to 1000.

Since the graph is a straight line, the data can be expressed by an equation of the form $A = C \cdot 10^{mn}$. The value of the vertical axis intercept at $n = 0$ is C. The value of m in the equation is the physical slope of the graph. The geometrical or measured slope of the line is $M = \dfrac{h}{a} \approx 0.63$. The physical slope is obtained by dividing the geometrical slope by 25, the horizontal axis scale factor:

$$m = \frac{M}{25} = \frac{0.63}{25} \approx 0.025$$

The equation which satisfies the data is

$$A \approx 200 \cdot 10^{0.025n}$$

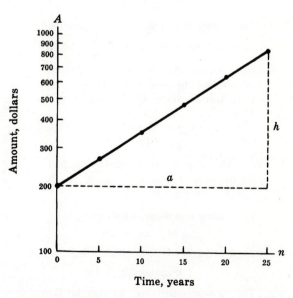

Fig. 9-13. The exponential function $A \approx 200 \cdot 10^{0.025n}$

Alternative Method

The values of C and m may be obtained algebraically. Select two points on the graph, say (10, 358) and (20, 642), and substitute the coordinates of each in the equation of an exponential function,

$$358 = C \cdot 10^{m \cdot 10} \qquad 642 = C \cdot 10^{m \cdot 20}$$

Taking logarithms of each side of the two equations:

$$\log 358 = \log C + 10m \qquad \log 642 = \log C + 20m$$

Subtracting the first equation from the second:

$$\log 642 - \log 358 = 10m$$

$$m = \frac{\log 642 - \log 358}{10} = \frac{2.8075 - 2.5539}{10} = 0.02536 \approx 0.025$$

Substituting $m = 0.02536$ into the logarithmic form of the first equation:

$$\log 358 = \log C + 0.2536$$
$$2.5539 = \log C + 0.2536$$
$$\log C = 2.3003$$
$$C \approx 200$$

9.32. The percentage of light T transmitted by a solution decreases with the concentration C of the solution. The data for a copper solution for 3800 angstroms are given in the following table.

C, mg/100 ml	0.61	1.21	1.82	2.42
T, %	70.0	51.0	33.5	22.8

Does the data satisfy an exponential equation? If so, determine the approximate functional equation.

Plot the concentration C along the horizontal axis of semilogarithmic paper and the transmission T along the vertical axis, as shown in Fig. 9-14. Label the vertical axis from 10 to 100.

Since the graph is a straight line on semilogarithmic graph paper, the data can be represented by an equation of the form:

$$T = K \cdot 10^{mC}$$

The geometrical slope from the measurements on the graph is $M = -\dfrac{h}{a} \approx -0.70$. However, the physical slope m is obtained by dividing the geometrical slope by the scale factor, which in this case is 2.5. Thus:

$$m = \frac{M}{2.5} = \frac{-0.70}{2.5} \cong -0.28$$

The value of C is obtained by substituting the point (1, 56) on the graph into the equation: $56 = K \cdot 10^{(-0.28)(1)}$.

Taking logarithms of both sides:

$$\log 56 = -0.28 + \log K$$
$$\log K = \log 56 + 0.27 = 1.7482 + 0.27 \approx 2.02$$
$$K \approx \text{antilog } 2.02 \approx 105$$

The equation satisfying the data is $T \approx 105 \cdot 10^{-0.27C}$.

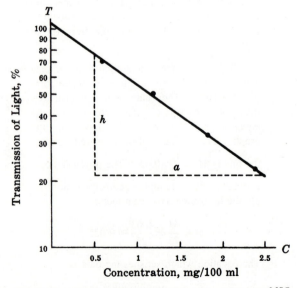

Fig. 9-14. The exponential function $T = 105 \cdot 10^{-0.27C}$

9.33. Water from a tank is allowed to discharge through a capillary. The volume of the water remaining in the tank is measured periodically. From the data shown in the following table determine the empirical equation relating the time t to volume V.

t, sec	V, liters	t, sec	V, liters
0	14.0	160	6.1
20	12.6	180	5.5
40	11.4	200	4.9
60	10.3	220	4.4
80	9.3	240	4.0
100	8.4	260	3.6
120	7.6	280	3.2
140	6.8	300	3.0

Plot the data on *two-cycle* semilogarithmic paper. Time is the independent variable and is plotted along the horizontal axis, which has a uniform scale. Volume is plotted along the vertical axis, which has two logarithmic scales. Since the graph is a straight line, its equation is of the form $V = C \cdot 10^{mt}$, where C is the V-intercept and m is the slope. The apparent geometrical slope in Fig. 9-15 is -0.46. However, since the grid is rectangular, the actual geometrical slope must be larger by the ratio of the rectangle length to its height, or $7/5$. Thus, the actual geometrical slope is

$$(-0.46)(7/5) \approx -0.64$$

Furthermore, the slope must be expressed in terms of the units on the horizontal axis; i.e., the geometrical slope must be divided by the scale factor used for the horizontal axis. Therefore, the physical slope is

$$m = -\frac{0.64}{280 \text{ seconds}} \approx -0.0023 \text{ sec}^{-1}$$

The V-intercept is 14. The equation of the graph is $V \approx 14 \cdot 10^{-0.0023t}$.

Also, the equation may be written to base e:

$$V \approx 14 e^{-(2.303)(0.0023)t}$$
$$\approx 14 e^{-0.053t}$$

9.34. Sunlight produces radioactive carbon-14 that is absorbed by all living organisms. By measuring the amount of carbon-14 present in an organism it is possible to

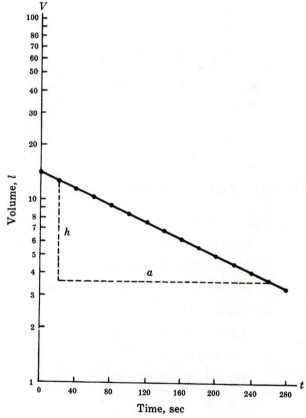

Fig. 9-15. The exponential function $V \approx 14 \cdot 10^{-0.0023t}$

approximate the time when an organism died. If this time can be approximated by the formula $t = 5700 \log_{(1/2)} C$ where C is the amount of C-14 present, how long ago did an organism die if the amount of C-14 present is 1/8 of the amount of C-14 when the animal was alive?

Substitute the values in the formula to obtain $t = 5700 \log_{(1/2)}(1/8)$. Calculate $\log_{(1/2)}(1/8)$. Use the conversion formula of section 9.11 to obtain $\log_{(1/2)}(1/8) = \log(1/8)/\log(1/2)$. Replace this quantity in the formula to estimate that the organism died about 17,100 years ago as indicated by the equation shown below.

$$t = 5700[\log(1/8)/\log(1/2)] \approx 17,100 \text{ years}$$

9.35. Assume that the amount of radium present in a compound is given by the formula $R = I_0 e^{-0.00043t}$. If the current amount of the compound I_0 is equal to 100 g, how much radium will remain in 8000 years?

Replace the values in the formula to obtain $R = 100e^{-(0.00043*8000)} = 3.21$ g.

9.36. Solve $2 \ln x = 1$ with a graphing calculator.

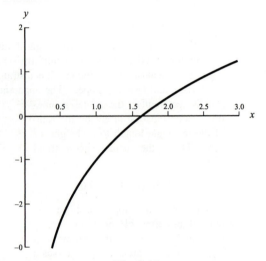

Fig. 9-16

Using an HP-38G:

(1) Define the function.

{LIB}FUNCTION{ENTER}EDIT

2{*}{SHIFT}{LN}{X,T,θ}{)}{−}1{ENTER}

(2) Plot the graph.

{PLOT}

(3) Find the intercept.

MENU FCN Root {ENTER}

$x \approx 1.65$

The graph of this equation is shown in Fig. 9-16.

9.37. Solve $2 \ln x = -x + 4$ using a calculator.

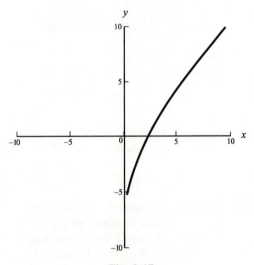

Fig. 9-17

Using an HP-38G:

(1) Define the function.

{LIB}FUNCTION{ENTER}EDIT

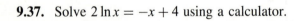

2{*}{SHIFT}{Ln}{X,T,θ}{)}{+}{X,T,θ}{−}4{ENTER}

(2) Plot the graph.

{PLOT}

(3) Find the intercept.

MENU FCN Root {ENTER}

$x \approx 2.22$

The graph is shown in Fig. 9.17.

9.38. Assume that a colony of bacteria grows according to the formula $N_t = N_0 e^{2t}$ where N_0 is the initial count of bacteria, t the time in hours and N_t the amount of bacteria after t hours.   How long it will take for the culture to quadruple its initial population?

   Substitute values in the formula to obtain: $4N_0 = N_0 e^{2t}$.   Simplify the equation, so that $e^{2t} = 4$.   Take natural logarithms at both sides to obtain $2t = \ln 4$ (remember that $\ln e = 1$).   Therefore, $t = (\ln 4)/2 \approx 0.69$ hours.

9.39. Radioactive decay provides an illustration of an exponential function with a negative exponent.   All radioactive elements disintegrate according to the equation $m = m_0 e^{-\lambda t}$, where m_0 is the initial mass of "parent" atoms, m is the mass of these atoms left after a time t, and λ is a constant characteristic of a given element.   For iodine-131, the value of λ is 0.087 per day.   If one starts with 5 µg of this element, the number of micrograms left after t days will be given by the equation $m = 5e^{-0.087t}$.

   Draw the graph of this function.   What does this curve illustrate?

   The curve illustrates what is known as exponential decay; i.e. as the value of t increases, the value of m decreases at a rate proportional to itself.   The graph of this function is shown in Fig. 9-18.

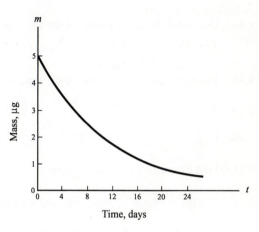

Fig. 9-18. A radioactive decay curve

Supplementary Problems

The answers to this set of problems are at the end of this chapter.

9.40. Express each number as a base raised to an integral power to compute the following:

(a) $\dfrac{(4)(16)(128)}{(2)(64)(256)}$ (b) $\dfrac{(125)(25)(5)}{(625)(125)}$

9.41. Write each of the following exponential equations in equivalent logarithmic form:

(a) $3^5 = 243$ (b) $2^{-2} = 1/4$ (c) $5^{-x} = 7$ (d) $10^5 = 100{,}000$

9.42. Write each of the following exponential equations in equivalent logarithmic form:

(a) $a^x = b$ (b) $\left(\dfrac{1}{2}\right)^x = c$ (c) $10^{-2} = 0.01$ (d) $7^0 = 1$

9.43. Write each of the following logarithmic equations in exponential form:

(a) $3 = \log_2 8$ (b) $5 = \log_3 N$ (c) $-2 = \log_7 x$ (d) $-3 = \log 0.001$

9.44. Write each of the following logarithmic equations in exponential form:

   (a) $\log_b x = y$ (b) $\log_5 p = -3$ (c) $\log_a 16 = -5$ (d) $\log_3 N = 1/2$

9.45. Determine the common logarithm of each of the following numbers:

   (a) 10   (b) 1   (c) 0.000001   (d) 0.0001   (e) 1,000,000   (f) 10^{50}

9.46. Why is it that $\log_b 1 = 0$, regardless of what base b is used?

9.47. Determine x if we know that $\log x$ is: (a) -2,   (b) 6,   (c) 5,   (d) -5,   (e) $1/2$, (f) $-1/2$,   (g) 0, (h) 1,   (i) -1.

9.48. Determine the characteristic of the logarithm of each of the following numbers:

   (a) 453.2   (b) 1230   (c) 5647   (d) 1200   (e) 5,463,000

9.49. Determine the characteristic of the logarithm of each of the following numbers:

   (a) 0.00243   (b) 0.0145   (c) 0.0000563   (d) 0.01456   (e) 0.9742

9.50. Find the mantissa of the logarithm of each of the numbers in Problem 9.48.

9.51. Find the mantissa of the logarithm of each of the numbers in Problem 9.49.

9.52. Use the results of Problem 9.48 and 9.50 to write the desired logarithms in two different forms.

9.53. Use the results of Problem 9.49 and 9.51 to write the desired logarithms in two different forms.

9.54. Find the antilogs of the following numbers:

   (a) $0.4520 + 3$ (b) $0.3200 + 2$ (c) $0.9453 + 5$ (d) $0.3870 + 4$

9.55. Find the antilogs of the following numbers:

   (a) $0.5430 - 2$ (c) $0.4752 - 4$ (e) $0.3428 - 10$

   (b) $0.6498 - 3$ (d) $7.4675 - 10$

9.56. Use logarithms to perform the indicated computations.

   (a) $(12.56)(0.1465)$ (f) $\dfrac{129,000}{43,500}$ (k) $\sqrt{7564}$ (or $7564^{1/2}$)

   (b) $(0.456)(12.79)$ (g) $(34.56)(13.3^5)(0.0013^4)$ (l) $\sqrt[3]{54.53}$ (or $54.53^{1/3}$)

   (c) $(0.0095)(0.0234)(56.3)$ (h) $(0.345^5)(123.8^5)$ (m) $\sqrt[3]{0.00354}$

   (d) $\dfrac{564.3}{54.6}$ (i) $\dfrac{(43.5)(0.4603)}{(3.002)(12.67)}$ (n) $\sqrt{\dfrac{427.4}{0.394}}$

   (e) $\dfrac{453.9}{32.76}$ (j) $\dfrac{(34.42^3)(4500)}{(4.56^2)(0.439)}$

9.57. The difference in intensity level of two sounds is measured in decibels according to the equation decibels $= 10 \log \dfrac{I_2}{I_1}$, where I_1 and I_2 are the intensities of the two sounds. Calculate the intensity level difference when $I_1 = 10^{-16}$ watt/cm^2 and $I_2 = 10^{-9}$ watt/cm^2.

9.58. In chemistry, the concentration of hydrogen ions in a solution may be expressed on a pH scale. The equation relating the hydrogen ion concentration H^+ to the pH of the solution is $pH = -\log H^+$, where H^+ is expressed in gram-ions per liter. Thus, a solution which has 10^{-6} g-ion per liter has a $pH = 6$. Conversely, if the given pH of a solution is 8, then the hydrogen ion concentration is 10^{-8} g-ion per liter.

(*a*) What is the pH of a solution which has a concentration of 10^{-3} g-ion per liter?

(*b*) What is the pH of a solution which has a concentration of 0.004 g-ion per liter? (*Hint*: Add the positive mantissa and the negative characteristic. Change the sign of the sum.)

(*c*) What is the hydrogen ion concentration of a solution whose pH is 12?

(*d*) What is the hydrogen ion concentration of a solution whose pH is 8.4? (*Hint*: Find the antilog of -8.4. Remember that the mantissa must be positive.)

9.59. Find: (*a*) $\ln 2$,   (*b*) $\ln 0.45$,   (*c*) $\ln 100$.

9.60. Find: (*a*) $\log_2 6$,   (*b*) $\log_5 10$,   (*c*) $\log_3 1000$.

9.61. Evaluate: (*a*) $\ln\left(\dfrac{4}{3}\right)^2$;   (*b*) $53.3 \ln 15/85$,   (*c*) $600(0.48) \ln 373/273$.

9.62. Write in exponential form: (*a*) $m = \log_2 43$,   (*b*) $\ln(N/N_0) = -\lambda t$.

9.63. Write in logarithmic form: (*a*) $i = 10e^{-t/c}$,   (*b*) $I = I_0 e^{-k/x}$,   (*c*) $A = A_0 e^{-et}$.

9.64. Evaluate: (*a*) $e^{-0.2}$,   (*b*) $18e^2$,   (*c*) $10^5 e^{-3.8}$.

9.65. (*a*) If $\ln R = 6.2538$, find R.   (*b*) If $\ln S = 1.704$, find S.

9.66. Solve each of the given equations for the unknown symbol:

(*a*) $3^x = 27$ (*f*) $(7.52)^{x-2} = 327$

(*b*) $2^{-y} = 1/64$ (*g*) $\log x + \log 10 = \log 5$

(*c*) $3^t - 3 = 81$ (*h*) $\log y^2 = 4$

(*d*) $\log(2x + 3) = 4$ (*i*) $\log(4x - 4) - \log(x + 2) = \log 2$

(*e*) $\log(p - 2) = 2.6021$ (*j*) $32(3^{x-2}) = 656$

9.67. If $M = m + 5 - 5 \log d$, find:

(*a*) M, for $m = 0.14$ and $d = 8.13$ (*b*) d, for $M = -3.2$ and $m = 1.22$

9.68. Decibels $= 10 \log(I_2/I_1)$.

(*a*) If $I_2 = 3I_1$, calculate the number of decibels.

(*b*) If $I_1 = 3I_2$, calculate the number of decibels.

9.69. Find I_2 from the equation: $45 = 10 \log \dfrac{10^{-8}}{I_2}$.

9.70. Solve for the unknown: $3 = 5e^{0.1t}$.

9.71. Solve for T: $530 = 3400 \ln \dfrac{T}{293}$.

9.72. If $p = 30e^{-0.0795h}$, find p for $h = 2$.

9.73. Given $I = I_0 e^{-kd}$. If $k = 2.0$, what should be the value of d, to make $I = I_0/2$?

9.74. $N/N_0 = e^{-\lambda t}$. For $t = 10$, $N/N_0 = 0.1633$. (a) Find λ. (b) For what t will $N/N_0 = 1/2$?

9.75. $I = I_0 e^{-(RL)t}$. Find I, if $I_0 = 10$, $R = 50$, $L = 0.1$, $t = 0.01$.

9.76. Write in equivalent logarithmic form: (a) $I = I_0 e^{-kx}$, (b) $AB^2 = C^{1/2}$.

9.77. Write in equivalent exponential form:

(a) $\ln N - \ln N_0 = -ct$ (b) $3 \log R - \dfrac{1}{2} \log P = \dfrac{2}{3} \log S$

9.78. (a) What would be the slope and the intercept on logarithmic graph paper of (1) $I = 0.6V^{3/2}$; (2) $d = 1.15\sqrt[3]{H}$;   (3) $N = 1850/d^2$;   (4) $V = 240T^{-0.30}$?

(b) Draw the graphs of (1), (2), (3), and (4) on log–log paper.

9.79. (a) Plot the following data on logarithmic paper.

L, m	0.915	0.610	0.306	0.153
P, sec	1.55	1.27	0.87	0.63

(b) Determine the equation connecting P and L.

9.80. The characteristics of a thyrite element are shown in the table below.

E, volts	20	25	30	35	40	45	50	55	60	65	70
I, mA	0.25	0.4	0.6	0.8	1.1	1.4	1.8	2.2	2.8	3.4	4.2

(a) Plot the data on logarithmic paper.

(b) Determine the equation of the graph in (a).

9.81. The power P transmitted by a steel shaft depends on the shaft diameter d according to the following table.

d, in.	1.5	2.0	2.5	3.0	3.5	4.0	4.5	5.0
P, hp	11	27	52	90	140	210	310	420

(a) Plot the data on log–log graph paper.

(b) Determine the equation of the line.

9.82. The osmotic pressure P of a sucrose solution depends on the concentration C of the solution. The data are reproduced in the table below.

C, mole/kg H_2O	1	2	3	4	5	6
P, atm	27.2	58.4	95.2	139	187	232

(a) Plot the data on logarithmic graph paper.

(b) Determine the equation between P and C.

9.83. The equivalent conductance L of a sulfuric acid solution depends on its concentration C.   The data for the solution at 25 °C are as follows:

C, equivalent/liter	0.005	0.01	0.05	0.1
L, mho cm^{-1} equivalent^{-1}	97.5	83.3	58.8	50.5

(a) Plot the data on log–log paper.

(b) Determine the equation relating L to C.

9.84. The deposit D in a bank savings account increases with time t as shown in the table below.

t, years	5	10	15	20
D, dollars	38.3	48.8	62.4	79.6

(a) Plot the data on semilogarithmic paper.

(b) Determine the equation of the graph in (a).

(c) What was the original amount deposited?

(d) How much will there be on deposit when 25 years has elapsed?

9.85. The activity I of a radioactive isotope decreases with time t.   The experimental data of an isotope are shown in the table below.

t, sec	0	20	40	60	80	100	120	140	160
I, counts/min	730	600	480	380	310	240	190	150	120

(a) Plot the above data on cartesian graph paper, with t as the independent variable.

(b) Plot the data on semilogarithmic paper.

(c) Determine the equation for the graph in (b).

(d) Write the equation with e as the base.

(e) If the equation in (d) were represented on semilogarithmic paper, what would be the physical slope of this graph?

(f) The half-life T of the radioactive substance is given by the relation $T = 0.693/m$, where m is the physical slope of graph of (e) on semilog paper.   Calculate the half-life of the given substance.

(g) The half-life of a radioactive substance is also defined as the time it takes for the activity of the sample to decrease to one-half its initial value.   (1) From the graph determine the time it took the given sample to decay from 700 counts/min to half that value.   (2) Repeat (1) for an initial count of 200 counts/min.

9.86. The absorption of beta rays by aluminum foil depends on the thickness of the foil.   The following data give the activity A of the beta rays as a function of foil thickness d.

d, cm	0	0.044	0.088	0.132	0.176	0.220	0.264	0.308
A, counts/min	15,500	11,300	8100	5900	4250	3000	2200	1550

(a) Plot the data on cartesian graph paper, with d as the independent variable.

(b) Plot the data on semilogarithmic paper.

(c) Determine the equation for the graph in (b).

(d) Write the equation with e as the base.

(e) What are the units of the physical slope?

9.87. A charged capacitor is allowed to discharge through a resistor. The deflections D of a galvanometer scale in the circuit are a measure of the residual charge as a function of time t. The data are recorded below.

t, sec	0	27.7	55.4	85.3	117.9	154
D, cm	45.3	42.3	39.3	36.3	33.3	30.3

t, sec	193.2	238.5	289.8	350.2	423.4	516.5
D, cm	27.3	24.3	21.3	18.3	15.3	12.3

(a) Plot a graph of the data on semilogarithmic paper.

(b) Determine the equation of the graph in (a).

(c) Write the equation with e as the base.

9.88. The inversion of sucrose causes a rotation of the plane of polarized light. The rotation θ as a function of time t for a solution is given by the following data.

t, min	8.65	19.6	31.0	40.0	48.3	58.3	66.7	75.0	86.7	96.6
θ, degrees	15.1	13.1	11.7	10.7	9.8	8.6	7.7	7.2	6.3	5.4

(a) Plot a graph on semilogarithmic paper.

(b) Determine the physical slope of the graph.

(c) Determine the equation of the graph.

9.89. Assume that a colony of bacteria has a population of 1.2 million individuals. If the colony grows at the rate of 1.2% per hour, estimate the time when the population reaches 5 millions.

9.90. Solve the logarithmic equation $3 \log x = 7 \log(x - 2)$ using a graphing calculator.

9.91. Solve the exponential equation $3^x = 2^x$ using a graphing calculator.

Answers to Supplementary Problems

9.40. (a) 1/4   (b) 1/5

9.41. (a) $\log_3 243 = 5$ (b) $\log_2 1/4 = -2$ (c) $\log_5 7 = -x$ (d) $\log 100,000 = 5$

9.42. (a) $\log_a b = x$ (b) $\log_{1/2} c = x$ (c) $\log 0.01 = -2$ (d) $\log_7 1 = 0$

9.43. (a) $2^3 = 8$ (b) $3^5 = N$ (c) $7^{-2} = x$ (d) $10^{-3} = 0.001$

9.44. (a) $b^y = x$　　(b) $5^{-3} = p$　　(c) $a^{-5} = 16$　　(d) $3^{1/2} = N$

9.45. (a) 1　　(b) 0　　(c) -6　　(d) -4　　(e) 6　　(f) 50

9.46. Because, by definition, $b^0 = 1$ for $b \neq 0$.

9.47. (a) 0.01　　(d) 0.00001　　(g) 1
(b) 1,000,000　　(e) $\sqrt{10}$　　(h) 10
(c) 100,000　　(f) $1/\sqrt{10}$　　(i) 0.1

9.48. (a) 2　　(b) 3　　(c) 3　　(d) 3　　(e) 6

9.49. (a) -3　　(b) -2　　(c) -5　　(d) -2　　(e) -1

9.50. (a) 0.6563　　(b) 0.0899　　(c) 0.7518　　(d) 0.0792　　(e) 0.7374

9.51. (a) 0.3856　　(b) 0.1614　　(c) 0.7505　　(d) 0.1632　　(e) 0.9887

9.52. (a) $0.6563 + 2$, 2.6563　　(c) $0.7518 + 3$, 3.7518　　(e) $0.7374 + 6$, 6.7374
(b) $0.0899 + 3$, 3.0899　　(d) $0.0792 + 3$, 3.0792

9.53. (a) $0.3856 - 3$, $7.3856 - 10$　　(c) $0.7505 - 5$, $5.7505 - 10$　　(e) $0.9887 - 1$, $0.9887 - 10$
(b) $0.1614 - 2$, $8.1614 - 10$　　(d) $0.1632 - 2$, $8.1632 - 10$

9.54. (a) 2831　　(b) 208.9　　(c) 881,700　　(d) 24,380

9.55. (a) 0.0349　　(b) 0.004465　　(c) 0.0002987　　(d) 0.002934　　(e) $2.202(10^{-10})$

9.56. (a) 1.840　　(d) 10.34　　(g) 0.00004109　　(j) $2.010(10^7)$　　(m) 0.1524
(b) 5.832　　(e) 13.86　　(h) $(1.421)(10^8)$　　(k) 86.97　　(n) 32.94
(c) 0.01252　　(f) 2.966　　(i) 0.5264　　(l) 3.792

9.57. 70

9.58. (a) 3　　(b) 2.3979　　(c) 10^{-12}　　(d) $(3.981)(10^{-9})$

9.59. (a) 0.6932　　(b) -0.7985　　(c) 4.606

9.60. (a) 2.585　　(b) 1.431　　(c) 6.288

9.61. (a) 0.5754　　(b) -92.44　　(c) 89.84

9.62. (a) $2^m = 43$　　(b) $e^{-\lambda t} = N/N_0$

9.63. (a) $-t/c = \ln(i/10)$　　(b) $-k/x = \ln(I/I_0)$　　(c) $-et = \ln(A/A_0)$

9.64. (a) 0.8187　　(b) 133　　(c) 2237

9.65. (a) 520　　(b) 5.495

9.66. (a) 3   (f) $2 + (\log 327)/(\log 7.52)$ or 4.87 approx.

(b) 6   (g) $\dfrac{1}{2}$

(c) $(\log 84)/(\log 3)$ or 4.033 approx.   (h) 100

(d) $4998\frac{1}{2}$ (i) 4

(e) 402   (j) $2 + (\log 41 - \log 2)/(\log 3)$ or 4.75 approx.

9.67. (a) 0.59   (b) 76.6

9.68. (a) 4.77   (b) -4.77

9.69. $(3.16)(10^{-13})$

9.70. -5.11

9.71. 342

9.72. 25.59

9.73. 0.346

9.74. (a) 0.1812   (b) 3.82

9.75. 9.51

9.76. (a) $\ln I = \ln I_0 - kx$, or $\log I = \log I_0 - kx \log e$

(b) $\log A + 2 \log B = \dfrac{1}{2} \log C$

9.77. (a) $N = N_0 e^{-ct}$ (b) $R^3 p^{-1/2} = S^{2/3}$, or $R^3 = p^{1/2} S^{2/3}$

9.78. (a) (1)   slope $= 3/2$, intercept $= 0.6$
           (2)   slope $= 1/3$, intercept $= 1.15$
           (3)   slope $= -2$, intercept $= 1850$
           (4)   slope $= -0.30$, intercept $= 240$

(b) Graphs on log–log paper with appropriate slopes and y-intercepts.

9.79. (a) See Fig. 9-19.   (b) $P \approx 1.6L^{0.49} \approx 1.6L^{1/2}$

9.80. (a) See Fig. 9-20.   (b) $I \approx 0.00033E^{2.2} \approx (3.3 \times 10^{-4})E^{2.2}$

9.81. (a) See Fig. 9-21.   (b) $P \approx 3.3d^{3.0}$

9.82. (a) See Fig. 9-22.   (b) $P \approx 27C^{1.2}$

9.83. (a) See Fig. 9-23.   (b) $L \approx 30C^{-0.22}$

9.84. (a) See Fig. 9-24.   (b) $D \approx 30 \cdot 10^{0.021t}$ (c) $30   (d) $100 approximately

9.85. (a) See Fig. 9-25.   (c) $I \approx (730)(10^{-0.0048t})$ (e) -0.011 (g) (1) 60 ± 5 sec;
                                                                                                          (2) $\approx 60 \pm 5$ sec

(b) See Fig. 9-26.   (d) $I \approx 730e^{-0.011t}$ (f) $T = 63$ sec

9.86. (*a*) See Fig. 9-27.   (*c*) $A \approx (16{,}000)(10^{-3.2d})$ (*e*) counts/min/cm

       (*b*) See Fig. 9-28.   (*d*) $A \approx 16{,}000e^{-7.4d}$

9.87. (*a*) See Fig. 9-29.   (*b*) $D \approx (45)(10^{-0.0011t})$ (*c*) $D \approx 45e^{-0.0025t}$

9.88. (*a*) See Fig. 9-30.   (*b*) -0.0049 degree/minute   (*c*) $\theta \approx (17)(10^{-0.0049t})$

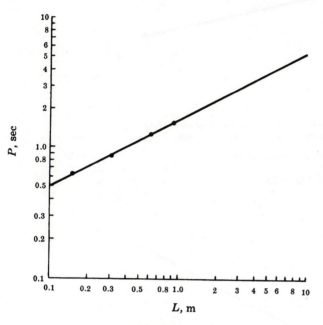

Fig. 9-19

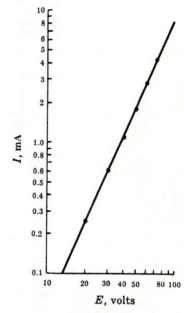

Fig. 9-20

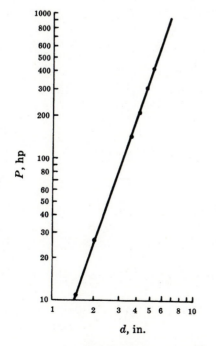

Fig. 9-21

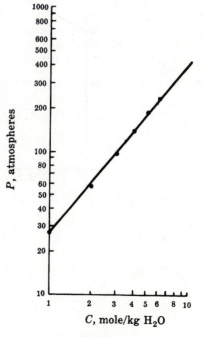

Fig. 9-22

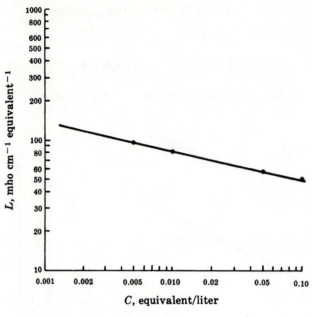

Fig. 9-23

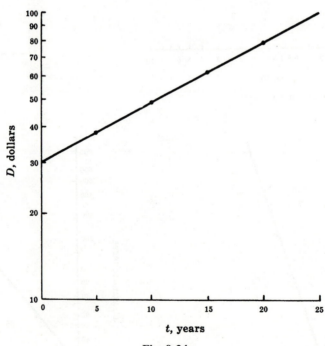

Fig. 9-24

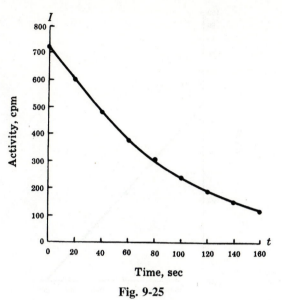

Fig. 9-25

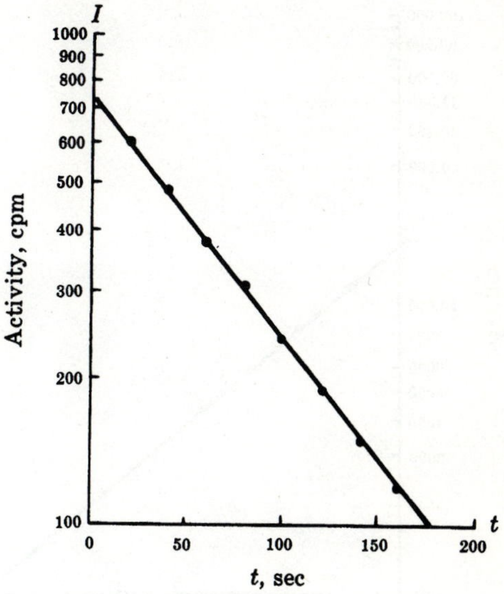

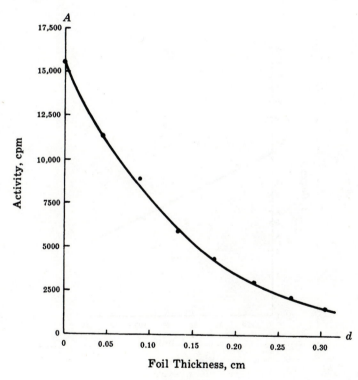

Fig. 9-26

Fig. 9-27

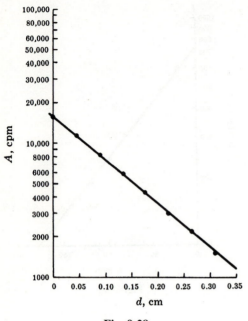

Fig. 9-28

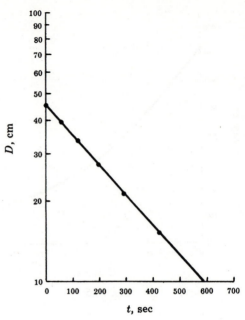

Fig. 9-29

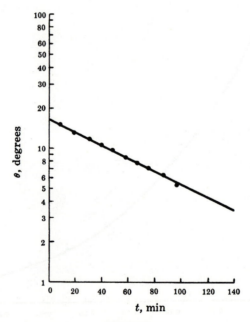

Fig. 9-30

9.89. About 119 hours or approximately 5 days.

9.90. The intersection point is (3.76, 1.72).

>   The sequence of steps to obtain this solution using a TI-82 is the following:

{Y =}

3{×}{LOG}{(}{X,T,θ}{)}{ENTER}   ← First curve: $3\log(x)$

7{×}{LOG}{(}{X,T,θ}{−}2{)}   ← Second curve: $7\log(x-2)$

{GRAPH}   ← "Using the standard window settings"

{2nd}{CALC}intersect{ENTER}

Use the directional keys to move the cross hair close to the intersection of the graphs before answering each of the following questions posed by the calculator.

First curve? {ENTER}

Second curve? {ENTER}

Guess? {ENTER}

9.91. Intersection occurs at (0, 1).

>   Using a HP-38G this solution can be obtained as follows:

{LIB}FUNCTION{ENTER}

EDIT

3{x^y}{X,T,θ}{ENTER}   ← This defines the function $y = 3^x$

2{x^y}{X,T,θ}{ENTER}   ← This defines the function $y = 2^x$

{PLOT}

MENU FCN Intersection {ENTER}

{ENTER}{ENTER}

Chapter 10

Quadratic Equations and Square Roots

Definitions

10.1 QUADRATIC EQUATION

A polynomial equation in which the highest power of the unknown is 2 is called *quadratic*.

EXAMPLES: $x^2 = 125;$ $t^2 - 4t - 3 = 0;$ $v^2 = 19.6d;$ $10n^2 + 24n = 12;$ $2x^2 - 17x + 21 = 0.$

A *second-degree equation in the unknown* is another name for a quadratic equation. The *general* or *standard form* of a quadratic equation is $ax^2 + bx + c = 0$, where a, b, and c are constants and $a \neq 0$.

A *complete quadratic equation* contains terms with nonzero coefficients of the second and first powers of the unknown.

EXAMPLE: $144 = 96t - 16t^2.$

A *pure* or *incomplete quadratic equation* is one in which the first-degree term has a zero coefficient.

EXAMPLE: $v^2 = 12.67.$

10.2 SOLUTION

To *solve* a quadratic equation is to determine the values of the unknown which will satisfy the equation. The numbers which satisfy a quadratic equation are called its *solutions* or *roots*. For example: when either $y = 3$ or $y = -1$ is substituted in $y^2 - 2y - 3 = 0$, the equation is satisfied. Thus, $y = 3$ and $y = -1$ are solutions or roots of this equation.

If the solutions of a quadratic equation are real numbers like 2, -5, $\sqrt{3}$, 1.75, then the roots are said to be *real*. If the equation is satisfied by numbers like $\sqrt{-2}$, $\sqrt{-4.75}$, then the roots are said to be *imaginary*. Such solutions belong to the *complex number system*, which is not discussed in this book.

Methods of Solving Quadratic Equations

10.3 PURE QUADRATIC EQUATIONS

To solve a pure quadratic equation, solve the equation for the square of the unknown and take the square root of each side.

EXAMPLE 10.1. Determine the value of h in the equation $4h^2 - 64 = 0$.

$$4h^2 = 64 \qquad \text{transpose 64 to the right side}$$
$$h^2 = 16 \qquad \text{divide both members by 4}$$
$$h = \pm\sqrt{16} \qquad \text{take the square root of each member}$$
$$h = \pm 4$$

The roots of the equation are $h = +4$ and $h = -4$.

Check: $4(+4)^2 \overset{?}{=} 64$ $4(-4)^2 \overset{?}{=} 64$

$\qquad\qquad 4(16) \overset{?}{=} 64$ $4(16) \overset{?}{=} 64$

$\qquad\qquad\quad 64 = 64$ $\qquad 64 = 64$

The two roots of a pure quadratic equation have the same absolute value but are opposite in sign.

10.4 QUADRATIC EQUATIONS BY FACTORING

This method makes use of the property of real numbers that if the product of two of them is 0, at least one of the numbers must be 0.   Expressed algebraically, if a and b are real numbers such that $ab = 0$, then either $a = 0$ or $b = 0$, or both.

EXAMPLE 10.2.   Solve the equation $x^2 - 5x + 6 = 0$.

The left member of this equation can be easily factored, and we obtain

$$(x - 3)(x - 2) = 0$$

Since $x - 2$ and $x - 3$ are representations of real numbers, it follows from the above property that $x - 3 = 0$ or $x - 2 = 0$.   The two solutions are then seen to be $x = 2$ and $x = 3$.

Check: $(2)^2 - 5(2) + 6 \overset{?}{=} 0$ $(3)^2 - (5)(3) + 6 \overset{?}{=} 0$

$\qquad\qquad 4 - 10 + 6 \overset{?}{=} 0$ $\qquad 9 - 15 + 6 \overset{?}{=} 0$

$\qquad\qquad\qquad 0 = 0$ $\qquad\qquad 0 = 0$

10.5 THE QUADRATIC FORMULA

After an equation has been reduced to the standard quadratic form, $ax^2 + bx + c = 0$, it may not be possible to factor the left member.   In such cases we are able to solve the equation by means of the quadratic formula:

$$x = \frac{-b \pm \sqrt{b^2 - 4ac}}{2a}$$

Note that one root is obtained by using $+\sqrt{b^2 - 4ac}$ in the formula; the other root results from $-\sqrt{b^2 - 4ac}$.

EXAMPLE 10.3.   Solve the equation $2x^2 - 3x + 1 = 0$.

The given equation is not a pure quadratic equation.   The equation is in standard form, with $a = 2$, $b = -3$, $c = 1$.   An application of the quadratic formula yields

$$x = \frac{-(-3) \pm \sqrt{(-3)^2 - (4)(2)(1)}}{(2)(2)} = \frac{3 \pm \sqrt{9 - 8}}{4} = \frac{3 \pm \sqrt{1}}{4}$$

$$= \frac{3 + 1}{4} = 1 \qquad \text{or} \qquad \frac{3 - 1}{4} = \frac{1}{2}$$

Therefore the two solutions are $x = 1$ and $x = 1/2$.

Check: Substitution in the given equation yields

$$2(1)^2 - 3(1) + 1 \overset{?}{=} 0 \qquad \text{and} \qquad 2(1/2)^2 - 3(1/2) + 1 \overset{?}{=} 0$$

$$2 - 3 + 1 \overset{?}{=} 0 \qquad\qquad 2(1/4) - 3(1/2) + 1 \overset{?}{=} 0$$

$$0 = 0 \qquad\qquad 1/2 - 3/2 + 1 \overset{?}{=} 0$$

$$0 = 0$$

The quantity $b^2 - 4ac$, which is under the radical sign in the quadratic formula, is called the *discriminant* of the equation and completely determines the nature of its solutions.

If the discriminant D is	*The roots are*
0	real and equal
> 0	real and unequal
< 0	imaginary

EXAMPLE 10.4.   Determine the nature of the solutions for:

(*a*) $4x^2 - 4x + 1 = 0$ (*b*) $x^2 - 2x - 3 = 0$ (*c*) $2x^2 + 3x + 2 = 0$

(*a*) Here $a = 4$, $b = -4$, $c = 1$.   The value of the discriminant is

$$D = (-4)^2 - (4)(4)(1) = 0$$

It follows that the two solutions of the equation are identical, and the single solution is a rational number.   An application of the quadratic formula gives

$$x = \frac{-(-4) \pm \sqrt{(-4)^2 - (4)(4)(1)}}{(2)(4)} = \frac{4 \pm \sqrt{16 - 16}}{8} = \frac{4 \pm 0}{8} = \frac{1}{2}$$

The two solutions are identical and have the value $1/2$.

Check: $4(1/2)^2 - 4(1/2) + 1 = 4(1/4) - 4(1/2) + 1 \overset{?}{=} 0$

$$1 - 2 + 1 \overset{?}{=} 0$$

$$0 = 0$$

(*b*) Here $a = 1$, $b = -2$, $c = 3$.   The value of the discriminant is

$$D = (-2)^2 - (4)(1)(-3) = 16$$

Since $16 > 0$, the two roots are real and unequal.   By applying the quadratic formula, we obtain

$$x = \frac{-(-2) \pm \sqrt{(-2)^2 - 4(1)(-3)}}{(2)(1)} = \frac{2 \pm \sqrt{16}}{2}$$

$$x = \frac{2 + 4}{2} \qquad \text{or} \qquad x = \frac{2 - 4}{2}$$

$$x = 3 \qquad \text{or} \qquad x = -1$$

Check: $(3)^2 - 2(3) - 3 \overset{?}{=} 0 \qquad (-1)^2 - 2(-1) - 3 \overset{?}{=} 0$

$$9 - 6 - 3 \overset{?}{=} 0 \qquad\qquad 1 + 2 - 3 \overset{?}{=} 0$$

$$0 = 0 \qquad\qquad 0 = 0$$

(*c*) Here, $a = 2$, $b = 3$, $c = 2$, and $D = 3^2 - 4(2)(2) = -7$.   The solutions of this equation are *not* real.   The use of the quadratic formula gives

$$x = \frac{-(3) \pm \sqrt{(3)^2 - (4)(2)(2)}}{(2)(2)} = \frac{-3 \pm \sqrt{9 - 16}}{4} = \frac{-3 \pm \sqrt{-7}}{4}$$

In *elementary science* problems this kind of answer to a problem usually indicates that the student has made an error in setting up the problem or in arithmetical computations.

10.6 USING A GRAPHING CALCULATOR TO SOLVE QUADRATIC EQUATIONS

Quadratic equations can also be solved using a graphing calculator. The graph of a quadratic function of the form $y = ax^2 + bx + c$ with $a \neq 0$ is called a *parabola*. Figures 10-1 through 10-3 show the graphs of some quadratic functions. Figure 10-1 shows a quadratic function whose graph intersects the x-axis at two different points. These two points ($x_1 = 1$; $x_2 = 4$) correspond to the two different and real solutions of the equation $x^2 - 5x + 4 = 0$. Notice that when the variable x is substituted by either of the two solutions, the y value of the quadratic function $y = x^2 - 5x + 4$ is equal to 0.

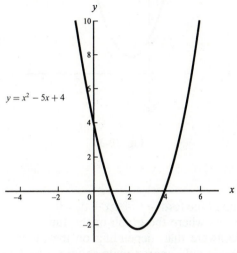

Fig. 10-1

Figure 10-2 shows a situation where the quadratic equation $x^2 - 6x + 9 = 0$ has two identical real solutions ($x_1 = x_2 = 3$). Notice that the graph of the function $y = x^2 - 6x + 9$ touches the x-axis only once.

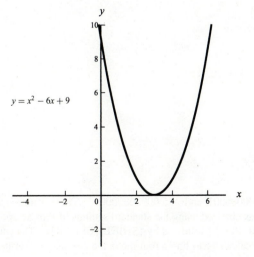

Fig. 10-2

Figure 10-3 shows the graph of a quadratic function whose graph does not intersect the x-axis. Notice that the equation $x^2 - 2x + 2 = 0$ has no real solutions.

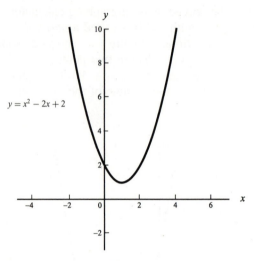

$y = x^2 - 2x + 2$

Fig. 10-3

Solving a quadratic equation of the form $ax^2 + bx + c = 0$ with $a \neq 0$ using a graphing calculator is equivalent to finding the points where the graph of the function $y = ax^2 + bx + c$ intersects the x-axis. The reader should be aware that, depending on the model of the calculator being used, a graphing calculator may provide only approximate solutions to the equation.

EXAMPLE 10.5. Solve $x^2 + 8x + 15 = 0$.

To solve this equation with an HP-38G graphing calculator we can proceed as follows.

(1) Define the function.

{LIB}<u>FUNCTION</u>{ENTER}

<u>EDIT</u>

{X,T,θ}{x^y}2{+}8{*}{X,T,θ}{+}15{ENTER}

(2) Plot the graph.

{PLOT}

(3) Find the x-intercepts.

<u>MENU</u> <u>FCN</u> <u>Root</u> {ENTER}

The intercepts are -3.00000000001 and -5.00000000001. These values can be rounded to -3 and -5 respectively. This graph was obtained using the standard settings of Plot Setup. If necessary, reset these values by pressing {SHIFT}{SETUP-PLOT} followed by {SHIFT}{CLEAR}. The calculator shows the root nearest to the crosshairs. To help the calculator to find a root move the crosshair closer to an intersect point. Figure 10-4 shows the graph of the function $y = x^2 + 8x + 15$.

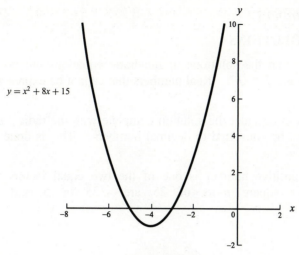

Fig. 10-4

EXAMPLE 10.6.   Solve $x^2 - 2x + 2 = 0$.

Using the TI-82 graphing calculator we can proceed as follows.

(1) Define the function.

{Y=}
{X,T,θ}{^}2{−}2{×}{X,T,θ}{+}2{ENTER}

(2) Plot the graph.

{GRAPH}   "If necessary, before pressing the GRAPH key, reset the settings of the Viewing Window to the standard values defined in the calculator guidebook."

The solutions of this equation are imaginary.   As Fig. 10-5 shows, the graph of the quadratic function $y = x^2 - 2x + 2$ does not intersect the x-axis.

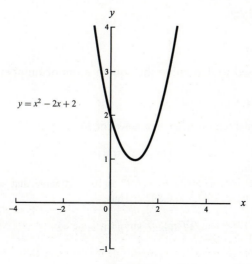

Fig. 10-5

Square Roots of Numbers

10.7   IRRATIONAL SOLUTIONS

It happens frequently that the solutions of quadratic equations are not integers or common fractions for example $\sqrt{2}$ and $3 + \sqrt{7}$.   Real numbers that cannot be expressed as ratios of integers are called *irrational*.

Though it is common to consider the solution complete with the radical sign in it, for practical applications the roots must be converted to decimal numbers.   This is done by *finding the square root*.

A square root of a positive number is one of the two equal factors whose product is the number.   Thus, the two square roots of 25 are $\sqrt{25}$ or 5 and $-\sqrt{25}$ or -5, for $5 \cdot 5 = (-5)(-5) = 25$.

10.8   TABLE OF SQUARE ROOTS

The easiest method of determining the square root of a number is to look it up in a table of square roots.   The table in the Appendix on page 464 gives the square roots of numbers from 1 to 150 to three decimal places.   The squares and cubes of these numbers are also given in this table.   For example, $\sqrt{7} = 2.646$; $\sqrt{41} = 6.403$; $\sqrt{149} = 12.207$; $68^2 = 4624$; $75^3 = 421{,}875$.

The table can also be used to determine the square roots of decimal fractions whose values fall between two integers in the table.   This process is called *linear interpolation*.

EXAMPLE 10.7.   Determine $\sqrt{52.8}$.

From the table, $\sqrt{53} = 7.280$; $\sqrt{52} = 7.211$.   Use the reasonable approximation that $\sqrt{52.8}$ is larger than $\sqrt{52}$ by 0.8 of the difference between $\sqrt{53}$ and $\sqrt{52}$.

$$\sqrt{53} = 7.280$$
$$\sqrt{52} = \underline{7.211}$$
$$\text{difference} = 0.069$$
$$(0.8)(0.069) = 0.0552 \approx 0.055$$

Thus, $\sqrt{52.8} \approx 7.211 + 0.055 \approx 7.266$.

Check: $(7.266)^2 \approx 52.795 \approx 52.8$.

The table may also be used to determine the square roots of numbers larger than 150.

EXAMPLE 10.8.   Determine $\sqrt{4750}$.

Write the number in exponential form with an *even* power of 10:

$$4750 = (47.50)(10^2)$$

Then, $\sqrt{4750} = \sqrt{(47.50)(10^2)} = \sqrt{47.50}\sqrt{10^2} = \sqrt{(47.5)}(10)$.   Assume that $\sqrt{47.5}$ is halfway between $\sqrt{47}$ and $\sqrt{48}$.   Interpolating from the table, $\sqrt{47.5} = 6.892$.   Thus,

$$\sqrt{4750} = (6.892)(10) = 68.92$$

Check: $(68.92)^2 \overset{?}{=} 4750$ or $4749.966 \approx 4750$.

10.9 ITERATIVE METHOD

It is possible to determine the approximate square root of a number by an *iterative process* which is usually less cumbersome than the direct procedure.   The square root obtained by this process will have the same number of significant figures as the given number, and this is consistent with scientific requirements.

One procedure for extracting the square root is:

1. Estimate the approximate square root of the given number.

2. Divide the given number by the estimated square root.

3. Calculate the arithmetic average of the results in steps 1 and 2.   This is an approximate square root.

4. If a more precise value is desired, repeat the process but use the result of step 3 as the divisor for step 2.

EXAMPLE 10.9.   Determine the square root of 28.5.

1. The approximate square root of 28.5 is estimated to be 5.00 because 5^2 is closer than 6^2 to 28.5.

2. Divide 28.5 by 5.00.

$$28.5 \div 5.00 = 5.70$$

3. Calculate the arithmetic average of 5.00 and 5.70.

$$\frac{5.00 + 5.70}{2} = \frac{10.70}{2} = 5.35$$

The approximate square root of 28.5 is 5.35.

Check: $(5.35)^2 = 28.5225$, which agrees in the first three digits with the given number.

4. If the given number had more digits, say 28.50, then the answer of step 3 would be used as the divisor, so that $28.50 \div 5.250 = 5.327$.   The arithmetic average of 5.350 and 5.327 would be the more precise square root.   Thus,

$$\frac{5.350 + 5.327}{2} = 5.339$$

Check: $(5.339)^2 = 28.5049$, which agrees in the first four digits with the given number.

10.10 TRADITIONAL METHOD

The method usually taught in schools is illustrated in the following example.

EXAMPLE 10.10.   Extract the square root of 556.25.

1. Place the decimal point above the square root sign and immediately above the decimal point in the radicand.

$$\sqrt{5\ 56.25}$$

2. Mark the radicand off in units of two digits, starting from the decimal point working left, and from the decimal point working right (the only time one digit comprises a full unit is when there is an odd number of digits to the left of the decimal point).

$$\sqrt{5\ 56.25}$$

3. Find the square root of the largest perfect square which is equal to or less than the first unit.

4. Write this perfect square below the first unit and its square root directly above the first *unit*.   *Subtract* this square from the first unit.

$$
\begin{array}{r}
2 \\
\sqrt{5\ 56.25} \\
\underline{4} \\
1
\end{array}
$$

5. *Bring down* the next unit and multiply the number in the square root by 20, to determine the trial divisor.

$$
2 \times 20 = 40 \qquad
\begin{array}{r}
2 \\
\sqrt{5\ 56.25} \\
\underline{4} \\
1\ 56
\end{array}
$$

6. Estimate the number of times the number (40) will divide into the dividend (156) and write this *estimation* (3) into the square root.

$$
40 \qquad
\begin{array}{r}
2\ 3 \\
\sqrt{5\ 56.25} \\
\underline{4} \\
1\ 56
\end{array}
$$

7. Add the estimation (3) to the trial divisor and multiply that sum (43) by the estimation (3).   If the product of these numbers is larger than the dividend, reduce the estimation.

$$
40 + 3 = 43 \qquad
\begin{array}{r}
2\ 3 \\
\sqrt{5\ 56.25} \\
\underline{4} \\
1\ 56 \\
1\ 29
\end{array}
$$

8. Subtract the (129) from the (156) and bring down the third unit (25).

$$
43 \qquad
\begin{array}{r}
2\ 3 \\
\sqrt{5\ 56.25} \\
\underline{4} \\
1\ 56 \\
\underline{1\ 29} \\
27\ 25
\end{array}
$$

9. Multiply the new square root (23) by 20.   Use this answer of 460 as a trial divisor.

$$
\begin{array}{r}
43 \\
(23)(20) = 460
\end{array}
\qquad
\begin{array}{r}
2\ 3 \\
\sqrt{5\ 56.25} \\
\underline{4} \\
1\ 56 \\
\underline{1\ 29} \\
27\ 25
\end{array}
$$

10. Estimate the number of times the trial divisor (460) will divide into the dividend (2725) and write this estimation into the square root.   Our estimation is 5.

$$
\begin{array}{r}
43 \\
460
\end{array}
\qquad
\begin{array}{r}
2\ 3\ .5 \\
\sqrt{5\ 56.25} \\
\underline{4} \\
1\ 56 \\
\underline{1\ 29} \\
27\ 25
\end{array}
$$

11. Add the estimation (5) to the new trial divisor (460). This results in 465. Multiply this sum (465) by the estimation (5). Subtract this product (2325) from the last figure (2725).

$$
\begin{array}{r}
2\ 3\ .\ 5 \\
\sqrt{5\ 56.25} \\
\hline
4 \\
\hline
1\ 56 \\
1\ 29 \\
\hline
27\ 25 \\
23\ 25 \\
\hline
\end{array}
$$

with 43 and $460 + 5 = 465$ shown to the left.

12. The approximate square root is then 23.5. This value can be checked by squaring 23.5 and adding the remainder 4.00 to give the original number 556.25.

Each *pair* of numbers in the radicand contributes *one* number in the square root. Thus twice as many decimal places are necessary in the radicand as are obtained in the square root.

To avoid confusion, go through the same steps with zero as with any other number, even though the multiplications and subtractions involved are trivial.

EXAMPLE 10.11. Determine the square root of 11,025.

Proceeding as in previous example, we get $\sqrt{11,025} = 105$.

$$
\begin{array}{r}
1\ 0\ \ 5\ . \\
\sqrt{1\ 10\ 25.} \\
\hline
1 \\
\hline
0\ 10 \\
0 \\
\hline
10\ 25 \\
10\ 25 \\
\hline
\end{array}
$$

with $20 + 0 = 20$ and $200 + 5 = 205$ shown to the left.

Check: $(105)^2 = 11,025$.

When the radicand is a decimal number less than 1, with zeros immediately following the decimal point, place one zero in the answer for each *pair* of zeros at the beginning of the number, and begin the formal extraction with the first nonzero pair.

EXAMPLE 10.12. Extract the square root of 0.00000351.

Pairing off zeros and digits to the right of the decimal point and proceeding as in previous examples, we obtain $\sqrt{0.00000351} \approx 0.00187$.

$$
\begin{array}{r}
0.\ 0\ \ 0\ \ 1\ \ 8\ \ 7 \\
\sqrt{0.00\ 00\ 03\ 51\ 00} \\
\hline
1 \\
\hline
2\ 51 \\
2\ 24 \\
\hline
27\ 00 \\
25\ 69 \\
\hline
\end{array}
$$

with $1 \times 20 = 20$, $20 + 8 = 28$, $(20)(18) = 360$, $360 + 7 = 367$ shown to the left.

Check: $(0.00187)^2 \stackrel{?}{=} 0.00000351$ $0.00000350 \approx 0.00000351$.

To find the square root of a fraction whose numerator and denominator are not both perfect squares, change the fraction to a decimal and find the square root by the above process.

EXAMPLE 10.13. Extract the square root of 5/7.

$$\sqrt{5/7} = \sqrt{0.714285} \approx 0.845$$

Check: $(0.845)^2 \overset{?}{=} 0.714285$ $0.714 \approx 0.714285.$

10.11 LOGARITHMIC METHOD

The square root of a number can be readily extracted by applying the power law of logarithms which was stated in Section 9.10. If N is the number, then $\sqrt{N} = N^{1/2}$ and $\log N^{1/2} = (1/2)\log N$. Thus, the procedure is to look up the logarithm of the number in the table of logarithms, divide the logarithm by 2 and look up the antilogarithm.

EXAMPLE 10.14. Extract $\sqrt{352}$.

$$\log\sqrt{352} = (1/2)(2.5465) = 1.2732 \qquad \sqrt{352} = \text{antilog } 1.2732 = 18.76$$

Check: $(18.76)^2 \overset{?}{=} 352$ $351.93 \approx 352.$

10.12 FINDING SQUARE ROOTS WITH A CALCULATOR

To find the square root of a number with the help of a calculator you can use two methods. The first method uses the $\sqrt{x}$ key. Here, depending on the model of your calculator, you may have to press the $\sqrt{x}$ key first and then the number or press the number first and then press the $\sqrt{x}$ key.

EXAMPLE 10-15. Determine $\sqrt{567}$ using a calculator.

When working with an HP-38G graphing calculator, press $\{\sqrt{x}\}567\{\text{ENTER}\}$. The display shows 23.8117617996.

If you are using a TI-82 graphing calculator, press $\{2^{\text{nd}}\}\{\sqrt{\ }\}567\{\text{ENTER}\}$. The display shows 23.8117618.

We can also find the square root of any given number by raising it to the 1/2 power. To do this, we use the x^y key of the HP-38G or use the $^\wedge$ key of the TI-82.

EXAMPLE 10-16. Find $\sqrt{567}$ with the help of a calculator.

If you are using an HP-38G use the following sequence: $567\{x^y\}\{()1\{/\}2()\}\{\text{ENTER}\}$.

Similarly, if you are using a TI-82 proceed as follows: $567\{^\wedge\}\{()1\{\div\}2()\}\{\text{ENTER}\}$.

Solved Problems

10.1. Solve for the unknown and check.

(a) $4.90t^2 = 39.2$

(b) $4 = \dfrac{v^2}{100}$

(c) $A = 4\pi r^2$, where $A = 225$ cm^2, $\pi = 3.14$

(d) $c^2 = 11^2 + 25^2$

(e) $v^2 = v_0^2 + 2ad$, where $v_0 = 60$; $a = -32$; $d = 18$

(a) $4.90t^2 = 39.2$

$t^2 = \dfrac{39.2}{4.9} = 8$ divide both members by 4.9

$t = \pm\sqrt{t^2} = \pm\sqrt{8}$ take the square root of each member

$t \approx \pm 2.83$ use the table, page 464, to evaluate $\sqrt{8}$

$t \approx 2.83$

Check: $(4.90)(2.83)^2 \overset{?}{=} 39.2$ $(4.90)(-2.83)^2 \overset{?}{=} 39.2$

$(4.90)(8.00) \overset{?}{=} 39.2$ $39.2 = 39.2$

$39.2 = 39.2$

(b) $4 = \dfrac{v^2}{100}$

$400 = v^2$ multiply both members by 100

$v = \pm\sqrt{v^2} = \pm\sqrt{400}$ take the square root of each member

$v = \pm 20$

Check: $4 \overset{?}{=} \dfrac{20^2}{100}$ $4 \overset{?}{=} \dfrac{(-20)^2}{100}$

$4 = 4$ $4 = 4$

(c) $A = 4\pi r^2$

$225 = 4(3.14)r^2$ substitute the known values

$225 = 12.56r^2$ carry out the multiplication

$\dfrac{225}{12.56} = r^2$ divide both members by 12.56

$r^2 \approx 17.91$ carry out the division

$r \approx \pm\sqrt{17.91}$ take the square root of each member

$r \approx \pm 4.23$ use the table on page 464 and interpolate between $\sqrt{17}$ and $\sqrt{18}$ or extract the square root by one of the other methods

$r \approx 4.23$ cm   the negative value of r has no physical meaning

Check: $225 \overset{?}{=} 4(3.14)(4.23)^2$

$225 \approx 224.7$

(d) $c^2 = 11^2 + 25^2$

$= 121 + 625$ look up the squares of 11 and 25 in the table on page 464

$= 746$

$c = \pm\sqrt{746}$ take the square root of each member

$= \pm 27.31$ extract the square root of 746

Check: $(27.31)^2 \overset{?}{=} 11^2 + 25^2$

$745.8 \approx 746$

(e) $v^2 = v_0^2 + 2ad$

$v^2 = 60^2 + 2(-32)(18)$ substitute the given values

$= 3600 - 1152$ carry out the indicated sperations

$= 2448$

$v = \pm\sqrt{2448}$ take the square root of each member

$\approx \pm 49.48$ extract the square root of 2448

Check: $(49.48)^2 \overset{?}{=} 60^2 + (-32)(18)$

$$\overset{?}{=} 2448$$

$$2448.2 \overset{?}{=} 2448$$

10.2. Determine the nature of the solutions in the following quadratic equations:

(a) $p^2 - 9p = 0$ (c) $6x^2 - x - 2 = 0$ (e) $3s^2 + 5s + 2 = 0$

(b) $t^2 - 2t + 1 = 0$ (d) $2r^2 - 5r = -4$

(a) $a = 1$ $b = -9$ $c = 0$

    $b^2 - 4ac = 9^2 - (4)(1)(0)$ evaluate the discriminant

            $= 81$

Since $81 > 0$, the solutions are real and unequal.

(b) $a = 1$ $b = -2$ $c = 1$

    $b^2 - 4ac = (-2)^2 - 4(1)(1) = 0$ evaluate the discriminant

Since the discriminant is zero, the solutions are real and equal.

(c) $a = 6$ $b = -1$ $c = -2$

    $b^2 - 4ac = (-1)^2 - 4(6)(-2)$ evaluate the discriminant

            $= 49$

Since the discriminant is > 0, the solutions are real and unequal.

(d) $2r^2 - 5r + 4 = 0$ put the equation in standard form

    $a = 2$ $b = -5$ $c = 4$

    $b^2 - 4ac = (-5)^2 - 4(2)(4)$ evaluate the discriminant

            $= -7$

Since $-7 < 0$, the solutions are imaginary.

(e) $3s^2 + 5s + 2 = 0$ put the equation in standard form

    $a = 3$ $b = -5$ $c = 2$

    $b^2 - 4ac = (5)^2 - 4(3)(2) = 1$ evaluate the discriminant

Since the discriminant is > 0, the solutions are real and unequal.

10.3. Solve by factoring and check.

(a) $x^2 - 4x + 4 = 0$ (c) $y^2 + 7y + 12 = 0$ (e) $3r^2 = 10r + 8$

(b) $4t^2 - 25 = 0$ (d) $3b + 5b^2 = 2$

(a) $x^2 - 4x + 4 = 0$ left member is the square of the binomial $x - 2$

    $(x - 2)(x - 2) = 0$

         $x - 2 = 0$ a factor must be zero

           $x = 2$ transpose -2

The two roots are equal.

$Check$: $2^2 - 4(2) + 4 \overset{?}{=} 0$

$4 - 8 + 4 \overset{?}{=} 0$

$0 = 0$

(b) $4t^2 - 25 = 0$ left member is the difference of two squares

$(2t + 5)(2t - 5) = 0$ factor the given equation

$2t + 5 = 0$ a factor must be zero

$t = -5/2$ solve for t

$2t - 5 = 0$ solve for the second value of t

$t = 5/2$

$Check$: $4(-5/2)^2 - 25 \overset{?}{=} 0$ $4(5/2)^2 - 25 \overset{?}{=} 0$

$4(25/4) - 25 \overset{?}{=} 0$ $4(25/4) - 25 \overset{?}{=} 0$

$0 = 0$ $0 = 0$

(c) $y^2 + 7y + 12 = 0$

$(y + 4)(y + 3) = 0$ factor the left member

$y + 4 = 0$ a factor must be zero

$y = -4$

$y + 3 = 0$ solve for the two values of y

$y = -3$

$Check$: $(-4)^2 + 7(-4) + 12 \overset{?}{=} 0$ $(-3)^2 + 7(-3) + 12 \overset{?}{=} 0$

$16 - 28 + 12 \overset{?}{=} 0$ $9 - 21 + 12 \overset{?}{=} 0$

$0 = 0$ $0 = 0$

(d) $3b + 5b^2 = 2$

$5b^2 + 3b - 2 = 0$ put equation in standard form

$(5b - 2)(b + 1) = 0$ factor the left member

$5b - 2 = 0$ a factor must be zero

$b = 2/5$ solve for the two values of b

$b + 1 = 0$

$b = -1$

$Check$: $5(2/5)^2 + (3)(2/5) \overset{?}{=} 2$ $5(-1)^2 + 3(-1) \overset{?}{=} 2$

$20/25 + 6/5 \overset{?}{=} 2$ $5 - 3 \overset{?}{=} 2$

$20/25 + 30/25 \overset{?}{=} 2$ $2 = 2$

$50/25 \overset{?}{=} 2$

$2 = 2$

(e) $3r^2 = 10r + 8$

$3r^2 - 10r - 8 = 0$ put equation in standard form

$(3r + 2)(r - 4) = 0$ factor

$3r + 2 = 0$ a factor must be zero

$r = -2/3$

$r - 4 = 0$ solve for the two values of r

$r = 4$

Check: $3(-2/3)^2 \overset{?}{=} 10(-2/3) - 8$ $3(4^2) \overset{?}{=} 10(4) + 8$

$+12/9 \overset{?}{=} -20/3 + 8$ $48 = 48$

$4/3 = 4/3$

10.4. Use the quadratic formula to solve the following equations:

(a) $x^2 + x - 6 = 0$ (c) $2x^2 - x + 1 = 0$ (e) $3q^2 - 12 = -1$

(b) $3r^2 + 5r = 2$ (d) $-3t^2 - 2t + 5 = 0$

To use the formula, each equation must be in the standard form, $ax^2 + bx + c = 0$. Substitute the values of a, b, and c in the formula

$$x = \frac{-b \pm \sqrt{b^2 - 4ac}}{2a}$$

(a) $a = 1$ $b = 1$ $c = -6$

$x = \dfrac{-1 \pm \sqrt{1^2 - 4(1)(-6)}}{2(1)}$ substitute in the formula

$x_1 = \dfrac{-1 + \sqrt{25}}{2} = 2$

$x_2 = \dfrac{-1 - \sqrt{25}}{2} = -3$

Check: $2^2 + 2 - 6 \overset{?}{=} 0$ $(-3)^2 + (-3) - 6 \overset{?}{=} 0$

$4 + 2 - 6 \overset{?}{=} 0$ $9 - 3 - 6 \overset{?}{=} 0$

$0 = 0$ $0 = 0$

(b) $3r^2 + 5r - 2 = 0$ put equation in standard form

$a = 3$ $b = 5$ $c = -2$

$r = \dfrac{-5 \pm \sqrt{5^2 - 4(3)(-2)}}{2 \cdot 3}$ substitute in the formula

$= \dfrac{-5 \pm \sqrt{49}}{6}$

$r_1 = \dfrac{-5 + 7}{6} = 1/3$

$r_2 = \dfrac{-5 - 7}{6} = -2$

Check: $3(1/3)^2 + 5(1/3) \overset{?}{=} 2$ $3(-2)^2 + 5(-2) \overset{?}{=} 2$

$3/9 + 5/3 \overset{?}{=} 2$ $12 - 10 \overset{?}{=} 2$

$3/9 + 15/9 \overset{?}{=} 2$ $2 = 2$

$2 = 2$

(c) $2x^2 - x + 1 = 0$

$a = 2$ $b = -1$ $c = 1$

$$x = \frac{-(-1) \pm \sqrt{(-1)^2 - 4(2)(1)}}{(2)(2)}$$ substitute in the formula

$$= \frac{1 \pm \sqrt{1 - 8}}{4}$$

$$= \frac{1 \pm \sqrt{-7}}{4}$$

The solutions are *not* real numbers since $\sqrt{-7}$ is not a real number.

(d) $-3t^2 - 2t + 5 = 0$

$a = -3 \quad b = -2 \quad c = 5$

$$t = \frac{-(-2) \pm \sqrt{(-2)^2 - 4(-3)(5)}}{2(-3)}$$ substitute in the formula

$$= \frac{2 \pm \sqrt{4 + 60}}{-6}$$

$$= \frac{2 \pm \sqrt{64}}{-6}$$

$$t_1 = \frac{2 + 8}{-6} = -5/3$$

$$t_2 = \frac{2 - 8}{-6} = 1$$

Check: $-3(-5/3)^2 - 2(-5/3) + 5 \overset{?}{=} 0$ $-3(1)^2 - 2(1) + 5 \overset{?}{=} 0$

$-75/9 + 10/3 + 5 \overset{?}{=} 0$ $-3 - 2 + 5 \overset{?}{=} 0$

$0 = 0$ $0 = 0$

(e) $3q^2 - 12q + 1 = 0$ put equation in standard form

$a = 3 \quad b = -12 \quad c = 1$

$$q = \frac{-(-12) \pm \sqrt{(-12^2) - 4(3)(1)}}{2(3)}$$ substitute in the formula

$$q = \frac{12 \pm \sqrt{144 - 12}}{6}$$

$$q_1 = \frac{12 + \sqrt{132}}{6}$$ look up $\sqrt{132}$ in the table on page 464

$$= \frac{12 + 11.48}{6} = \frac{23.48}{6}$$

$$= 3.91$$

$$q_2 = \frac{12 - \sqrt{132}}{6}$$

$$= \frac{12 - 11.48}{6} = \frac{0.52}{6}$$

$$= 0.0867$$

Check: $3(3.91)^2 - 12(3.91) + 1 \overset{?}{=} 0$ $3(0.0867)^2 - 12(0.0867) + 1 \overset{?}{=} 0$

$3(15.28) - 46.92 + 1 \overset{?}{=} 0$ $0.0225 - 1.040 + 1 \approx 0$

$45.84 - 46.92 + 1 \overset{?}{=} 0$ $-0.0175 \approx 0$

$-0.08 \approx 0$

10.5. If a 40-cm tube closed at one end is thrust, open and downward, 10 cm into mercury, the pressure of air inside the tube depresses the mercury level by h cm according to the equation $h^2 + 106h - 760 = 0$.   Solve for h in the equation.

In this equation $a = 1$, $b = 106$, $c = -760$.   Application of the quadratic formula gives:

$$h = \frac{-106 \pm \sqrt{(+106)^2 - 4(1)(-760)}}{2} = \frac{-106 \pm \sqrt{11,236 + 3040}}{2}$$

$$= \frac{-106 \pm \sqrt{14,276}}{2} \approx \frac{-106 \pm 119.5}{2}$$

$$h \approx \frac{13.5}{2} \text{ cm} \approx 6.75 \text{ cm} \quad \text{and} \quad h \approx \frac{-225.5}{2} \text{ cm} \approx -112.75 \text{ cm}$$

Check: $(6.75)^2 + 106(6.75) - 760 \overset{?}{=} 0$

$$45.6 + 714 - 760 \overset{?}{=} 0$$

$$759.6 - 760 \approx 0$$

In many physical science problems one of the solutions is frequently discarded because it has no physical meaning.   In the above problem, $-225.5/2$ or -112.75 cm could not be a solution since the entire tube is only 40 cm long.

10.6. Solve and check the equation: $\dfrac{50}{d^2} = \dfrac{150}{(2-d)^2}$.

$\dfrac{1}{d^2} = \dfrac{3}{(2-d)^2}$	divide both sides by 50
$(2-d)^2 = 3d^2$	product of means equals product of extremes
$4 - 4d + d^2 = 3d^2$	remove parentheses
$4 - 4d - 2d^2 = 0$	transpose $3d^2$ to left side
$2 - 2d - d^2 = 0$	divide both members by 2
$-d^2 - 2d + 2 = 0$	put in standard form

$$d = \frac{-(-2) \pm \sqrt{(-2)^2 - (4)(-1)(2)}}{2(-1)} \qquad \text{use quadratic formula}$$

$$= \frac{2 \pm \sqrt{12}}{-2} = \frac{2 \pm 3.464}{-2}$$

$$d = +0.732 \quad \text{and} \quad d = -2.732.$$

Check: $\dfrac{50}{(0.732)} \overset{?}{=} \dfrac{150}{(2 - 0.732)^2}$ and $\dfrac{50}{(-2.732)^2} \overset{?}{=} \dfrac{150}{(4.732)^2}$

$$\frac{50}{0.537} \overset{?}{=} \frac{150}{1.61} \qquad\qquad\qquad \frac{50}{7.46} \overset{?}{=} \frac{150}{22.4}$$

$$93.2 = 93.2 \qquad\qquad\qquad\qquad 6.70 = 6.70$$

10.7. A problem on the reflection of light leads to the equation:

$$10 = \frac{(n-1)^2}{(n+1)^2} \, 100$$

where n is the index of refraction relative to air.   Determine n.

$$10(n+1)^2 = (n-1)^2(100)$$ simplify the equation

$$10(n^2 + 2n + 1) = (n^2 - 2n + 1)100$$ expand the two binomials

$$10n^2 + 20n + 10 = 100n^2 - 200n + 100$$ simplify

$$90n^2 - 220n + 90 = 0$$

$$9n^2 - 22n + 9 = 0$$

$$n = \frac{22 \pm \sqrt{(22)^2 - 4(9)(9)}}{(2)(9)}$$ substitute in the quadratic formula

$$= \frac{22 \pm \sqrt{484 - 324}}{18}$$

$$= \frac{22 \pm \sqrt{160}}{18} = \frac{22 \pm 12.6}{18}$$

$$n = \frac{34.6}{18} \text{ and } \frac{9.4}{18}, \text{ or } n = 1.92 \text{ and } n = 0.52.$$

Mathematically both solutions are correct; but the solution 0.52 has no physical meaning, as the index of refraction cannot be less than 1.

Check: $10 \overset{?}{=} \dfrac{(1.92 - 1)^2}{(1.92 + 1)^2}(100) = \dfrac{(0.92)^2}{(2.92)^2}(100)$

$10 \overset{?}{=} \dfrac{84.6}{8.53}$

$10 \approx 9.92$

10.8.   The distance d in feet covered by a freely falling object is given by the equation $d = 16t^2$, where t is the time of fall in seconds.   How long will it take for an object to fall 9 feet?

$$9 = 16t^2$$ substitute 9 for s

$$\frac{9}{16} = t^2$$ divide both sides by 16

$$\pm\sqrt{9/16} = t$$ take the square root of both sides

$$\pm 3/4 = t$$ since 9 and 16 are perfect squares, take the square root of each and divide

The time of fall is 3/4 second.   The negative root has no physical meaning.

10.9.   In the equation $a = \dfrac{v^2}{R}$, if $a = 4$ and $R = 500$, calculate v.

$$4 = \frac{v^2}{500}$$ substitute 4 for a and 500 for R

$$2000 = v^2$$ multiply both sides by 500

$$\pm\sqrt{2000} = v$$ take the square root of both sides

$$44.7 \approx v$$

Check: $4 \overset{?}{=} \dfrac{(44.7)^2}{500} = \dfrac{1998}{500}$

$4 \approx 3.996$

10.10. Solve for R: $\dfrac{100}{250} = \dfrac{(6400)^2}{R^2}$.

$$100R^2 = (250)(6400)^2 \qquad \text{cross-multiply}$$

$$R^2 = \frac{(250)(6400)^2}{100} \qquad \text{divide both sides by 100 and multiply}$$

$$R^2 = (2.5)(6.4 \times 10^3)^2 \qquad \text{use exponential notation}$$

$$R = \pm\sqrt{(2.5)(6.4 \times 10^3)^2} \qquad \text{take the square root of both sides}$$

$$R = \pm\sqrt{2.5}\sqrt{(6.4 \times 10^3)^2} \qquad \text{take the square root of each factor}$$

$$R = \pm(1.58)(6.4 \times 10^3)$$

$$R = \pm 1.01 \times 10^4$$

Check: $\quad \dfrac{100}{250} \overset{?}{=} \dfrac{(6400)^2}{(1.01 \times 10^4)^2}$

$\qquad\qquad 0.40 \approx 0.402$

10.11. Solve for t: $(t + 0.80)^2 = 2.5t + t^2$.

$$t^2 + 1.6t + 0.64 = 2.5t + t^2 \qquad \text{expand the binomial}$$

$$t^2 - t^2 + 1.6t - 2.5t + 0.64 = 0 \qquad \text{transpose}$$

$$-0.9t + 0.64 = 0 \qquad \text{simplify}$$

$$t = \frac{0.64}{0.9} \qquad \text{solve for } t$$

$$t = 0.71$$

Check: $\quad (0.71 + 0.80)^2 \overset{?}{=} 2.5(0.71) + (0.71)^2$

$\qquad\qquad 2.28 \overset{?}{=} 1.775 + 0.504$

$\qquad\qquad 2.28 \approx 2.279$

10.12. In the equation $v^2 = v_0^2 + 2as$, $v = 50$; $a = -10$; $s = 150$.   Compare the value of v_0.

$$50^2 = v_0^2 + 2(-10)(150) \qquad \text{substitute the given values}$$

$$2500 = v_0^2 - 3000 \qquad \text{simplify}$$

$$v_0^2 = 5500 \qquad \text{solve for } v$$

$$v_0 = \pm\sqrt{5500} \qquad \text{take the square root of each side}$$

$$\approx 74.16$$

$$\approx 74.2$$

Check: $\quad (50)^2 \overset{?}{=} (74.2)^2 + 2(-10)150$

$\qquad\qquad 2500 \overset{?}{=} 5506 - 3000$

$\qquad\qquad 2500 \approx 2506.$

10.13. Solve: $\dfrac{0.20}{(0.50 - d)^2} = \dfrac{0.5}{d^2}$.

$$0.20d^2 = 0.5(0.25 - d + d^2) \qquad \text{cross-multiply and expand the binomial}$$
$$0.20d^2 = 0.125 - 0.5d + 0.5d^2 \qquad \text{simplify}$$
$$(0.20d^2 - 0.5d^2) + 0.5d - 0.125 = 0 \qquad \text{put equation in standard form}$$
$$-0.3d^2 + 0.5d - 0.125 = 0$$
$$-30d^2 + 50d - 12.5 = 0 \qquad \text{multiply by 100}$$
$$a = -30 \quad b = +50 \quad c = -12.5$$
$$d = \frac{-(+50) \pm \sqrt{(+50)^2 - 4(-30)(-12.5)}}{2(-30)} \qquad \text{use quadratic formula}$$
$$d_1 = 0.306 \quad \text{and} \quad d_2 = 1.360 \qquad \text{simplify}$$

Check: $\dfrac{0.20}{(0.50 - 0.306)^2} \overset{?}{=} \dfrac{0.5}{(0.306)^2} \qquad 5.31 \approx 5.34 \qquad \dfrac{0.20}{(0.50 - 1.360)^2} \overset{?}{=} \dfrac{0.5}{(1.360)^2} \qquad 0.270 \approx 0.270$

10.14. Solve the following equations for the indicated symbol:

(a) $S = (1/2)at^2$ for t

(b) $2\pi fL = \dfrac{1}{2\pi fC}$, for f

(c) $hf = W + (1/2)mv^2$, for v

(d) $m = m_0/\sqrt{1 - v^2/c^2}$, for v

(e) $A = \pi(R_1^2 - R_2^2)$, for R_2

(a) $S = (1/2)at^2$

$$2S = at^2 \qquad \text{multiply both sides by 2}$$
$$2S/a = t^2 \qquad \text{divide both sides by } a$$
$$\pm\sqrt{2S/a} = t \qquad \text{extract the square root of each side}$$
$$t = \pm\sqrt{2S/a}$$

(b) $2\pi fL = \dfrac{1}{2\pi fC}$

$$4\pi^2 f^2 LC = 1 \qquad \text{multiply both sides by } 2\pi fC$$
$$f^2 = \frac{1}{4\pi^2 LC} \qquad \text{divide both sides by } 4\pi^2 LC$$
$$f = \pm\sqrt{\frac{1}{4\pi^2 LC}} \qquad \text{take the square root of each side}$$
$$f = \pm\frac{1}{2\pi}\frac{1}{\sqrt{LC}} \qquad \text{take } \frac{1}{2\pi} \text{ outside the radical sign since } \sqrt{\frac{1}{4\pi^2}} = \frac{1}{2\pi}$$

(c) $hf = W + (1/2)mv^2$

$$hf - W = (1/2)mv^2 \qquad \text{transpose } W$$
$$\frac{2(hf - W)}{m} = v^2 \qquad \text{multiply both sides by 2 and divide by } m$$
$$\pm\sqrt{\frac{2(hf - W)}{m}} = v \qquad \text{extract the square root of each side}$$
$$v = \pm\sqrt{\frac{2(hf - W)}{m}}$$

(d) $m = m_0 \Big/ \sqrt{1 - \dfrac{v^2}{c^2}}$

$m\sqrt{1 - \dfrac{v^2}{c^2}} = m_0$ multiply both sides by $\sqrt{1 - \dfrac{v^2}{c^2}}$

$\sqrt{1 - \dfrac{v^2}{c^2}} = \dfrac{m_0}{m}$ divide both sides by m

$1 - \dfrac{v^2}{c^2} = \dfrac{m_0^2}{m^2}$ square both sides

$\dfrac{-v^2}{c^2} = \dfrac{m_0^2}{m^2} - 1$ transpose 1

$\dfrac{v^2}{c^2} = 1 - \dfrac{m_0^2}{m^2}$ multiply both sides by -1

$v^2 = c^2\left(1 - \dfrac{m_0^2}{m^2}\right)$ multiply both sides by c^2

$v = \pm\sqrt{c^2\left(1 - \dfrac{m_0^2}{m^2}\right)} = \pm c\sqrt{1 - \dfrac{m_0^2}{m^2}}$ extract the square root of each side

(e) $A = \pi(R_1^2 - R_2^2)$

$\dfrac{A}{\pi} = R_1^2 - R_2^2$ divide both sides by π

$\dfrac{A}{\pi} - R_1^2 = -R_2^2$ transpose R_1^2

$R_1^2 - \dfrac{A}{\pi} = R_2^2$ multiply both sides by -1

$\pm\sqrt{R_1^2 - \dfrac{A}{\pi}} = R_2$ extract the square root of each side

$R_2 = \pm\sqrt{R_1^2 - \dfrac{A}{\pi}}$

Note: Since the above five formulas deal with intrinsically positive physical quantities, the negative solutions have no physical meaning in any case.

10.15. Using the table of square roots and interpolation, determine the square roots of the following to three decimal places: (*a*) 36.4, (*b*) 118.2, (*c*) 80.35, (*d*) 7.9, (*e*) 12.08.

(*a*)
$$\sqrt{37} \approx 6.083$$
$$\sqrt{36} = 6.000$$
$$\text{difference} = 0.083$$
$$(0.4)(0.083) = 0.0332$$
$$\sqrt{36.4} \approx 6.000 + 0.0332$$
$$\approx 6.033$$

Check: $6.033^2 \overset{?}{=} 36.4$
$$36.4 \approx 36.4$$

(*b*)
$$\sqrt{119} = 10.909$$
$$\sqrt{118} = 10.863$$
$$\text{difference} = 0.046$$
$$(0.2)(0.046) = 0.0092$$
$$\sqrt{118.2} \approx 10.863 + 0.0092$$
$$\approx 10.872$$

Check: $(10.872)^2 \overset{?}{=} 118$
$$118.2 \approx 118$$

(c)
$$\sqrt{81} = 9.000$$
$$\sqrt{80} = 8.944$$
$$\text{difference} = 0.056$$
$$(0.35)(0.056) = 0.0196$$
$$\sqrt{80.35} = 8.944 + 0.0196$$
$$\approx 8.964$$

Check $(8.964)^2 \overset{?}{=} 80.35$
$$80.35 \approx 80.35$$

(e)
$$\sqrt{13} = 3.606$$
$$\sqrt{12} = 3.464$$
$$\text{difference} = 0.142$$
$$(0.08)(0.142) = 0.0114$$
$$\sqrt{12.08} = 3.464 + 0.0114$$
$$\approx 3.475$$

Check: $(3.475)^2 \overset{?}{=} 12.08$ $12.08 = 12.08.$

(d)
$$\sqrt{8} = 2.828$$
$$\sqrt{7} = 2.646$$
$$\text{difference} = 0.182$$
$$(0.9)(0.182) = 0.1638$$
$$\sqrt{7.9} = 2.646 + 0.1638$$
$$\approx 2.810$$

Check: $(2.810)^2 \overset{?}{=} 7.9$
$$7.896 \approx 7.9$$

10.16. Extract the square root of the following numbers by the iterative method to at least two significant figures: (a) 96.8,   (b) 165,   (c) 5.73,   (d) 18.3,   (e) 0.00792.

(a) The nearest approximate square root of 96.8 is 10 because 10^2 is closer to 96 than 9^2.   Divide 96.8 by 10.0:

$$96.8 \div 10.0 = 9.68$$

Calculate the arithmetic average of 10 and 9.68:

$$\frac{10 + 9.68}{2} = \frac{19.68}{2} = 9.84$$

Thus, $\sqrt{96.8} \approx 9.84$.

Check: $(9.84)^2 \overset{?}{=} 96.8$ $96.825 \approx 96.8.$

(b) The nearest approximate square root of 165 is 13.   Thus,

$$165 \div 13 = 12.69 \quad \text{and} \quad \frac{13 + 12.69}{2} = \frac{25.69}{2} = 12.84$$
$$\sqrt{165} \approx 12.84$$

Check: $(12.84)^2 \overset{?}{=} 165$ $164.99 \approx 165.$

(c) The nearest approximate square root of 5.73 is 2.   Thus,

$$5.73 \div 2 = 2.86 \quad \text{and} \quad \frac{2 + 2.86}{2} = \frac{4.86}{2} = 2.43$$
$$\sqrt{5.73} \approx 2.43$$

Check: $(2.43)^2 \overset{?}{=} 5.73$ $5.90 \approx 5.73.$

(d) The nearest approximate square root of 18.3 is 4.   Thus,

$$18.3 \div 4 = 4.575 \quad \text{and} \quad \frac{4 + 4.575}{2} = \frac{8.575}{2} \approx 4.288$$
$$\sqrt{18.3} \approx 4.29$$

Check: $(4.29)^2 \overset{?}{=} 18.3$ $18.4 \approx 18.3.$

(*e*) Round off the nonzero digits to one digit. Pair off sets of two digits to the right of the decimal point: $0.00792 \cong 0.008000$.

   The first digit of the square root will be 0. Estimating the square root of 80, $\sqrt{80} \approx 9$, since $9 \cdot 9 = 81$.

   Using the estimated square root as the divisor, $80 \div 9 = 8.889$.

   Averaging 9.000 and 8.889, we get $\dfrac{9.000 + 8.889}{2} = 8.944$.

   Therefore the approximate square root of 0.00792 is 0.0894.

Check: $(0.0894)^2 = 0.00798$, which agrees in the first two digits with the given number. Higher precision could be obtained by using 8.9 or 8.94 instead of 9 as the divisor.

10.17. Extract the square root of the following by the traditional method (to three nonzero digits):

(*a*) 2.5761,   (*b*) 0.006950,   (*c*) 115.83

(*a*)
$$
\begin{array}{r}
1.\,6\ 0\ 5 \\
\sqrt{2.576100}
\end{array}
$$

$$
\begin{array}{r}
1
\end{array}
$$

$$
\begin{array}{rr}
 & 1\ 57 \qquad\qquad 1 \times 20 = 20 \\
 & 1\ 56 \qquad\qquad\quad +6 \\
\hline
160 \times 20 = 3200 \qquad\qquad & 161\ 00 \qquad\qquad\quad 26 \\
+5 \qquad\qquad & 160\ 25 \\
\hline
3205 \\
\sqrt{2.5761} \approx 1.605
\end{array}
$$

Check: $(1.605)^2 \stackrel{?}{=} 2.5761$ $2.5760 \approx 2.5761$.

(*b*)
$$
\begin{array}{r}
0.\,0\ 8\ 3\ 3 \\
\sqrt{0.00695000}
\end{array}
$$

$$
\begin{array}{rr}
 & 64 \\
\hline
8 \times 20 = 160 & 550 \\
+3 & 489 \\
\hline
163 & 6100 \\
83 \times 20 = 1660 & 4989 \\
\hline
+3 & 1101 \\
\hline
1663
\end{array}
$$

$\sqrt{0.00690} \approx 0.0833$

Check: $(0.0833)^2 \stackrel{?}{=} 0.00690$ $0.00694 \approx 0.00690$.

(*c*)
$$
\begin{array}{r}
1\,0\,.7\ 6 \\
\sqrt{115.8300}
\end{array}
$$

$$
\begin{array}{rr}
 & 1 \\
\hline
1 \times 20 = \ \ 20 & 15\ 83 \\
10 \times 20 = 200 & 14\ 49 \\
\hline
+7 & 1\ 3400 \qquad 107 \times 20 = 2140 \\
\hline
207 & 1\ 2876 \qquad\qquad\quad +6 \\
 & \qquad\qquad\qquad\qquad 2146
\end{array}
$$

$\sqrt{115.83} \approx 10.8$

Check: $(10.8)^2 \stackrel{?}{=} 115.83$ $116.6 \approx 115.8$.

10.18. Use logarithms to extract the square root of: (*a*) 5882,   (*b*) 60.45,   (*c*) 0.09604.

(*a*) $\log \sqrt{5882} = (1/2) \log 5882$.   Use the table of logarithms.

$$\log 5882 = 3.7695$$

$$(1/2) \log 5882 = \frac{3.7695}{2} = 1.8848$$

$$\sqrt{5882} = \text{antilog } 1.8848 \approx 76.7$$

Thus, $\sqrt{5882} \approx 76.7$.

Check: $(76.7)^2 \overset{?}{=} 5882$ $5883 \approx 5882$.

(*b*) $\log \sqrt{60.45} = (1/2) \log 60.45$.

$$(1/2) \log 60.45 = \frac{1.7814}{2} = 0.8907$$

$$\sqrt{60.45} = \text{antilog } 0.8907 = 7.775 \approx 7.78$$

Check: $(7.78)^2 \overset{?}{=} 60.45$ $60.5 \approx 60.45$.

(*c*) $\log \sqrt{0.09604} = (1/2) \log 0.09604$.

$$(1/2) \log 0.09604 = \frac{8.9852 - 10}{2}$$

To make it easier to evaluate the antilogarithm of the quotient, add 10 to 8 and subtract 10 from -10. Thus,

$$(1/2) \log 0.09604 = \frac{18.9825 - 20}{2} = 9.4912 - 10$$

$$\sqrt{0.09604} = \text{antilog } (9.4912 - 10) = 0.3098 \approx 0.31$$

Check: $(0.31)^2 \overset{?}{=} 0.09604$ $0.0961 \approx 0.09604$.

10.19. Solve the equation $-x^2 + 2x + 7 = 0$ using a graphing calculator.

Using an HP-38G graphing calculator you may proceed as indicated below.

(1)   Define the function.

{LIB} <u>FUNCTION</u> {ENTER}

{−x}{X,T,θ}{x^y}2{+}2{*}{X,T,θ}{+}7{ENTER}

(2)   Plot the graph.

{PLOT}

(3)   Find the *x*-intercepts.

<u>MENU</u> <u>FCN</u> <u>Root</u> {ENTER}

The intercepts are $x_1 = -1.82$ and $x_2 = -3.82$.   These values are the corresponding solutions of the equation $-x^2 - 2x + 7 = 0$.   To obtain the second solution or root it may be necessary to move the crosshairs along the curve so that it is closer to the other *x*-intercept.   For this particular example, we adjusted the viewing window and set the YRNG range from -2 to 10.

Figure 10-6 shows the graph of the function $y = -x^2 + 2x + 7$.   Notice that the parabola opens downward since the sign of the coefficient of x^2 is negative.

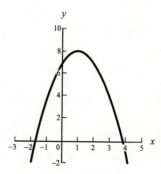

Fig. 10-6

10.20. Solve the equation $2x^2 - 8x + 5 = 0$ using a graphing calculator.

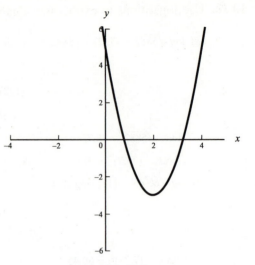

Using a TI-82 graphing calculator you may proceed as follows.

(1) Define the function.

   {Y=}

   2{×}{X,T,θ}{^}2{−}8{×}{X,T,θ}{+}5{ENTER}

(2) Draw the graph.

   {GRAPH}

(3) Find the x-intercept.

   {2nd}{CALC}root{ENTER}

The intercepts are $x_1 = 3.22$ and $x_2 = 0.8$.

Fig. 10-7

   Use the directional keys to select a lower bound, an upper bound and an "educated guess" of the intersect. Repeat this process for every intersect.

   Figure 10-7 shows the graph of the function $y = 2x^2 - 8x + 5$.

10.21. Solve the equation $2x^2 + 3x + 5 = 0$ using a graphing calculator.

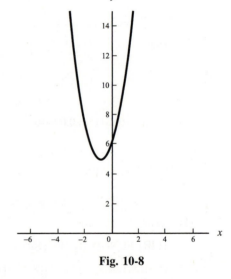

   Using an HP-38G graphing calculator you may proceed as indicated below.

(1) Define the function.

   {LIB} FUNCTION {ENTER}

   2{*}{X,T,θ}{x^y}2{+}3{*}{X,T,θ}{+}5{ENTER}

(2) Plot the graph.

   {PLOT}

Fig. 10-8

(3) Find the x-intercepts.

   There are no x-intercepts. The roots of the equation are imaginary. Notice that since there are no intercepts only an extremum (lowest value) is displayed. Figure 10-8 shows that the graph of the function $y = 2x^2 + 3x + 5$ does not intercept the x-axis.

Supplementary Problems

The answers to this set of problems are at the end of this chapter.

10.22. Determine the nature of the solutions of each of the following quadratic equations:

   (a) $2x^2 - 3x + 4 = 0$ (c) $4y^2 - 8y + 16 = 0$ (e) $5v^2 + 6v - 4 = 0$

   (b) $5m^2 + 10m - 1 = 0$ (d) $t^2 + 2t - 5 = 0$

10.23. Solve each of the following equations by factoring and check the solutions:

(a) $2x^2 + 3x - 2 = 0$ (c) $y^2 - y - 2 = 0$ (e) $4r^2 + 20r + 25 = 0$

(b) $3x^2 + 5x - 2 = 0$ (d) $4t^2 - 9 = 0$

10.24. Use the quadratic formula to solve each of the equations in Problem 10.23.

10.25. Either factor or use the quadratic formula to solve the following equations, if the solutions are real numbers:

(a) $3x^2 - 4x + 1 = 0$ (c) $5y^2 + 4y - 5 = 0$ (e) $r^2 - 4r + 2 = 0$

(b) $4t^2 - 16 = 0$ (d) $3x^2 + 3x - 1 = 0$

10.26. Find the real solutions of the following equations and check:

(a) $3p^2 - p + 1 = p^2 + 2p - 3$ (c) $4 - 2m + 2m^2 = 2m + m^2 + 2$ (e) $\dfrac{2v - 4}{v - 2} = \dfrac{v - 4}{v}$

(b) $4x^2 + 2x - 5 = 2x^2 - 2x - 3$ (d) $(f + 3)^2 = 225 - f^2$

10.27. Use the quadratic formula to solve the following equations for x:

(a) $x^2 - 2ax + b = 0$ (c) $rx^2 - x + s = 0$ (e) $s = x + kx^2$

(b) $2x^2 - \sqrt{3b}x + 2a = 0$ (d) $cx^2 + 2dx = 3c$

10.28. Solve the following equations and check the solutions:

(a) $n^2 + 2n + 1 = 11n^2 - 22n + 11$ (c) $525 = 100L + L^2$ (e) $81 = 100x + (100x)x + 100$

(b) $25(15 - d)^2 = 100d$ (d) $2/(20 - d)^2 = 6/d^2$

10.29. The final velocity v of a freely falling body from a height h m near the surface of the earth is given by $v = \sqrt{19.6h}$ m/sec (neglect friction).   Calculate v for $h = 2.50$ m.

10.30. The time required for a volume of gas at constant pressure and temperature to diffuse through a small opening is given by $t = 10.75\sqrt{32/18}$ min.   Calculate the diffusion time for the gas.

10.31. The wavelength of a resonant circuit is $\lambda = 1885\sqrt{0.000125}$ m.   Calculate the wavelength.

10.32. Solve the following for the unknown and check.   The negative solution has no physical meaning.

(a) $1.5t^2 = 29.4$

(b) $a = \dfrac{v^2}{R}$, where $a = 4$ and $R = 1000$

(c) $v^2 = v_0^2 + 2aS$, where $v = 0$; $v_0 = 44$; $S = -20$

(d) $480 = 64t + 16t^2$

(e) $h_1 + \dfrac{v_1^2}{2g} = \dfrac{v_2^2}{2g}$, where $h_1 = 60$; $g = 32.2$; $v_2 = 63$

10.33. Solve and check.   Discard the negative root.

(a) $2 = \dfrac{d^2}{(20 - d)^2}$ (b) $\dfrac{3}{4} = \dfrac{4000^2}{R^2}$

10.34. Solve for r_2 and for λ in (a) and (b) respectively:

   (a) $2GMr_1 = 2GMr_2 - r_1r_2v^2$ (b) $Ve = \dfrac{1}{2}m\left(\dfrac{h}{m\lambda}\right)^2$

10.35. Calculate m in the equation

$$m = \frac{m_0}{\sqrt{1 - v^2/c^2}}$$

   where $m_0 = 1$; $v = (1.5)(10^8)$; $c = (3)(10^8)$.

10.36. In the equation of Problem 10.35, what should be the value of v if $m = 2m_0$?

10.37. Determine and check the approximate square root of the following by using the table on page 464 and interpolating: (a) 34.75,   (b) 129.4,   (c) 0.0056,   (d) 8.25,   (e) 48,200.

10.38. Repeat Problem 10.37 by the iterative method.

10.39. Repeat Problem 10.37 by using logarithms.

10.40. Determine the approximate square root of each of the following by the traditional method correct to three significant figures.   In each case, check your result.

   (a) 342.35   (b) 654,395.28   (c) 0.005643   (d) 48,657.36   (e) 12.047

10.41. The graphs of figures 10-9 through 10-11 correspond to quadratic functions of the form $y = ax^2 + bx + c$ with $a \neq 0$.   What can be said about the roots of their corresponding quadratic equations $ax^2 + bx + c = 0$ where $a \neq 0$?

(a)

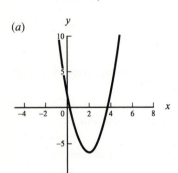

Fig. 10-9

(b)

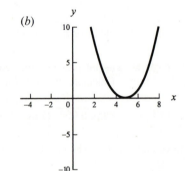

Fig. 10-10

(c)

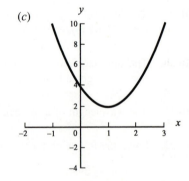

Fig. 10-11

10.42. Solve the equation $-3x^2 - 2x + 1 = 0$ with the help of a graphing calculator.

10.43. What are the sum and product of the roots of $7x^2 - 3x - 2 = 0$?   Use a graphing calculator to find the roots.   Draw the graph of the function $y = 7x^2 - 3x - 2$.

10.44. Does the graph of the function $y = 3x^2 - x - 2 = 0$ open up or down?   Graph this function using a graphing calculator.   In the function given above, the number 2 is called the independent term, what does this term represent?

Answers to Supplementary Problems

10.22. (*a*) not real   (*b*) real, irrational, unequal   (*c*) not real   (*d*) real, irrational, unequal

(*e*) real, irrational, unequal

10.23. (*a*) $1/2, -2$ (*b*) $1/3, -2$ (*c*) $-1, 2$ (*d*) $\pm 3/2$ (*e*) $-5/2, -5/2$

10.24. Same as Problem 10.23.

10.25. (*a*) $1/3, 1$ (*b*) ± 2 (*c*) $(-2 \pm \sqrt{29})/5$ (*d*) $(-3 \pm \sqrt{21})/6$ (*e*) $2 \pm \sqrt{2}$

10.26. (*a*) not real   (*b*) $-1 \pm \sqrt{2}$ (*c*) $2 \pm \sqrt{2}$ (*d*) $9, -12$ (*e*) $2, -4$

10.27. (*a*) $a \pm \sqrt{a^2 - b}$ (*c*) $(1 \pm \sqrt{1 - 4rs})/2r$ (*e*) $(-1 \pm \sqrt{1 + 4ks})/2k$

(*b*) $(\sqrt{3b} \pm \sqrt{3b - 16a})/4$ (*d*) $(-d \pm \sqrt{d^2 + 3c^2})/c$

10.28. (*a*) $(6 \pm \sqrt{11})/5$ (*b*) $9, 25$ (*c*) $5, -105$ (*d*) $30 \pm 10\sqrt{3}$ (*e*) $-1/2 \pm \sqrt{6}/10$

10.29. 7.0 m/sec

10.30. 14 min 20 sec

10.31. 21.1 m

10.32. (*a*) $t = 4.43$ (*b*) $v = 63.2$ (*c*) $a = 48.4$ (*d*) $t = 3.8$ (*e*) $v_1 = 10.25$

10.33. (*a*) $d = 68.28, 11.72$ (*b*) 4620

10.34. (*a*) $r_2 = r_1 \left/ \left(1 - \dfrac{r_1 v^2}{2GM} \right) \right.$ (*b*) $\lambda = h/(\sqrt{2mVe})$

10.35. $m \approx 1.15$

10.36. $v \approx (2.6)(10^8)$

10.37. (*a*) 5.89   (*b*) 11.38   (*c*) 0.075   (*d*) 2.87   (*e*) 220

10.38. Same as for Problem 10.37.

10.39. Same as for Problem 10.38.

10.40. (*a*) 18.5   (*b*) 809   (*c*) 0.0751   (*d*) 221   (*e*) 3.47

10.41. (*a*) two different solutions, both real

(*b*) two identical solutions, both real

(*c*) no real solution

10.42. Using a TI-82 graphing calculator:

(1) Define the function.

{Y =}

{(−)}3{×}{X,T,θ}{^}2{−}2{×}{X,T,θ}{+}1{ENTER}

(2) Plot the graph.

{GRAPH}

(3) Find the x-intercepts.

There are two intercepts $x_1 = -1$ and $x_2 = 0.333$. It may be necessary to use the zoom key to help t calculator locate the intercepts. Figure 10-12 shows the graph of the function $y = -3x^2 - 2x +$

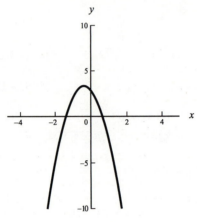

Fig. 10-12

10.43. sum ≈ 0.43, product ≈ -0.29. Figure 10-13 shows the graph of the corresponding equation.

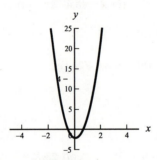

Fig. 10-13

10.44. The parabola opens upward since the coefficient of x^2 is positive. The independent term represents the intercept; that is, the point where the graph crosses the y-axis. Figure 10-14 shows the graph of this functic

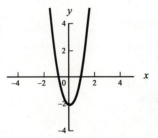

Fig. 10-14

Chapter 11

Essentials of Plane Geometry

Basic Concepts

11.1 TERMS

Point, line, and *plane* are the three undefined terms in Euclidean geometry.

Euclidean space may be identified with the set (collection) of all physical points.

Euclidean geometry is a study of the space properties of idealized physical objects and their interrelations.

A *geometrical solid* is a limited enclosed portion of Euclidean space.

Solid geometry is a study of the spatial properties of idealized geometrical figures such as cubes, cylinders, and spheres.

Surfaces are the boundaries of solids.   If a straight line joining any two points in a surface lies wholly within it, the surface is a *plane surface* or a *plane*.   A surface which is not plane is called *curved*.

Plane geometry deals with the properties of figures in a plane: angles, triangles, circles, parallelograms, etc.

An *axiom* or *postulate* is a general statement accepted as true without proof.

A *theorem* is a statement which requires proof.

11.2 LINES

Certain familiar sets of points in a plane are called *lines*.   It will be understood that the term "line" will mean a straight line, extending indefinitely far in two directions.

A *segment* is a part of a line included between two points on the line.   A segment may be defined more precisely as the set of points of a line between and including two points known as its *endpoints*.

A *ray* is a portion of a line bounded in one direction by a point but extending indefinitely in the other direction.   Using the set terminology, a ray is a set of points consisting of an endpoint and all the points of a line in one direction from the endpoint.

A line, a segment, and a ray are represented graphically in Fig. 11-1.

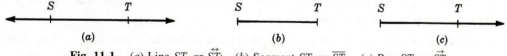

Fig. 11-1.   (*a*) Line *ST*, or $\overleftrightarrow{ST}$;   (*b*) Segment *ST*, or $\overline{ST}$;   (*c*) Ray *ST*, or $\overrightarrow{ST}$

11.3 SYMBOLS

A number of symbols used frequently in geometry are reproduced below.

$\angle$	angle	$''$	second of arc
$\angle\!\!\!s$	angles	$\overline{AB}$	line segment AB
$\parallel$	parallel; parallel to	$\overset{\frown}{CD}$	arc CD
$\parallel\!s$	parallels	$\circ$	degree
$\perp$	perpendicular; perpendicular to	$\sim$	similar
$\perp\!s$	perpendiculars	$\cong$	congruent
$\triangle$	triangle	$\therefore$	therefore
$\triangle\!\!\!s$	triangles	$\neq$	not equal
$\square$	rectangle	$\approx$ or $\doteq$	approximately equal
$\square$	parallelogram	$\equiv$	identically equal
$\odot$	circle	$>$	greater than
$\circledS$	circles	$<$	less than
$'$	minute of arc	$\llcorner$	right angle

11.4 SELECTED GEOMETRICAL FACTS

The following facts are used frequently in plane geometry.

1. The shortest distance between two points is the length of the line segment joining them.

2. Only one straight line segment can be drawn between two points.

3. A straight line segment can be extended indefinitely in both directions.

4. A geometric figure can be moved in the plane without any effect on its size or shape.

5. At any point in a straight line only one perpendicular to the line can be erected in a plane containing the given line.

6. Two straight lines in the same plane intersect in not more than one point.

7. Two straight lines in the same plane are either parallel or intersecting.

8. Through a given external point only one line can be drawn parallel to a given line.

9. Two lines parallel to a third line are parallel to each other.

10. The shortest distance from a point to a line is the length of the perpendicular drawn from the point to the line.

Angles

11.5 TYPES

An *angle* is a plane geometrical figure formed by two rays extending from a common point of the plane. In Fig. 11-2, the rays *AB* and *AC* are called *sides* of the angle *BAC*; point *A* is the *vertex*. Ray *AC* is the *right side* of the angle.

If ray *AB* coincides with ray *AC*, then the angle *BAC* is said to be a zero angle. One complete revolution by a ray describes an angle of 360 degrees (360°). If side *AB* is one quarter of a revolution from side *AC*, then *AB* is perpendicular to *AC* and angle *BAC* in Fig. 11-3 is a *right angle*, or a 90° angle. If both sides of an angle are oppositely directed along a straight line, the angle is called a *straight angle*, or a 180° angle, for example $\angle CAD$.

An angle of smaller size than a right angle is called *acute*, for example $\angle BAC$ in Fig. 11-4. An angle larger than a right angle is called *obtuse*, for example $\angle CAD$. Two angles are said to be *adjacent* if they have a common vertex and a common side, for example $\angle BAC$ and $\angle BAD$.

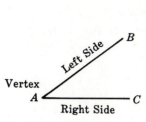

Fig. 11-2. Components of
an angle

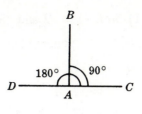

Fig. 11-3. Right and
straight angles

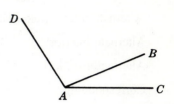

Fig. 11-4. Acute, obtuse, and
adjacent angles

If the sum of two angles measures 180°, the angles are *supplementary*, as ∠1 and ∠2 in Fig. 11-5(*a*). If the sum of two angles measures 90°, the angles are *complementary*, as ∠1 and ∠2 in

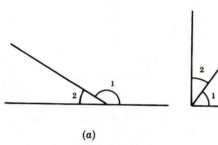

Fig. 11-5. (*a*) Supplementary angles, (*b*) Complementary angles, (*c*) Vertical angles

Fig. 11-5 (*b*). *Vertical angles* are nonadjacent angles formed by two intersecting straight lines, as ∠1 and ∠2, as well as ∠3 and ∠4, in Fig. 11-5(*c*).

When two parallel lines are intersected by a third line, the angles shown in Fig. 11-6 are formed.

Alternate interior: ∡3 and 6, ∡4 and 5.

Alternate exterior: ∡1 and 8, ∡2 and 7.

Interior: ∡4 and 6, ∡3 and 5.

Exterior: ∡1 and 7, ∡2 and 8.

Corresponding: ∡1 and 5, ∡2 and 6, ∡3 and 7, ∡4 and 8.

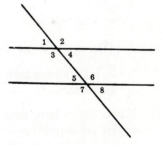

Fig. 11-6.   Exterior and
interior angles

A *central* angle has two radii of a circle for its sides and the center of the circle for its vertex, for example ∠1 in Fig. 11-7. An angle *inscribed* in a circle has its vertex on the circle, for example ∠2. It is assumed that the reader is familiar with the use of a protractor for measuring angles.

An angle is called *positive* if it is measured counterclockwise from its right side; *negative*, if clockwise. Angles of Euclidean plane geometry are usually considered positive.

11.6 THEOREMS ON EQUAL ANGLES

Two angles are said to be equal if they have equal measures. Some useful theorems on equal angles are:

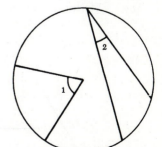

Fig. 11-7.   Central and
inscribed angles

1. Vertical angles are equal. In Fig. 11-5(c), $\angle 1 = \angle 2$ and $\angle 3 = \angle 4$.

2. Alternate interior, alternate exterior, and corresponding angles are equal. In Fig. 11-6,

   Alternate interior: $\angle 3 = \angle 6$ and $\angle 4 = \angle 5$.

   Alternate exterior: $\angle 1 = \angle 8$ and $\angle 2 = \angle 7$.

   Corresponding: $\angle 1 = \angle 5$, $\angle 2 = \angle 6$, $\angle 3 = \angle 7$, $\angle 4 = \angle 8$.

3. If the right and left sides of two angles are respectively parallel to each other, the angles are equal. In Fig. 11-8(a), $\angle 1 = \angle 2$.

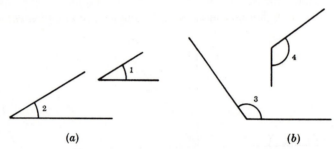

(a)   (b)

Fig. 11-8. (a) Angles with parallel sides, (b) Angles with perpendicular sides

4. If two angles have their right sides mutually perpendicular and their left sides mutually perpendicular, the angles are equal. In Fig. 11-8(b), $\angle 3 = \angle 4$.

5. The angles which are opposite equal sides of a triangle are equal.

11.7 THEOREMS ON SUPPLEMENTARY ANGLES

1. If two angles have their sides parallel, right side to left side and left side to right side, the angles are supplementary. In Fig. 11-9, $\angle 1 + \angle 2 = 180°$.

2. If two angles have their sides perpendicular, right side to left side and left side to right side, the angles are supplementary. In Fig. 11-10, $\angle 3 + \angle 4 = 180°$.

3. Two parallel lines cut by a third line determine pairs of supplementary angles which are:

   (a) Interior angles on the same side. In Fig. 11-6, $\angle 4 + \angle 6 = 180°$ and $\angle 3 + \angle 5 = 180°$.

   (b) Exterior angles on the same side. In Fig. 11-6, $\angle 1 + \angle 7 = 180°$ and $\angle 2 + \angle 8 = 180°$.

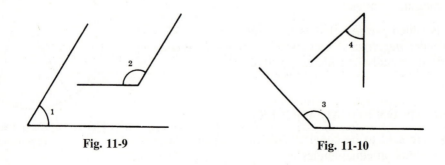

Fig. 11-9   Fig. 11-10

11.8   THE DEGREE MEASURE

The degree is subdivided into sixty minutes; the minute is subdivided into sixty seconds.   In symbols,

$$1° = 60'   1' = 60''$$

Thus, the degree measure is not a decimal system.   Therefore, the arithmetical operations with measures of angles become more involved in many cases.

EXAMPLE 11.1.   Add 24°38'45'' and 19°29'36''.

Write the two measures under each other.   Add *separately* the degrees, minutes, and seconds.

$$
\begin{array}{ccc}
24° & 38' & 45'' \\
19° & 29' & 36'' \\
\hline
43° & 67' & 81'' \\
\end{array}
$$

The sum of the seconds is $45'' + 36'' = 81'' = 1' + 21''$.   Add the 1 to the sum of the minutes.   That is, $38' + 29' + 1 = 68'$.   This value can be rewritten as $68' = 1° + 8'$.   Add 1° to the sum of the degrees.   That is, $24° + 19° + 1° = 44°$.   Thus, the final result is 44°8'21''.

EXAMPLE 11.2.   Subtract 8°54'15'' from 24°32'19''.

Write the two measures under each other.

$$
\begin{array}{ccc}
24° & 32' & 19'' \\
8° & 54' & 15'' \\
\end{array}
$$

Since 54' is larger than 32', "borrow" 1° from 24° and add the equivalent 60' to 32'.   Then subtract *separately* degrees, minutes, and seconds.

$$
\begin{array}{cccl}
23° & 92' & 19'' & \\
8° & 54' & 15'' & \\
\hline
15° & 38' & 4'' & \textit{Ans.} \\
\end{array}
$$

EXAMPLE 11.3.   Multiply 89°52'47.5'' by 2.

Multiply each of the three numbers by 2 and simplify the answer.

$$
\begin{array}{cccl}
89° & 52' & 47.5'' & \\
 & \times 2 & & \\
\hline
178° & 104' & 95'' & 104' = 1°44' \\
\text{or}\quad 178° & & & 95'' = 1'35'' \\
 & 1°44' & & \\
 & & 1'35'' & \\
\end{array}
$$

Thus, 179°45'35'' is the required answer.

If the degree subdivisions are expressed as decimal fractions, then the operations with angle measures are the same as for decimal numbers.

EXAMPLE 11.4.   Divide 115°44'00'' by 3.

Convert 44' to a decimal fraction of a degree; or $\frac{44}{60} \approx 0.73$.   Perform the division, $\frac{115.73}{3} = 38.58$.   The result is 38.58°.

11.9   ANGLE OPERATIONS WITH A CALCULATOR

A calculator can also be used to carry out operations with angles expressed in the format Degree–Minutes–Seconds.   However, before doing any operation, most calculators require that the MODE be set to Degrees.

EXAMPLE 11.5.   Add $24°38'45''$ and $19°29'36''$ using a calculator.

Using a TI-82 graphing calculator

In this calculator, degrees, minutes, and seconds are expressed using the minute mark.   That is, the angle $24°38'45''$ is expressed as $24'38'45'$.   In the example that follows, the selection of this option within the ANGLE MENU is indicated as follows: {2nd}{ANGLE}2.   Notice that the 2 is underlined.

(1)   Set the mode.

  {MODE}Degree {ENTER}   ← "It may be necessary to use the directional arrows (◄, ►, ▲, ▼) to highlight the degree option before
  {CLEAR}   pressing ENTER."

(2a) Enter the first angle.

  24{2nd}{ANGLE}2   38{2nd}{ANGLE}2   45{2nd}{ANGLE}2

  {+}

(2b) Enter the second angle.

  19{2nd}{ANGLE}2   29{2nd}{ANGLE}2   36{2nd}{ANGLE}2

  {ENTER} ← "Pressing {ENTER} carries out the addition operation."

(3)   Display the answer in the format Degrees–Minutes–Seconds

The display shows an answer of 44.13916667.   To express this value in the format Degrees–Minutes–Seconds use the option DMS of the ANGLE menu.   Therefore, after the answer has been displayed, key in the sequence {2nd}{ANGLE}4 {ENTER}   ← "DMS is selection 4."

The result appears now as $44°8'21''$.

Using an HP-38G

Notice that, in this calculator, Degrees–Minutes–Seconds are expressed in the format

D.MMSS.

(1)   Set the mode.

  {SHIFT}{MODES}Degrees{ENTER}

  {HOME}

Before entering the angles, express them in "decimal."   That is, each angle should be in the format X.X (number of degrees with a decimal fraction).   To do this, select the function HMS → pressing the following sequence of keys: {MATH}{►}{▼}{▼}{▼}{▼}{ENTER}

The display shows HMS → (

(2a) Enter the first angle.

  24.3845{)}{ENTER}.   ← "Use the key {.} to enter the period"

Store this angle in variable A.

  STO ► {A . . . Z}{A}{ENTER}   ← "The A variable is also the HOME key."

(2b) Enter the second angle.

  {MATH}{►}{▼}{▼}{▼}{▼}{ENTER}

The display shows HMS → (

19.2936{)}{ENTER}   ← "Use the key {.} to enter the period"

Store this angle in variable B.

STO ►{A...Z}{B}{ENTER}   ← "The variable B is also the SIN key."

(3) Add the angles and express the result in the DD.MMSS format.

To add the angles and express the result in the DD.MMSS format use the function → HMS.

Select this function keying in the sequence {MATH}{►}{▼}{▼}{▼}{▼}{▼}{▼}{▼}{ENTER}

The display shows → HMS(

Recall variable A.

{A...Z}{A}

{+}

Recall variable B.

{A...Z}{B}

{)}{ENTER}

The result is expressed as 44.0821.   That is, 44°08′21″.

11.10   AN ABBREVIATED METHOD TO CARRY OUT ARITHMETIC OPERATIONS WITH ANGLES USING A CALCULATOR

EXAMPLE 11.6.   Add 16°14′18″ and 18°12′11″ using a calculator.

Express all degree subdivisions as decimal fractions.   That is, express minutes and seconds as fractions of degrees.   Remember that $1° = 60′$ and that $1° = 3600″$.   Enter the expression using extra parentheses if necessary.   The algebraic expression to be entered is $(16 + (14/60) + (18/3600)) + (18 + (12/60) + (11/3600))$. This is illustrated below using an HP-38G calculator.   Key in the following sequence

{(}16 {+}{(}14{/}60{)}{+}{(}18{/}3600{)}{)}{+}{(}18{+}{(}12{/}60{)}{+}{(}11{/}3600{)}{)} {ENTER}

The result obtained is 34.4413888889.   The integer part represents the degrees: 34°.   Subtract the degrees.

Multiply the difference by 60 to obtain the minutes.   In this case, multiply 0.4413888889 by 60 to obtain 26.483333334.   The integer part represents the minutes: 26′.   Subtract the minutes.   Multiply the difference by 60 to obtain the seconds.   In this case, multiply 0.483333334 by 60 to obtain 29.00000004.   The integer part represents the seconds.

The final result is 34°26′29″.

Triangles

11.11   PARTS OF A TRIANGLE

A *triangle* is a plane figure bounded by three line segments.   The segments are called the *sides* of the triangle.   A point which is common to two sides is a *vertex*.

An *altitude* is the perpendicular segment drawn from any vertex to the opposite side.   The side to which an altitude is drawn is called the *base* for this particular orientation.   A triangle with vertices A, B, C will be referred to as △ABC.

In a right triangle the sides which include the right angle are called *legs*.   The side which lies opposite the right angle is the *hypotenuse*.   The two legs of a right triangle may also be seen to be two of its altitudes.

11.12   TYPES OF TRIANGLES

If no two sides of a triangle are equal, it is called *scalene* [Fig. 11-11(*a*)].   An *isosceles* triangle has two equal sides [Fig. 11-11(*b*)].   All the three sides of an *equilateral* triangle are equal [Fig. 11-11(*c*)].   In an *acute* triangle each of the angles is smaller than 90° [Fig. 11-11(*a*)].   All the angles in an equiangular triangle are equal, as are also its sides [Fig. 11-11(*c*)].   An *obtuse* triangle has one angle greater than 90° [Fig. 11-11(*d*).   One of the angles in a *right* triangle is 90° [Fig. 11-11(*e*)].

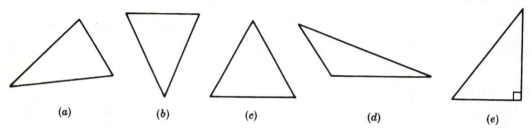

Fig. 11-11.   (*a*) Scalene acute triangle, (*b*) isosceles triangle, (*c*) equilateral or equiangular triangle, (*d*) obtuse triangle, (*e*) right triangle

11.13   CONGRUENCE OF TRIANGLES

If two triangles can be made to coincide by superposition, they are said to be congruent ($\cong$).   The following theorems are stated without proof.

Two triangles are congruent if:

1. Two sides and the included angle of one are equal respectively to two sides and the included angle of the other.   In Fig. 11-12,

$$AB = A'B' \qquad BC = B'C' \qquad \angle B = \angle B'$$

therefore, $\triangle ABC \cong \triangle A'B'C'$.

2. Two angles and the included side of one are equal respectively to two angles and the included side of the other.   In Fig. 11-12, $\angle A = \angle A'$, $\angle C = \angle C'$, $AC = A'C'$; thus, $\triangle ABC \cong \triangle A'B'C'$.

3. The three sides of one are equal respectively to the three sides of the other.   In Fig. 11-12, $AB = A'B'$, $BC = B'C'$, $AC = A'C'$.   Hence, $\triangle ABC \cong \triangle A'B'C'$.

It follows from the definition that corresponding parts of congruent triangles are equal.   In other words, if two triangles are congruent, then equal sides lie opposite equal angles; also, equal angles lie opposite equal sides.

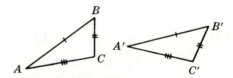

Fig. 11-12.   Congruent triangles

11.14   SIMILARITY OF TRIANGLES

Two geometrical figures are similar ($\sim$) if the dimensions of one are in a constant ratio to the corresponding dimensions of the other.   For example, every part of a photographic enlargement is

similar to the corresponding part in the negative. Following are some theorems on the similarity of triangles.

Two triangles are similar if:

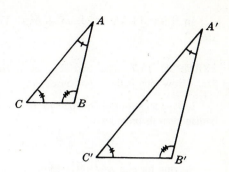

Fig. 11-13. Similar triangles

1. Two angles of one are equal respectively to two angles
    of the other. In Fig. 11-13.

$$\angle A = \angle A', \qquad \angle B = \angle B'$$

    therefore, $\triangle ABC \sim \triangle A'B'C'$.

2. Two sides of one are proportional to two sides of the
    other, and the included angles are equal. In Fig.
    11-13, $AB/AC = A'B'/A'C'$, $\angle A = \angle A'$; consequently,
    $\triangle ABC \sim \triangle A'B'C'$.

3. Their corresponding sides are in the same ratio. In Fig. 11-13, if:

$$AB/A'B' = AC/A'C' = BC/B'C', \qquad \text{then} \qquad \triangle ABC \sim \triangle A'B'C'$$

4. The three angles of one are equal to the three angles of the other.

5. Their corresponding sides are respectively parallel or perpendicular.

Two right triangles are similar if:

6. The hypotenuse and a leg of one are respectively proportional to the hypotenuse and a leg of the
    other.

7. An acute angle of one is equal to an acute angle of the other.

If two triangles are similar, then: (*a*) their angles are respectively equal, and (*b*) their
corresponding sides are proportional.

11.15   OTHER THEOREMS ON TRIANGLES

1. The sum of the angles in a triangle is $180°$, or is equal
    to two right angles.

2. In every triangle the sum of any two sides is greater
    than the third side; the difference of any two sides is
    less than the third side.

3. The angles opposite equal sides are equal, and
    conversely.

4. The largest angle in a triangle lies opposite the largest
    side, and conversely.

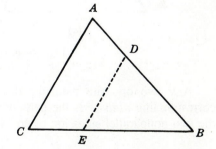

Fig. 11-14.   Division of triangle sides
into proportional segments

5. A line parallel to one side of a triangle divides the other
    two sides into proportional segments, and conversely.   In Fig. 11-14, if *DE* is parallel to *AC*, then

$$AD/DB = CE/EB$$

11.16   THE PYTHAGOREAN THEOREM

This is a theorem which has numerous applications.   It may be stated as follows.

Theorem.   The square upon the hypotenuse of a right triangle is equal in area to the sum of the squares upon the two
legs.

In Fig. 11-15, $c^2 = a^2 + b^2$. Thus, area 1 = area 2 + area 3.

EXAMPLE 11.7 The two legs of a right triangle are 12 m and 5 m. How long is the hypotenuse?

Let the length of the hypotenuse be c m. Then an application of the Pythagorean theorem gives:

$$c^2 = 12^2 + 5^2 = 144 + 25 = 169$$

Solving for c, $c = \sqrt{169} = 13$ m.

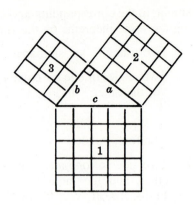

Fig. 11-15. Pythagorean theorem: area 1 = area 2 + area 3

Quadrilaterals

11.17 TERMS

A *quadrilateral* is a plane figure bounded by four line segments.

A *parallelogram* is a quadrilateral in which opposite sides are parallel to each other.

A *trapezoid* is a quadrilateral with only one pair of parallel opposite sides [Fig. 11-16(*a*)].

A *rhombus* is a parallelogram with all sides equal [Fig. 11-16(*b*)].

A *rectangle* is a parallelogram with four right angles [Fig. 11-16(*c*)].

A *square* is a rectangle with all sides equal.

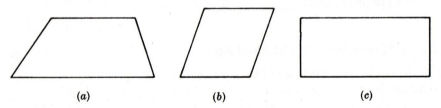

(*a*)   (*b*)   (*c*)

Fig. 11-16. (*a*) Trapezoid, (*b*) rhombus, (*c*) rectangle

Any two opposite parallel sides are called *bases* of a trapezoid or a parallelogram. The corresponding *altitude* is the perpendicular distance between the bases. In an *isosceles* trapezoid the two nonparallel sides are of equal length.

11.18 THEOREMS ON PARALLELOGRAMS

A quadrilateral is a parallelogram if any one of the following statements is true:

1. Two sides are equal and parallel.

2. Each interior angle is equal to the interior opposite angle.

3. The diagonals bisect each other.

4. The opposite sides are equal in length.

The converse of each of the above theorems is also true. Thus, in a parallelogram the opposite angles are equal, and the opposite sides are of equal length.

A diagonal of a parallelogram divides it into two congruent triangles.

Two angles with a common side in a parallelogram are supplementary. Thus, the measure of the sum of its four angles is $180° + 180° = 360°$.

Some properties of special parallelograms are:

1. In a rhombus, the diagonals are perpendicular to each other.

2. In a rhombus, the diagonals bisect the angles.

3. In a rectangle, the square of the length of a diagonal is equal to the sum of the squares of the lengths of two adjacent sides.

4. In a rectangle the two diagonals are equal.

5. A square, being a parallelogram, a rhombus, and a rectangle, has all the properties listed above.

Circles and Arcs

11.19   TERMS

A *circle* is a closed plane curve with all its points equidistant from some fixed point. The fixed point is called the *center* of the circle.

The linear measure or length of a circle is its *circumference*, although the word is frequently used as a synonym for the points of the circle itself.

A *radius* of a circle is a line segment connecting the center of the circle with any point on it; however, the word is often used to mean the linear measure or length of this segment.

A *secant* is any line passing through two points on the circle, as *AB* in Fig. 11-17.

A *tangent* is a line which has exactly one point in common with the circle, for example *CD* in Fig. 11-17.

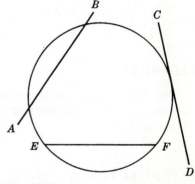

Fig. 11-17.   Secant, tangent, and chord

A *chord* is a line segment connecting two points on the circle, for example *EF* in Fig. 11-17. A *diameter* is a chord through the center of the circle.

An *arc* is a part of a circle. Half of a circle is a *semicircle*. An arc equal to one-fourth of a circle is called a *quadrant*.

A *sector* is a plane figure bounded by a circular arc and two radii drawn to the art's extremities, for example *ABCA* in Fig. 11-18.

A *segment* is a plane region bounded by an arc of a circle and its connecting or subtending chord, for example region *DEFD* in Fig. 11-18. A perpendicular joining the midpoints of the segment's arc and chord is called a *sagitta*; it is shown as *EG* in Fig. 11-18.

A central angle and its intercepted arc may both be measured in degrees and are of equal magnitude.

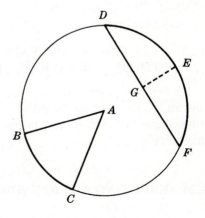

Fig. 11-18.   Sector, segment, and sagitta

11.20   THEOREMS ON CIRCLES AND ARCS

1. In a circle (or equal circles), equal chords subtend equal arcs, and conversely.

2. In a circle (or equal circles), equal chords are equidistant from the circle's center, and conversely.

3. A line through the center of a circle and perpendicular to a chord bisects the chord and its arc, and conversely.

4. A tangent to a circle is perpendicular to the radius at the point of contact, and conversely.

5. Two arcs of a circle, intercepted between parallel lines, are of equal length.

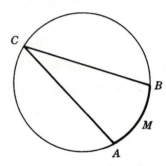

6. An inscribed angle is measured by one-half of its intercepted arc.   In Fig. 11-19, if arc AMB is one-sixth of a circle, then the arc is measured by $1/6 \times 360°$, or $60°$.   The inscribed angle ACB has $1/2 \times 60°$, or $30°$.

7. One and only one circle can be passed through three points not in a straight line.

Fig. 11-19.   Inscribed angle

Constructions

A number of useful constructions can be made by using only a compass and a straightedge (rule without a scale).

11.21   BISECTING A LINE SEGMENT (Fig. 11-20)

Spread the compass more than one-half of the given line segment AB.   With A and B as centers draw two arcs which intersect at C and D.   A line through C and D will bisect the given line segment at P.

11.22   BISECTING AN ANGLE (Fig. 11-21)

From the vertex B of the given angle ABC draw an arc intersecting the sides at D and E.   With D and E as centers, and with equal radii, draw two arcs intersecting at F.   The line through F and B bisects the angle.

11.23   ERECTING A PERPENDICULAR TO A GIVEN LINE FROM A GIVEN POINT NOT ON THE LINE (Fig. 11-22)

Let O be the point from which a perpendicular is to be dropped to the line AB.   With O as center draw an arc intersecting the given line at D and E.   With D and E as centers and with the same radius draw the arcs st and pr intersecting at F.   The line through O and F is the desired perpendicular.

11.24   ERECTING A PERPENDICULAR TO A LINE AT A GIVEN POINT ON THE LINE (Fig. 11-23)

The construction procedure can be seen in Fig. 11-23.

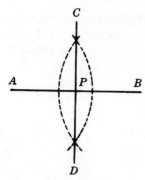

Fig. 11-20.   Bisection of a
              segment

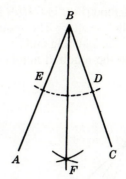

Fig. 11-21.   Bisection of
              an angle

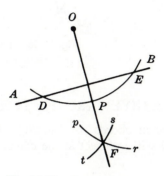

Fig. 11-22.   Perpendicular to a line
              from an external point

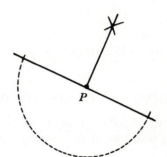

Fig. 11-23.   Perpendicular from
              a point on a line

11.25   DRAWING A PARALLEL TO A GIVEN LINE (Fig. 11-24(a) and 11-24(b))

Given line ON draw two perpendicular lines AB and CD. With A and C as centers, and with equal radii, draw two arcs intersecting at F and E respectively. The line through F and E is parallel to the line ON.

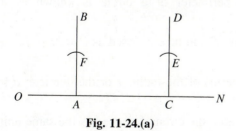

Fig. 11-24.(a)

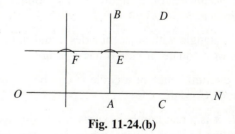

Fig. 11-24.(b)

11.26   DIVIDING A GIVEN LINE SEGMENT INTO A GIVEN NUMBER OF EQUAL PARTS (Fig. 11-25)

(a)   Draw CD parallel to the given line segment AB.   Lay off on CD the required number of equal segments of arbitrary length.   Draw lines through AO and BN until they intersect at P.   Draw lines from P through the end points of the equal segments on CD.   These lines will subdivide AB into the required number of equal segments on AB.

(*b*)   A second method [Fig. 11-24(*b*)] is to draw the line *AC* at some convenient angle. Line *BD* is drawn parallel to *AC*. The required number of equal segments are laid off on *AC* and on *BD*. The corresponding end points of the segments are joined by lines which divide the given segment into the required number of equal parts.

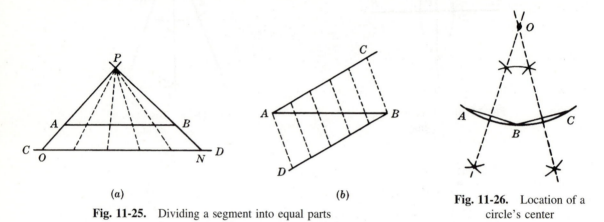

(a)   (b)

Fig. 11-25. Dividing a segment into equal parts   **Fig. 11-26.** Location of a circle's center

11.27 FINDING THE CENTER OF A CIRCLE FROM A GIVEN ARC (Fig. 11-26)

Through three points *A*, *B*, *C* on the arc draw the adjacent chords *AB* and *BC*. Construct perpendicular bisectors of the chords. When extended, they will intersect at the center *O* of the circle.

Some of the above constructions may be made more quickly and conveniently by using a protractor; for example, the constructions shown in Figs. 11-21, 11-22, and 11-23.

Perimeters and Areas of Plane Figures

11.28 PERIMETERS

The *perimeter* of a figure is its boundary. The perimeter of a circle is known as its *circumference*.

In a rectangle with adjacent sides *a* and *b*, the perimeter *P* can be expressed as $P = 2a + 2b$. A rhombus or a square of side *a* has $P = 4a$.

The circumference of a circle *C* can be expressed in terms of the radius *r* or the diameter *d* by $C = 2\pi r$ or $C = \pi d$.

Since π is a real number expressing a dimensionless ratio, the circumference has the same units as the radius or the diameter, when the above formulas are used.

The perimeters of similar figures are in the same ratio as the lengths of their corresponding linear parts.

11.29 AREAS

The area of a plane figure is the area of the surface enclosed by the figure, expressed in square units, such as square feet or square meters.

Triangle

The area A of a triangle may be calculated from either of two formulas:

$$A = (1/2)hb$$

where h is the altitude to the base b, or

$$A = \sqrt{s(s-a)(s-b)(s-c)}$$

where a, b, c, are the three sides of the triangle and $s = (1/2)(a+b+c)$.

EXAMPLE 11.8.   Calculate the area of triangle with sides 6, 8, and 10 cm.

Calculate s to obtain

$$s = \frac{6+8+10}{2} = 12$$

Substitute in the formula for the area in terms of sides:

$$A = \sqrt{12(12-6)(12-8)(12-10)} = \sqrt{(12)(6)(4)(2)}$$
$$= \sqrt{576} = 24$$

The area of the triangle is $24\,\text{cm}^2$.

To calculate the area of this triangle using an HP-38G graphing calculator use the following steps.

(1)   Calculate the value of s and save it in a variable.   Use extra parentheses if necessary.

{(}6{+}8{+}10{)}{/}2{ENTER}   ← "Extra parentheses ensure that the numerator is the sum of the given sides."

STO ▶ {A...Z}{S}{ENTER}   ← "Save result in variable S."

(2)   Substitute s in the formula for the area.   To do this key in the following sequence.

{$\sqrt{x}$}{(}{(}{A...Z}{S}{*}{(}{(}{A...Z}{S}{−}6{)}{)}{*}{(}{(}{A...Z}{S}{−}8{)}{)}{*}{(}{(}{A...Z}{S}{−}10{)}{)}{)}
{ENTER}

Notice that variable s was retrieved by pressing the sequence {A...Z}{S}

With a TI-82 calculator the area of the triangle can be calculated as follows.

(1)   Calculate the value of s and save it in memory.

{(}6{+}8{+}10{)}{÷}2{ENTER}{STO▶}{ALPHA}{S}{ENTER}   ← "Save result in variable S."

(2)   Substitute s in the formula for the area.   To do this key in the following sequence.

{2^{nd}}{$\sqrt{}$}{(}{(}{2^{nd}}{RCL}{ALPHA}{S}{ENTER}{×}{(}{(}{2^{nd}}{RCL}{ALPHA}{S}{ENTER}{−}6{)}{)}{×}{(}{(}{2^{nd}}{RCL}
{ALPHA}{S}{ENTER}{−}8{)}{)}{×}{(}{(}{2^{nd}}{RCL}{ALPHA}{S}{ENTER}{−}10{)}{)}{)}{ENTER}

Variable s was retrieved using the sequence {2^{nd}}{RCL}{ALPHA}{S}.

Parallelogram, Trapezoid and Square

The area A of a parallelogram is given by $A = bh$, where b is one of the sides and h is an altitude to that side.   In a rectangle, each of the sides a is an altitude to an adjacent side b, and the area is $A = ab$.   The area of a square is given by $A = a^z$ since all the sides are equal.

The area of a rhombus or a square can also be obtained if its two diagonals d_1 and d_2 are known:

$$A = \frac{1}{2}(d_1 d_2)$$

The area of a trapezoid is found from the expression:

$$A = \frac{1}{2}(a + b)h$$

where a and b are the lengths of the two parallel sides and h is the distance between them.

EXAMPLE 11.9.   Compute the area of a trapezoid with parallel sides of 12.40 and 15.90 inch with a distance of 3.20 inch between them.

$$A = \frac{(12.40 + 15.90)3.20}{2} \approx 45.28$$

The area to the correct number of significant figures is 45.3 in^2.

Circle

The area of a circle may be calculated from either of the following two formulas:

$$A = \pi r^2 \approx 3.142\,r^2, \text{ where } r \text{ measures the radius of the circle}$$

$$A = \frac{\pi d^2}{4} \approx 0.785 d^2, \text{ where } d \text{ measures the diameter of the circle}$$

Sector of a Circle

The area of a sector may be calculated from the formula

$$A = \frac{\pi r^2 \theta}{360}$$

where r is the radius of the sector and θ is the number of degrees in the central angle subtended by the sector, as shown in Fig. 11-27.

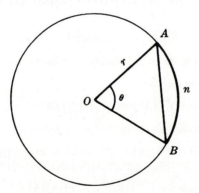

Fig. 11-27.   Sector and segment areas

EXAMPLE 11.10.   Find the area of a sector whose radius is 30 cm and whose arc subtends 20°.

Substitution in the above formula gives:

$$A = \frac{(3.142)(30)^2(20)}{360} = \frac{(3.142)(900)(20)}{360} = (3.142)(50)$$

$A \approx 157$, or the area of the sector is approximately 157 cm^2.

To calculate the area of the sector of this circle using a TI-82 calculator proceed as follows:

$\{(\}\{2^{nd}\}\{\Pi\}\{\times\}30\{^\wedge\}2\{\times\}20\{)\}\{\div\}360\{ENTER\} \approx 157\,cm^2$.

Segment of a Circle

The area of the segment AnB in Fig. 11-26 is obtained by subtracting the area of the triangle AOB from the area of the sector $AOBnA$.

Areas of Similar Figures

The areas of any two similar figures are in the same ratio as the squares of the lengths of any two corresponding linear parts.

EXAMPLE 11.11.   The corresponding sides of two similar triangles are in the ratio $3:2$.   What is the ratio of their areas?

Since the ratio of the areas is equal to the square of the ratio of the sides, we can write:

$$\frac{A_1}{A_2} = \left(\frac{3}{2}\right)^2 = \frac{9}{4} = 2.25$$

Thus, the area of the larger triangle is 2.25 times the area of the smaller triangle.

EXAMPLE 11.12.   The diameter of a disk is 7.5 cm.   How does its area compare with that of a disk 15 cm in diameter?

The areas of two disks are in the same ratio as the square of the ratio of their diameters.   Therefore:

$$\frac{A_1}{A_2} = \left(\frac{7.5}{15}\right)^2 = \left(\frac{1}{2}\right)^2 = \frac{1}{4}$$

The area of the smaller disk is one-fourth the area of the larger.

Solved Problems

ANGLES

11.1   (a) What name is given to angles related as $\angle 1$ and $\angle 2$ in Fig. 11.28?

(b) If $\angle 2$ is $34°$, how large is $\angle 3$?

(c) How large is $\angle 1$?

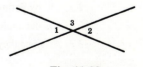

Fig. 11-28

(a) Vertical

(b) $\angle 2 + \angle 3 = 180°$

$34° + \angle 3 = 180°$

$\angle 3 = 180° - 34° = 146°$

(c) $\angle 1 = 34°$, since vertical angles are equal.

11.2   The ratio of the magnitudes of two complementary angles is $2:1$. How large are the angles?

   Let one of the unknown angles be $\angle A$ and the other $\angle B$.   Then, $\dfrac{\angle A}{\angle B} = \dfrac{2}{1}$ or $\angle A = 2\angle B$.   Also, by definition $\angle A + \angle B = 90°$.

$$2\angle B + \angle B = 90° \qquad \text{by substitution}$$
$$3\angle B = 90°$$
$$\angle B = 30° \qquad \text{and} \qquad \angle A = 2\angle B = 60°$$

11.3.   In Fig. 11-29, line segments AP and BE are parallel to each other; line segments CP and DP are perpendicular to each other; also $PE \perp EB$.

   (*a*) Which pairs of angles are equal?

   (*b*) Which pairs of adjacent angles are complementary?

   (*c*) Which pairs of labeled angles are supplementary?

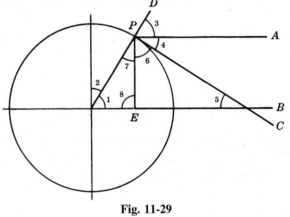

Fig. 11-29

   (*a*) $\angle 1 = \angle 3$ corresponding)

   $\angle 2 = \angle 7$ (alternate interior)

   $\angle 4 = \angle 5$ (alternate interior)

   $\left. \begin{array}{l} \angle 1 = \angle 6 \\ \angle 2 = \angle 5 \\ \angle 3 = \angle 6 \\ \angle 5 = \angle 7 \end{array} \right\}$ (mutually perpendicular sides)

   (*b*) 1, 2; 3, 4; 4, 6; 6, 7 by definition; 5, 6; 1, 7 in right $\triangle$

   (*c*) none

11.4   In Fig. 11-30, side AB is parallel to side DE; side BC is parallel to EF. $\angle 2$ and $\angle 4$ are right angles.   Which pairs of angles are equal to each other?

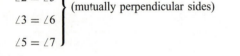

   $\angle 2 = \angle 4$, by definition

   $\angle 3 = \angle 6$, the angles have mutually parallel sides

   $\angle 3 = \angle 5$, since they are complements of equal angles

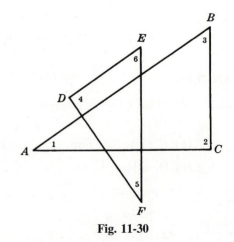

Fig. 11-30

11.5   In Fig. 11-31, AB is parallel to CD, and both lines are perpendicular to BD.   Also, BE is parallel to DF and perpendicular to DG.

   (*a*) List the pairs of equal angles.

   (*b*) List some pairs which are complementary.

   (*c*) List some pairs which are supplementary.

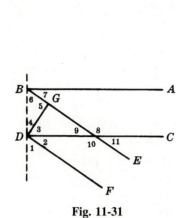

Fig. 11-31

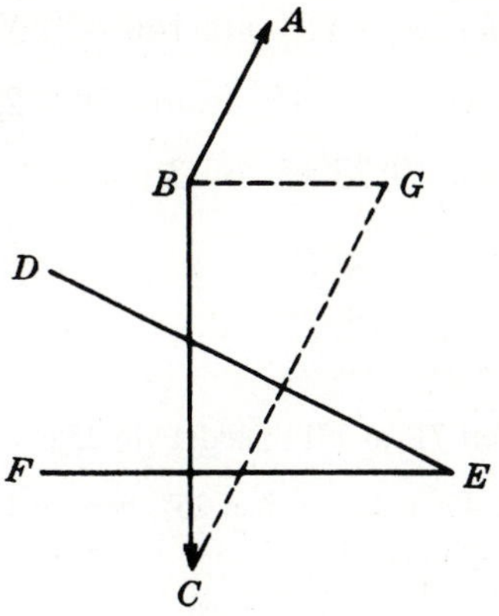

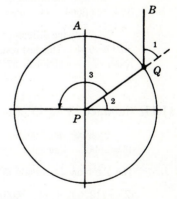

Fig. 11-32

(*a*) 2, 9; 1, 3; 3, 6; 1, 6; 2, 4; 8, 10; 9, 11; 7, 9; 7, 11; 2, 7

(*b*) 1, 2; 2, 3; 3, 4; 3, 9; 4, 6; 6, 7; 3, 7; 1, 4; 2, 6; and others

(*c*) 7, 8; 8, 9; 8, 11; 9, 10; 10, 11; 7, 10; 2, 8; 2, 10; and others

11.6   In Fig. 11-32, $AB \perp DE$, $BC \perp FE$.

(*a*) What is the relationship between $\angle ABC$ and $\angle DEF$?   Why?

(*b*) If $\angle DEF$ is 15°, how large is $\angle ABC$?

(*c*) If CG is drawn parallel to BA, how large is $\angle BCG$?   Why?

(*a*) Supplementary; theorem on supplementary angles. $\angle DEF$ and $\angle ABC$ are supplementary since their sides
are mutually perpendicular right side to left side and left side to right side.

(*b*) Thus, $\angle ABC + \angle DEF = 180°$

$$\angle ABC + 15° = 180°$$

$$\angle ABC = 165°$$

(*c*) 15°; theorem on equal angles because of perpendicular sides.   If $AB \perp DE$, then $CG \perp DE$ also.

11.7   In Fig. 11-33, $\angle 1 = 40°$.   What is (*a*) the measure of
$\angle 2$?   (*b*) if $\angle 3$?

(*a*) $\angle APQ = \angle 1 = 40°$, corresponding $\angle s$

$$\angle 2 = 90° - \angle APQ = 90° - 40° = 50°$$

(*b*) $\angle 3 = 180° - \angle 2 = 180° - 50° = 130°$

11.8   (*a*) Convert 18°27′ into minutes.

(*b*) Convert 3800″ into degrees, minutes, and seconds.

(*a*) 1° = 60′ 18° = (18)(60) = 1080′

$$18°27' = 1080' + 27' = 1107'$$

(*b*) 1′ = 60″ $\dfrac{3800''}{60} = 63'20''$

1° = 60′ 63′ = 1°3′   The answer is 1°3′20″

Fig. 11-33

11.9   (a)   Add $118°29'25''$ and $140°42'229''$.

       (b)   Subtract $72°3'45''$ from $138°4'25''$.

       (c)   Divide $19°23'15''$ by 3.

(a)
$$\begin{array}{rrr} 118° & 29' & 25'' \\ 140° & 42' & 29'' \\ \hline 258° & 71' & 54'' \end{array}$$

Convert $71'$ to $1°11'$; add $1°$ to $258°$.   The desired sum is $259°11'54''$.

(b)   Since $45''$ is larger than $25''$, borrow $1'$ from $4'$ and add it to $25''$.   Subtract:

$$\begin{array}{rrr} 138° & 3' & 85'' \\ 72° & 3' & 45'' \\ \hline 66° & 0' & 40'' \end{array}$$

The answer is $66° 40''$.

(c)   Divide each part by 3.

$$(19° + 23' + 15') \div 3 = 6\tfrac{1}{3}° + 7\tfrac{2}{3}' + 5''$$

$$\tfrac{1}{3}° = (1/3)(60') = 20' \qquad \tfrac{2}{3}' = (2/3)(60'') = 40''$$

Combining like units, the quotient becomes $6°27'45''$.

11.10.   Add $45°20'13''$ and $12°15'18''$ using a calculator.

Express all degree subdivisions as decimal fractions. Remember that $1° = 60'$ and that $1° = 3600''$.   Enter the expression using extra parentheses if necessary.

The algebraic expression to be entered is

$$(45 + (20/60) + (13/3600)) + (12 + (15/60) + (18/3600))$$

In a calculator, key in the following sequence:

{(}45{ + }{(}20{/}60{)}{ + }{(}13{/}3600{)}{)}{ + }{(}12{ + }{(}15{/}60{)}{ + }
{(}18{/}3600{)}{)}{ENTER}

The result is 57.5919444444.   The integer part represents the degrees.   That is, $57°$.   Subtract the degrees.

Multiply the difference by 60 to obtain the minutes.   That is, multiply 0.5919444444 by 60.   The result is 35.516666664.   The integer part represents the minutes.   That is, $35'$.   Subtract the minutes.

Multiply the difference by 60 to obtain the seconds.   That is, multiply 0.516666664 by 60.   The result is 30.9999994.   This last result can be rounded to 31.   The integer part represents the seconds.   Therefore, the result of the addition is $57°35'31''$.

11.11.   Subtract $23°12'45''$ from $142°16'25''$ using a calculator.

Express all degree subdivisions as decimal fractions. Remember that $1° = 60'$ and that $1° = 3600''$.   Enter the expression using extra parentheses if necessary.

The algebraic expression to be entered is

$(23 + (12/60) + (45/3600)) - (142 + (16/60) + (25/3600))$

Enter this expression in a calculator as follows:

{(}23{ + }{(}12{/}60{)}{ + }{(}45{/}3600{)}{)}{ - }{(}142{ + }{(}16{/}60{)}{ + }
{(}25{/}3600{)}{)}{ENTER}

The result is 119.061111111.   The integer part represents the degrees.   That is, $119°$.   Subtract the degrees.

Multiply the difference by 60 to obtain the minutes.   That is, multiply 0.061111111 by 60.   The result is 3.66666666.   The integer part represents the minutes.   That is, $3'$.   Subtract the minutes.

Multiply the difference by 60 to obtain the seconds.   That is, multiply 0.66666666 by 60.   The result is 39.9999996.   This result can be rounded to $40''$.   Therefore, the result is $119°3'40''$.

TRIANGLES AND PYTHAGOREAN THEOREM

11.12. In Fig. 11-34, $XY \perp XZ$.

(*a*) What type of $\triangle$ is XYZ?

(*b*) What names are given to parts XZ and XY of the $\triangle$?

(*c*) What is ZY called?

(*d*) If $XZ = XY$, how many degrees are there in $\angle Z$?   In $\angle Y$?

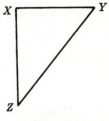

Fig. 11-34

(*a*) right   (*b*) legs   (*c*) hypotenuse

(*d*) $\angle Z + \angle Y = 180° - 90° = 90°$.   Equal angles are opposite equal sides:

$$\angle Z = \angle = \frac{90°}{2} = 45°$$

11.13. (*a*) What type of triangle is $\triangle ABC$ in Fig. 11-35?

(*b*) Are any two angles in $\triangle ABC$ equal?   If so, which ones?

(*c*) A perpendicular line segment is dropped from B to AC.   What is the length of this line segment?

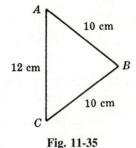

Fig. 11-35

(*a*) Isosceles.

(*b*) Yes; A and C.

(*c*) The segment will be $\perp$ to AC and will bisect it.   The segment will be one of the legs in a triangle with the other leg 6 cm in length and a 10-cm hypotenuse.   Applying the Pythagorean theorem, with x the length of the unknown segment,

$$x^2 + 6^2 = 10^2$$
$$x^2 = 10^2 - 6^2 = 100 - 36 = 64$$
$$x = \sqrt{64} = 8$$

and the length of the segment is 8 cm.

11.14. In triangle PQR in Fig. 11-36, $PQ = QR = PR$, $QS \perp PR$.

(*a*) What type of triangle is $\triangle PQR$?   (*b*) How many degrees are there in $\angle 1$?   in $\angle 2$?

(*a*) equilateral   (*b*) 60°; 30°, since $\overline{QS}$ bisects $\angle PQR$.

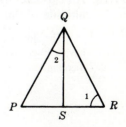

Fig. 11-36

11.15. In triangle UVW in Fig. 11-37, $UV \perp VW$.   Calculate the length of UW.

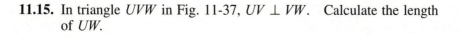

By the Pythagorean theorem,

$$\overline{UW}^2 = 6^2 + 11^2 = 36 + 121 = 157$$

$$\overline{UW} = \sqrt{157} \approx 12.5 \text{ inch}$$

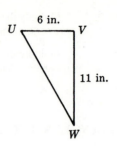

11.16. In Fig. 11-38, $FE \perp ED$, $FC \perp BC$, $AB \perp BC$, $FE \parallel AC$.

(*a*) What is the relationship between $\triangle ABC$ and $\triangle FED$?

(*b*) $AB = 3$ ft; $BC = 10$ ft. If $FE = 1.5$ ft, how long is ED? FD?

(*a*) $\triangle ABC \sim \triangle FED$ by the theorem on similar triangles: their corresponding angles are equal.

(*b*) In similar triangles, the corresponding sides are proportional.

$$\frac{\overline{AB}}{\overline{BC}} = \frac{\overline{FE}}{\overline{ED}} \qquad \text{and} \qquad \frac{3}{10} = \frac{1.5}{\overline{ED}}$$

$$3\,\overline{ED} = 15$$

$$\overline{ED} = 5$$

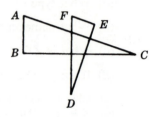

Since $\triangle FED$ is a right triangle, $\overline{ED}^2 + \overline{FE}^2 = \overline{FD}^2$ and $5^2 + 1.5^2 = \overline{FD}^2$

$$27.25 = \overline{FD}^2$$

$$\overline{FD} = \sqrt{27.25} \approx 5.2$$

The length of ED is 5 ft; of FD, 5.2 ft.

11.17. A 25-foot ladder is placed against a wall. The base of the ladder is 12 feet from the wall on level ground. How high up the wall does the ladder reach?

The ladder, wall, and ground form a right triangle with the ladder as the hypotenuse. By the Pythagorean theorem,

$$25^2 = 12^2 + h^2$$

$$625 = 144 + h^2$$

$$625 - 144 = h^2$$

$$\sqrt{481} = h$$

$$h \approx 21.9$$

The ladder reaches 21.9 feet up the wall.

11.18. An isosceles triangle has equal sides each 30 cm and a base 20 cm in length. Determine the altitude of this triangle.

The altitude will divide the isosceles triangle into two congruent right triangles. Each of the triangles will have a 30-cm hypotenuse and a 10-cm leg; the unknown altitude a will be the second leg. By the Pythagorean theorem,

$$a^2 + 10^2 = 30^2$$

$$a^2 = 30^2 - 10^2$$

$$a = \sqrt{800} \approx 28.3$$

The length of the altitude is 28.3 cm approximately.

11.19. Is the triangle with sides of lengths 8, 14, and 17 units, a right triangle?

> A right triangle must satisfy the Pythagorean equation.
>
> $$8^2 + 14^2 \overset{?}{=} 17^2$$
>
> $$64 + 196 \overset{?}{=} 289$$
>
> $$260 \neq 289$$

Therefore the given triangle is not a right triangle.

QUADRILATERALS

11.20. In the rectangle $ABCD$ shown in Fig. 11-39, $AC = 80$ cm, $AD = 100$ cm.

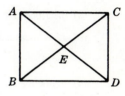

(*a*) Calculate the length of AB.

(*b*) How long is EC?

Fig. 11-39

(*a*) $AB = CD$. By the Pythagorean theorem,

$$\overline{AD}^2 = \overline{CD}^2 + \overline{AC}^2$$

$$\overline{CD} = \sqrt{\overline{AD}^2 - \overline{AC}^2} = \sqrt{100^2 - 80^2}$$

$$= \sqrt{3600} = 60$$

> The length of AB is 60 cm.

(*b*) $EC = (1/2)BC$, since the diagonals of a rectangle bisect each other. $BC = AD = 100$ cm, since the diagonals of a rectangle are equal in length. Thus, $EC = (1/2)100 = 50$, or EC is 50 cm in length.

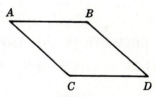

11.21. In Fig. 11-40, $ABCD$ is a parallelogram.

(*a*) If $\angle BAC = 40°$, how large is $\angle BDC$?

(*b*) How large is $\angle ABD$?

(*c*) If $AB = AC$, what name is given to the figure?

Fig. 11-40

(*a*) 40°. In a $\square$ an interior angle = opposite interior angle.

(*b*) 140°. The angles with a common side in a $\square$ are supplementary.

(*c*) rhombus

11.22. The diagonal of a square is 10 inches in length. Compute the length of each side.

> The diagonal forms the hypotenuse of a right triangle. Let the length of each side by S. Then,
>
> $$S^2 + S^2 = 10^2$$
>
> $$2S^2 = 100$$
>
> $$S^2 = 50$$
>
> $$S = \sqrt{50} \approx 7.07$$

The length of each side is approximately 7.07 inch.

CIRCLES AND ARCS

11.23. In Fig. 11-41,

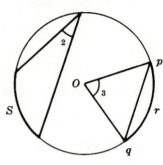

    (a) $\angle 2$ may be referred to as what kind of angle?

    (b) If $\angle 2 = 30°$, how many arc degrees are there in arc S?

    (c) What name may be given to a line segment such as pq?

    (d) What name may be given to a figure such as pqr?

    (e) If $\angle 3 = 72°$, what fraction of the circumference is arc prq?

Fig. 11-41

    (a) inscribed

    (b) $\angle 2$ is measured by $\frac{1}{2}S$.   Thus S is measured by $2(\angle 2)$, or $60°$.

    (c) chord

    (d) segment

    (e) prq: circumference $:: 72° : 360°$

$$\frac{72°}{360°} = \frac{1}{5} = 0.2$$

    $\therefore$ arc prq is equal to $1/s$ or 0.2 of the circumference.

11.24. What angle does a diameter make with the tangent to a circle at the outer end of the diameter?

    Since the diameter lies along the radius, the angle at the point of tangency is $90°$.

PERIMETERS AND AREAS

11.25. In $\square PQRS$ in Fig. 11-42, $PQ = 10$ m, $QR = 21$ m.

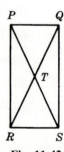

    (a) Calculate the length of QS.

    (b) Compute the area of $PQRS$.

    (c) What is the length of PT?

    (a) $QS = PR$

        From the Pythagorean theorem,

$$\overline{QS} = \overline{PR} = \sqrt{21^2 - 10^2} = \sqrt{341} \approx 18.5$$

        The length of QS is about 18.5 m.

Fig. 11-42

    (b) $A = (PQ)(PR) \approx (10)(18.5) \approx 185$

        The area of the rectangle is $185\,\text{m}^2$ approximately.

    (c) $\overline{PT} = \frac{1}{2}\overline{PS} = \frac{1}{2}\overline{QR} = (1/2)21 = 10.5.$ The length of PT is 10.5 m.

11.26. (a) Calculate the area of a triangle with sides 5 cm, 12 cm, 13 cm.

(b) Calculate the length of the altitude to the 13-cm side.

$$A = \sqrt{s(s-a)(s-b)(s-c)}$$

$$s = \frac{a+b+c}{2} = \frac{5+12+13}{2} = 15$$

$$A = \sqrt{15(15-5)(15-12)(15-13)} = \sqrt{15(10)(3)(2)} = \sqrt{900} = 30$$

The area is $30\,\text{cm}^2$.

(b) $A = (1/2)(\text{altitude})(\text{base}) = \dfrac{1}{2}\text{hb}$

$$30 = \frac{1}{2}(h)(13)$$

$$h = 60/13 \approx 4.62$$

The altitude is 4.62 cm approximately.

11.27. Given a circle of 20-mm radius: (a) compute the circumference of the given circle, and (b) calculate the area of the circle.

(a) $C = 2\pi r \approx 2(3.14)(20) = 125.6$

The circumference is 126 mm approximately.

(b) $A = \pi r^2 \approx (3.14)(20^2) = 1260$

The area is $1260\,\text{mm}^2$ approximately.

11.28. A sector subtends an arc of $30°$ in a circle with a diameter of 2.50 m. Calculate (a) the arc length, and (b) the sector area.

(a) Let the arc length be L and the diameter be D. Then, $L : \pi D :: 30° : 360°$, from which

$$L = \frac{(\pi D)\cancel{(30)}}{\underset{12}{\cancel{360}}} \approx \frac{(3.14)(2.50)}{12} \approx 0.654$$

The arc length is 0.654 m approximately.

(b) $A = \dfrac{\pi r^2 \theta}{360°}$, where $\theta = 30°$ and $r = 1.25$ m.

$$A \approx \frac{(3.14)(1.25^2)\cancel{30}}{\underset{12}{\cancel{360}}} \approx 0.409$$

The area of the sector is $0.409\,\text{m}^2$ approximately.

11.29. The altitude of a trapezoid is 2.5 cm. One of the parallel sides is 10 cm; the other is 18 cm. Compute the area of the trapezoid.

$$A = (1/2)(a+b)h$$
$$A = (1/2)(10+18)(2.5) = (1/2)(28)(2.5)$$
$$A = 35$$

The area of the trapezoid is $35\,\text{cm}^2$.

PERIMETERS AND AREAS OF SIMILAR FIGURES

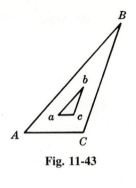

11.30. The linear dimensions of $\triangle ABC$ in Fig. 11-43 are four times as large as the corresponding dimensions of $\triangle abc$. Determine the ratio of their areas.

The areas are proportional to the squares of the linear dimensions. Thus,

$$\frac{\text{area } \triangle ABC}{\text{area } \triangle abc} = \frac{(\overline{AB})^2}{(\overline{ab})^2}$$

$$= \left(\frac{\overline{AB}}{\overline{ab}}\right)^2 = \left(\frac{4}{1}\right)^2 = 16$$

Fig. 11-43

The ratio is $16:1$.

11.31. Given a circle of 50-cm diameter:

(*a*) What would the radius of another circle have to be to produce a circle whose circumference is three times that of the given circle?

(*b*) How does the area of a circle 5 cm in diameter compare with the area of the given circle?

(*a*) The circumferences C of circles are directly proportional to their diameters d.

$$C_1 : C_2 :: d_1 : d_2$$
$$C_1 : 3C_1 :: 50 : d_2, \quad \text{since } C_2 = 3C_1$$
$$150C_1 = d_2 C_1$$
$$d_2 = 150$$
$$r_2 = 150/2 = 75, \quad \text{since } r_2 = d_2/2$$

The required radius is 75 cm.

(*b*) The areas of circles are proportional to the squares of their diameters.

$$A_2 : A_1 :: 5^2 : 50^2$$
$$A_2 : A_1 :: 25 : 2500$$
$$\frac{A_2}{A_1} = \frac{25}{2500} = \frac{1}{100} = 0.01$$

The area of the 5-cm circle is 0.01 as large as the area of the 50-cm circle.

11.32. The adjacent sides of a parallelogram are 10 cm and 15 cm in length. In a similar parallelogram the corresponding sides are 8 cm and 12 cm in length.

(*a*) What is the ratio of their perimeters?

(*b*) What is the ratio of their areas?

(*a*) The perimeters are in the same ratio as the lengths of the corresponding sides:

$$P_1 : P_2 :: 10 : 8$$

The ratio of the first parallelogram's perimeter to that of the second is $10:8$ or $5:4$.

(*b*) The areas are in the ratio of the squares of the corresponding sides:

$$A_1 : A_2 :: 10^2 : 8^2$$

The ratio of the first parallelogram's area to that of the second is $100:64$ or $25:16$.

Supplementary Problems

The answers to this set of problems are at the end of this chapter.

11.33. If one of two complementary angles is 15°, how large is the other?

11.34. If one of two supplementary angles is 45°, how large is the other?

11.35. The ratio of the magnitudes of two supplementary angles is 4 : 5. How large are the two angles?

11.36. Two straight lines intersect so that one of the angles formed is 75°. How large are the other three angles?

11.37. In Fig. 11-44:

  (a) Identify the types of the angles *ZYX*, *ZYW*, *ZYV*.

  (b) What is point *Y* called?

  (c) What is the line *YW* called if ∠*VYW* = ∠*WYZ*?

  (d) If ∠*ZYX* measured counterclockwise is 150°, how large is the associated angle measured clockwise from *ZY*?

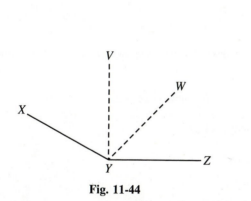

Fig. 11-44   Fig. 11-45

11.38. In Fig. 11-45, *AB*∥*CD*, *FG*∥*HI*, *BE* ⊥ *AB*, *KD* ⊥ *BJ*, *DJ* ⊥ *HI*. List the pairs of equal angles among those that are numbered.

11.39. In Fig. 11-46, *W* ⊥ *AB*, *N* ⊥ *AC*, *F*∥*AC*. If *ACB* is 65°, how large is (a) ∠1, (b) ∠2?

11.40. In Fig. 11-47, V_a ⊥ R_a, V_b ⊥ R_b.

  (a) How do angles 1 and 2 compare in size?   (b) Give a reason for your answer in (a).

11.41. Add: (a) 120°45′22″ and 16°17′38″,   (b) 149°17′35″ and 210°42′25″.

11.42. Subtract: (a) 12°30′50″ from 42°12′30″,   (b) 182°56′42″ from 310°7″.

11.43. Convert: (a) 17′37″ into seconds,   (b) 197′ into a decimal number of degrees.

11.44. In △*PQR* in Fig. 11-48, *PQ* ⊥ *QR*.

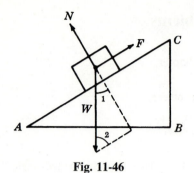
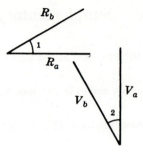
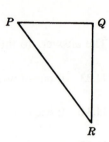

| Fig. 11-46 | Fig. 11-47 | Fig. 11-48 |

(a) How many degrees are there in $\angle QPR + \angle QRP$?

(b) What name is given to PR?

(c) What name is given to $\angle PQR$?

(d) If $PQ = QR$, what type is $\triangle PQR$?

11.45. In $\triangle XYZ$ in Fig. 11-49, $XY = YZ = ZX = 20$ cm.

(a) What type is $\triangle XYZ$?

(b) How many degrees are there in each of the interior angles?

(c) A perpendicular line segment is dropped from Z to XY.

    (1) What is the length of this line segment?

    (2) How many degrees will there be in each of the angles of the two triangles formed by the line segment in (1) above?

11.46. (a) What type is $\triangle XYZ$ in Fig. 11-49?   (b) If $\angle X = 70°$, how large is $\angle Y$? $\angle Z$?

11.47. In $\triangle STV$ in Fig. 11-50, $SV \perp VT$. Calculate the length of ST.

11.48. In Fig. 11-52, $AB \| DE$, $EF \perp DF$, $AC \perp DE$, $BC \perp DF$.

(a) What is the relationship between $\triangle EFD$ and $\triangle ABC$?

(b) $DF = 50$ cm, $EF = 20$ cm. If $AB = 8$ m, how long is DE? AC? BC?

11.49. In the triangles of Fig. 11-53, $v_a = v_b$, $R_1 = R_2$, $R_1 \perp v_b$, $R_2 \perp v_a$.

(a) What is the relationship between the $\triangle$?   (c) To which angles is $\angle 3$ equal?

(b) What type are these triangles?   (d) To which angles is $\angle B$ equal?

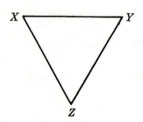
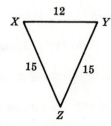
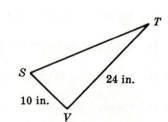

| Fig. 11-49 | Fig. 11-50 | Fig. 11-51 |

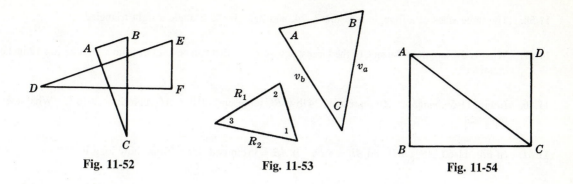

Fig. 11-52   **Fig. 11-53**   **Fig. 11-54**

11.50. In Fig. 11-54:

(a) If *AB* represents 9 lb-f and *BC* represents 12 lb-f, what does *AC* represent?

(b) If *AC* represents 100 mph and *DC* represents 50 mph, what does *AD* represent?

11.51. What is the dimension of the indicated altitude in the triangle of Fig. 11-55?

11.52. The two triangles in Fig. 11-56 are similar.   If *W* represents a force of 60 lb-f, what force is represented (a) by *T*; (b) by *R*?

11.53. Triangles *ABC* and *A'B'C'* in Fig. 11-57 have their respective sides parallel to each other.   *AB* = 4 ft; *BC* =5 in.   If *B'D'* = 2*B'C'* and represents a force of 40 lb-f, what force is represented by *A'B'*?   (Figure 11-57 is not drawn to scale.)

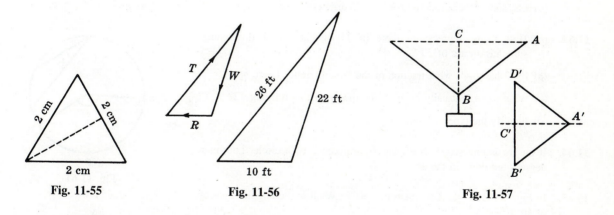

Fig. 11-55   **Fig. 11-56**   **Fig. 11-57**

11.54. In Fig. 11.58, *AB* ⊥ *BC*, *EF* ⊥ *BC*, *DF* ⊥ *AC*.

(a) Which pairs of angles are equal?

(b) Express the relationships between the sides as 3 pairs of ratios.

11.55. The two legs of a right triangle have lengths 3.0 ft and 8.0 ft.   What is the length of the hypotenuse?

11.56. The hypotenuse of a right triangle is 0.25 m long; one of the legs is 0.15 m in length.   Compute the length of the other leg.

11.57. A certain telephone pole is 80 feet high.   A brace is run from a point 3/4 of the way up the pole to a point 20 feet from the base of the pole.   How long is the brace?

11.58. The three sides of a triangle measure 7, 24, and 25. Is the triangle a right triangle?

11.59. What is the length of the side of the largest square beam that can be cut out from a circular log 12 inches in diameter?

11.60. In Fig. 11-59, $mn \| pq$. AB represents 24 lb-f; BC represents 40 lb-f; AD represents 30 lb-f. What does DE represent?

11.61. In Fig. 11-60, $AB \perp AD$ and $AB = CD$. If $AB = 35$ cm and $AD = 55$ cm, how long is AC?

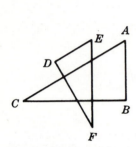

Fig. 11-58

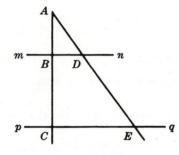

Fig. 11-59

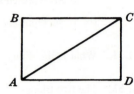

Fig. 11-60

11.62 The sides of a given quadrilateral are equal to each other but its diagonals are unequal.

    (*a*) What is the name of such a figure (it is known that its opposite sides are parallel)?

    (*b*) List some other properties of this quadrilateral.

11.63. (*a*) How many degrees are there in $\angle 1$ of Fig. 11-61 if the angle subtends an arc of 20°?

    (*b*) If $\angle 2$ has 144°, what fraction of the circumference is arc *mpn*?

    (*c*) What name is usually given to the line segments forming $\angle 2$? $\angle 1$?

    (*d*) Identify the type of figure *Ompn*.

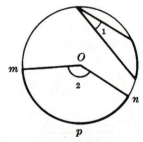

Fig. 11-61

11.64. If an arc is one-sixteenth of the circumference of a circle, how many degrees are there in the arc?

11.65. In Fig. 11-62, angle ABC measures 50°. What is the angular measure of arc AXC?

11.66. In Fig. 11-62, are BD measures 42°. What is the magnitude of angle BCD?

11.67. What angle does a radius make with the tangent at the outer end of the radius?

11.68. How many degrees are there in a central angle subtended by a semicircle? By a quadrant?

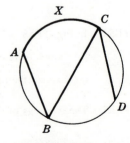

Fig. 11-62

11.69. How many degrees are there in an arc which is subtended by a chord equal to a radius?

11.70. The side of a rhombus is 2.5 inch. Calculate the perimeter.

11.71. If the diagonal of a square is 8 cm, compute its perimeter.

11.72. Calculate the circumference of a circle whose diameter is 0.40 m.

11.73. If the circumference of a circle is 62.8 inch, compute its radius.

11.74. A triangle has an altitude of 32 mm and a base of 56 mm.   Compute the area of the triangle.

11.75. Find the area of a rhombus whose diagonals are 3.0 inch and 1.6 inch.

11.76. One of the legs of a rectangle is 6 cm.   The diagonal is 10 cm.   Calculate the area of the rectangle.

11.77. Compute the area of a circle with a diameter of 50 mm.

11.78. Calculate the area of a sector subtending an arc of 22.5° in the circle of Problem 11.77.

11.79. Calculate the ratio of the area of a circle 1.0 m in diameter to the area of a square 1.0 m on a side.

11.80. In Fig. 11-63, $pq \perp rs$, $pq = 5$ mm, $rs = 3$ mm.   Calculate the shaded area.

11.81. In $ABCD$ in Fig. 11-64, $AC = 14$ cm, $BD = 31$ cm, $GH = 4.2$ cm, $AE = EB$, $CF = FD$, $AC \| BD$.

    (a) What name is given to a figure like $ABCD$?

    (b) How long is EF?

    (c) Calculate the area of $ABCD$.

11.82. In Problem 11.80 what is:

    (a) the ratio of the circumference of the larger circle to that of the smaller?

    (b) the ratio of the area of the smaller circle to that of the larger?

11.83. If the triangle ABC is cut out of a rectangular piece of brass sheet as shown in Fig. 11-64, what percent of the sheet is wasted?

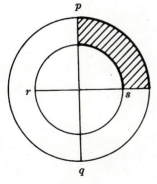

Fig. 11-63

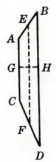

Fig. 11-64

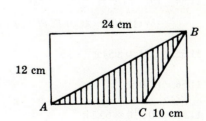

Fig. 11-65

Answers to Suplementary Problems

11.33. 75°

11.34. 135°

11.35. 80°; 100°

11.36. 105°, 75°, 105°

11.37. (*a*) obtuse; acute; right   (*b*) vertex   (*c*) bisector   (*d*) 210°

11.38. 1, 4; 2, 3; 3, 8; 7, 8; 1, 6; 4, 6; 2, 5; 3, 5; 3, 7

11.39. 25°, 65°

11.40. (*a*) they are equal   (*b*) theorem on equal angles

11.41. (*a*) 137°3′   (*b*) 360°

11.42. (*a*) 29°41′40″   (*b*) 127°3′25″

11.43. (*a*) 1023″   (*b*) 3.28° approximately

11.44. (*a*) 90°   (*b*) hypotenuse   (*c*) right   (*d*) isosceles; right

11.45. (*a*) equilateral   (*b*) 60°   (*c*) (1) $10\sqrt{3}$; (2) 30°, 60°, 90°

11.46. (*a*) isosceles   (*b*) 70°; 40°

11.47. 26 inches

11.48. (*a*) they are similar   (*b*) $10\sqrt{29} \approx 53.9\,\text{cm}$; 20 cm; $4\sqrt{29} \approx 21.5\,\text{cm}$

11.49. (*a*) they are similar   (*b*) isosceles   (*c*) C (*d*) $A, \angle 1, \angle 2$

11.50. (*a*) 15 lb   (*b*) $50\sqrt{3} \approx 86.7\,\text{mph}$

11.51. $\sqrt{3} \approx 1.732\,\text{cm}$

11.52. (*a*) 70.9 lb   (*b*) 27.3 lb

11.53. 192 lb

11.54. (*a*) $\angle EDF = \angle ABC,\ \angle FED = \angle BAC,\ \angle DFE = \angle ACB$

      (*b*) $AB : DE :: AC : EF;\ BC : DF :: AB : ED;\ BC : DF :: AC : EF$

11.55. $\sqrt{73} \approx 8.54\,\text{ft}$

11.56. 0.20 m

11.57. $20\sqrt{10} \approx 63.2\,\text{ft}$

11.58. yes

11.59. $6\sqrt{2} \approx 8.5$ in.

11.60. 50 lb

11.61. $5\sqrt{170} \approx 65.2$ cm

11.62. (a) rhombus
(b) diagonals bisect each other and related angles; opposite angles are equal; adjacent angles are supplementary

11.63. (a) $10°$ (b) 144/360, 2/5, or 0.4   (c) radii; chords   (d) sector

11.64. $22.5°$

11.65. $100°$

11.66. $21°$

11.67. $90°$

11.68. $180°$; $90°$

11.69. $60°$

11.70. 10 in.

11.71. $16\sqrt{2} \approx 22.6$ cm

11.72. $0.4\pi \approx 1.26$ m

11.73. $\dfrac{31.4}{\pi} \approx 10$ in.

11.74. 896 mm^2

11.75. 2.4 in^2

11.76. 48 cm^2

11.77. $625\pi \approx 1960$ mm^2

11.78. $\dfrac{625\pi}{16} \approx 122$ mm^2

11.79. $\pi : 4$ or $0.79 : 1$ approximately

11.80. π or 3.14 mm^2 approximately

11.81. (a) trapezoid   (b) 22.5 cm   (c) 94.5 cm^2

11.82. (a) 5 : 3   (b) 9 : 25

11.83. 71% approximately

Chapter 12

Solid Figures

Most of the applications of solid geometry to elementary science and technology require only a knowledge of the names, shapes, and components of regular solid figures and the ability to compute their areas and volumes.

Terms

12.1 POLYHEDRONS

A *polyhedral angle* is formed when three or more planes meet at a common point called the *vertex*. The intersections of the planes are called *edges*. The sections of the planes between the edges are called *faces*. Shown in Fig. 12-1 is a polyhedral angle with five faces and five edges.

A *polyhedron* is a solid bounded by planes.

A *regular polyhedron* has faces which are congruent regular polygons and equal polyhedral angles. Three out of five possible regular polyhedrons are shown in Fig. 12-2.

Two types of polyhedrons are of special interest in science: the prism and the pyramid.

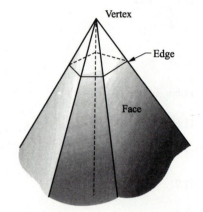

Fig. 12-1. Polyhedral angle

A *prism* is a polyhedron two faces of which are congruent polygons in parallel planes and whose other faces are formed by planes through the corresponding sides of the polygons. The congruent polygons are the *bases* of the prism, while the other faces are the *lateral faces* of the prism.

(*a*) Tetrahedron   (*b*) Hexahedron   (*c*) Octahedron

Fig. 12-2. Regular polyhedrons

If the lateral edges are perpendicular to the bases, the prism is a *right prism*. A right triangular prism is shown in Fig. 12-3.

A *regular prism* has bases which are regular polygons.

The *altitude* of a prism is the perpendicular distance between its bases.

A *parallelepiped* is a prism whose bases are parallelograms. This solid and its altitude are shown in Fig. 12-4.

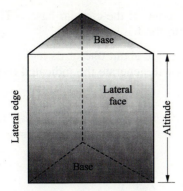

Fig. 12-3. Right triangular prism

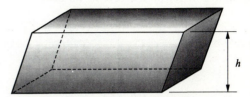

Fig. 12-4. Parallelepiped and its altitude, *h*

A *rectangular solid* or *right parallelepiped* is a parallelepiped of which all the faces are rectangles. A *cube* is a right parallelepiped of which all the six faces are congruent squares.

A *pyramid* (Fig. 12-5) is a polyhedron of which one face is a polygon, called its *base* (*ABCD* in Fig. 12-5), and the other lateral faces are triangles meeting at a common point, the *vertex* (*O*). The perpendicular from the vertex to the base is the *altitude* of the pyramid (*OP*).

A *regular* or *right* pyramid has a regular polygon as its base, and its altitude *h* meets the base at its center. The *slant height* of a regular pyramid is the altitude of one of the lateral faces. The pyramid in Fig. 12-5 is a *square right pyramid* with the base *ABCD* a square and the slant height *OS*.

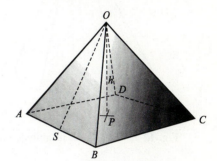

Fig. 12-5. Regular pyramid, its altitude, and slant angle

12.2 CYLINDERS

A *cylinder* is bounded by two parallel planes and by the surface generated by a line segment rotating parallel to itself. The sections of the parallel planes are the *bases* of the cylinder.

A *right circular cylinder* can be generated by revolving a rectangle about one of its sides. Thus, if rectangle *ABCD* in Fig. 12-6 is revolved about *AD* as an axis, a right circular cylinder is generated with circles for its bases.

The *axis* of a right circular cylinder is the line joining the centers of the circular bases. The *altitude* of a cylinder is the perpendicular distance between its bases.

Fig. 12-6. Right circular cylinder with its altitude and generating rectangle

12.3 CONES

A *cone* is a solid formed by joining with straight-line segments the points of a closed curve to a point outside the plane of the curve, as shown in Fig. 12-7. The closed curve (*ABCA*) is the *base* of the cone; the right (*V*) is the *vertex* of the cone. The joining lines are the *elements*. The *axis* of a right circular cone is the line segment joining the vertex to the midpoint of the circular base. The *altitude* of a cone is the perpendicular *VP* drawn from the vertex to the base.

A *right circular cone* may be generated by revolving a right triangle about one of its legs as an axis. In Fig. 12-8 revolution of triangle *ABC* about *AB* as an axis will generate the indicated right circular cone, with a circle as its base.

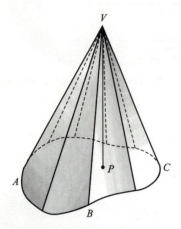

Fig. 12-7. Cone with its altitude (*VP*)

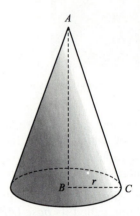

Fig. 12-8. Right circular cone and its generating triangle

The *slant height* of a right circular cone is the length of one of its elements (*AC* in Fig. 12-8). The *lateral area* of the cone is the area of the conical surface, exclusive of the base.

12.4 SPHERES

A *spherical surface* is a curved surface all points of which are equidistant from the point called the *center*.

A *sphere* is a solid bounded by a closed spherical surface. In Fig. 12-9 the center of the sphere is *O*; *OA* is a radius; *AB* is a diameter; *CD* is a chord.

A *great circle* of the sphere is any circle made by a plane passing through the center of the sphere (*ADBA*, Fig. 12-9).

A *small circle* is a circle made by a plane cutting the sphere but not through its center (*EHFE*, Fig. 12-9). A *minor arc* is any circular arc on the sphere whose length is less than one-half the circumference of the circle.

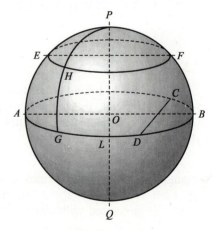

Fig. 12-9. Sphere

12.5 SIMILAR SOLIDS

Two solids are said to be *similar* if all their corresponding dimensions are in the same ratio. For example, in Fig. 12-10 if the altitude and radius of the smaller right circular cylinder are one-half as large as the altitude and radius of the larger cylinder, then the two cylinders are similar.

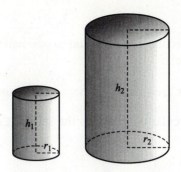

Fig. 12-10. Similar cylinders

Areas of Solids

12.6   AREA OF A PRISM

The lateral area is the sum of the areas of the lateral faces.   The lateral area S of a right prism equals the product of the base perimeter p and the altitude h:

$$S = ph$$

The total area A of any prism is the sum of the lateral area S and the areas of the two bases, B_1 and B_2:

$$A = S + B_1 + B_2$$

12.7   AREA OF A PYRAMID

The lateral area S of a regular pyramid is equal to one-half the product of its slant height L and the base perimeter P:

$$S = \frac{1}{2}LP$$

The total area A of any pyramid is the sum of the lateral area S and the area of the base B:

$$A = S + B$$

12.8   AREA OF A CYLINDER

The lateral area S of a right circular cylinder equals the product of the circumference C of its base and its altitude h:

$$S = Ch = 2\pi rh \qquad \text{where } r = \text{radius of the base}$$

The total area A of a right circular cylinder equals the sum of the lateral area S and the area of the two bases $2B$:

$$A = S + 2B$$
$$= 2\pi rh + 2\pi r^2 \qquad \text{where } r = \text{radius of base}$$
$$= 2\pi r(h + r)$$

12.9   AREA OF A CONE

The lateral area S of a right circular cone is equal to one-half the product of the circumference C of its base and the slant height L:

$$S = \frac{1}{2}CL = \pi r L \qquad \text{where } r = \text{radius of base}$$

The total area A of a right circular cone equals the sum of the lateral area S and the area of the base B:

$$
\begin{aligned}
A &= S + B \\
  &= \pi r L + \pi r^2 \qquad \text{where } r = \text{radius of base} \\
  &= \pi r (L + r)
\end{aligned}
$$

12.10   AREA OF A SPHERE

The area A of a sphere of radius r is four times the surface area of a great circle of the sphere:

$$A = 4\pi r^2$$

12.11   AREAS OF SIMILAR SOLIDS

The lateral and total surface areas of similar figures are to each other as the squares of their corresponding linear dimensions. For example, the area of the smaller cylinder considered in Section 12.5 is $\left(\frac{1}{2}\right)^2$ or one-fourth the corresponding area of the larger cylinder.

Volumes of Solids

12.12   VOLUME OF A PRISM

The volume V of any prism equals the product of its base area B and its altitude h:

$$V = Bh$$

The volume of a rectangular parallelepiped is the product of its three linear dimensions: $V = abc$, where a, b, c are the lengths of three intersecting edges.

12.13   VOLUME OF A PYRAMID

The volume V of any pyramid is equal to one-third the product of its base area B and altitude h:

$$V = \frac{1}{3}Bh$$

12.14   VOLUME OF A CYLINDER

The volume of any cylinder is the product of its base area B and its altitude h:

$$V = Bh$$

The volume V of a circular cylinder with a base radius r and altitude h is given by the formula

$$V = \pi r^2 h$$

12.15   VOLUME OF A CONE

The volume V of any cone is equal to one-third the product of the base area B and the altitude h:

$$V = \frac{1}{3}Bh$$

The volume V of a circular cone with a base radius r is given as a special case by the formula

$$V = \frac{1}{3}\pi r^2 h$$

12.16   VOLUME OF A SPHERE

The volume V of a sphere of a radius r is given by the formula

$$V = \frac{4}{3}\pi r^3$$

12.17   VOLUMES OF SIMILAR SOLIDS

The volumes of similar solids are to each other as the cubes of their corresponding linear dimensions.   The volume of the smaller cylinder considered in Section 12.5 is $\left(\dfrac{1}{2}\right)^3$ or one-eighth the volume of the larger cylinder.

Solved Problems

12.1.   Given a cube whose edges are 5 cm, calculate (*a*) the total area of the cube, and (*b*) the volume of the cube.

(*a*) The given cube is a rectangular solid with six square faces.   The area a of each face is

$$a = 5^2 = 25 \text{ cm}^2$$

The total area is six times the area of a single face.

$$A = 6a = (6)(25) = 150$$

The area is 150 cm^2.

(*b*) The volume V equals the product of the lengths of three equal intersecting edges.

$$V = (5)(5)(5) = 5^3 = 125$$

The volume is 125 cm^3.

With an HP38G calculator these computations can be carried out as follows:

The area of each face is: 5{x^y}2{ENTER}.   The result is 25.

The total area is: {*}6{ENTER}   ← "The previous result is multiplied by 6."

The volume V is: 5{x^y}3{ENTER}.   The result is 125.

12.2. Given the rectangular solid shown in Fig. 12-11, calculate the solid's (*a*) lateral area, (*b*) total area, and (*c*) volume.

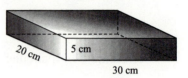

    (*a*) The base perimeter *p* of the solid is

$$p = 20 + 30 + 20 + 30 = 100 \text{ cm}$$

    and the altitude *h* is 5 cm. The lateral area *S* is then

$$S = ph = (100)(5) = 500 \quad \text{or} \quad S \text{ is } 500 \text{ cm}^2$$

Fig. 12-11

    (*b*) The area of each base is $A_1 = A_2 = (20)(30) = 600$. The total area $A = S + A_1 + A_2 = 500 + 600 + 600$, or *A* is 1700 cm².

    (*c*) The volume is the product of the lengths of three intersecting edges:

$$V = (30)(20)(5) = 3000$$

The volume of the solid is 3000 cm³.

With an HP-38G calculator these computations can be carried out as follows:

    The perimeter is: 20{+}30{+}20{+}30{ENTER}.   ← "The result is 100."

    The lateral area *S* is: {*}5{ENTER}.   ← "The previous result is multiplied by 5."

    Store the lateral area into the variable S: Keying the following sequence <u>STO▶</u> {A...Z}{S}{ENTER}

    The area of each base is: 20{*}30{ENTER}   ← "The result is 600."

    The total area *S* is: {*}2 + {A...Z}{S}{ENTER}   "The previous result is multiplied by 2 and added to *S* (the latter value was recalled from the memory variable *S*). The result is 1700."

    The volume of the solid is: 20{*}30{*}5{ENTER}.   ← "The result is 3000."

12.3. The volume of a rectangular prism is 540 in³. If the length is 6 inches and the height is 18 inches, determine (*a*) the width of the prism, (*b*) the area of the smallest face.

    (*a*) A rectangular prism is also a rectangular parallelepiped or solid. The volume $V = lwh$, where l = length, w = width, h = height. Substitute the given values in the formula and solve for *w*:

$$540 = (6)(w)(18) = 108w$$

$$w = \frac{540}{108} = 5, \text{ or the width is 5 inches}$$

    (*b*) The smallest faces are those of area $A = lw$.

$$A = (6)(5) = 30 \quad \text{or} \quad A \text{ is } 30 \text{ in}^2$$

12.4. A triangular prism has a base area of 0.45 m² and an altitude of 0.07 m. Calculate the volume of the prism.

    The volume is the product of the base and the altitude.

$$V = Bh = (0.45)(0.07) = 0.0315$$

    The volume is 0.0315 m².

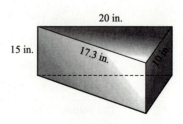

12.5. The dimensions of a right triangular prism are given in Fig. 12-12. The angle between the 17.3-inch and the 10-inch edges is 90°. Calculate the prism's (*a*) lateral area, (*b*) total area, (*c*) volume.

Fig. 12-12

(a) The perimeter of the base of the prism is

$$P = 17.3 + 10 + 20 = 47.3 \text{ in.}$$

The altitude h is 15 in.   The lateral area is

$$S = Ph = (47.3)(15) = 709.5 \quad \text{or} \quad S \text{ is } 710 \text{ in}^2 \text{ approximately}$$

(b) The two prism bases are congruent right triangles, and the area of one triangle is

$$B_1 = (1/2)(17.3)(10)$$

The area of the two bases is therefore

$$B_1 + B_2 = (1/2)(17.3)(10) + (1/2)(17.3)(10) = 173 \quad \text{or} \quad \text{the area is } 173 \text{ in}^2$$

The total area of the prism is

$$A = S + B_1 + B_2 \approx 710 + 173 = 883 \quad \text{or} \quad \approx 883 \text{ in}^2$$

(c) The volume of the prism is

$$V = Bh = (1/2)(17.3)(10)(15) \approx 1298 \quad \text{or} \quad V \text{ is } 1298 \text{ in}^2 \text{ approximately}$$

12.6. A square pyramid has a base 5 cm on an edge and an altitude of 12 cm.   Calculate the volume of the pyramid.

The area of the base is $B = (5)^2 = 25$.   The volume of the pyramid is

$$V = (1/3)Bh = (1/3)(25)(12) = 100 \quad \text{or} \quad V \text{ is } 100 \text{ cm}^3$$

12.7. A triangular pyramid has a base 10 in. on each side and an altitude of 8 in.   Calculate the volume of the pyramid.

The area of the base is

$$B = \sqrt{s(s-a)(s-b)(s-c)} = \sqrt{15(15-10)(15-10)(15-10)}$$
$$= \sqrt{(15)(5)(5)(5)} \approx 43.3$$

The volume of the pyramid is

$$V = Bh \approx (43.3)(8) \approx 346, \quad \text{or} \quad V \text{ is } 346 \text{ in}^3 \text{ approximately}$$

12.8. A square pyramid has a base edge of 1.2 m and a lateral edge 1.0 m as shown in Fig. 12-13.   Calculate the pyramid's   (a) lateral area,   (b) total area,   (c) volume.

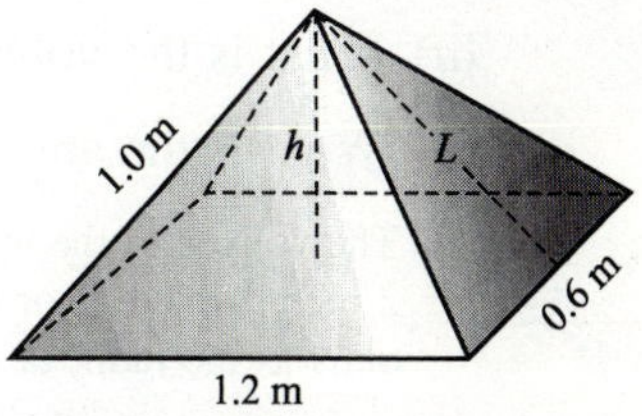

Fig. 12-13

(a) Calculate the slant height from the Pythagorean theorem:

$$L = \sqrt{1.0^2 - 0.6^2} = 0.8$$

The base perimeter is $P = 4(1.2) = 4.8$.   The lateral area is

$$S = (1/2)LP = (1/2)(0.8)(4.8) = 1.92 \quad \text{or} \quad S \text{ is } 1.92 \text{ m}^2$$

(b) The area of the base is $B = (1.2)^2 = 1.44$.   The total area of the pyramid is $A = S + B = 1.92 + 1.44 = 3.36$, or A is 3.36 m^2.

(c) The altitude is calculated by the Pythagorean theorem:

$$h = \sqrt{0.8^2 - 0.6^2} = \sqrt{0.64 - 0.36} = \sqrt{0.28} \approx 0.531$$

The volume of the pyramid is

$$V = (1/3)Bh = (1/3)(1.44)(0.53) \approx 0.254 \quad \text{or} \quad V \text{ is } 0.25 \text{ m}^3 \text{ approximately}$$

12.9. A right circular cylinder has a base with a radius of 4.6 cm and an altitude of 8.5 cm. Calculate the cylinder's (*a*) lateral area, (*b*) total area, (*c*) volume.

(*a*) The circumference of the base is

$$C = 2\pi r = 2(3.14)(4.6) \approx 28.9$$

The lateral area is

$$S = Ch \approx (28.9)(8.5) \approx 246 \quad \text{or} \quad S \text{ is } 246 \text{ cm}^2 \text{ approximately}$$

(*b*) The area of each base is

$$B = \pi r^2 \approx (3.14)(4.6)^2 \approx 66.4$$

The total area is $A = S + 2B \approx 246 + 2(66.4) \approx 379$, or A is 379 cm^2 approximately.

(*c*) The volume is

$$V = \pi r^2 h \approx 3.14(4.6)^2(8.5) \approx 560$$

Alternatively, $V = Bh \approx (66.4)(8.5) \approx 560$, or V is 560 cm^3 approximately.

12.10. What is the radius of a right circular cylinder with a volume of 628 in^3 and an altitude of 5 inches?

Substitute the given values in $V = \pi r^2 h$.

$$628 = (3.14)(r^2)(5)$$

Solve for r^2.

$$r^2 = \frac{628}{(3.14)(5)} = 40$$

$$r = \sqrt{40} \approx 6.3 \quad \text{or} \quad r \text{ is } 6.3 \text{ in. approximately}$$

12.11. A cylindrical tube 9 feet long has an inside diameter of 2 inches and an outside diameter of 3 inches.

(*a*) What is the volume of the metal in the tube?

(*b*) What is the area of the outside surface?

(*a*) The volume of the metal is the difference between the volumes of right circular cylinders with 3-inch and 2-inch diameters, or radii of 1.5 and 1.0 in. The altitude is the pipe lengths, which must be in the *same units* as the radii, or $h = (9)(12) = 108$ in. Thus,

$$\begin{aligned} V_{\text{metal}} &= \pi(1.5^2)(108) - \pi(1.0^2)(108) \\ &= \pi(1.08)(1.5^2 - 1.0^2), \text{ by factoring} \\ &\approx (3.14)(108)(2.25 - 1.0) \approx 424 \end{aligned}$$

or V_{metal} is 424 in^3 approximately.

(*b*) The area of the outide surface is

$$S = Ch = 2\pi rh \approx 2(3.14)(1.5)(108) \approx 1017 \quad \text{or} \quad S \text{ is } 1017 \text{ in}^2 \text{ approximately}$$

With an HP-38G calculator these computations can be carried out by keying in the sequnce shown below.

{SHIFT}{π}{*}1.5{x^y}2{*}108{−}{SHIFT}{π}{*}1.02{x^y}2{*}108{ENTER}

The display shows 424.1150008235.   This result can be rounded to 424.

12.12. A right circular cone (see Fig. 12-14) has a 15-cm radius and a 20-cm altitude.   Determine the cone's   (a) lateral area,   (b) total area,   (c) volume.

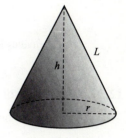

Fig. 12-14

(a) The cone's altitude, radius, and slant height form a right triangle.   From the Pythagorean theorem,

$$L = \sqrt{h^2 + r^2} = \sqrt{20^2 + 15^2} = 25 \quad \text{or} \quad L \text{ is 25 cm}$$

The lateral area is $S = \pi r L = 3.14(15)(25) \approx 1180$, or S is $1180 \, \text{cm}^2$ approximately.

(b) The total area is $A = S + B = \pi r L + \pi r^2 = \pi r(L + r)$.

$$A \approx 3.14(15)(15 + 25) \approx 1880 \quad \text{or} \quad A \text{ is } 1880 \text{ cm}^2 \text{ approximately}$$

(c) The volume is $V = \pi r^2 h/3$.

$$V = 3.14(15^2)(20)/3 \approx 14{,}100/3 \approx 4700 \quad \text{or} \quad V \text{ is } 4700 \text{ cm}^3 \text{ approximately}$$

12.13. A right circular cone (see Fig. 12-14) has a volume of $2.512 \, \text{ft}^3$.   If the altitude is $0.250 \, \text{ft}$, calculate the radius of the base.

Substitute the given values in the formula $V = \dfrac{\pi r^2 h}{3}$ and solve for r.

$$2.52 = 3.14(r^2)(0.25)/3$$

$$r^2 = \frac{(2.512)(3)}{3.14(0.25)} \approx 9.60$$

$r \approx \sqrt{9.60} \approx 3.1$, and the radius is 3.1 ft approximately.

12.14. A sphere has a radius of $5 \, \text{cm}$.   Calculate the sphere's   (a) area and   (b) volume.

(a) The area is $A = 4\pi r^2 \approx 4(3.14)(5^2) \approx 314$, or A is $314 \, \text{cm}^2$ approximately.

(b) The volume is $V = (4/3)\pi r^3 \approx (4/3)(3.14)(5^3) \approx 523$, or V is $523 \, \text{cm}^3$ approximately.

12.15. Determine the radius of a sphere whose surface area is $2464 \, \text{ft}^2$.

Substitute the given value in $A = A\pi r^2$.

$$2464 = 4(3.14)r^2$$

Solve for r.

$$R = \sqrt{\frac{2464}{4(3.14)}} \approx 14 \quad \text{or} \quad \text{the radius is 14 ft approximately}$$

12.16. Determine the diameter of a sphere whose volume is $113.14 \, \text{in}^3$.

Substitute the given value in $V = 4/3\pi r^3$.

$$113.14 = (4/3)(3.14)r^3$$

Then,

$$r^3 = \frac{113.14}{(4/3)(3.14)} \approx 27$$
$$r \approx \sqrt[3]{27} = 3$$

The radius is approximately 3 inches.   The diameter is approximately 6 inches.

12.17. If the edge of a cube is doubled, what will be the effect on:

(*a*) the diagonal of a face,   (*b*) the area,   (*c*) the volume

(*a*) The diagonal $d = \sqrt{S^2 + S^2} = \sqrt{2S^2} = \sqrt{2}S$, where $S =$ length of edge.   Therefore, d varies directly as the length of the edge.   If S is doubled, d is also doubled.

(*b*) The area will be quadrupled (2^2), since it varies as the square of the edge length.

(*c*) The volume will increase 2^3 or 8 times, since it varies as the cube of the edge.

12.18. A can has the greatest capacity for a fixed amount of metal if its diameter is made equal to its height.   If such a can is to hold $6280\,\text{cm}^3$, what should be its height and diameter?

Substitute $h = 2r$ and $V + 6280$ in $V = \pi r^2 h$

$$6280 = 3.14(r^2)2r$$
$$r^3 = \frac{6280}{6.28} = 1000$$
$$r = \sqrt[3]{1000} = 10$$

The radius is 10 cm.   The diameter and height should each be 20 cm.

12.19. The radii of two similar right cylinders are 2 m and 6 m.

(*a*) What is the ratio of their altitudes?

(*b*) What is the ratio of their total surface areas?

(*c*) What is the ratio of their volumes?

(*a*) In similar solids all corresponding linear dimensions are in the *same* ratio.   Since the ratio of the radii is 2 : 6 or 1 : 3, the ratio of the altitudes is also 1 : 3.

(*b*) In similar solids the areas vary as the *square* of the linear dimensions.   Thus, the ratio of the surface area is $\left(\frac{1}{3}\right)^2 : 1$, or $\frac{1}{9} : 1$, or 1 : 9.

(*c*) In the two similar cylinders the volumes will be in the same ratio as the *cubes* of the linear dimensions.   Thus, the ratio is $\left(\frac{1}{3}\right)^3 : 1$, or $\frac{1}{27} : 1$, or 1 : 27.

12.20. A hexagonal prism has a base that measures 12 inches on a side.   The altitude of the prism is 40 inches.   Determine the prism's

(*a*) lateral areas,   (*b*) total area,   and   (*c*) volume.

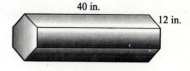

The bases of a hexagonal prism are regular hexagons or six-sided polygons of equal sides, as shown in Fig. 12-15.

(a) The lateral area is $S = Ph = (6)(12)(40) = 2880$. S is $2880\,\text{in}^2$.

Fig. 12-15

(b) In the given prism, each base can be divided into six equilateral triangles with 12-inch sides. By the Pythagorean theorem, the altitude of each triangle is $\sqrt{12^2 - 6^2} = \sqrt{108} \approx 10.39$.

The area of each triangle $\approx (1/2)(10.39)(12) \approx 62.3$.

The area of each base $= (6)(62.3) \approx 374$.

The total area of the prism is

$$A = S + B + B \approx 2880 + 2(374) = 3628 \quad \text{or} \quad A \text{ is } 3630\,\text{in}^2 \text{ approximately}$$

(c) The volume of the prism is

$$V = Bh \approx (374)(40) = 14{,}960 \quad \text{or} \quad V \text{ is } 15{,}000\,\text{in}^3 \text{ approximately}$$

Supplementary Problems

The answers to this set of problems are at the end of this chapter.

12.21. What different names can be given to the solids represented by (a) Fig. 12-16(a) and (b) Fig. 12-16(b).

12.22. (a) How many faces has
  (1) a triangular prism?
  (2) a hexagonal prism?
  (3) a tetrahedron?

(b) How many edgtes has
  (1) a quadrangular prism?
  (2) an octagonal pyramid
  (3) a hexahedron?

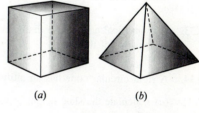

(a)   (b)

Fig. 12-16

12.23. A cube is 8 cm on an edge.

(a) Calculate the surface area of the cube.

(b) Calculate its volume.

(c) Calculate the length of a diagonal.

12.24. Three intersecting edges of a rectangular parallelepiped are 4, 8, and 12 inches.

(a) Calculate the volume.

(b) Calculate the surface area.

(c) Calculate the length of a diagonal.

12.25. The linear dimensions of a certain cube are three times the dimensions of another cube.

(a) What is the value of the ratio, volume of larger cube : volume of smaller cube?

(b) What is the value of the ratio, area of smaller cube : area of larger cube?

12.26. A right pyramid has a square base 6 cm on an edge; the lateral edges are 5 cm long.

    (*a*) Calculate the slant height.   (*c*) Calculate the lateral area.

    (*b*) Calculate the altitude   (*d*) Calculate the volume.

12.27. A prism and a pyramid have bases and altitudes which are respectively equal.   What is the ratio of their volumes?

12.28. The base of any pyramid is 0.6 m^2 in area.   The altitude is 0.8 m.   A plane parallel to the base cuts 0.3 m from the vertex.

    (*a*) Compute the area of the section by the plane.

    (*b*) Compute the volume of the bottom portion of the resulting figure.

12.29. The volume of a cylinder is 1200 cm^3; its altitude is 15 cm.   Calculate the area of the base.

12.30. What is the locus of all points in space equidistant from a point?

12.31. A right cylinder has a base with a diameter of 6 inches.   The height of the cylinder is 10 inches.

    (*a*) Calculate the lateral area of the cylinder.

    (*b*) Calculate the total area of the cylinder.

    (*c*) Compute the volume of the cylinder.

12.32. The volume of a right circular cylinder is 3142 in^3.   The altitude is 10 inches.

    (*a*) Compute the radius of the cylinder   (*b*) Calculate the lateral area.

12.33. A cylindrical can whose diameter equals its height contains 785.5 cm^3 of a liquid.   Calculate the linear dimensions of the can.

12.34. A right circular cone has an altitude of 40 cm and a diameter of 60 cm.

    (*a*) Calculate the slant height of the cone.   (*c*) Calculate the total surface area.

    (*b*) Calculate the lateral area.   (*d*) Calculate the volume.

12.35. An isosceles triangle with a base of 0.4 m and an altitude of 0.3 m is revolved about a line through the vertex and perpendicular to the base.

    (*a*) What is the name of the solid generated by the revolving triangle?

    (*b*) Calculate the volume generated by the triangle.

12.36. A right circular cone has an altitude of 24 inches and a base radius of 10 inches.

    (*a*) Identify the intersection figure formed by a plane passing through the vertex and perpendicular to the base.

    (*b*) What are the linear dimensions of the figure in (*a*)?

    (*c*) Identify the intersection figure formed by a plane perpendicular to the axis.

    (*d*) Calculate the area of the section in (*c*) if it is 4.8 in. below the vertex.

12.37. A sphere has a diameter of 20 cm.

    (*a*) Calculate the volume of the sphere.   (b) Calculate the surface area.

12.38. The radius of a sphere is 10 inches. Find the area of a small circle whose plane lies 6 inches from the center of the sphere.

12.39. A sphere has a surface area of $2512 \, \text{cm}^2$. What is the diameter of the sphere?

12.40. The radii of two similar right circular cylinders are 5 and 2.5 in.

   (*a*) What is the ratio of the total surface area of the larger cylinder to the total surface area of the smaller cylinder?

   (*b*) What is the ratio of the volume of the smaller cylinder to the volume of the larger?

   (*c*) What is the ratio of their altitudes?

12.41. For the solid shown in Fig. 12-17, compute: (*a*) the total surface area, (*b*) the volume.

12.42. The volume of a certain cone is 64 times the volume of a similar cone. Evaluate the following ratios.

   (*a*) base area of smaller cone : base area of larger cone

   (*b*) altitude of smaller cone : altitude of larger cone

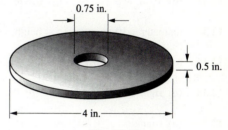

Fig. 12-17

12.43. The altitude of a right octagonal prism is 20 cm. Each edge of the base is 5 cm. Calculate the lateral area of the prism.

12.44. If the diameter of a sphere is reduced by a factor of 3, what happens to its:

   (*a*) surface area   (*b*) volume?

12.45. Show that the volume V of a circular cylinder can be expressed approximately by the formula:

$$V = 0.785 \, d^2 h, \qquad \text{where } d = \text{the diameter of the base} \quad \text{and} \quad h = \text{altitude}$$

12.46. Show that the total area S of a right circular cone can be written

$$S = \frac{\pi d}{4}(d + 2L), \qquad \text{where } d = \text{base diameter of the cone} \quad \text{and} \quad L = \text{slant height}$$

12.47. What is the maximum number of points in which a line can intersect a sphere?

Answers to Supplementary Problems

12.21. (*a*) Cube; hexahedron; regular polyhedron; regular solid or rectangular parallelepiped; quadrangular prism, (*b*) regular square pyramid; right square pyramid.

12.22. (*a*) (1) 5;   (2) 8;   (3) 4   (*b*) (1) 12;   (2) 16;   (3) 12

12.23. (*a*) $384 \, \text{cm}^2$ (*b*) $512 \, \text{cm}^3$ (*c*) $\sqrt{192} = 8\sqrt{3} \approx 13.9 \, \text{cm}$

12.24. (*a*) $384 \, \text{in}^3$ (*b*) $352 \, \text{in}^2$ (*c*) $4\sqrt{14}$, or 15 in.

12.25. (*a*) 27 : 1   (*b*) 1/9 or 0.111 or 1 : 9

12.26. (*a*) 4 cm   (*b*) $\sqrt{7}$ or 2.65 cm   (*c*) 48 cm^2 (*d*) $12\sqrt{7}$ or 31.8 cm^3

12.27. $V_{\text{prism}}/V_{\text{pyramid}} = 3$

12.28. (*a*) 0.0844 m^2 (*b*) 0.152 m^3

12.29. 80 cm^2

12.30. sphere

12.31. (*a*) $60\pi \approx 188$ in^2 (*b*) $69\pi \approx 245$ in^2 (*c*) $90\pi \approx 283$ in^3

12.32. (*a*) 10 in. approx.   (*b*) 628 in^2 approx.

12.33. diameter = 10 cm approx.;   height = 10 cm approx.

12.34. (*a*) 50 cm   (*b*) 4713 cm^2 (*c*) 7540 cm^2 approx.   (*d*) 37,700 cm^3 approx.

12.35. (*a*) right circular cone   (*b*) 0.0126 m^3 approx.

12.36. (*a*) triangle   (*b*) base = 20 in.;   altitude = 24 in.   (*c*) circle   (*d*) 12.6 in^2 approx.

12.37. (*a*) 4190 cm^3 approx.   (*b*) 1257 cm^2 approx.

12.38. 64π in$^2 \approx 201$ in^2

12.39. $20\sqrt{2} \approx 28.3$ cm

12.40. (*a*) 4 : 1   (*b*) 1 : 8   (*c*) 2 : 1

12.41. (*a*) 31.7 in^2 (*b*) $247\pi/128 \approx 6.06$ in^3

12.42. (*a*) 1 : 16   (*b*) 1 : 4

12.43. 800 cm^2

12.44. (*a*) reduced to 1/9 of former value   (*b*) reduced to 1/27 of former value

12.45. *Hint:* Substitute $r = d/2$ in $V = \pi r^2 h$.

12.46. *Hint:* Let $r = d/2$ in $A = 2\pi r(L + r)$

12.47. two

Chapter 13

Trigonometric Functions

Basic Concepts

13.1   TRIGONOMETRY

Trigonometry means "the measuring of triangles."

Plane trigonometry is concerned with plane angles and triangles.   The subject also deals with special functions of angles without reference to triangles.

To *solve a triangle* means to determine the measure of its unknown sides or angles from the given triangle parts.

13.2   RATIOS IN SIMILAR TRIANGLES

In Fig. 13-1, perpendiculars are dropped from one side of the acute angle *KAM* to its other side to form right triangles *BAC*, *B'AC'* and *B''AC''*.   Since these triangles are similar, their corresponding sides are in direct proportion.   This means that

$$\frac{BC}{AC} = \frac{B'C'}{AC'} = \frac{B''C''}{AC''} \quad \text{and} \quad \frac{AC}{AB} = \frac{AC'}{AB'} = \frac{AC''}{AB''}$$

Other proportions may also be written.

The ratios do *not* depend on the size of the triangle as long as angle *KAM* remains unchanged.

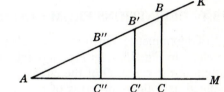

Fig. 13-1.   Similar right triangles

Trigonometric Functions

13.3   DEFINITIONS FROM TRIANGLE RATIOS

The relationships between an acute angle of a right triangle and its sides are given by six ratios.   For angle θ of triangle *ABC* (shown in Fig. 13-2) the "side opposite" is a, the "side adjacent" is b, and the hypotenuse is c.

The six *trigonometric functions* of angle θ are defined in terms of the sides of the triangle, as follows:

$$\text{sine } \theta = \frac{\text{side opposite}}{\text{hypotenuse}} = \frac{a}{c}$$

$$\text{cosine } \theta = \frac{\text{side adjacent}}{\text{hypotenuse}} = \frac{b}{c}$$

$$\text{tangent } \theta = \frac{\text{side opposite}}{\text{side adjacent}} = \frac{a}{b}$$

$$\text{cosecant } \theta = \frac{\text{hypotenuse}}{\text{side opposite}} = \frac{c}{a}$$

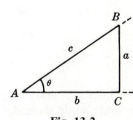

Fig. 13-2

$$\text{secant } \theta = \frac{\text{hypotenuse}}{\text{side adjacent}} = \frac{c}{b}$$

$$\text{cotangent } \theta = \frac{\text{side adjacent}}{\text{side opposite}} = \frac{b}{a}$$

The above functional values are written in abbreviated form as $\sin \theta$, $\cos \theta$, $\tan \theta$, $\csc \theta$, $\sec \theta$, and $\cot \theta$.

The three functions most commonly used in elementary science and technology are sine, cosine, tangent. Students are urged strongly to memorize the above definitions of these three functions.

EXAMPLE 13.1. Determine the six trigonometric functions of $\angle LMN$ (Fig. 13-3).

The side opposite $\angle LMN$ is 3; the side adjacent to $\angle LMN$ is 4; the hypoteneuse is 5.

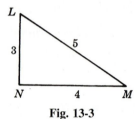

Substituting the given values in the definitions:

$$\sin \theta = 3/5 = 0.600 \qquad \csc \theta = 5/3 \approx 1.67$$
$$\cos \theta = 4/5 = 0.800 \qquad \sec \theta = 5/4 = 1.25$$
$$\tan \theta = 3/4 = 0.750 \qquad \cot \theta = 4/3 \approx 1.33$$

Fig. 13-3

13.4 DEFINITIONS FROM COORDINATES

Trigonometric functions may also be defined in terms of the coordinates of a point on a circle. In Fig. 13-4, let P be a point on a circle of radius r. Rectangular axes are drawn with the origin at the center O of the circle. The trigonometric functions of the indicated angle θ may be defined as follows in terms of the coordinates x and y, the angle θ being said to be in *standard position*:

$$\sin \theta = \frac{\text{ordinate}}{\text{radius}} = \frac{y}{r}$$

$$\cos \theta = \frac{\text{abscissa}}{\text{radius}} = \frac{x}{r}$$

$$\tan \theta = \frac{\text{ordinate}}{\text{abscissa}} = \frac{y}{x}$$

$$\csc \theta = \frac{\text{radius}}{\text{ordinate}} = \frac{r}{y}$$

$$\sec = \frac{\text{radius}}{\text{abscissa}} = \frac{r}{x}$$

$$\cot \theta = \frac{\text{abscissa}}{\text{ordinate}} = \frac{x}{y}$$

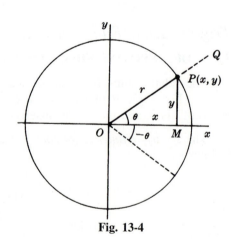

Fig. 13-4

Since x, y, and r may be considered to measure the sides of $\triangle POM$, it is clear that for the case of an acute angle these definitions are equivalent to those given earlier for a right triangle.

EXAMPLE 13.2. A circle has a radius of 26 units. Determine the trigonometric functions for the angle between the x-axis and the radius through the point (10, 24).

The radius is 26; the abscissa is 10; the ordinate is 24.   Substituting the given values in the definitions of the functions:

$$\sin \theta = 24/26 \cong 0.923 \qquad \csc \theta = 26/24 \cong 1.083$$
$$\cos \theta = 10/26 \cong 0.385 \qquad \sec \theta = 26/10 = 2.600$$
$$\tan \theta = 24/10 = 2.400 \qquad \cot \theta = 10/24 \cong 0.417$$

13.5.   TRIGONOMETRIC TABLES

The numerical values of the sine, cosine, and tangent for angles from $0°$ to $90°$ are tabulated in the Appendix, page 467.   As will be discussed later, the values of the cosecant, secant, and cotangent functions may also be read from these tables.

EXAMPLE 13.3.   Determine   (a) $\cos 14°$,   (b) $\tan 59°$,   (c) $\sin 7°$.

Locate the given angle in the column labeled "Degrees." Move horizontally to the value in the column for the required function.   The desired values are

$$\cos 14° = 0.9703 \qquad \tan 59° = 1.664 \qquad \sin 7° = 0.1219$$

EXAMPLE 13.4.   Determine the unknown angle if:

(a) $\tan \theta = 0.700$ (b) $\sin \phi = 0.961$ (c) $\cos D = 0.934$

The values are given to three decimal places; those in the table have four decimals.   Select from the appropriate column the value which is nearest the given value and read off the corresponding angle.   To the nearest degree, the desired angles are

(a) $35°$ (b) $74°$ (c) $21°$

The symbol ∞ for the tangent of $90°$ means that the function is not defined.

If greater precision is required, then *linear interpolation* may be used.   The method assumes a direct proportion between the difference in two angle values and the difference in the values of their trigonometric functions.

EXAMPLE 13.5.   Determine $\sin 30.4°$.

From the table on page 467,

$$\sin 31° = 0.5150$$
$$\sin 30° = 0.5000$$
$$\text{difference} = 0.0150$$

The desired value is 0.4 of the difference between 0.5000 and 0.5150.   Thus

$$(0.4)(0.0150) = 0.0060$$
$$\sin 30.4° = 0.5000 + 0.0060 = 0.5060$$

EXAMPLE 13.6.   Determine θ if $\tan \theta = 1.083$.

From the table on page 467.

$$\tan 48° = 1.111$$
$$\tan 47° = 1.072$$
$$\text{difference} = 0.039$$

The difference between $\tan 47°$ and $\tan \theta$ is $1.083 - 1.072 = 0.011$.

Therefore, $\theta = \dfrac{0.011}{0.039}$ of the distance between $47°$ and $48°$.   Carrying out the division and adding the result to $47°$, $\theta = 47.28° \approx 47.3°$.

If the answer is to be expressed in degrees and minutes, the decimal fraction is multiplied by 60.   Thus $(0.28)(60) = 16.8 \approx 17$, and the desired answer is $\theta = 47°17'$.

13.6 TRIGONOMETRIC FUNCTIONS OF SPECIAL ANGLES

Several angles are of frequent occurrence in scientific applications.   These angles are $0°$, $30°$, $45°$, $60°$, $90°$, $180°$, $270°$, and $360°$.   Their trigonometric functions can be calculated by simple methods.

The functions of $45°$ may be obtained by reference to the isosceles right triangle of Fig. 13-5.   If each of the two legs is 1 unit long, then the hypotenuse is $\sqrt{2}$ units long.   The trigonometric functions of $45°$ are then obtained by application of the definitions of Section 13.3:

$$\sin 45° = \frac{1}{\sqrt{2}} = \frac{\sqrt{2}}{2} \approx 0.707$$

$$\cos 45° = \frac{1}{\sqrt{2}} = \frac{\sqrt{2}}{2} \approx 0.707$$

$$\tan 45° = \frac{1}{1} = 1$$

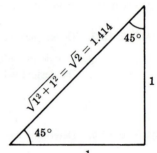

Fig, 13-5.   Calculating trigonometric functions of $45°$

The functions of $30°$ and $60°$ may be computed in like manner by reference to the equilateral triangle of Fig. 13-6.   Let the sides of the triangle be 2 units long.   A perpendicular bisector from a vertex to the opposite side will result in two adjacent $30°$–$60°$–$90°$ triangles, each of sides $2, \sqrt{3}, 1$.   From the first definitions of the trigonometric functions we obtain:

$$\sin 30° = 1/2 = 0.500 \qquad \sin 60° = \frac{\sqrt{3}}{2} \approx 0.866$$

$$\cos 30 = \frac{\sqrt{3}}{2} \approx 0.866 \qquad \cos 60° = 1/2 = 0.500$$

$$\tan 30° = \frac{1}{\sqrt{3}} \approx 0.577 \qquad \tan 60° = \frac{\sqrt{3}}{1} \approx 1.732$$

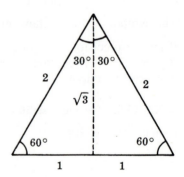

Fig. 13-6.   Calculating trigonometric functions of $30°$ and $60°$

13.7 CALCULATING TRIGONOMETRIC FUNCTIONS WITH A CALCULATOR

A calculator can also be used to obtain decimal approximation of the values of the trigonometric functions.   However, it is necessary to make sure that the MODE of the calculator has been set correctly.   There are generally two options: degrees and radians.   For all examples of this chapter, make sure that the mode of your calculator has been set to degrees.

EXAMPLE 13.7.   Determine $\sin \theta$, $\cos \theta$, and $\tan \theta$ where $\theta = 37.6°$.

Using an HP-38G calculator proceed as indicated below.

(1)　Set the mode to Degrees.

　　　{SHIFT}{MODES}<u>CHOOS</u> <u>Degrees</u>{ENTER} ← It may be necessary to use the directional keys to highlight degrees before pressing ENTER.

　　　{HOME}

(2)　Calculate the functions.

　　　{SIN}37.6{)}{ENTER}　　The answer is 0.610145163901.　This answer can be rounded to 0.6101.

　　　{COS}37.6{)}{ENTER}　　The answer is 0.792289643355.　This answer can be rounded to 0.7923.

　　　{TAN}37.6{)}{ENTER}　　The answer is 0.770103672336.　This answer can be rounded to 0.7701.

EXAMPLE 13.8.　Determine the θ angle if $\sin \theta = 0.961$.

　　To calculate the angle θ with a TI-82 graphing calculator key in the sequence

$\{2^{nd}\}\{SIN^{-1}\}0.961\{ENTER\}$

The answer is 73.94569193.　This value can be rounded to 74°.

　　Using an HP-38G proceed as follows

$\{SHIFT\}\{ASIN\}0.961\{)\}\{ENTER\}$

The answer is 73.9456919257.　This value can be rounded to 74°.

13.8　COFUNCTIONS

　　It can be seen from Fig. 13-6 that the side opposite 30° is the side adjacent 60°.　Therefore, $\sin 30° = \cos 60°$.　Similarly $\csc 30° = \sec\ 60°$, and $\tan 30° = \cot 60°$.　Such pairs of functions which have the property that their values for complementary angles are equal are called *cofunctions*.　Hence sine and cosine, secant and cosecant, and tangent and cotangent are pairs of cofunctions, as their names indicate.

EXAMPLE 13.9.　Determine $\cot 18°$.

$$\cot 18° = \tan(90° - 18°) = \tan 72°$$

From the table on page 467, $\tan 72° = 3.078$.　Thus, $\cot 18° = 3.078$.

13.9　SIGNS OF TRIGONOMETRIC FUNCTIONS

　　Since abscissas and ordinates may be positive or negative, it follows from the definitions of Section 13.4 that the values of the generalized trigonometric functions are also signed numbers.　It is to be noted that the radius is always positive in these definitions.

EXAMPLE 13.10.　Determine the numerical values of $\sin 120°$, $\cos 120°$, and $\tan 120°$.

　　Let the terminal side of 120° in standard position intersect a unit circle at point P, as shown in Fig. 13-7.　Then the coordinates of P will have the same absolute values as for the case of 60°.　However, the abscissa is negative.　Therefore, from Section 13.4 we have

$$\sin 120° = \sin 60° \approx 0.866$$
$$\cos 120° = -\cos 60° = -0.500$$
$$\tan 120° = -\tan 60° \approx -1.732$$

The Roman numerals in Fig. 13-7 refer to the four *quadrants* of the circle.

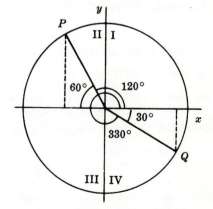

Fig. 13-7.　Determining the signs of trigonometric functions

EXAMPLE 13.11.   Determine the numerical values of $\sin 330°$, $\cos 330°$, and $\tan 330°$.

It is seen from Fig. 13-7 that the coordinates of Q will have the same absolute values as the coordinates of the point associated with $30°$.   However, the ordinate of Q is negative.   Therefore, from the definitions of the three basic functions, we get:

$$\sin 330° = -\sin 30° = -0.500$$
$$\cos 330° = \cos 30° \approx 0.866$$
$$\tan 330° = -\tan 30° \approx -0.577$$

13.10 TRIGONOMETRIC IDENTITIES

Relationships between trigonomeric functions are called *trigonometric identities*.   The reader can verify the validity of the identities shown below using the definitions of Section 13.4.

$$\sin \theta = 1/\csc \theta \qquad \cos \theta = 1/\sec \theta \qquad \tan \theta = 1/\cot \theta$$
$$\csc \theta = 1/\sin \theta \qquad \sec \theta = 1/\cos \theta \qquad \cot \theta = 1/\tan \theta$$

Some additional and useful trigonometric identities are:

$$\sin^2 \theta + \cos^2 \theta = 1 \qquad 1 + \tan^2 \theta = \sec^2 \theta \qquad 1 + \cot^2 \theta = \csc^2 \theta$$

where $\sin^2 \theta$, $\cos^2 \theta$ and $\tan^2 \theta$ stand for $(\sin \theta)^2$, $(\cos \theta)^2$, and $(\tan \theta)^2$, respectively.

EXAMPLE 13.12.   Determine $\csc 21°$, $\sec 21°$ and $\cot 21°$ using a calculator.

Since there are no function keys to calculate these values directly from the calculator, use the definition of these trigonometric functions:

$$\csc 21° = 1/\sin 21°$$

To calculate $\csc 21°$ using a TI-82 calculator key in the sequence

1{÷}{SIN}21{ENTER}

The result is 2.79042811.   This value can be rounded to 2.7904. Therefore, $\csc 21° = 2.7904$

$$\sec 21° = 1/\cos 21°$$

Calculate this value by keying in

1{/}{COS}21{ENTER}

The result is 1.071144994.   This value can be rounded to 1.0711.   Therefore, $\sec 21° = 1.0711$

$$\cot 21° = 1/\tan 21°$$

Calculate this value by keying in

1{÷}{TAN}21{ENTER}

The result is 2.605089065.   This value can be rounded to 2.6051. Therefore, $\cot 21° = 2.6051$

EXAMPLE 13.13.   Determine $\sec 35°$ given $\tan 35° = 0.70$.

Use the identity

$$1 + \tan^2 \theta = \sec^2 \theta, \text{ to obtain } \sec \theta = \sqrt{1 + \tan^2 \theta}$$

Replace $\tan 35°$ in the previous equation to obtain

$$\sec \theta = \sqrt{1 + \tan^2 35} = \sqrt{1 + (0.70)^2} = 1.2206.$$

EXAMPLE 13.14.   Determine $\sin 42°$ given $\tan 42° = 0.900$.

Determine $\cot 42°$ using the identity

$$\cot \theta = 1/\tan \theta$$

Therefore, $\cot 42° = 1/\tan 42°$.

Replace $\tan 42°$ by its value to obtain

$$\cot 42° = 1/0.900 = 1.111.$$

Use the trigonometric identity

$$1 + \cot^2 42° = \csc^2 42°$$

to obtain the value of $\csc 42°$.   That is,

$$\csc 42° = \sqrt{1 + \cot^2 42°} = \sqrt{1 + (1.111)^2} = \sqrt{2.234} = 1.4948.$$

Finally, substitute this value in $\sin 42° = 1/\csc 42°$ to obtain that $\sin 42° = 1/1.4948 = 0.669.$

Solved Problems

13.1.   From the right triangle shown in Fig. 13-8, write out the six trigonometric functions of angle A.

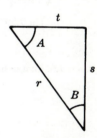

Fig. 13-8

   s is side opposite.

   t is side adjacent.

   r is the hypotenuse.

From the definitions of the functions,

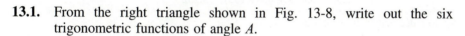

$$\sin A = s/r \qquad \cos A = t/r \qquad \tan A = s/t$$
$$\csc A = r/s \qquad \sec A = r/t \qquad \cot A = t/s$$

13.2.   Referring to the triangle in Fig. 13-9, determine to three significant figures the functions:

   (*a*) $\tan \phi$,   (*b*) $\tan \theta$,   (*c*) $\sin \theta$,   (*d*) $\sec \phi$,   (*e*) $\cot \phi$.

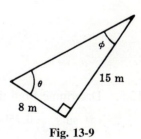

Fig. 13-9

   (*a*) $\tan \phi = \dfrac{\text{side opposite}}{\text{side adjacent}} = \dfrac{8}{15} \approx 0.533$

   (*b*) $\tan \phi = \dfrac{\text{side opposite}}{\text{side adjacent}} = \dfrac{15}{8} \approx 1.88$

   (*c*) Use the Pythagorean theorem to determine the hypotenuse H.

   $H^2 = 8^2 + 15^2 = 289$

   $H = \sqrt{289} = 17$

   $\sin \theta = \dfrac{\text{side opposite}}{\text{hypotenuse}} = \dfrac{15}{17} \approx 0.882$

   (*d*) $\sec \phi = \dfrac{\text{side hypotenuse}}{\text{side adjacent}} = \dfrac{17}{15} \approx 1.13$

(e) $\cot \phi = \dfrac{\text{side adjacent}}{\text{side opposite}} = \dfrac{15}{8} \approx 1.88$

13.3. Determine the measures of θ and ϕ in Problem 13.2 to the nearest degree.

Referring to the table on page 467, and using the values of the functions, we obtain $\theta = 62°$ and $\phi = 28°$.

13.4. Determine from the table on page 467, the values of the following functions:

(a) $\cos 19.3°$ (b) $\tan 72.9°$ (c) $\sin 27°50'$

(a) $\cos 19° = 0.9455$
        $\cos 20° = \underline{0.9397}$
    difference $= \overline{0.0058}$

$\cos 19.3° = \cos 19° - 0.3 \times$ (the difference between $\cos 20°$ and $\cos 19°$), since the cosine decreases as the angle increases.

$$\cos 19.3° = 0.9455 - 0.3(0.0058) \approx 0.9455 - 0.0017 \approx 0.9438$$

(b) $\tan 73° = 3.271$
        $\tan 72° = \underline{3.078}$
    difference $= \overline{0.193}$

$\tan 72.9° = \tan 72° + 0.9 \times$ (the difference between $\tan 72°$ and $\tan 73°$).

$$\tan 72.9° = 3.078 + 0.9(0.193) \approx 3.252$$

(c) First convert $50'$ to a decimal fraction of a degree.

$$\frac{50}{60} = 0.833$$

$$\sin 27°50' \approx \sin 27.83°$$

$$\sin 28° = 0.4695$$
$$\sin 27° = \underline{0.4540}$$
$$\text{difference} = \overline{0.0155}$$

$$\sin 27.83° = 0.4540 + (0.83)(0.0155) \approx 0.4669$$
$$\sin 27°50' \approx 0.4669$$

13.5. Determine from the given functions the measures of the angles to the nearest $0.1°$.

(a) $\sin P = 0.7590$ (b) $\tan \phi = 3.2505$ (c) $\cos R = 0.1612$

Refer to the table on page 467.

(a) $\sin 50° = 0.7660$ $\sin P = 0.7590$
        $\sin 49° = \underline{0.7547}$ $\sin 49° = \underline{0.7547}$
    difference $= \overline{0.0113}$ difference $= \overline{0.0043}$

Angle P is $\dfrac{0.0043}{0.0113}$ of the distance between $49°$ and $50°$.

Angle P is $49° + 0.38° = 49.38° \approx 49°23'$.

(b) $\tan 73° = 3.271$ $\tan \phi = 3.250$
        $\tan 72° = \underline{3.078}$ $\tan 72° = \underline{3.078}$
    difference $= \overline{0.193}$ difference $= \overline{0.172}$

Angle ϕ is $\dfrac{0.172}{0.193}$ of the distance between $72°$ and $73°$.

Angle ϕ is $72° + 0.89° \approx 72.9° \approx 72°54'$.

(c) $\cos 80° = 0.1736$ $\cos 80° = 0.1736$
       $\cos 81° = \underline{0.1564}$ $\cos R = \underline{0.1612}$
  difference $= \overline{0.0172}$ difference $= \overline{0.0124}$

Angle R is $\dfrac{0.0124}{0.0172}$ of the distance between $80°$ and $81°$.

Angle R is $80° + 0.72° \approx 80.72° \approx 80°43'$.

13.6. In which quadrants is the

  (a) sine negative?   (c) cosine negative?
  (b) tangent positive?   (d) tangent negative and cosine positive?

The signs of the abscissas and ordinates are shown in Fig. 13-10.   The radius is always positive.

(a) From the definition, $\sin\theta = y/r$; it will be negative if the ordinate is negative.   This condition is satisfied for quadrants III and IV only.

(b) Since $\tan\theta = y/x$, the positive value of the ratio requires that y and x be both positive or both negative.   This occurs in quadrants I and III.

(c) Since $\cos\theta = x/r$, its value will be negative for negative values for the abscissa.   The cosine is negative in quadrants II and III.

(d) The tangent is negative in quadrants II and IV; the cosine is positive in quadrants I and IV.   Therefore, the tangent is negative and the cosine positive in quadrant IV only.

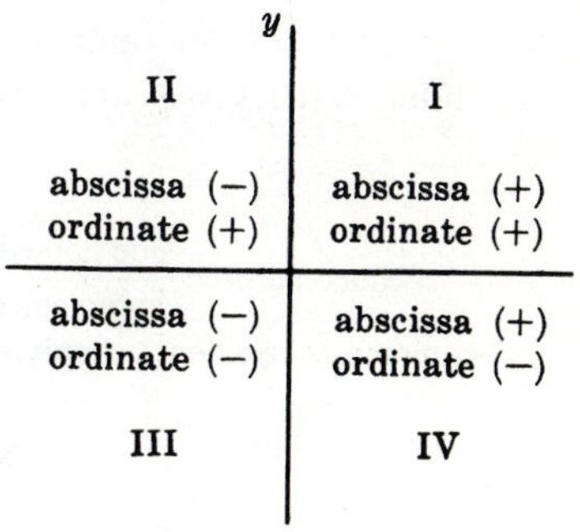

Fig. 13-10

13.7. Find the smallest positive angle that satisfies:

  (a) $\sin 20° = \cos\theta$ (c) $\sec 37° = \csc M$
  (b) $\tan 85° = \cot\phi$ (d) $\cos 41° = \sin T$

The functions of the unknown angles are cofunctions of the given angles.   The unknown angles are complementary to the given angles.

  (a) $\theta = 90° - 20° = 70°$ (c) $M = 90° - 37° = 53°$
  (b) $\phi = 90° - 85° = 5°$ (d) $T = 90° - 41° = 49°$

13.8. Express the given functional values in terms of angles smaller than $90°$:

  (a) $\sin 150°$ (b) $\cos 205°$ (c) $\sin 590°$ (d) $\tan 315°$

Sketch the approximate positions of the terminal sides of the given central angles shown intercepting a circle of unit radius (Fig. 13-11).

(a) $\sin 150°$ will have the same value as $\sin 30°$ since for both angles the ordinates of the points on the circle are positive.   Thus, $\sin 150° = \sin 30°$.

(b) cos 205° will have the same absolute value as cos 25°, since the abscissas of the intercept points on the circle have the same absolute values. But the abscissa for 205° is negative. Thus cos 205° = − cos 25°.

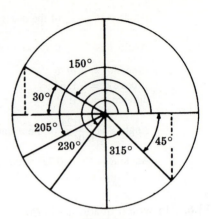

Fig. 13-11

(c) For angles greater than 360°, subtract 360°, or 720°, or any multiple of 360°, so as to make the resultants angle smaller than 360°.

$$590° − 360° = 230°$$

Then,

$$\sin 590° = \sin 230° \quad \text{and} \quad |\sin 230°| = |\sin 50°|$$

But since the ordinate of the intercept point is negative $\sin 230° = − \sin 50°$.

(d) tan 315° and tan 45° have the same absolute values. With ordinates opposite in sign, we have tan 315° = − tan 45°.

13.9. If $\sin \phi = 0.80$ and ϕ is an acute angle, determine without using trigonometric tables (a) $\cos \phi$ and (b) $\tan \phi$.

$$\sin \phi = 0.80 = 80/100 = 8/10 = 4/5$$

Since $\sin \phi = \dfrac{\text{side opposite}}{\text{hyptotenuse}}$, by letting side opposite be 4 units and the hypotenuse be 5 units the given equation is satisfied. Applying the Pythagorean theorem,

$$\text{side adjacent} = \sqrt{(\text{hypotenuse})^2 − (\text{side opposite})^2}$$

(a) $\cos \phi = \dfrac{\text{side adjacent}}{\text{hypotenuse}} = 3/5 = 0.60$ (b) $\tan \phi = \dfrac{\text{side opposite}}{\text{side adjacent}} = 4/3 \approx 1.33$

13.10. Prove that $\csc \theta$ is the reciprocal of $\sin \theta$ if $\theta \neq n\pi$.

Using the definitions of $\csc \theta$ and $\sin \theta$ in terms of the coordinates on a circle,

$$\csc \theta = \frac{r}{y} \qquad \sin \theta = \frac{y}{r}$$

Therefore, $$\csc \theta = \frac{1}{\sin \theta}, \quad \text{if } \sin \theta \neq 0$$

But $\sin \theta$ is zero only when θ is an integral multiple of π.

13.11. Which of the following statements are possible and which are not?

(a) $\cos \theta = 3$ (c) $\tan \theta = 1$ (e) $\tan \theta$ is undefined

(b) $\sin \theta = 0$ (d) $\sin \theta = 2$ (f) $\sin \theta = 1/3$ and $\csc \theta = −3$

(b), (c), (e) are possible; the others are not. For part (f) consider the solution of Problem 13.10. Since $\csc \theta = 1/\sin \theta$, both must have the same sign.

13.12. Determine the six trigonometric functions of the angle whose terminal side passes through the point $(−5, 12)$. (See Fig. 13-12.)

The angle θ is in the second quadrant.

$$r^2 = x^2 + y^2 = (-5)^2 + (12)^2 = 169$$
$$r = 13$$

From the definitions of the functions,

$$\sin \theta = 12/13 \qquad \cos \theta = -5/13 \qquad \tan \theta = -12/5$$
$$\csc \theta = 13/12 \qquad \sec \theta = -13/5 \qquad \cot \theta = -5/12$$

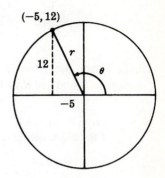

Fig. 13-12

13.13. Determine the angle in Problem 13.12.

Use the value of $\sin \theta$, $\cos \theta$, or $\tan \theta$ and the trigonometric tables. For instance,

$$\sin \theta = 12/13 \approx 0.9231$$
$$\theta \approx 180° - 67.4° = 112.6°$$

13.14. Determine the sine, cosine, and tangent of (a) 90° and (b) 180°.

Using a circle with unit radius, the coordinates of the intercept point of the terminal side of the angle are $(0, 1)$ for 90° and $(-1, 0)$ for 180° (Fig. 13-13).

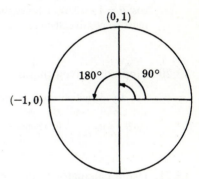

Fig. 13-13

(a) $\sin 90° = \dfrac{1}{1} = 1$

$\cos 90° = \dfrac{0}{1} = 0$

$\tan 90° = \dfrac{1}{0}$, not defined.

(b) $\sin 180° = \dfrac{0}{1} = 0 \quad \cos 180° = \dfrac{-1}{1} = -1 \quad \tan 180° = \dfrac{0}{1} = 0$

13.15. Determine the value of $\cos \theta$ if $\tan \theta = -4/3$ and $\sin \theta$ is negative.

Since $\tan \theta = y/x$ and $r^2 = x^2 + y^2$, we may write $r = \sqrt{4^2 + 3^2} = 5$. From Fig. 13-14, $\tan \theta$ is negative in quadrants II and IV; $\sin \theta$ is negative in quadrants III and IV. Thus θ must have its terminal side in quadrants IV, where the abscissa is positive. Thus, $\cos \theta = 3/5$.

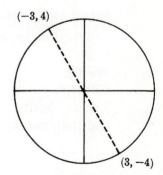

Fig. 13-14

13.16. In the triangle of Fig. 13-15, what function of ϕ is represented by the ratio

(a) 20.0/34.6,   (b) 34.6/40.0,   (c) 40.0/20.0?

(a) $\cot \phi$ (b) $\sin \phi$ (c) $\sec \phi$

13.17. In the triangle of Fig. 13-15, evaluate the following ratios:

(a) $\dfrac{1}{\tan \theta}$ (b) $\dfrac{1}{\sin \theta}$ (c) $\dfrac{1}{\cos \theta}$

Fig. 13-15

(a) $\dfrac{1}{20/34.6} = \dfrac{34.6}{20.0} = 1.73$ (b) $\dfrac{1}{20/40} = \dfrac{40.0}{20.0} = 2.00$ (c) $\dfrac{1}{34.6/40} = \dfrac{40.0}{34.6} \approx 1.16$

13.18. From the triangle in Fig. 13-16 determine $\cos R$ and $\cot R$ in terms of m and n.

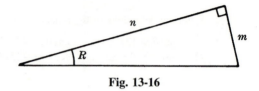

Fig. 13-16

Let $h = $ unknown hypotenuse. From the Pythagorean theorem, $h = \sqrt{m^2 + n^2}$. From the definitions of the trigonometric functions:

$$\cos R = \dfrac{n}{\sqrt{m^2 + n^2}} \quad \text{and} \quad \cot R = \dfrac{n}{m}$$

13.19. Determine $\cos \theta$ if $\sin \theta = 0.6$ and $0° \leq \theta \leq 90°$.

Use the identity $\sin^2 \theta + \cos^2 \theta = 1$ obtain that $\cos^2 \theta = 1 - \sin^2 \theta$. Replace $\sin \theta$ by its value to obtain $\cos^2 \theta = 1 - (0.6)^2$. Solve this equation for $\cos \theta$ and obtain that $\cos \theta = \pm 0.8$. Since $0° \leq \theta \leq 90°$ $\cos \theta$ must be positive, therefore, $\cos \theta = 0.8$.

13.20. Find $\cot \theta$ given that $\cos \theta = 0.6$ and $0° \leq \theta \leq 90°$.

By definition $\sec \theta = 1/\cos \theta$. Therefore, $\sec \theta = 1/0.6 \approx 1.667$. Replace this value in the identity $1 + \tan^2 \theta = \sec^2 \theta$; solve for $\tan \theta$ and obtain that $\tan \theta = 1.333$. By definition $\cot \theta = 1/\tan \theta$. Replace $\tan \theta$ by its value to obtain that $\cot \theta = 0.751$.

13.21. Use a calculator to determine $\sin \theta$, $\cos \theta$ and $\tan \theta$ where $\theta = 25°$.

Using an HP-38G proceed as indicated below.

Make sure that the MODE is set to degrees. If not, key in the following sequence:

{SHIFT}{MODES}<u>CHOOS</u> <u>Degrees</u>{ENTER} ← It may be necessary to use the directional keys to highlight degrees.

{HOME}
{SIN}25{)}{ENTER}   The result is 0.422618261741. This result can be rounded to 0.423.
{COS}25{)}{ENTER}   The result is 0.906307787037. This result can be rounded to 0.906.
{TAN}25{)}{ENTER}   The result is 0.466307658155. This result can be rounded to 0.466.

Supplementary Problems

The answers to this set of problems are at the end of this chapter.

13.22. For the triangle of Fig. 13-17, determine each of the following:

(a) $\sin \phi$ (b) $\tan \theta$ (c) $\cos \phi$ (d) $\csc \theta$

13.23. From the right triangle shown in Fig. 13-18, determine:

(a) $\sin \theta$ (b) $\cos \phi$ (c) $\tan \theta$ (d) $\cot \phi$ (e) θ (f) ϕ

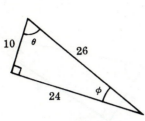

Fig. 13-17

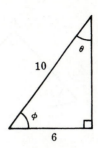

Fig. 13-18

13.24. Using the table of trigonometric functions in the Appendix, page 467, determine:

(a) $\tan 34.2°$ (b) $\cos 19°20'$ (c) $\sin 48.4°$ (d) $\cos 81.3°$ (e) $\tan 74°15'$

13.25. Express (if possible) the given values as functions of acute angles:

(a) $\tan 270°$ (b) $\sin 210°$ (c) $\cos 150°$ (d) $\sin 305°$ (e) $\tan 217°$ (f) $\cos 475°$

13.26. Determine the acute angle θ in each of the following cases:

(a) $\cos \theta = \sin 20°$ (c) $\sin 82° = \cos \theta$ (e) $\cot 70° = \tan \theta$

(b) $\tan 63° = \cot \theta$ (d) $\csc 53° = \sec \theta$ (f) $\cos 38°25' = \sin \theta$

13.27. In which quadrants is   (a) the cosine negative?   (b) the tangent positive?   (c) the sine negative? (d) the sine negative and the cosine positive.   (e) the cosine negative and the tangent positive? (f) the sine negative and the tangent negative?

13.28. If $\cos \theta = 5/13$ and θ is an acute angle, determine the other trigonometric functions of θ without using trigonometric tables.

13.29. Prove that $\cos \phi$ is the reciprocal of $\sec \phi$.

13.30. Which of the following statements are not possible?

(a) $\sin \theta = 1.500$ (c) $\tan \theta = 10^6$ (e) $\cos \theta$ is positive and $\tan \theta$ is negative

(b) $\cos \theta = 0.01$ (d) $\cos \theta = 1.05$ (f) $\cot \theta = 2$ and $\tan \theta = 1/2$

13.31. The terminal side of an angle intersects a circle with center at the origin at the point $(1, -2)$.   Determine the six trigonometric functions of the angle.

13.32. Without using trigonometric tables, determine if possible the sine, cosine, and tangent of   (a) 270° and (b) 360°.

13.33. (a) Referring to Fig. 13-5, determine   (1) $\csc 45°$;   (2) $\cot 45°$;   (3) $\sec 45°$.

(b) Referring to Fig. 13-6, determine   (1) $\sec 60°$;   (2) $\cot 30°$;   (3) $\csc 60°$.

13.34. If $\sin \theta = -4/5$ and the tangent is positive, determine   (a) $\cos \theta$ and   (b) $\cot \theta$.

13.35. Determine θ in Problem 13.34   (a) to the nearest 0.1° and   (b) to the nearest 5'.

13.36. In the triangle of Fig. 13-19, what functions of θ are represented by the ratios:

(a) $\dfrac{BC}{AB}$?   (b) $\dfrac{AC}{BC}$?   (c) $\dfrac{AC}{AB}$?

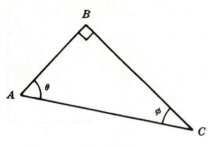

Fig. 13-19

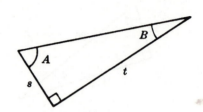

Fig. 13-20

13.37. From Fig. 13-20, express $\sin A$, $\tan A$, $\cos B$, and $\csc B$ in terms of s and t.

13.38. Prove $\dfrac{\sin \theta}{\cos \theta} = \tan \theta$ if $\cos \theta \neq 0$.

13.39. Find $\sin \theta$ and $\cos \theta$ given that $\cot \theta = 3$.

13.40. Using a calculator verify that $\cos 230° \cos 310° - \sin(-50) \sin(-130) = -1$.

13.41. Using a calculator find $\dfrac{\tan 30° + \cos 45° + 4}{\sin^2 25° + \cos^2 25}$

13.42. Using a calculator find $\dfrac{\sin 60° + \cos 45° + 1}{\cot 30° + 5}$

Answers to Supplementary Problems

13.22. (a) $10/26 \approx 0.385$ (b) $24/10 = 2.40$ (c) $24/26 \approx 0.923$ (d) $26/24 \approx 1.08$

13.23. (a) $6/10 = 0.600$ (c) $6/8 = 0.750$ (e) $\theta \approx 36.9° \approx 37°$

        (b) $6/10 = 0.600$ (d) $6/8 = 0.750$ (f) $\phi \approx 53.1° \approx 53°$

13.24. (a) 0.6796 (b) 0.9436 (c) 0.7478 (d) 0.1513 (e) 3.546

13.25. (a) undefined   (b) $-\sin 30°$ (c) $-\cos 30°$ (d) $-\sin 55°$ (e) $\tan 37°$ (f) $-\cos 65°$

13.26. (a) $70°$ (b) $27°$ (c) $8°$ (d) $37°$ (e) $20°$ (f) $51°35'$

13.27. (a) II, III   (b) I, III   (c) III, IV   (d) IV   (e) III   (f) IV

13.28. $\sin \theta = 12/13$, $\tan \theta = 12/5$, $\cot \theta = 5/12$, $\sec \theta = 13/5$, $\csc \theta = 13/12$

13.29. $1/(\sec \phi) = \dfrac{1}{r/x} = x/r = \cos \phi$

13.30. (a), (d)

13.31. $\sin \theta = -2\sqrt{5}$ $\cos \theta = 1/\sqrt{5}$ $\tan \theta = -2$ $\csc \theta = -\sqrt{5}/2$ $\sec \theta = \sqrt{5}$ $\cot \theta = -1/2$

13.32.   (a) $\sin 270° = -1$ $\cos 270° = 0$ $\tan 270°$ is not defined

   (b) $\sin 360° = 0$ $\cos 360° = 1$ $\tan 360° = 0$

13.33.   (a) (1) $\sqrt{2}$ or 1.414;   (2) 1;   (3) $\sqrt{2}$ or 1.414

   (b) (1) 2;   (2) $3/\sqrt{3} = \sqrt{3} = 1.732$;   (3) $2/\sqrt{3}$ or 1.155

13.34.   (a) $-3/5$ (b) 3/4

13.35.   (a) 233.1°   (b) 233°10′

13.36.   (a) $\tan \theta$ (b) $\csc \theta$ (c) $\sec \theta$

13.37. $\sin A = t/\sqrt{s^2 + t^2}$ $\tan A = t/s$

   $\cos B = t/\sqrt{s^2 + t^2}$ $\csc B = (\sqrt{s^2 + t^2})/s$

13.38. $\dfrac{y/r}{x/r} = y/x = \tan \theta$ (if $\cos \theta \neq 0$)

13.39. $\sin \theta \approx \dfrac{\sqrt{10}}{10}$ $\cos \theta \approx \dfrac{3\sqrt{10}}{10}$

13.40. -1

13.41. ≈ 5.284

13.42. ≈ 0.3822

Chapter 14

Solution of Triangles

Right Triangles

There are many scientific studies in which right triangles play a leading role. The solution of these and other triangles is especially important. To solve a right triangle is to determine three of its unknown parts from either (1) one known side and one known acute angle, or (2) two known sides. The methods of solution are best illustrated by examples.

14.1 ONE KNOWN SIDE AND ONE KNOWN ACUTE ANGLE

EXAMPLE 14.1. Three distances are represented by a right triangle. The distance of 12.8 ft is represented by the hypotenuse, which makes an angle of 65° with one of the legs of the triangle. Determine the values of the third angle and of the two unknown distances.

Sketch and label the triangle as in Fig. 14-1. The orientation of the triangle is immaterial. Angle θ is complementary to 65°, or $\theta = 90° - 65° = 25°$. Since d_2 is the side opposite 65°, by applying the definition of the sine function we can write

$$\sin 65° = \frac{d_2}{12.8}$$

from which

$$d_2 = (12.8)(\sin 65°)$$
$$= (12.8)(0.906) \approx 11.6$$

Hence, d_2 has a value of 11.6 ft. Similarly,

$$\sin 25° = \frac{d_1}{12.8}$$
$$d_1 = (12.8)(\sin 25°) = (12.8)(0.423) \approx 5.41$$

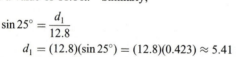

Fig. 14-1

and so d_1 has a value of 5.41 ft. The solution can be checked graphically by drawing the triangle to scale or applying the Pythagorean theorem:

$$12.8^2 \overset{?}{=} 11.6^2 + 5.41^2$$
$$163.84 \overset{?}{=} 134.56 + 29.27$$
$$163.84 \overset{?}{=} 163.83$$

Before calculating these values using a calculator, make sure that the MODE of the calculator has been set to Degrees. To calculate the value of d_2, key in 12.8{*}{SIN}65{ENTER}. The result is 11.6007396741. This value can be rounded to 11.6.

Likewise, the value of d_1 can be calculated by keying in 12.8{*}{SIN}25{ENTER}. The result is 5.40951375028. This value can be rounded to 5.41.

14.2 TWO KNOWN SIDES

EXAMPLE 14.2. In a right triangle one leg is 2.10 m long; the hypotenuse is 3.50 m long.

(*a*) Determine the length of the second leg.

(*b*) Determine the angles of the triangle.

Sketch and label the triangle. Identify the known and unknown parts, as shown in Fig. 14-2. To find L, use the Pythagorean theorem:

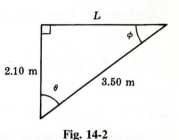

Fig. 14-2

$$3.50^2 = 2.10^2 + L^2$$
$$L^2 = 3.50^2 - 2.10^2$$
$$L = \sqrt{3.50^2 - 2.10^2}$$
$$= \sqrt{12.25 - 4.41}$$
$$= \sqrt{7.8400} = 2.80$$

Therefore the unknown leg is 2.80 m long. From the definitions of trigonometric functions:

$$\cos\theta = \frac{2.10}{3.50} = 0.600$$

In the table of trigonometric functions one finds $\cos 53° \approx 0.602$ and $\cos 54° \approx 0.588$. Since 0.602 is very close to 0.600, $\theta \approx 53°$. One could also interpolate and get a more precise value of θ. Thus, the difference between $\cos 53°$ and $\cos 54°$ is $0.602 - 0.588$, or 0.014; the difference between $\cos 53°$ and $\cos\theta$ is $0.602 - 0.600$, or 0.002. Therefore, 0.002 corresponds to 0.002/0.014 of 1°, or 0.14° in excess of 53°. The value of θ to the nearest tenth of a degree is then 53.1°.

The angle ϕ is the complement of θ. Therefore, $\phi = 90° - 53° = 37°$, to the nearest degree, or $\phi = 90 - 53.1 = 36.9°$, to the nearest tenth of a degree.

To calculate the square root of a difference of squares, use extra parentheses to make sure that the square root is applied to the entire difference and not to the first term only. Using a TI-82 calculator, calculate the value of L by keying in {2nd}{$\sqrt{}$}{(}3.50{^}2{−}2.10{^}2{)}{ENTER}. The result is 2.8. Calculate the value of the angle θ by keying in {2nd}{COS^{-1}}0.6{ENTER}. The result is 53.13010235. This value can be rounded to 53.1°.

Oblique Triangles

Since an oblique triangle contains no right angle, it is not possible to solve this type of triangle by the methods of the previous section. We shall state two laws which are applicable to the solution of any triangle.

14.3 THE LAW OF SINES

The law of sines states that *in any triangle the lengths of the sides are proportional to the sines of the opposite angles.* Thus, in Fig. 14-3:

$$\frac{a}{\sin A} = \frac{b}{\sin B} = \frac{c}{\sin C}$$

EXAMPLE 14.3. Three distances are represented by the sides of a triangle. One of the distances is 200 cm. The other two sides make angles of 30° and 45° with the given side. Determine the unknown distances.

Sketch and label the triangle of distances as shown in Fig. 14-4. Since we know that the sum of the angles in a triangle is 180°, we can determine the unknown angle θ.

$$30° + 45° + \theta = 180° \quad \text{or} \quad \theta = 180° - 30° - 45° = 105°$$

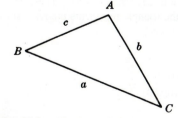

Fig. 14-3. The law of sines:
$a/\sin A = b/\sin B = c/\sin C$

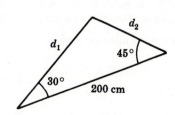

Fig. 14-4

Applying the law of sines, we can write

$$\frac{d_1}{\sin 45°} = \frac{200}{\sin 105°}$$

Solving for d_1, we obtain:

$$d_1 = \frac{200}{\sin 105°}\sin 45°$$

But
$$\sin 105° = \sin 75° = 0.9659 \approx 0.966$$

Substituting the numerical values of $\sin 105°$ and $\sin 45°$ in the equation for d_1, we get

$$d_1 \approx \frac{200}{0.966}(0.707) \approx 146$$

Therefore, $d_1 \approx 146$ cm.   Similarly, we can write

$$\frac{d_2}{\sin 30°} = \frac{200}{\sin 105°}$$
$$d_2 = \frac{200}{\sin 105°}\sin 30°$$
$$d_2 = \frac{200}{0.966}(0.500) \approx 104$$

Therefore, $d_2 = 104$ cm.   The solution may be checked by drawing the diagram to scale.

To calculate the value of d_2 with a calculator, use extra parentheses to ensure that $\sin 30°$ multiplies the entire fraction instead of the denominator. With a TI-82 calculator, key in {(}200{÷}{sin}105°{)}{×}SIN}30{ENTER}. The result is 103.527618. This value can be rounded to 104°.

14.4   THE LAW OF COSINES

The law of cosines states that *the square of any triangle side is equal to the sum of the squares of the other two sides minus twice the product of these sides and the cosine of the included angle.* Thus, expressed in terms of the sides and angles given in triangle ABC of Fig. 14-5, the cosine law is

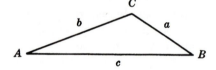

Fig. 14-5.   Oblique triangle

$$c^2 = a^2 + b^2 - 2ab \, \cos C$$

Two similar expressions can be written:

$$a^2 = b^2 + c^2 = 2bc \, \cos A$$
$$b^2 = a^2 + c^2 = 2ac \, \cos B$$

If triangle ABC is a right triangle with $C = 90°$, then $\cos C = \cos 90° = 0$. The equation for c^2 then reduces to the Pythagorean relation, $c^2 = a^2 + b^2$.

The law of cosines makes it possible to determine one side of a triangle if the other two sides and the angle between them are known.

EXAMPLE 14.4.   A force of 100 newtons makes an angle of 60° with a force of 80 newtons. These two forces may be represented by the adjacent sides of a triangle. The third side of the triangle represents the resultant force R as shown in Fig. 14-6. Determine the magnitude and direction of the resultant.

Applying the law of cosines, we get

$$R^2 = 100^2 + 80^2 - 2(100)(80)\cos 60°$$
$$= 10{,}000 + 6400 - (16{,}000)(0.500)$$
$$= 16{,}400 - 8000$$
$$= 8400$$
$$R = 92$$

The magnitude of the resultant is 92 newtons.   To find angle θ, we use the law of sines and get

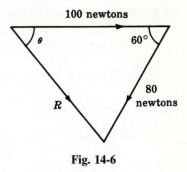

Fig. 14-6

$$\frac{92}{\sin 60°} = \frac{80}{\sin \theta}$$

$$\sin \theta = \frac{80}{92} \sin 60° = \frac{80}{92}(0.866) \approx 0.753$$

Hence, from the trigonometric tables we find $\theta \approx 49°$.   The resultant makes an angle of 49° with the 100-newton force.

To calculate the value of $\sin \theta$ using a calculator use extra parentheses to ensure that $\sin 60°$ multiplies the entire fraction instead of the denominator.   With an HP-38G key in {(}80{/}92{)}{*}{SIN}60{)}{ENTER}.   The result is 0.753065568508.   This value can be rounded to 0.753.

Solved Problems

14.1.   In the right triangles shown in Fig.14-7, express the labeled unknown side in terms of the given parts.

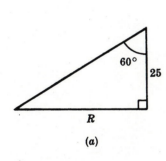

(a)

(b)

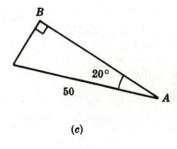

(c)

Fig. 14-7

(a) $\dfrac{R}{25} = \tan 60°$ by definition

   $R = 25 \tan 60°$ multiply both sides of the equation by 25

(b) $\dfrac{12}{t} = \sin 30°$ by definition

   $12 = t \sin 30°$ multiply both sides by t

   $t = \dfrac{12}{\sin 30°}$ divide both sides by $\sin 30°$

   Alternate Solution: $\dfrac{t}{12} = \csc 30°$ and $t = 12 \csc 30°$

(c) $\dfrac{\overline{AB}}{50} = \cos 20°$ by definition

   $\overline{AB} = 50 \cos 20°$ multiply both sides by 50

14.2. For the triangle of Fig. 14-8 express   (*a*) *p* as a function of *r* and ϕ;   (*b*) *r* as a function of *p* and θ;   (*c*) *q* as a function of *p* and θ;   (*d*) *p* as a function of *q* and ϕ.

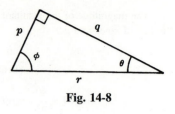

Fig. 14-8

(*a*) $\dfrac{p}{r} = \cos\phi$

   $p = r\cos\phi$

(*b*) $\dfrac{r}{p} = \csc\theta$

   $r = p\csc\theta$

(*c*) $\dfrac{q}{p} = \cot\theta$

   $q = p\cot\theta$

(*d*) $\dfrac{p}{q} = \cot\phi$

   $p = q\cot\phi$

14.3. In the triangle of Fig. 14-9, solve for all the unknown parts of the triangle.

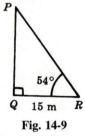

Fig. 14-9

$$\frac{\overline{PQ}}{\overline{QR}} = \tan 54°$$

$$\overline{PQ} = \overline{QR}\,\tan 54°$$

$$= (15)(1.376) \qquad \text{by substitution of } \tan 54° = 1.376$$
$$\qquad\qquad\qquad \text{from the table, page 467}$$

$$\approx 20.6$$

$\overline{PQ}$ is 20.6 m approximately.

$$\frac{15}{\overline{PR}} = \cos 54°$$

$$15 = \overline{PR}\,\cos 54°$$

$$\overline{PR} = \frac{15}{\cos 54°} = \frac{15}{0.5878} \approx 25.5$$

$\overline{PR}$ is 25.5 m approximately.

$$\angle QPR = 90° - 54° = 36°$$

$\angle QPR$ is 36°.

Check: $\overline{QR}^2 + \overline{PQ}^2 = \overline{PR}^2$

$$(15)^2 + (20.6)^2 \stackrel{?}{=} (25.5)^2$$

$$225 + 424.4 \stackrel{?}{=} 650.2$$

$$649.4 \approx 650.2$$

14.4. In a right triangle, the side opposite an angle of 42° is 23.5 ft long.   Solve for all the unknown parts of the triangle.

Sketch the triangle with the given values as shown in Fig. 14-10.   Denote the hypotenuse by *h* and the unknown leg by *s*.

$$\frac{23.5}{h} = \sin 42° \quad \text{and} \quad h = \frac{23.5}{\sin 42°} = \frac{23.5}{0.6691} \approx 35.1$$

The hypotenuse is 35.1 ft approximately.

$$\frac{s}{h} = \cos\theta$$

$$s = h\cos\theta = 35.1\cos 42° = (35.1)(0.7431) \approx 26.1$$

The unknown leg is 26.1 ft approximately.

The unknown acute angle is $(90 - 42)°$, or 48°.

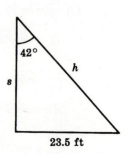

23.5 ft

Fig. 14-10

Check: $(23.5)^2 + (26.1)^2 \overset{?}{=} (35.1)^2$

$$552.2 + 681.2 \overset{?}{=} 1232$$

$$1233.4 \overset{?}{=} 1232$$

14.5. A kite string makes an angle of 50° with level ground. If the string is 65 m long, how high is the kite above ground?

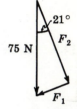

Sketch the triangle with the given values as shown in Fig. 14-11, with H as the required height

$$\frac{H}{65} = \sin 50°$$

$$H = 65 \ \sin 50° = 65(0.766) \approx 49.8$$

Fig. 14-11

The kite is approximately 49.8 m above ground.

14.6. Given the right triangle of forces shown in Fig. 14-12. Calculate F_1 and F_2.

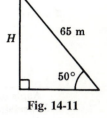

$$\frac{F_1}{75} = \sin 21°$$

$$F_1 = 75 \ \sin 21° = 75(0.3584) \approx 26.9$$

F_1 is approximately 26.9 N.

Fig. 14-12

$$\frac{F_2}{75} = \cos 21°$$

$$F_2 = 75 \cos 21° = 75(0.9336) \approx 70$$

F_2 is approximately 70 N.

Check: $(26.9)^2 + (70.0)^2 \overset{?}{=} (75)^2$

$$724 + 4900 \overset{?}{=} 5625$$

$$5624 \approx 5625$$

14.7. When the angle of elevation of the sun is 59°, a building casts a shadow 85 feet long. How high is the building?

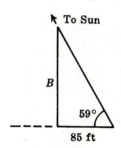

The *angle of elevation* is the angle between the horizontal and the upward line of sight to the observed point. Sketch the triangle as shown in Fig. 14-13.

Let the height of the building be denoted by B.

$$\frac{B}{85} = \tan 59° \quad \text{and} \quad B = 85 \ \tan 59° = 85(1.664) \approx 141$$

Fig. 14-13

The building is approximately 141 feet in height.

14.8. A tower is 35 meters above sea level. The angle of depression of a boat as seen from the tower is 14°. How far is the boat from the tower?

The *angle of depression* is the angle between the horizontal and the observed point which is below the horizontal.   Represent the given data by a triangle as shown in Fig. 14-14.   Let d be the desired distance.

$$\frac{d}{35} = \tan(90 - 14)^\circ = \tan 76^\circ$$

$$d = 35 \ \tan 76^\circ = 35(4.011) \approx 140.4$$

The boat is approximately 140 m from the tower.

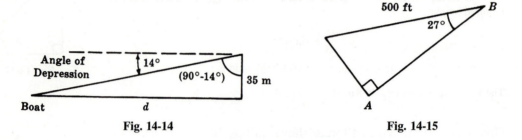

Fig. 14-14   Fig. 14-15

14.9.   In the right triangle of Fig. 14-15, calculate the distance AB.

$$\frac{\overline{AB}}{500} = \cos 27^\circ$$

$$\overline{AB} = 500 \ \cos 27^\circ = 500(0.8910) \approx 446$$

The distance AB is approximately 446 feet.

14.10.   In Fig. 14-16, an observer at point 1 on a river bank sees a tree T on the opposite shore.   The observer walks along the shore until the tree makes an angle of 45° with his path.   (*a*) If the distance between points 1 and 2 is 50 meters, how wide is the river? (*b*) What is the distance r between point 2 and the tree?

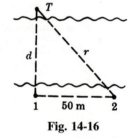

Fig. 14-16

(*a*) The legs of a 45°–45°–90° triangle are equal.   Thus, the width of the river is 50 meters.   Alternatively,

$$\frac{d}{50} = \tan 45^\circ$$

$$d = 50 \ \tan 45^\circ = 50(1.000) = 50$$

(*b*)
$$\frac{50}{r} = \cos 45^\circ$$

$$r = \frac{50}{\cos 45^\circ} = \frac{50}{0.7071} \approx 71$$

The distance r is approximately 71 meters.

14.11.   The two legs of a right triangle are 50 cm and 70 cm long.   Determine the length of the hypotenuse and the measures of the acute angles.

Applying the Pythagorean theorem,

$$H^2 = 50^2 + 70^2 = 7400$$

$$H = \sqrt{7400} \approx 86$$

The hypotenuse is 86 cm approximately.   Let θ be the angle opposite the longer leg.

$$\tan \theta = \frac{70}{50} = 1.40$$

From the tables on page 467, $\theta \approx 54.5°$.   The other acute angle is $(90 - 54.5)°$ or $35.5°$.

14.12. The velocities of a plane and wind are shown in Fig. 14-17.   Calculate the magnitude of the velocity represented by R and the angle ϕ.

From the Pythagorean theorem,

$$R^2 = 80^2 + 300^2 = 96,400$$

$$R = \sqrt{96,400} \approx 310$$

$$\tan \phi = \frac{80}{300} \approx 0.2667$$

$$\phi \approx 14.9°$$

The magnitude of R is approximately 310 km/hr.   The angle ϕ is 14.9°.

14.13. A wire supporting a pole is fastened to it 15 feet from the ground and to the ground 10 feet from the pole.   Calculate the length of the wire and the angle it makes with the pole.

Let L be the wire length and θ the unknown angle.   Applying the Pythagorean theorem,

$$L^2 = 10^2 + 15^2 = 325$$

$$L = \sqrt{325} \approx 18.0$$

$$\tan \theta = \frac{10}{15} \approx 0.667$$

$$\theta \approx 33.7°$$

The supporting wire is approximately 18.0 ft long and makes an angle of 33.7° with the pole.

14.14. In the parallelogram shown in Fig. 14-18, calculate the values of R and P.

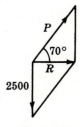

Fig. 14-18

Oppposite sides of a parallogram are equal.   In the upper triangle, one leg is R and the other is 2500.

$$\frac{2500}{R} = \tan 70°$$

$$R = \frac{2500}{\tan 70°} \approx \frac{2500}{2.748} \approx 910$$

$$\frac{2500}{P} = \sin 70°$$

$$P = \frac{2500}{\sin 70°} \approx \frac{2500}{0.9397} \approx 2660$$

14.15. From the data in Fig. 14-19 calculate F and ϕ.

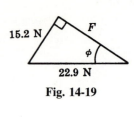

Fig. 14-19

From the Pythagorean theorem, $(22.9)^2 = (15.2)^2 + F^2$.

$$F^2 = (22.9)^2 - (15.2)^2 \approx 524.4 - 231.0 \approx 293.4$$

$$F \approx \sqrt{293.4} \approx 17.1$$

$$\sin \phi = \frac{15.2}{22.9} \approx 0.664$$

$$\phi \approx 41.6°$$

The approximate magnitude of F is 17.1 N and that of ϕ is 41.6°.

14.16. (*a*) Determine the angle between a diagonal of a rectangle and one of the longer sides.   The length and width of the rectangle are 16 cm and 30 cm respectively.

(*b*) Determine the length of a diagonal.

Let the unknown angle be θ and the unknown diagonal D, as shown in Fig. 14-20.

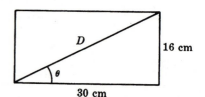

Fig. 14-20

(*a*) $\tan \theta = \dfrac{16}{30} = \dfrac{8}{15} \approx 0.533$

$$\theta = 28°$$

The angle is 28° approximately.

(*b*) $\dfrac{16}{D} = \sin \theta$

$$D = \frac{16}{\sin \theta} = \frac{16}{\sin 28°} \approx \frac{16}{0.470} \approx 34$$

The diagonal is about 34 cm in length.

Check: $16^2 + 30^2 \overset{?}{=} 34^2$

$$256 + 900 \overset{?}{=} 1156$$

$$1156 = 1156$$

14.17. The isosceles triangle shown in Fig. 14-21 has sides 2.0 meter long and a 2.4-meter base.

(*a*) Determine the altitude of the triangle.

(*b*) Determine the measure of the vertex angle.

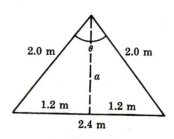

Fig. 14-21

Denote the altitude by a and the vertex angle by θ.   The altitude bisects the base into segments of 1.2 m each and the vertex angles into equal angles $\theta/2$.

(*a*) $(2)^2 = (1.2)^2 + a^2$

$$a^2 = 2^2 - (1.2)^2 = 4 - 1.44 = 2.56$$

$$a = \sqrt{2.56} = 1.6, \text{ and the altitude is 1.6 m in length.}$$

(*b*)

$$\sin\left(\frac{\theta}{2}\right) = \frac{1.2}{2.0} = 0.600$$

$$\frac{\theta}{2} \approx 37°$$

$$\theta \approx 74°$$

The measure of the vertex angle is 74°.

14.18. A *gradient* is a ratio of the vertical distance to the horizontal distance.   The gradient may also be stated in percent or in degrees.   A highway has a gradient of 0.087.

(*a*) Express this gradient G in percent and in degrees.

(*b*) How many miles would an automobile rise vertically if it traveled 2 miles along this highway?

(*a*) The gradient in percent is $G = (0.087)(100) = 8.7\%$

$$\tan G = 0.087$$
$$G \approx 5°$$

(*b*) The 2 miles is along the hypotenuse of a triangle, the hypotenuse making a 5° angle with the horizontal.   If the vertical rise is R, then

$$\frac{R}{2} = \sin \theta$$
$$R = 2 \sin \theta = 2 \sin 5° = 2(0.08716) \approx 0.174$$

The rise is 0.174 mile, or 920 feet.

14.19. Given the triangle shown in Fig. 14-22, calculate the values of the unknown parts.

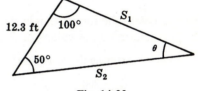

Fig. 14-22

The sum of the angles in a triangle is 180°.

$$50 + 100 + \theta = 180°$$
$$\theta = 180 - 150 = 30°$$

Applying the law of sines,

$$\frac{S_2}{\sin 100°} = \frac{12.3}{\sin 30°}$$

$$S_2 \ \sin 30° = 12.3 \ \sin 100° \qquad \text{by cross-multiplication}$$

$$S_2 = 12.3 \ \frac{\sin 100°}{\sin 30°} \qquad \text{by solving for } S_2$$

But $\sin 100° = \sin(180 - 100)° = \sin 80°$.

$$S_2 = 12.3 \ \frac{0.9848}{0.5000} \approx 24.2 \qquad \text{by substituting numerical values of the sines}$$

Side S_2 is 24.2 ft approximately.   Using the law of sines once more,

$$\frac{S_1}{\sin 50°} = \frac{12.3}{\sin 30°} \quad \text{and} \quad S_1 = 12.3 \ \frac{\sin 50°}{\sin 30°} = 12.3 \ \frac{0.7660}{0.5000} \approx 18.8$$

Side S_1 is approximately 18.8 feet.

14.20. Given a triangle with a side 5.0 m long opposite an angle of 30° and adjoining a side 8.0 m in length.   Calculate the unknown parts of the triangle.

A sketch of the triangle (Fig. 14-23) shows that two solutions are possible.   Using the law of sines,

$$\frac{\sin 30°}{5.0} = \frac{\sin \theta}{8.0}$$

$$\sin \theta = 8.0 \frac{\sin 30°}{5.0} = 8.0 \ \frac{0.5000}{5.0} = 0.800$$

But $\sin 53° \approx 0.800$ and $\sin 127° \approx 0.800$.   Thus, $\theta_1 \approx 53°$ and $\theta_2 \approx 127°$.

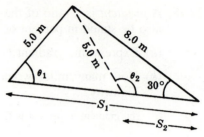

Fig. 14-23

With θ_1 and $30°$ as two angles, the third angle is $180° - (53 + 30)°$, or $97°$. Thus,

$$\frac{\sin 97°}{S_1} = \frac{\sin 30°}{5.0} \qquad \text{by the law of sines}$$

$$S_1 = 5.0\,\frac{\sin 97°}{\sin 30°} \qquad \text{solving for } S_1$$

But $\sin 97° = \sin(180 - 97)° = \sin 83° = 0.9926$.

$$S_1 = 5.0\,\frac{0.9926}{0.5000} \approx 9.9$$

Side S_1 is 9.9 m approximately. With θ_2 and $30°$ as two angles, the third angle is $180° - (30 + 127)°$, or $23°$.

Repeating the above procedure,

$$\frac{\sin 23°}{S_2} = \frac{\sin 30°}{5.0}$$

$$S_2 = 5.0\,\frac{\sin 23°}{\sin 30°} = 5.0\,\frac{0.3907}{0.5000} \approx 3.9$$

Side S_2 in the second solution is 3.9 m approximately.

14.21. Given the force triangle shown in Fig. 14-24, determine the values of R and θ.

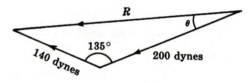

Fig. 14-24

Applying the cosine law,

$$R^2 = 200^2 + 140^2 - 2(200)140\,\cos 135°$$

But $\qquad \cos 135° = -\cos(180 - 135)° = -\cos 45°$

or, $\qquad R^2 = 200^2 + 140^2 + 2(200)(140)\cos 45°$

Note that the plus sign results from subtracting the negative cosine term.

$$R^2 = 40,000 + 19,600 + 56,000(0.707) \approx 99,190$$

$$R \approx \sqrt{99,190} \approx 315$$

The force R is 315 dynes approximately. Using the law of sines,

$$\frac{\sin \theta}{140} = \frac{\sin 135°}{315}$$

$$\sin 135° = \sin(180 - 135)° = \sin 45° = 0.707$$

$$\sin \theta = 140\,\frac{0.707}{315} \approx 0.314$$

$$\theta \approx 18.3°$$

The angle θ is 18.3° approximately.

14.22. For the triangle of Fig. 14-25, express V in terms of the given quantities.

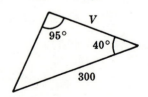

Fig. 14-25

The angle opposite V is $180° - (40 + 95)°$, or $45°$.

Applying the law of sines,

$$\frac{\sin 45°}{V} = \frac{\sin 95°}{300}$$

Solving for V:

$$V = 300 \, \frac{\sin 45°}{\sin 95°} = 300 \frac{0.7071}{0.9962} \approx 213$$

14.23. For the triangle of Fig. 14-26, $x = 6$, $y = 15$, and $\phi = 145°$. Determine the value of z.

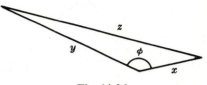

Fig. 14-26

Using the law of cosines,

$$z^2 = 6^2 + 15^2 - 2(6)(15)(\cos 145°)$$
$$z = \sqrt{6^2 + 15^2 - 2(6)(15)(\cos 145°)}$$
$$= \sqrt{36 + 225 - 2(6)(15)(-0.8192)}$$
$$\approx \sqrt{261 + 147.5} \approx \sqrt{408.5}$$
$$\approx 20.2$$
$$\approx 20$$

To calculate the value of z using a calculator, use extra parentheses to ensure that the square root of the entire expression is calculated. Using an HP-38G calculator key in

$\{\sqrt{x}\}\{(\}6\{x^y\}2\{+\}15\{x^y\}2\{-\}\{(\}2\{*\}6\{*\}15\{*\}\{COS\}145\{)\}\{)\}\{ENTER\}$.

The result is 20.2100808502. This value can be rounded to 20.

Supplementary Problems

The answers to this set of problems are at the end of this chapter.

14.24. Determine the values of the unknown parts for the triangle in Fig. 14-27.

14.25. One leg of a right triangle is 30 cm long. The angle adjacent to this leg is 50°.

(*a*) Determine the length of the other leg   (*b*) Determine the length of the hypotenuse.

14.26. Using the data of Fig. 14-28, determine F_1 and F_2.

14.27. From the data of the Fig. 14-29, determine V_a and V_b.

14.28. If in the representation of Fig. 14-30, $A = 60$ cm/sec^2 and $B = 150$ cm/sec^2, determine C, x, and y.

14.29. From the data of Fig. 14-31, determine H and θ.

14.30. If a vertical pole 18 feet high casts a horizontal shadow 30 feet long, determine the angle of elevation of the sun.

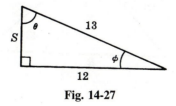

Fig. 14-27

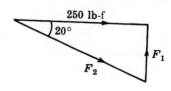

Fig. 14-28

14.31. In the representation of Fig. 14-32, $a_1 = 10$ ft/sec^2, $a_2 = 4$ ft/sec^2.   Determine a_3, θ and ϕ.

14.32. In the representation of Fig. 14-33, $R = 1200$ ohms and $X = 1600$ ohms.   Determine Z and ϕ.

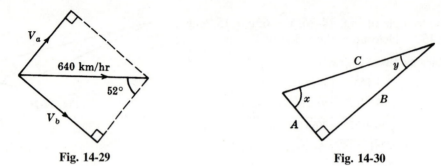

Fig. 14-29   Fig. 14-30

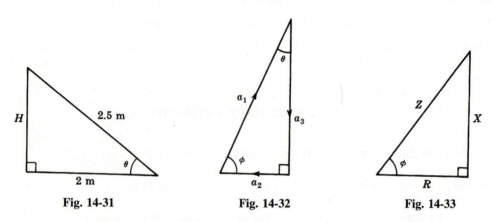

Fig. 14-31   Fig. 14-32   Fig. 14-33

14.33. In each triangle shown in Fig. 14-34, express the unknown labeled part in terms of the given parts.

14.34. For the triangle of Fig. 14-35, express

    (*a*) q as a function of r and ϕ (*c*) p as a function of r and θ

    (*b*) q as a function of p and θ (*d*) r as a function of p and ϕ

14.35. One leg of a right triangle is 20 inches long; the angle opposite this leg measures 40°.

    (*a*) Determine the length of the other leg.   (*b*) Determine the length of the hypotenuse.

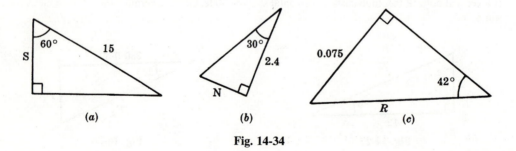

(*a*)   (*b*)   (*c*)

Fig. 14-34

14.36.   Solve for the unknown parts of the triangle in Fig. 14-36.   Check the solution.

14.37.   Solve for the unknown parts of the triangle shown in Fig. 14-37.   Check the solution.

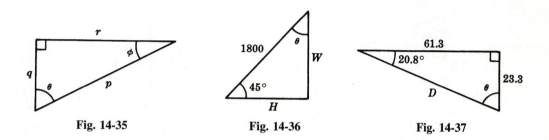

Fig. 14-35   Fig. 14-36   Fig. 14-37

14.38.   Solve for the unknown parts of the triangle shown in Fig. 14-38.

14.39.   The gradient of a road is 0.12.   Express the gradient   (*a*) in percent and   (*b*) in degrees.   (*c*) If a truck rises 250 meters vertically, what distance did it travel along the road?

14.40.   The angle of depression of a boat from the top of a tower is 50°.   If the tower is 175 yards above sea level, how far away is the boat from the tower?

14.41.   For the triangle of Fig. 14-39, express R in terms of the given parts.

14.42.   In Fig. 14-40, $M = 100, n = 50, \theta = 140°$.   Express p in terms of the given parts.

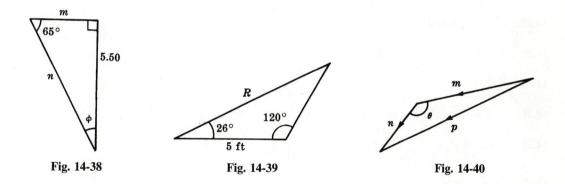

Fig. 14-38   Fig. 14-39   Fig. 14-40

14.43.   From the data of Fig. 14-41, determine T.

14.44.   In the triangle shown in Fig. 14-42, $V_1 = 10$ ft/sec, $V_3 = 7.5$ ft/sec, and $\theta = 35°$.   Determine ϕ.

14.45.   One side of a triangle is 40 cm long.   The adjacent angles are 73° and 49°.   Determine the other two sides of the triangle.

14.46.   Two sides of an oblique triangle are 3 m and 5 m long.   The included angle is 23°.   Determine the third side of the triangle.

14.47.   In the triangle shown in Fig. 14-43, determine R and θ.

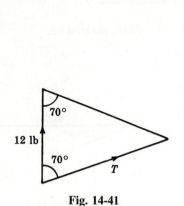

Fig. 14-41

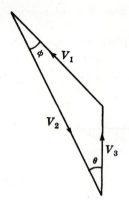

Fig. 14-42

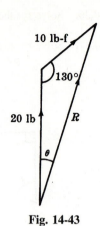

Fig. 14-43

Answers to Supplementary Problems

14.24. $S = 5$ $\phi \approx 22.6°$ $\phi \approx 67.4°$

14.25. (a) 35.8 cm   (b) 46.7 cm

14.26. $F_1 \approx 91.0$ lb $F_2 \approx 266$ lb-f

14.27. $V_a \approx 394$ km/hr $V_b \approx 504$ km/hr

14.28. $C \approx 162$ cm/sec^2 $x \approx 68.2°$ $y \approx 21.8°$

14.29. $H = 1.5$ m $\theta \approx 36.9°$

14.30. $\theta \approx 31°$

14.31. $a_3 = 9.2$ ft/sec^2 $\theta = 23.6°$ $\phi = 66.4°$

14.32. $z = 2000$ ohms $\phi \approx 53.1°$

14.33. (a) $S = 15 \ \cos 60°$ (b) $N = 2.4 \ \tan 30°$ (c) $R = 0.075 \ \csc 42°$ or $R = \dfrac{0.075}{\sin 42°}$

14.34. (a) $q = r \tan \phi$ (b) $q = p \ \cos \theta$ (c) $p = r \ \csc \theta$ (d) $r = p \ \cos \phi$

14.35. (a) 23.8 in.   (b) 31.1 in.

14.36. $\theta = 45°$ $H \approx 1270$ $W \approx 1270$

14.37. $\theta = 69.2°$ $D \approx 65.6$

14.38. $\phi = 25°$ $m \approx 2.56$ $n \approx 6.07$

14.39. (a) 12%   (b) 6.8°   (c) 2098 m ≈ 2100 m

14.40.   147 yards

14.41. $5 \sin 120° / \sin 34°$

14.42. $p = \sqrt{100^2 + 50^2 - 2(100)(50)(\cos 140°)} = \sqrt{100^2 + 50^2 + 2(100)(50)(\cos 40°)}$

14.43. $T \approx 17.5$ lb

14.44. $\phi \approx 25.5°$

14.45.   45.1 cm   35.6 cm

14.46.   2.53 m

14.47. $R \approx 27.5$ lb-f $\theta \approx 16.2°$

Chapter 15

Vectors

Basic Concepts

15.1 TERMS

A *scalar quantity* is one which has magnitude or size and is independent of direction.   It is measured by a number or *scalar*.   Examples of scalar quantities are: time, length, speed, mass.   A number and a measurement unit completely describe a scalar measure.

A *vector quantity* is one which has magnitude *and* direction.   It is measured by things called *vectors*.   Examples of vector quantities are: velocity, force, acceleration, weight, momentum.   A plane vector may be specified by *two* numbers and a measurement unit.

15.2 VECTOR REPRESENTATION

A common symbol for a vector in elementary science is a boldface letter or a letter with a small arrow above it; for example, **A** or **a**; $\vec{A}$ or $\vec{a}$.

The magnitude of a vector is represented by the absolute value symbol or by a letter in italics.   Thus, $|\vec{A}|$, $|\mathbf{A}|$, or A represents the magnitude of vector $\vec{A}$ of **A**.

A vector in science may be represented graphically by a directed line segment or an arrow.   The length of the arrow is drawn to an arbitrary scale to indicate the magnitude of the vector.   The direction of the arrow is the direction of the vector.

EXAMPLE 15.1.   Represent graphically a velocity of 15 m/sec, 30° north of east.

Choose a convenient scale.   Use the conventional map directions.   With the scale shown in Fig. 15-1, the required vector is an arrow 3 cm long inclined 30° to the eastward direction.

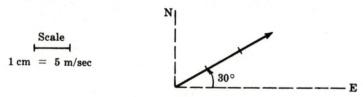

Fig. 15-1.   Vector drawn to scale

15.3 EQUALITY OF VECTORS

Vectors are equal if they have equal magnitudes and the same direction.   Vectors are unequal if they differ in magnitude, in direction, or in both.   Equal and unequal vectors are shown in Fig. 15-2.

322

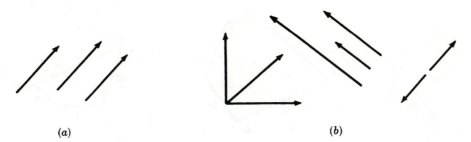

(a)   (b)

Fig. 15-2.   (a) Equal vectors (b) Unequal vectors

Vector Addition

15.4   GRAPHICAL OR GEOMETRICAL METHOD

The rule for determining the sum of vectors **A** and **B** is:

Rule.   Draw vector **A**.   Draw vector **B** with its tail at the head of vector **A**.   The vector drawn from the tail of **A** to the head of **B** is defined as the sum **C** or *resultant* of the two vectors.

The method is illustrated in Fig. 15-3(a).   The vector sum does not depend on the order of addition as shown in Fig. 15-3(b).   Thus, vector addition is *commutative*:

$$\mathbf{A} + \mathbf{B} = \mathbf{C} = \mathbf{B} + \mathbf{A}$$

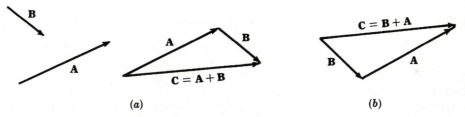

(a)   (b)

Fig. 15-3.   Addition of two vectors: (a) $\mathbf{A} + \mathbf{B} = \mathbf{C}$,   (b) $\mathbf{B} + \mathbf{A} = \mathbf{C}$

If the triangle in Fig. 15-3(b) is moved so that the resultant **C** coincides with its equal in the triangle of Fig. 15-3(a), a parallelogram is formed.   Therefore, an alternative rule for adding two vectors is possible.

The *parallelogram rule* is:

Rule.   To add two vectors, draw them from a common point.   Complete the parallelogram.   The sum is the vector along the diagonal drawn from the common point.

This rule is illustrated in Fig. 15-4.

The rule for addition can be extended to three or more vectors by the *polygon method*.   The vectors are added tail to head in succession.   The vector drawn from the tail of the first to the head of the last vector is the vector sum or resultant.   The method is commutative: the vectors may be added in any order, as shown in Fig. 15-5.

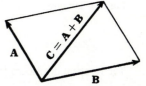

Fig. 15-4.   The parallelogram method

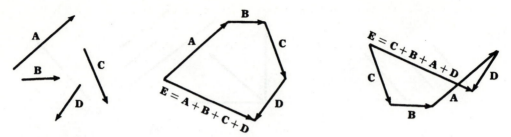

Fig. 15-5. The polygon method of vector addition

15.5   ANALYTICAL METHOD

The addition of vectors may also be carried out by the use of algebraic and trigonometric techniques.

(a) Parallel Vectors

To determine the magnitude of the sum of two vectors having the same direction, add their magnitudes.   The direction of the resultant is the same as that of the given vectors.

EXAMPLE 15.2.   Add a displacement of 8 miles west and a displacement of 6 miles west.

The magnitude of the vector sum is $(8+6)$ or 14 miles.   The direction is west.   The resultant displacement is 14 miles west.

To obtain the magnitude of the sum of two parallel vectors having opposite directions, subtract the smaller magnitude from the larger.   The direction of the resultant is that of the vector with the larger magnitude.

EXAMPLE 15.3.   Add a displacement 6 miles west and a displacement 8 miles east.

The magnitude of the vector sum is $(8 - 6)$ or 2 miles.   The direction is east.   The resultant displacement is 2 miles east.

(b) Perpendicular Vectors

The magnitude of the sum of two perpendicular vectors is the square root of the sum of the squares of their magnitudes.   The direction of the resultant is the direction of the diagonal of a rectangle with the two vectors as adjacent sides.

EXAMPLE 15.4.   Add displacements of 6 miles west and 8 miles south (see Fig. 15-6).

Let the resultant displacement be represented by R.   Then, the Pythagorean theorem can be applied to calculate the magnitude of R as

$$R = \sqrt{6^2 + 8^2} = \sqrt{100} = 10$$

The magnitude of the vector sum is 10 miles.

Let θ be the direction which the resultant makes with the west direction.   Then,

$$\tan \theta = 8/6 = 1.333$$
$$\theta \approx 53°$$

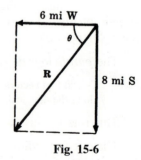

Fig. 15-6

The resultant displacement **R** is 10 miles, 53° south of west.

(c) Vectors at an Arbitrary Angle

The magnitude and direction of the resultant of two vectors at any angle may be obtained by an application of the law of cosines and the law of sines.

EXAMPLE 15.5. Add displacements of 6 miles west and 8 miles 60° north of east.

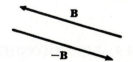

Sketch the sum of the given vectors as shown in Fig. 15-7 with **R** as the resultant. It is not necessary to make the drawing to scale.

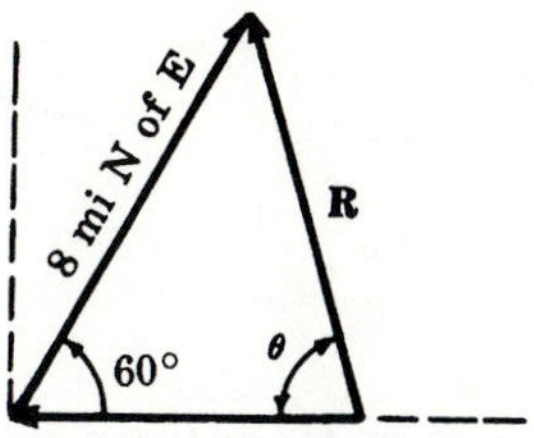

Apply the law of cosines to determine the magnitude of **R**.

$$R = \sqrt{6^2 + 8^2 - 2(6)(8)(\cos\ 60°)}$$
$$= \sqrt{100 - 96(0.500)} = \sqrt{1000 - 48} = \sqrt{52}$$
$$R \approx 7.2$$

Fig. 15-7

Apply the law of sines to determine θ, the direction of the resultant.

$$\frac{\sin\ \theta}{8} = \frac{\sin\ 60°}{7.2}$$

$$\sin\ \theta = \frac{8(0.866)}{7.2} \approx 0.962$$

$$\theta \approx 74°$$

The resultant displacement **R** is 7.2 miles, 74° north of west.

To calculate the value of R using a calculator, make sure that the MODE of the calculator has been set to Degrees. With an HP-38G graphing calculator key in the following sequence:

$\{\sqrt{x}\}\{()6\{x^y\}2\{+\}8\{x^y\}2\{-\}\{()2\{*\}6\{*\}8\{*\}\{cos\}60\{()\}\}\{()\}\{ENTER\}$

The display shows 7.21110255093. This value can be rounded to 7.2.

To calculate the value of $\sin\ \theta$ key in

$\{()8\{*\}0.866\{()\}\{/\}7.2\{ENTER\}$

The result is 0.962222222222. The value can be rounded to 0.962. Given that $\sin\ \theta = 0.962$, the angle θ can be calculated as follows:

$\{SHIFT\}\{ASIN\}0.962\{()\}\{ENTER\}$

The result is 74.1541924502. This angle can be rounded to 74°.

Vector Subtraction

15.6 THE NEGATIVE OF A VECTOR

The *negative of a vector* is a vector of equal magnitude but opposite in direction to the given vector. For example, in Fig. 15-8 vector $-\mathbf{B}$ is the negative of vector **B**.

Fig. 15-8. A vector and its negative

15.7 SUBTRACTION OF VECTORS

The operation of subtraction $\mathbf{A} - \mathbf{B}$ is defined as $\mathbf{A} + (-\mathbf{B})$.

Thus, the rule for subtracting one vector from another is:

Rule.   To subtract vector **B** from vector **A**, reverse the direction of **B** and add $-\mathbf{B}$ to **A**. The graphical method of vector subtraction is illustrated in Fig. 15-9.

Fig. 15-9. Vector subtraction

Vector Resolution

15.8   VECTOR RESOLUTION OR DECOMPOSITION

To *resolve* a vector is to determine two vectors whose sum is the given vector.   Thus, vector resolution is the reverse of vector addition.

Vector components are the vectors into which an initial vector has been resolved.

Though a vector can be resolved along any two directions, in most applications the standard directions are at right angles to each other.

Rectangular components of a vector are its components along perpendicular directions.

The rule for determining graphically the rectangular components follows:

Rule.   Through the tail of the vector draw two perpendicular direction lines. From the arrowhead drop perpendiculars on the direction lines.   The rectangular component vectors are drawn from the tail of the initial vector to the perpendiculars along the given directions.

If the standard directions are the *x*- and *y*-axes, as in Fig. 15-10(*a*), then the components A_x, A_y are called the *x*-component and the *y*-component, respectively.   The rectangular components of vector **F** in Fig. 15-10(*b*) are along a direction parallel to an incline ($F_{\parallel}$) and a direction perpendicular to it ($F_{\perp}$).

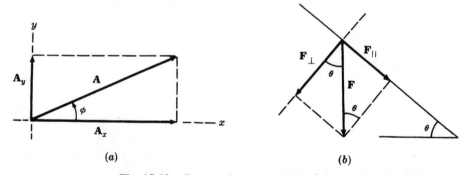

(*a*)   (*b*)

Fig. 15-10.   Rectangular components of a vector

15.9   RELATIONSHIPS BETWEEN A VECTOR AND ITS COMPONENTS

From Fig. 15-10(*a*) it is seen that the two rectangular components form the adjacent sides of a parallelogram with the given vector as its diagonal.   From the trigonometry of the right triangle,

$$\frac{A_x}{A} = \cos \phi \quad \text{or} \quad A_x = A \cos \phi \qquad \text{and} \qquad \frac{A_y}{A} = \sin \phi \quad \text{or} \quad A_y = A \sin \phi$$

The relationship between the vector magnitudes follows from the Pythagorean theorem,

$$A^2 = A_x^2 + A_y^2$$

Similarly, from Fig. 15-10(b),

$$\frac{F_\perp}{F} = \cos\theta \quad \text{or} \quad F_\perp = F\cos\theta$$

$$\frac{F_\parallel}{F} = \sin\theta \quad \text{or} \quad F_\parallel = F\sin\theta$$

and
$$F^2 = F_\perp^2 + F_\parallel^2$$

EXAMPLE 15.6. A car travels at 50 mi/hr in a direction 60° north of west. Calculate the car's velocity components in the northerly and westerly directions.

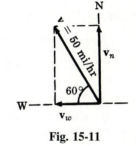

Fig. 15-11

Sketch the directions and the car's velocity **v** as shown in Fig. 15-11. The sketch need not be drawn to scale. Let v_w be the magnitude of the westerly component of **v** and v_n the magnitude of the northerly component.

$$\frac{v_w}{v} = \cos 60°$$

$$v_w = v\cos 60° = 50(0.500) = 25$$

The westerly component of the car's velocity is 25 mi/hr, or $v_w = 25$ mi/hr W.

$$\frac{v_n}{v} = \sin 60°$$

$$v_n = v\sin 60° = 50(0.866) = 43.3$$

The northerly component of the car's velocity is 43.3 mi/hr or $v_n = 43.3$ mi/hr N.

Vector Addition with Components

Vector addition is greatly simplified by the use of rectangular components.

15.10   THE COMPONENTS OF VECTOR SUMS

The x-component, R_x, of the resultant of several vectors is the sum of their x-components:

$$R_x = A_x + B_x + C_x + \cdots$$

The y-component, R_y of the resultant of several vectors is the sum of their y-components:

$$R_y = A_y + B_y + C_y + \cdots$$

15.11   VECTOR SUM FROM COMPONENTS

If we know R_x and R_y, then the magnitude of the resultant, R, is $\sqrt{R_x^2 + R_y^2}$. The direction of **R** is determined from trigonometry.

EXAMPLE 15.7. A force of 100 lb-f applied at the origin makes an angle of 30° with the positive x-axis and acts to the right. Another force of 75 lb-f makes an angle of $-120°$ with the positive x-axis and acts to the left. Calculate the resultant force by using rectangular components.

Sketch a diagram with the data as shown in Fig. 15-12. Let $\mathbf{F_1}$ be the 100-lb force; $\mathbf{F_2}$, the 75-lb force; and **R**, the resultant.

Then,

$$F_{1x} = F_1 \cos 30° \approx 100(0.866) \approx 86.6$$
$$F_{1y} = F_1 \sin 30° \approx 100(0.500) \approx 50.0$$
$$F_{2x} = F_2 \cos(-120°) \approx 75(-0.500) = -37.5$$
$$F_{2y} = F_2 \sin(-120°) \approx 75(-0.866) = -64.9$$
$$R_x = F_{1x} + F_{2x} \approx 86.6 - 37.5 \approx 49.1$$
$$R_y = F_{1y} + F_{2y} \approx 50 - 64.9 \approx -14.9$$
$$R = \sqrt{R_x^2 + R_y^2} \approx \sqrt{(49.1)^2 + (-14.9)^2}$$
$$\approx \sqrt{2632.82}$$
$$\approx 51.3$$
$$\tan\theta = \frac{-14.9}{49.1} \approx -0.303$$
$$\theta \approx -17°$$

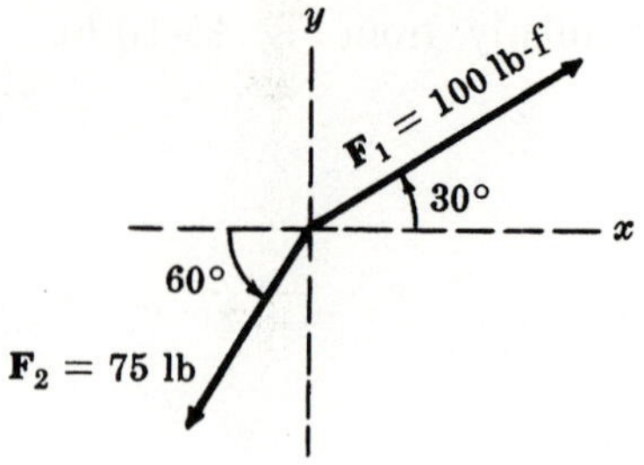

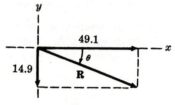

Fig. 15-12

Fig. 15-13

The resultant is a force of 51.3 lb, at an angle of $-17°$ with the positive x-axis (see Fig. 15-13).

To calculate the value of R using a calculator, make sure that the square root is applied to the whole expression $R_x^2 + R_y^2$. Therefore, you may have to use extra parentheses. With an HP-38G calculator key in the sequence

$\{\sqrt{x}\}\{()49.1\{x^y\}2\{+\}\{()\{-x\}14.9()\}\{x^y\}2()\}\{ENTER\}$

The display shows 51.3110124632. This value can be rounded to 51.3.

To calculate the value of the angle θ key in

$\{SHIFT\}\{ATAN\}\{-x\}0.303()\}\{ENTER\}$

The result of this operation is -16.8568085435. This value can be rounded to $-17°$.

Solved Problems

15.1. Represent graphically a displacement 45 mi/hr, 20° south of west.

The scale and the vector are shown in Fig. 15-14.

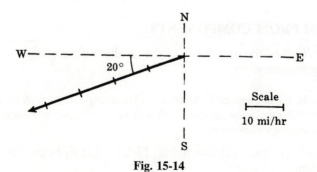

Fig. 15-14

15.2. Which of the following measures are vectors and which are scalars?

(a) $24.7 \, \text{cm}^3$ (d) $1.05 \, \text{sec}$ (g) $70.0 \, \text{km/hr}$, northeast

(b) $9.80 \, \text{m/s}^2$, downward   (e) $8.3 \, \text{lb-f}$, along the y-axis

(c) $15.0 \, \text{ft}^2$ (f) $2.7 \, \text{g/cm}^3$

(a) volume—scalar   (c) area—scalar   (e) force—vector   (g) velocity—vector

(b) acceleration—vector   (d) time—scalar   (f) density—scalar

15.3. A downward force of 25 lb-f is applied at 40° to the horizontal.

(a) Represent the given force by a vector.

(b) Describe the negative of the force vector.

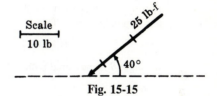

(a) Draw a horizontal direction. Lay off an angle of 40°. Choose a convenient scale to draw 25 lb-f along the 40° direction as shown in Fig. 15-15.

(b) An upward force of 25 lb-f at 40° to the horizontal.

Fig. 15-15

15.4. In Problem 15.3 determine the horizontal and vertical components of the given force.

Sketch the given vector and draw through its tail the horizontal and vertical directions, as shown in Fig. 15-16. Drop perpendiculars from the arrow head onto the two directions.

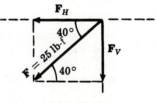

The magnitude of the horizontal component is

$$F_H = F \cos 40° = 25(0.766) \approx 19.1$$

The magnitude of the vertical component is

$$F_V = F \sin 40° = 25(0.6428) \approx 16.1$$

Fig. 15-16

The two components are $F_H = 19.1 \, \text{lb}$, to the left

$$F_V = 16.1 \, \text{lb}, \text{ downward}$$

Check: $25^2 \stackrel{?}{=} 19.1^2 + 16.1^2$

$$625 \approx 624$$

15.5. The *azimuth* of a point is the angle measured in a circle clockwise from north to the point. Determine the resultant velocity of the following velocities: 295 mi/hr, azimuth 215°; 45 mi/hr, azimuth 215°.

The two given velocities have the same direction. The magnitude V of the resultant is the sum of the two velocity magnitudes:

$$V = 295 + 45 = 340$$

The resultant velocity is

$$\mathbf{V} = 340 \text{ mi/hr, azimuth } 215°$$

15.6. In Problem 15.5, subtract the second velocity from the first.

Reverse the direction of the second velocity and add it to the first. The magnitude V of the resultant velocity is the difference between the two given magnitudes:

$$V = 295 - 45 = 250$$

The direction of the vector difference is the same as the direction of the larger velocity. The resultant is

$$\mathbf{V} = 250 \text{ mi/hr, azimuth } 215°$$

15.7. Add by the parallelogram method the vectors shown in Fig. 15-17.

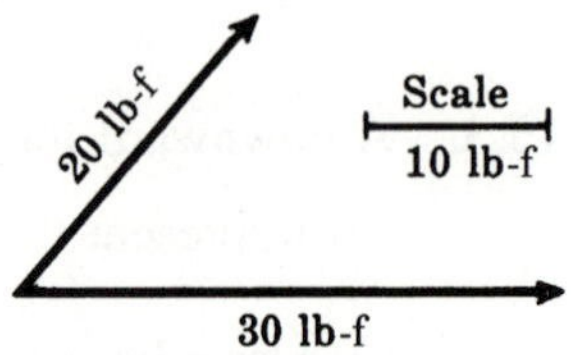

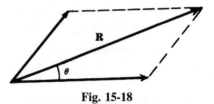

Fig. 15-17

Complete the parallelogram with the two given vectors as adjacent sides as shown in Fig. 15-18. Draw a diagonal from the common point to the opposite corner. Using the scale to which the vectors were drawn, determine the magnitude of the resultant. The length of $\mathbf{R}$ is 4.5 scale units approximately. Thus

$$|\mathbf{R}| \approx (4.5)(10) \approx 45$$

The angle θ is measured with a protractor.

$$\theta \approx 18°$$

$$\mathbf{R} = 45 \text{ lb-f, } 18° \text{ with the horizontal and to the right}$$

Fig. 15-18

15.8. Add graphically by the polygon method the vectors shown in Fig. 15-19.

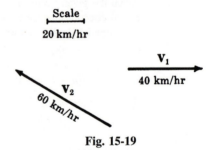

Fig. 15-19

From the arrow of the vector $\mathbf{V}_1$ drawn the vector $\mathbf{V}_2$ in the given direction. Join the tail of the vector $\mathbf{V}_1$ to the arrowhead of the vector $\mathbf{V}_2$ to obtain the sum or resultant as shown in Fig. 15-20. Measure the resultant and determine its magnitude from the scale used in the drawing.

$$R \approx 34 \text{ km/hr}$$

Measure the angle ϕ with a protractor.

$$\phi \approx 110°$$

The resultant of the two vectors is

$$\mathbf{R} \approx 34 \text{ km/hr, } 110° \text{ counterclockwise from } \mathbf{V}_1$$

15.9. Add the force vectors shown in Fig. 15-21 by the polygon method.

From the arrowhead of $\mathbf{F}_1$ draw $\mathbf{F}_2$; from the arrowhead of $\mathbf{F}_2$ draw $\mathbf{F}_3$. Join the tail of $\mathbf{F}_1$ to the arrowhead of $\mathbf{F}_3$; this is the resultant $\mathbf{R}$. The method is shown in Fig. 15-22(a). Since the process is

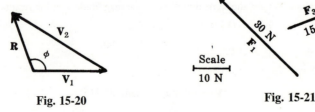

Fig. 15-20 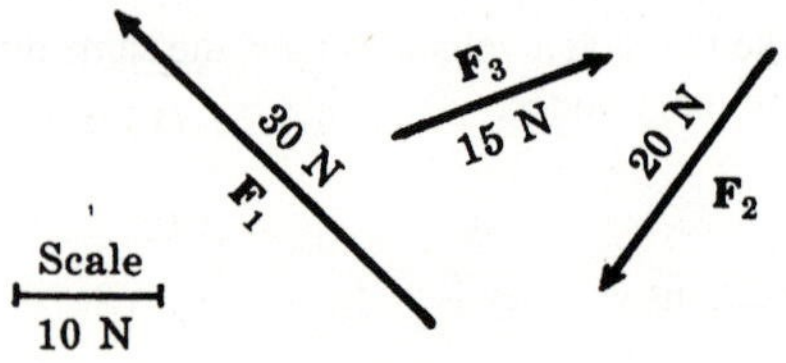   Fig. 15-21

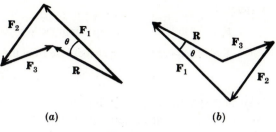

Fig. 15-22

commutative, the sum does not depend on the order in which the vectors are added.   In Fig. 15-22(*b*), the sequence of addition was F_3, F_2 and F_1.

Measure the resultant.   Its value from the scale used for the drawing is

$$R \approx (2.0)(10) \approx 20 \text{ N}$$

Measure the angle θ with a protractor.

$$\theta \approx 17°$$

The resultant force is

$$\mathbf{R} \approx 20 \text{ N},\ 17° \text{ counterclockwise from } \mathbf{F}_1$$

15.10. Determine the resultant of the displacements 7.0 miles to the north and 9.0 miles to the west.

A sketch of the given displacements is shown in Fig. 15-23.   The two displacements are at 90° to each other.   The magnitude of the resultant is obtained by the Pythagorean theorem.

$$D = \sqrt{7.0^2 + 9.0^2} = \sqrt{130} \approx 11.4$$

If θ is the direction angle of the resultant,

$$\tan \theta = 7/9 \approx 0.778$$
$$\theta \approx 38°$$

The resultant displacement is:

$$\mathbf{D} \approx 11.4 \text{ mi},\ 38° \text{ north of west}$$

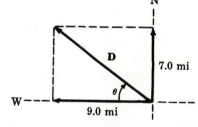

Fig. 15-23

15.11. Determine the resultant of a 10.0 lb-f horizontal force acting to the left and a 150. lb-f vertical force acting upward.

The geometrical layout of the forces is shown in Fig. 15-24.   The forces are perpendicular to each other.   From the Pythagorean theorem, the magnitude of the resultant is

$$F = \sqrt{(10.0)^2 + (15.0)^2} = \sqrt{325} \approx 18.0$$

Let the direction of the resultant be θ.

$$\tan \theta = 15.0/10.0 = 1.50$$
$$\theta = 56.3°$$

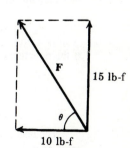

Fig. 15-24

The resultant of the two forces is

$$\mathbf{F} \approx 18.0 \text{ lb-f}, 56.3° \text{ clockwise from the horizontal force}$$

or

$$\mathbf{F} \approx 18.0 \text{ lb-f}, 33.7° \text{ counterclockwise from the vertical force}.$$

15.12. The velocity of a boat is 20 km/hr in a direction 50° north of east. The wind velocity is 5 km/hr from the west. The resultant velocity of the boat can be represented by the side of a triangle in which the two given velocities are the other sides. Determine the resultant velocity **V** of the boat.

Sketch and label the triangle of velocities as shown in Fig. 15-25. Applying the law of cosines, we have

$$V^2 = 20^2 + 5^2 - 2(20)(5) \cos 130°$$

But $\cos 130° = -\cos 50° = -0.643$

Hence $V^2 = 400 + 25 + 200(0.643)$

$$= 425 + 129 = 554$$

$$V = \sqrt{554} \approx 23.5$$

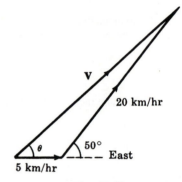

Fig. 15-25

The magnitude of the velocity of the boat is 23.5 km/hr. To determine the direction of the resultant velocity, we use the law of sines to compute angle θ.

$$\frac{20}{\sin \theta} = \frac{23.5}{\sin 130°}$$

From which we get

$$\sin \theta = \frac{20}{23.5} \sin 130° = 0.8511 \sin 50°$$

$$\approx 0.6519$$

$$\theta \approx 40.7°$$

The direction of the boat is then 40.7° north of east.

15.13. Solve Problem 15.12 by the method of rectangular components.

Resolve each velocity into rectangular components. Thus, the components of the boat's velocity are

$$|\mathbf{V}_{\text{north}}| = 20 \sin 50° = (20)(0.766) \approx 15.3$$
$$|\mathbf{V}_{\text{east}}| = 20 \cos 50° = (20)(0.6428) \approx 12.9$$

The wind velocity is 5 km/hr eastward. Add the eastward components to get the eastward component of the resultant: $V_{\text{east}} = 12.9 + 5 \approx 17.9$. From the Pythagorean theorem, the magnitude of the resultant is

$$V = \sqrt{(17.9)^2 + (15.3)^2} \approx 23.5 \approx 24$$

If the direction of the resultant is θ in Fig. 15-26, then

$$\tan \theta = \frac{15.3}{17.9} \approx 0.854$$

$$\theta \approx 41°$$

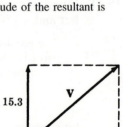

The boat's direction is 41° north of east. The value of the resultant is the same as in Problem 15.12, but the computations are greatly simplified by the use of rectangular components.

Fig. 15-26

15.14. Determine the resultant from the diagram of Fig. 15-27.

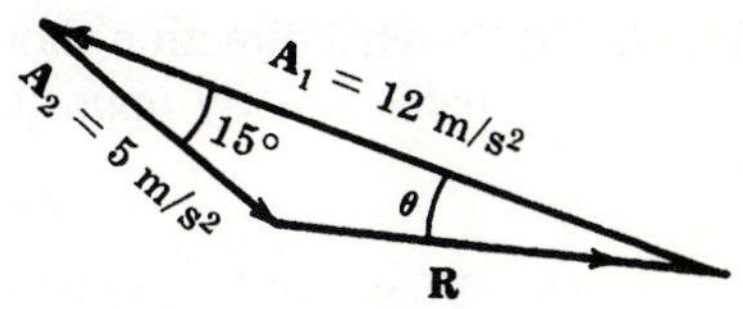

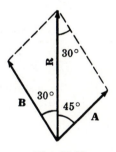

Fig. 15-27

Here it is more convenient to use the law of cosines than the method of rectangular components.

We have

$$R^2 = 12^2 + 5^2 - 2(12)(5)(\cos 15°) \approx 53.09$$

$$R \approx \sqrt{53.09} \approx 7.3$$

Applying the law of sines,

$$\frac{\sin \theta}{5} = \frac{\sin 15°}{7.3}$$

$$\sin \theta = 5\frac{\sin 15°}{7.3} \approx \frac{5(0.259)}{7.3} \approx 0.177$$

$$\theta \approx 10°$$

The resultant is

$$\mathbf{R} = 7.3 \text{ m/s}^2, \ 25° \text{ counterclockwise from the } \mathbf{A}_2 \text{ vector}$$

15.15. A force **A** makes an angle of 45° with the vertical; a force **B** makes an angle of 30° with the vertical. The vector sum of **A** and **B** is 20 newtons, vertically upward. Determine the magnitudes of **A** and **B** if they are on the opposite sides of the resultant.

Let the resultant be **R**. From the geometry of the vectors in Fig. 15-28, the angle opposite **R** is 180° − (30° + 45°), or 105°. Applying the law of sines,

$$\frac{\sin 105°}{20} = \frac{\sin 30°}{A}$$

$$A = 20\frac{\sin 30°}{\sin 105°} \approx 20\frac{0.500}{0.966} \approx 10.4$$

Similarly,

$$\frac{\sin 105°}{20} = \frac{\sin 45°}{B}$$

$$B = 20\frac{\sin 45°}{\sin 105°} \approx 20\frac{0.707}{0.966} \approx 14.6$$

Fig. 15-28

The magnitude of force **A** is 10.4 newtons; of force **B**, 14.6 newtons.

15.16. A river flows northward. The velocity of the current is 5 km/hr. A boat with a speed of 20 km/hr starts directly across the river from the west bank. What is the resultant velocity of the boat?

The resultant velocity **v** of the boat is the vector sum of the boat velocity $\mathbf{v}_b$ and the current velocity $\mathbf{v}_c$, as shown in Fig. 15-29.

$$v = \sqrt{20^2 + 5^2} = \sqrt{425} \approx 20.6$$

$$\tan \theta = 5/20 = 0.250$$

$$\theta \approx 14°$$

The boat's velocity is $\mathbf{v} \approx 20.6$ km/hr, 14° north of east.

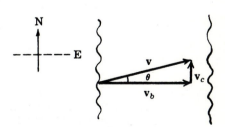

Fig. 15-29

15.17. (*a*) In what direction should the boat in Problem 15.16 head so as to reach a point on the east bank directly opposite the starting point?

(*b*) What will be the boat's easterly speed?

(*a*) The resultant velocity **v** must now have a direction eastward as shown in Fig. 15-30.

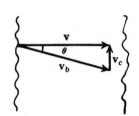

$$\sin \theta = \frac{v_c}{v_b} = 5/20 = 0.250$$

$$\theta \approx 14.5°$$

(*b*) The boat's easterly speed is the magnitude of its resultant velocity. From the Pythagorean theorem,

$$v = \sqrt{20^2 - 5^2} = \sqrt{375} \approx 19.4$$

Fig. 15-30

Thus the boat's speed is 19.4 km/hr.

15.18. Resolve a downward force of 40 lb-f into a component parallel to a 30° incline and a component perpendicular to it.

The geometry of the inclined plane and the forces is shown in Fig. 15-31. The angle between **F** and the perpendicular component $F_\perp$ is 30°, since the two vectors are perpendicular to the two sides of the angle.   From the definitions of cosine and sine,

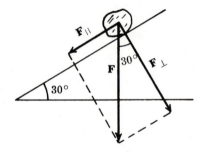

$$\frac{F_\perp}{F} \cos 30° \quad \text{or} \quad F_\perp = F \cos 30° = 40(0.866) \approx 34.6$$

$$\frac{F_\parallel}{F} \sin 30° \quad \text{or} \quad F_\parallel = F \sin 30° = 40(0.500) = 20.0$$

The magnitudes of the two components are 34.6 lb-f and 20.0 lb-f.

Fig. 15-31

Check: $40^2 \stackrel{?}{=} (34.6)^2 + (20.0)^2$

$$1600 \approx 1597$$

15.19. Determine by the method of rectangular components the resultant of the following displacements: 60 miles, east; 50 miles, 45° west of north; 30 miles, 20° east of south.

Sketch a diagram of the given displacements as in Fig. 15-32.

Resolve each vector into its rectangular components and tabulate the results as shown below.   Let the eastward and northward directions be positive.

Displacement	Horizontal component	Vertical component
60 mi, E	$60 \cos 0° = 60$	$60 \sin 0° = 0$
50 mi, 45° W of N	$50 \cos 135° \approx -35.4$	$50 \sin 135° \approx 35.4$
30 mi, 20° E of S	$30 \cos 290° \approx 10.3$	$30 \sin 290° \approx -28.2$
	sum $D_h \approx 34.9$	sum $D_v \approx 7.2$

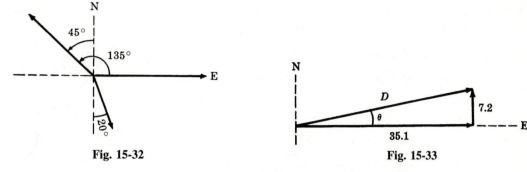

Fig. 15-32   Fig. 15-33

The sum of the horizontal components is 34.9, which is the horizontal component of the resultant; the sum of the vertical components is 7.2, which is the vertical component of the resultant.

From the Pythagorean theorem,

$$D = \sqrt{(34.9)^2 + (7.2)^2} \approx 35.6$$

From Fig. 15-33,

$$\tan \theta = \frac{7.2}{35.6} \approx 0.202$$
$$\theta \approx 11.5°$$

The resultant displacement **D** is 35.6 mi, 11.5° north of east.

15.20. The momentum vector $\mathbf{M}_1$ is 125 kg · m/sec at $+50°$ to the $+x$-axis. The momentum vector $\mathbf{M}_2$ is along the $-y$-axis as shown in Fig. 15-34. What should be the magnitude of $\mathbf{M}_2$ so that the sum of $\mathbf{M}_1$ and $\mathbf{M}_2$ will be along the $+x$-axis?

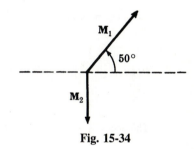

Fig. 15-34

The y-component of the resultant is to be made zero. Therefore the magnitude of the vector $\mathbf{M}_2$ must be equal to the magnitude of the y-component of $\mathbf{M}_1$. Thus,

$$M_2 = M_1 \sin 50° = 125(0.766) \approx 95.8$$

The magnitude of $\mathbf{M}_2$ is 95.8 kg · m/sec.

15.21. What momentum must be added to the momenta in Problem 15.20 so that the resultant is zero?

The added momentum must be equal in magnitude and opposite in direction to the horizontal component of $\mathbf{M}_1$. Therefore,

$$M_3 = M_1 \cos 50° = 125(0.643) \approx 80.4$$

The required momentum is 80.4 kg · m/sec along the $-x$-axis.

15.22. The velocity of a plane is 180 mph, 30° west of north. The wind is 60 mph from 30° south of east. Determine the resultant velocity of the plane.

The relationship of the vectors is shown in Fig. 15-35.   From the
law of cosines, the magnitude of the resultant is

$$V = \sqrt{(180)^2 + (60)^2 - 2(180)(60)\cos 150°}$$
$$= \sqrt{32,400 + 3600 - 21,600(-0.866)} = \sqrt{54,700} \approx 234$$

Applying the law of sines,

$$\frac{\sin \phi}{60} = \frac{\sin 150°}{234}$$

$$\sin \phi = 60\frac{\sin 150°}{234} = 60\frac{(0.50)}{234} \approx 0.128$$

$$\phi \approx 7.4°$$

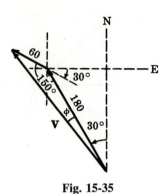

Fig. 15-35

The direction of **V** is $(30 + 7.4)°$ or $37.4°$ west of north.   The resultant
velocity of the plane is 234 mph, $37.4°$ west of north.

15.23. Check Problem 15.22 by the method of rectangular components.

Let the north and east directions be positive and resolve the given velocities into rectangular components.

$$\begin{array}{ll}
\text{north component of plane velocity} = 180\cos 30° = 180(0.866) \approx 156 \\
\text{north component of wind velocity} = \ \ 60\cos 60° = \ \ 60(0.500) = \ \ 30 \\
\hline
\text{sum} \approx 186
\end{array}$$

$$\begin{array}{ll}
\text{west component of plane velocity} = 180\sin 30° = 180(0.500) = \ \ 90 \\
\text{west component of wind velocity} = \ \ 60\sin 60° = \ \ 60(0.866) \approx \ \ 52 \\
\hline
\text{sum} \approx 142
\end{array}$$

From the Pythagorean theorem, the magnitude of the resultant velocity is

$$V = \sqrt{(186)^2 + (142)^2} = \sqrt{54,760} \approx 234$$

$$\tan \theta = \frac{142}{186} \approx 0.763$$

$$\theta \approx 37.4°$$

The check is within computational precision.

Supplementary Problems

The answers to this set of problems are at the end of this chapter.

15.24.   Which of the following measures are vectors and which scalars?

    (a) 280 cm/sec   (c) 12°25′   (e) 1.75 meters westward

    (b) 75 newtons, upward   (d) 4.8 ft/sec^2 at 30° to $+x$-axis   (f) 1.75 kilograms

15.25.   Represent graphically a downward force of 25 lb-f at 60° with the negative x-axis.

15.26.   A body has a velocity whose component along an axis inclined 30° to the horizontal is 3.80 m/sec to the
right.   Describe the component along the same axis of a velocity equal in magnitude but opposite in
direction to that of the body.

15.27. In Fig. 15-36, determine the rectangular components of the given acceleration.

15.28. Add graphically by the polygon method the vectors shown in Fig. 15-37.

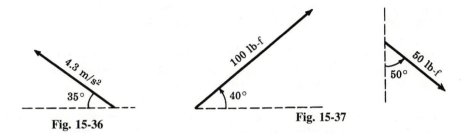

Fig. 15-36

Fig. 15-37

15.29. In the representation of Fig. 15-38, F = 15 newtons and $\theta = -40°$.

(*a*) Determine the horizontal component of **F**.

(*b*) Determine the vertical component of **F**.

15.30. The rectangular components of a velocity vector are:

$$v_x = -5.0 \text{ ft/sec} \quad \text{and} \quad v_y = 8.0 \text{ ft/sec}$$

Determine the magnitude and direction of the velocity.

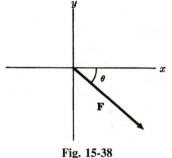

Fig. 15-38

15.31. Two forces making an angle of 60° with each other are applied at one point in a body. The magnitudes of the forces are 100 lb-f and 80 lb-f. Determine graphically by the parallelogram method (*a*) the magnitude of the resultant, (*b*) the direction it makes with the 100-lb-f force.

15.32. A boy walks 3 km westward; then he walks 2 km in a direction 40° east of north and stops. Determine the boy's displacement from his point of departure.

15.33. Determine the resultant of a momentum of 6.0 kg · m/sec downward and a momentum of 12.0 kg · m/sec along a horizontal to the right.

15.34. Subtract graphically a force of 15 newtons at 45° with the +*x*-axis from a force of 10 newtons at 60° from the +*y*-axis.

15.35. The speed of a boat in still water is 8 mi/hr. A river current has a velocity 3 mi/hr eastward. Find the boat's land velocity if its water velocity is (*a*) eastward, (*b*) westward, (*c*) northward.

15.36. In Problem 15.35, (*a*) what direction should the boat be steered to reach a point on the south shore directly opposite starting point and (*b*) what will be the boat's resultant speed?

15.37. Determine by the method of rectangular components the resultant of the following forces: 30 newtons to the right at 45° with the +*x*-axis, 20 newtons to the right at −30° with +*x*-axis.

15.38. Resolve a downward force of 500 dynes into (*a*) a component perpendicular to an inclined plane with an angle of 20° with the horizontal and (*b*) a component parallel to the plane.

15.39. The velocity of a plane in still air is 140 mph, 54° north of east. If a 30-mph wind comes up from the south, what will be the resultant velocity of the plane? Solve by the analytical method.

15.40. Solve Problem 15.39 by using rectangular components.

15.41. Calculate the magnitudes of the forces $\mathbf{F}_1$ and $\mathbf{F}_2$ in Fig. 15-39.   The figure is a parallelogram.

15.42. The resultant of two displacements is 25 km 30° west of south.   One of the displacements is 20° east of south, the other displacement is 60° west of south.   Determine the magnitudes of the two displacements.

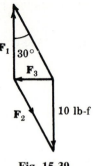

15.43. The components of a body's acceleration are $4.5\,\text{m/sec}^2$ along the positive y-axis and $3.0\,\text{m/sec}^2$ along the negative x-axis.   What additional acceleration vector should be added so that the body's acceleration is zero.

Fig. 15-39

15.44. Determine the resultant of the following forces: $\mathbf{F}_1 = 50\,\text{lb-f}$, vertically downward; $\mathbf{F}_2 = 25\,\text{lb-f}$, upward at 150° to the horizontal; $\mathbf{F}_3 = 43.3\,\text{lb-f}$, upward at 60° to the horizontal.

15.45. The resultant of three velocities has the components $R_x = -12\,\text{m/sec}$; $R_y = 6\,\text{m/sec}$.   Determine the resultant.

15.46. In Problem 15.45, two of the velocities have the components $A_x = 6\,\text{m/sec}$, $A_y = -2\,\text{m/sec}$; $B_x = 4\,\text{m/sec}$, $B_y = 4\,\text{m/sec}$.   What are the components of the third velocity, C_x and C_y?

Answers to Supplementary Problems

15.24. vectors: (b), (d), (e); scalars: (a), (c), (f).

15.25. See Fig. 15-40.

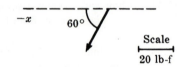

15.26. $3.80\,\text{m/sec}$ to the left at 210° (or $-150°$ with the horizontal to the right)

Fig. 15-40

15.27. horizontal: $3.5\,\text{m/s}^2$ to the left
vertical: $2.5\,\text{m/s}^2$ upward

15.28. See Fig. 15-41.

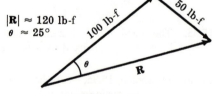

15.29. (a) 11.5 newton to the right
(b) 9.64 newton downward

15.30. $v \approx 9.4\,\text{ft/sec}$, 122° from $+x$-axis
or 58° from $-x$-axis

Fig. 15-41

15.31. See Fig. 15-42.

15.32. 2.30 km, 48.2° W of N

15.33. $13.4\,\text{kg} \cdot \text{m/sec}$ downward to the right at 26.6° with the horizontal

15.34. See Fig. 15-43.

15.35. (a) 11 mi/hr eastward (b) 5 mi/hr westward (c) 8.5 mi/, 69.5° north of east

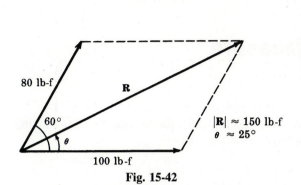

Fig. 15-42   Fig. 15-43

15.36. (a) 22° west of south   (b) 7.4 mi/hr

15.37. 40.1 newtons to the right at 16.2° with the $+x$-axis

15.38. (a) 4700 dynes   (b) 1700 dynes

15.39. 165 mph, 60.1° north of east

15.40. Same as Problem 15.39.

15.41. $F_1 = 10$ lb-f $F_2 \approx 11.5$ lb-f $F_3 \approx 5.8$ lb-f

15.42. 12.7 km, 20° east of south   19.4 km, 60° west of south

15.43. 5.4 m/sec², −56.3° or 303.7° with $+x$-axis

15.44. 0

15.45. 13.4 m/sec, 153.4° or −206.6° from the $+x$-axis

15.46. $C_x = -22$ m/sec, $C_y = 4$ m/sec

Chapter 16

Radian Measure

The Radian

In addition to the degree, there is another very useful unit for measuring angles—the radian. This unit is measure is important in theoretical developments.

16.1 DEFINITION

A *radian* is the central angle included between two radii of a circle which intercept on the circumference an arc equal in length to a radius. In Fig. 16-1, angle *AOB* is 1 radian.

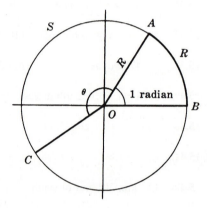

16.2 BASIC EQUATION

In Fig. 16-1, if angle θ intercepts an arc *ASC*,

$$\frac{S}{\theta \text{ radian}} = \frac{R}{1 \text{ radian}}$$

from which $S = R\theta$ and $\theta = S/R$. In this equation, θ is a radian measure, while S and R must be in the same units of length. We sometimes write θ^r to indicate that θ is in radians.

Fig. 16-1

EXAMPLE 16.1. If an arc length is 24 cm and the circle radius is 60 cm, what is the radian measure of the subtended central angle?

Substitute the given values in $\theta = S/R$

$$\theta = \frac{24 \text{ cm}}{60 \text{ cm}} = \frac{2}{5} = 0.40$$

The measure of the angle is 0.40 radian.

16.3 RELATIONSHIP BETWEEN THE RADIAN AND DEGREE MEASURES

Since the circumference of a circle has a length of $2\pi R$, where R is the radius, the circumference subtends a central angle of $\theta = \dfrac{2\pi R}{R} = 2\pi$ radians. The circumference also subtends 360°. Hence, 2π radian = 360° and

$$1 \text{ radian} = \frac{360°}{2\pi} = \frac{180°}{\pi} \approx \frac{180°}{3.14} \approx 57.3°$$

Also,

$$1° = \frac{2\pi \text{ radian}}{360} \approx \frac{6.28}{360} \approx 0.0175 \text{ radian}$$

EXAMPLE 16.2 Express 75° in radians.

Let θ be the desired radian measure of 75°.   Then we can set up the proportion

$$\frac{2\pi}{\theta} = \frac{360}{75}$$

Therefore,

$$\theta = 2\pi \frac{75}{360} \approx 0.417\pi \approx (0.417)(3.14) \approx 1.31^r$$

Check: $(75°)(0.0174 \, \text{rad/deg}) \approx 1.31 \, \text{rad}$

To calculate the value of θ using an HP-38G graphing calculator, key in the sequence

2{*}{SHIFT}{π}{*}{(}75{/}360{)}{ENTER}

The display shows 1.30899693899.   This value can be rounded to 1.31.

EXAMPLE 16.3.   Express 9.41 radians in degrees.

Let ϕ be the desired degree measure of the given angle.   We can set up the proportion

$$\frac{9.41}{2\pi} = \frac{\phi}{360}$$

Therefore, $\phi = 360\dfrac{9.41}{2\pi} \approx 360\dfrac{9.41}{6.28} \approx 539°$

Check: $(9.41 \, \text{rad})(57.3 \, \text{deg/rad}) \approx 539°$

To find the value of ϕ with a TI-82 graphing calculator, key in

360{x}{(}9.41{÷}{(}2{x}{2nd}{π}{)}{)}{ENTER}

The result is 539.1532852.   This result can be rounded to 539°.

16.4   EVALUATING TRIGONOMETRIC FUNCTIONS OF AN ANGLE USING A CALCULATOR

To work with angles expressed in radians, make sure that the MODE of the calculator is set to Radians.   For all the problems in this chapter make sure that your calculator has been set to Radian mode.

EXAMPLE 16.4.   Determine $\sin \pi/4$ and $\cos 3\pi/4$.

Using an HP-38G calculator follow the steps indicated below.

(1)   Set the Mode to Radians.

{SHIFT}{MODES}

CHOOS{▾}Radians{ENTER}

{HOME}

(2)   Calculate $\sin \pi/4$.

{SIN}{SHIFT}{π}{/}4{)}{ENTER}

The display shows 0.707106781187.   This value can be rounded to 0.7071.

(3)   Calculate $\cos 3\pi/4$.

{COS}{(}3{*}{SHIFT}{π}{)}{/}4{)}{ENTER}

Always use extra parentheses to ensure that the denominator divides the entire numerator.   The result is −0.707106781185.

EXAMPLE 16.5.   Determine $\cot(5\pi/3)$ and $\sec(13\pi/6)$.

Since there are no function keys to calculate directly the cotangent and secant of an angle, it is necessary to use the trigonometric definitions of these two functions.   Do not forget to use extra parentheses if you are not sure of the order in which the calculator carries out arithmetic operations.   By definition, $\cot(5\pi/3) = 1/\tan(5\pi/3)$, therefore, to calculate the value of $\cot(5\pi/3)$ using a TI-82 calculator proceed as indicated below.

(1)   Set the Mode to Radians.

{MODE}{˅}{˅}{ENTER}

{CLEAR}

(2)   Calculate $\cot(5\pi/3)$.

1{÷}{TAN}{(}{(}{(}5{×}{2ⁿᵈ}{π}{)}{)}{÷}3{)}{ENTER}

The display shows -0.5773502692.   This result can be rounded to -0.5774.

By definition, $\sec(13\pi/6) = 1/\cos(13\pi/6)$, therefore, to calculate the value of $\sec(13\pi/6)$ proceed as follows:

1{÷}{COS}{(}{(}{(}13{×}{2ⁿᵈ}{π}{)}{)}{÷}6{)}{ENTER}

The result is 1.15400538.   This value can be rounded to 1.1540.

EXAMPLE 16.6.   Find the angle in radians whose cosine is equal to 0.3.

Using an HP-38G graphing calculator key in {SHIFT}{ACOS}0.3{)}{ENTER}.   The display shows 1.26610367278.   This value can be approximated to 1.27 radians.

Using a TI-82 calculator key in {2ⁿᵈ}{COS⁻¹}0.3{ENTER}.   The display shows 1.266103673.   This value can be rounded to 1.27 radians.

Angular Speed and Velocity

16.5   ANGULAR SPEED

In Fig. 16-1, let point A move to position C in time t; then the distance covered is the arc length S; the angle swept out by radius OA is θ^{r}.

Average angular speed ω is then defined as the change in angle per unit time, or

$$\omega = \theta/t$$

ω is the Greek letter Omega.

The units of ω will depend on the units of θ and t.   If θ is in radians and t is in seconds, then ω is in radians per second.

EXAMPLE 16.7.   Determine the angular speed in radians per minute and radians per second of a centrifuge which makes 100,000 revolutions in 5 min.

Since one revolution is equivalent to 2π radians, the angle swept out by the centrifuge is $(100,000)(2\pi)$ radians.   Therefore the angular speed is

$$\omega = \frac{(100,000)2\pi}{5}\frac{\text{radian}}{\text{min}} = 40,000\pi \text{ radian/min} \approx 126,000 \ \text{radian/min}$$

Also, substituting 60 sec for 1 min,

$$\omega = \frac{40,000\pi}{60}\frac{\text{radian}}{\text{sec}} \approx 667\pi \text{ radian/sec} \approx 2090 \text{ radian/sec}$$

16.6 ANGULAR VELOCITY

Angular velocity is angular speed in a specified direction.   Rotation in a counterclockwise direction is said to be positive; clockwise rotation is called negative.   For example, an angular velocity of 5.25 rad/sec clockwise may be written as $\omega = 5.25$ rad/sec clockwise, or $\omega = -5.25$ rad/sec.

16.7 ANGULAR AND LINEAR SPEEDS

In Fig. 16-1 the average linear speed v of point A along the arc S is $v = S/t$.   But $S = R\theta$, and so by substitution $v = R\theta/t$. We have previously seen that $\omega = \theta/t$.   Therefore, $v = R\omega$ expresses the relationship between linear (circumferential) and angular speeds.   Thus, linear speed = radius $\times$ angular speed.   It is important to note that v and R must involve the same units of length.

EXAMPLE 16.8.   The angular speed of a grindstone is 1800 revolutions per minute.   The diameter of the grindstone is 10 inches.   (*a*) Calculate the linear speed of a point on the rim of the grindstone.   (*b*) Express the answer in ft/sec.

(*a*) The angular speed must be first expressed in radian/minute.   Since 1 rev $= 2\pi^r$,

$$1800 \left(\frac{\text{rev}}{\text{min}}\right) \left(\frac{2\pi \text{ rad}}{1 \text{ rev}}\right) = 3600\pi \text{ rad/min}$$

Substitute in the equation $v = R\omega$:

$$v = (5)(3600\pi) = 18{,}000\pi \text{ in/min} \approx 56{,}500 \text{ in/min}$$

(*b*) 1 ft $= 12$ in.; 1 min $= 60$ sec

$$v = 56{,}500 \frac{\text{in}}{\text{min}} \frac{1 \text{ ft}}{12 \text{ in}} \frac{1 \text{ min}}{60 \text{ sec}} \approx 78.5 \text{ ft/sec}$$

Small-Angle Approximations

16.8 TRIGONOMETRIC FUNCTIONS OF SMALL ANGLES

For the angle θ (radians) in Fig. 16-2, $\theta = S/R$; $\sin\theta = y/R$; $\tan\theta = y/x$. We note that as the angle becomes smaller and smaller the length of arc S differs less and less from the length of the leg y of the right triangle; also the length of the leg x approximates closer and closer the length of radius R.   The smaller the angle the better are these approximations.   Thus, for *small angles* we have $S \approx y$ and $R \approx x$, and the following approximations are possible:

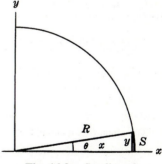

Fig. 16-2.   Small angle

$$\sin\theta = \frac{y}{R} \approx \frac{S}{R} = \theta$$

$$\tan\theta = \frac{y}{x} \approx \frac{S}{R} = \theta$$

from which, $\sin\theta \approx \tan\theta$.   Also,

$$\cos\theta = \frac{x}{R} \approx \frac{R}{R} = 1$$

The approximations are quite good for angles up to $10°$, or about 0.17 radians.

EXAMPLE 16.9. Determine the value of sin 2° and tan 2° without using trigonometric tables.

Convert 2° to radians.

$$1° = 0.0175^r \qquad \text{and} \qquad 2° = (2)(0.0175)^r = 0.0350^r$$

Since 2° is a small angle, $\sin \theta \approx \theta^r$. Therefore, sin 2° ≈ 0.0350. Also, tan 2° ≈ 0.0350. The trigonometric table gives sin 2° = 0.03490 and tan 2° = 0.03492.

To calculate the values of these small angles using a TI-82 calculator proceed as follows:

(1) Make sure that the MODE of the calculator is set to Radians.

{MODE}{▼}{▼}<u>Radian</u>{ENTER}

{CLEAR}

(2) Calculate sin 2°.

{SIN}{(}{(}{(}2{×}{2nd}{π}{)}{)}{÷}180{)}{ENTER}

The result is 0.3348994967. This value can be rounded to 0.0349.

(3) Calculate tan 2°.

{TAN}{(}{(}{(}2{×}{2nd}{π}{)}{)}{÷}180{)}{ENTER}

The result is 0.0348994967. This value can be rounded to 0.0349.

Notice that there may be slight differences between the values obtained with a calculator and the values obtained using a trigonometric table.

How much error occurs when the above small-angle approximation is used instead of the trigonometric function for calculation of distances? For an angle as large as 2°, the error is only about one part in 20,000, i.e. about 0.005 percent!

16.9 APPLICATIONS

Small-angle approximations are especially useful in optics and in astronomy.

EXAMPLE 16.10. A set of binoculars is advertised as having a field of 375 feet at 1000 yards.

(a) Calculate the subtended angle.

(b) What is the maximum size in feet of an object that can be seen in its entirety by the binoculars at a distance of 2 miles?

(a) Convert the distance R to feet:

$$R = 1000 \; \cancel{yd} \cdot 3 \frac{ft}{\cancel{yd}} = 3000 \text{ ft}$$

The angle θ is relatively small. Thus,

$$\theta \approx \frac{375^r}{3000} \approx 0.125^r \qquad (0.125)(57.3)° \approx 7.2°$$

(b) $\qquad\qquad R = (2)(5280) = 10{,}560 \text{ ft} \qquad S \approx R\theta \approx (10{,}560)(0.125) \approx 1320 \text{ ft}$

The largest dimension of the object is 1320 ft.

EXAMPLE 16.11. The moon subtends an angle of 31′ at its mean distance from the earth of 240,000 miles. Determine the diameter of the moon.

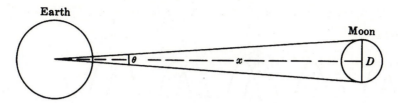

Fig. 16-3

In Fig. 16-3 (not to scale), $\theta = 31'$, $x = 240{,}000$ miles, and D is the unknown diameter.   Since θ is a small angle, we can write $\theta \approx D/x$ or $D \approx x\theta$.   To change the given angle to radian measure, it is convenient first to express the angle in degrees: $\theta^{\mathrm{r}} = \dfrac{31}{60}$ deg.   But 1 radian ≈ 57.3 deg.   Hence,

$$\theta \approx \frac{31}{(60)(57.3)} \approx 0.0090$$

so that $\theta \approx 0.0090^{\mathrm{r}}$.

An application of the formula $\theta \approx D/x$ then gives

$$0.0090 \approx \frac{D}{240{,}000}$$

Hence, $D \approx (0.0090)(240{,}000) \approx 2200$.   We have shown that the diameter of the moon is approximately 2200 miles.

Graphs of Trigonometric Functions

16.10   PERIODICITY

A *period function* is a function whose values repeat themselves at regular increments of the independent variable.   Thus, if there is a constant P such that $f(\theta + P) = f(\theta)$, then $f(\theta)$ is a periodic function and P is called its *period*.

The six trigonometric functions are periodic because they retain their values when the angle is increased or decreased by an integral multiple of 360°.   Thus,

$$\sin 30° = \sin(30° + 360°) = \sin(30° + 720°) = \sin(30° + n360°)$$

where n is an integer.   Similarly, $\cos(50° + n360°) = \cos 50°$.   This is also true for the cosecant, secant, tangent, and cotangent.

The period of the sine, cosine, cosecant, and secant functions is $2\pi^{\mathrm{r}}$ or 360°.   For the tangent and cotangent the period is π^{r} or 180°.   Thus,

$$\tan(30° + 180°) = \tan(30° + 360°) = \tan(30° + 540°)$$

16.11   GRAPHS OF $y = \sin\theta$ AND $y = \cos\theta$

If the values of θ are plotted along the horizontal axis and $\sin\theta$ along the vertical axis, the points $(\theta, \sin\theta)$ lie on a smooth curve shown in Fig. 16-4 (solid).   For the graph to be meaningful, we must known whether θ is the degree measure or the radian measure of an angle.   This is usually clear from the markings on the horizontal (θ) scale.

The graph of the function $y = \cos\theta$ is also shown in Fig. 16-4 (dashed).   It can be seen that the graphs of the sine and cosine functions have identical shapes, but the cosine curve is displaced $\pi/2$ units to the left with respect to the sine curve.   The two curves are called *sinusoids*; such curves are of great help in our description and understanding of all vibratory motion and wave propagation.

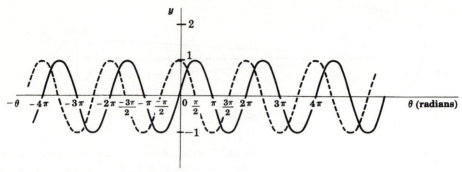

Fig. 16-4. The functions $y = \sin \theta$ (solid) and $y = \cos \theta$ (dashed)

The *amplitude* of a sinusoid is one-half the difference between its maximum and minimum values. The amplitude of either sinusoid in Fig. 16-4 is

$$(1/2)[1 - (-1)] = 1$$

16.12   GRAPH of $y = c \sin k\theta$

Let c and k be any two positive numbers. The graph of $y = c \sin k\theta$ will be similar in shape to the graph of $y = \sin \theta$; however, the amplitude is c times as great, and the period is $1/k$ times as great, as for the graph of $y = \sin \theta$.

EXAMPLE 16.12.   Construct the graph of $y = 2 \sin 3\theta$

The amplitude of the function is 2; the period of the function is

$$\frac{360°}{3} = 120° \qquad \text{or} \qquad \frac{2\pi}{3} \text{ radian}$$

The required graph is shown in Fig. 16-5 (solid). The graph of $y = \sin \theta$ (dashed) is shown for comparison.

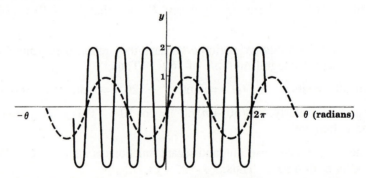

Fig. 16-5. The functions $y = \sin \theta$ (dashed) and $y = 2 \sin 3\theta$ (solid)

To construct the graph of a trigonometric function using a calculator, first, make sure that the MODE of the calculator has been set to Radians. Using an HP-38G graphing calculator proceed as indicated below.

(1)   Set the Mode to Radians.

{SHIFT}{MODES}CHOOS{˅}Radians{ENTER}

{HOME}

(2)   Define the function.

{LIB}FUNCTION{ENTER}

EDIT 2{*}{SIN}3{*}{X,T,θ}{)}{ENTER}

(3)   Set the range.

{SHIFT}{SETUP-PLOT}

XRNG: 0{ENTER}   2{*}{SHIFT}{π}{ENTER}

YRNG: {−×}2{ENTER}   2{ENTER}

XTICK: {SHIFT}{π}{/}2{ENTER}   YTICK: 0.5{ENTER}

(4)   Plot the graph.

{PLOT}

The user may have to ZOOM to see different parts of the graph.

16.13   GRAPHS OF OTHER TRIGONOMETRIC FUNCTIONS

The graph of $y = \tan \theta$ is shown in Fig. 16-6.   The graphs of other trigonometric functions can be found in any trigonometry textbook but are of less interest for scientific applications.

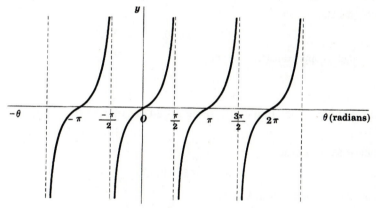

Fig. 16-6.   The function $y = \tan \theta$

Solved Problems

16.1.   Express the following angles in radians: (*a*) 30° and   (*b*) 135°.

Set up a proportion $\dfrac{\theta^r}{\phi^\circ} = \dfrac{2\pi^r}{360^\circ}$, where ϕ is the degree measure of the angle in the problem.

(*a*) $\dfrac{\theta^r}{30^\circ} = \dfrac{2\pi^r}{360^\circ}$

$$\theta^r = \frac{(30^\circ)(2\pi^r)}{360^\circ} = \frac{\pi^r}{6} \approx \frac{3.14^r}{6} \approx 0.523^r$$

(*b*) $\dfrac{\theta^r}{135^\circ} = \dfrac{2\pi^r}{360^\circ}$

$$\theta^r = \frac{(135^\circ)(2\pi^r)}{360^\circ} = \frac{3\pi^r}{4} \approx 2.36^r$$

16.2. Express the following angles in degrees: (*a*) 4.17^r and   (*b*) $(-5\pi/6)^r$.

$$1^r \approx 57.3°$$

(*a*) $4.17^r \approx (4.17)(57.3°) \approx 239°$

Alternatively,

$$\frac{4.17^r}{2\pi^r} = \frac{\phi°}{360°}$$

$$\phi° = \frac{(360°)(4.17)}{2\pi} \approx 239°$$

(*b*) $2\pi^r = 360°$ $\pi^r = 180°$

$$\phi° = \frac{-5(180°)}{6} = -150°$$

16.3. A central angle of $42.0°$ is inscribed in a circle of radius 18.0 cm.   Calculate the length of the intercepted arc using radian measure.

Convert $42.0°$ to radians:

$$42.0° \approx (42.0)(0.0174^r) \approx 0.731^r$$

Substitute the known values in $S = R\theta$:

$$S = (18.0)(0.731) \approx 13.2$$

The arc length is approximately 13.2 cm.

16.4. A central angle of $65°$ in a circle subtends an arc of 3.84 ft.   What is the length of the circle radius?

Convert $65°$ to radians:

$$65° \approx (65)(0.0174^r) \approx 1.13^r$$

Substitute the known values in $S = R\theta$:

$$3.84 = (R)(1.13) \qquad R = \frac{3.84}{1.13} \approx 3.40$$

The radius is approximately 3.40 ft long.

16.5. Using small-angle approximations determine the following.

(*a*) tan $3°$ (*b*) sin $1.3°$

(*c*) What is the percent difference between the calculated and table values for tan $3°$?

(*d*) Interpolate the table value for sin $1.3°$ and compute the percent difference with the calculated value.

(*a*) Convert $3°$ to radians.

$$\frac{\theta^r}{3°} = \frac{2\pi}{360}$$

$$\theta = \frac{\pi}{60} \approx 0.0523$$

$$\tan 3° \approx \theta \approx 0.0523$$

(b) Convert 1.3° to radians.

$$(1.3)(0.0174^{\text{r}}) \approx 0.0226^{\text{r}}$$

$$\sin 1.3° \approx \theta^{\text{r}} \approx 0.0226$$

(c) From the table, page 467, $\tan 3° = 0.05241$.

$$\text{percent absolute difference} \approx \frac{0.0523 - 0.05241}{0.05241} \times 100\% \approx 0.21\%$$

(d) From the table, page 467,

$$\sin 2° = 0.03490$$
$$\sin 1° = \underline{0.01745}$$
$$\text{difference} = 0.01745$$

$$(0.3)(\text{difference}) = (0.3)(0.01745) \approx 0.00524$$

$$\sin 1.3° = 0.01745 + 0.00524 \approx 0.02269$$

$$\text{percent absolute difference} \approx \frac{0.0226 - 0.02269}{0.02269} \times 100\% \approx 0.4\%$$

16.6. The length of a pendulum is 108 cm. An arc described by the pendulum bob is 12 cm. Determine the angle of the pendulum's motion in radians and degrees.

Substitute the known quantities in the equation $S = R\theta$.

$$12 = 108\theta$$

$$\theta = \frac{12}{108}\,\text{radian} = \frac{1}{9}\,\text{radian} \approx 0.111\,\text{radian}$$

$$\theta = \frac{1}{9}\left(\frac{360}{2\pi}\right)^{\circ} = \frac{20°}{\pi} \approx 6.37°$$

16.7. A car is traveling at 60 mph. What is the angular speed of the car's wheels in radians per second if the diameter of a tire is 30 in.?

The linear speed of the car is 60 mph, and it must be converted to inches per second.

$$1\text{ mile} = 5280\text{ feet} \qquad 1\text{ ft} = 12\text{ in.} \qquad 1\text{ hr} = (60)(60)\text{ sec} = 3600\text{ sec}$$

$$v = \frac{(\cancel{60})(5280)(12)}{(\cancel{60})(60)}\,\frac{\text{in}}{\text{sec}}$$
$$= 1056\text{ in/sec} \approx 1060\text{ in/sec}$$

The angular speed of a wheel of the car is obtained by substitution in $v = R\omega$. Thus,

$$1060 = 15\omega$$

$$\omega = \frac{1060}{15} \approx 70.7\text{ radian/sec}$$

16.8. The asteroid Eros has a horizontal geocentric parallax of 44″ at the distance of its nearest approach to the earth. Determine this distance, if the diameter of the earth is 12,800 km. The horizontal geocentric parallax of Eros is the angle subtended by the semidiameter of the earth as seen from Eros.

In Fig. 16-7,

$$R = \frac{12,800}{2} = 6400\text{ km}$$

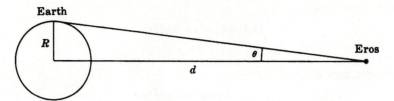

Fig. 16-7. Horizontal geocentric parallax (not to scale)

θ is the radian measure of an angle of 44 seconds.   The required distance is d.   Since θ is small,

$$\theta \approx \frac{R}{d} \qquad \text{or} \qquad d \approx \frac{R}{\theta}$$

To convert 44″ to radians, we proceed as follows.

$$\theta^r \approx \frac{44}{3600} \deg \approx \frac{44}{(3600)(57.3)} \text{ radian} \approx 0.000213 \text{ radian}$$

$$\theta \approx 0.000213$$

By substitution,

$$d \approx \frac{6400}{0.000213} \approx 30{,}000{,}000 = (3.0)(10^7)$$

The distance of Eros is $(3.0)(10^7)$ km, approximately.

16.9. A motor shaft is rotating at 1500 rpm, clockwise.   Determine the rotational velocity of the shaft in   (a) radians per second and   (b) degrees per hour.

(a) 1 rev $= 2\pi$ rad and 1 min $= 60$ sec.

$$1500 \text{ rpm} = 1500 \frac{\text{rev}}{\text{min}} \cdot 2\pi \frac{\text{rad}}{\text{rev}} \cdot \frac{1 \text{ min}}{60 \text{ sec}} = 50\pi \text{ rad/sec} \approx 157 \text{ rad/sec}$$

(b) 1 rev $= 360°$ and 1 hr $= 60$ min.

$$1500 \text{ rpm} = 1500 \frac{\text{rev}}{\text{min}} \cdot \frac{360°}{1 \text{ rev}} \cdot \frac{60 \text{ min}}{1 \text{ hr}} \approx 32{,}400{,}000 \text{ deg/hr}$$

16.10. A mosquito sits on a record rotating at 45 rpm.   If the mosquito is 7.5 cm from the center of the record, what is its linear speed through the air?

Convert 45 rpm; to rad/sec.   1 rev $= 2\pi$ rad   and   1 min $= 60$ sec.

$$\omega = 45 \frac{\text{rev}}{\text{min}} \cdot \frac{2\pi \text{ rad}}{1 \text{ rev}} \cdot \frac{1 \text{ min}}{60 \text{ sec}} \approx 1.5\pi \text{ rad/sec} \approx 4.71 \text{ rad/sec}$$

Substitute the known values in the equation $v = R\omega$:

$$v = (7.5)(4.71) \approx 35.3$$

The linear speed of the mosquito is approximately 35.3 cm/sec.

16.11. The drive wheel of a train is 6 feet in diameter.   The train is traveling at 60 mph.   Calculate the angular speed of the drive wheel in   (a) rad/sec and   (b) rev/min.

Convert the train speed v to ft/sec.

(a) $1 \text{ mi} = 5280 \text{ ft}$ and $1 \text{ hr} = 3600 \text{ sec}$.

$$v = 60 \frac{\text{mi}}{\text{hr}} \cdot \frac{5280 \text{ ft}}{1 \text{ mi}} \cdot \frac{1 \text{ hr}}{3600 \text{ sec}} = 88 \text{ ft/sec}$$

Substitute the known values in the equation $v = R\omega$:

$$88 = 3\omega$$

$$\omega = \frac{88}{3} \approx 29.3 \text{ rad/sec}$$

(b) $1 \text{ rev} = 2\pi \text{ rad}$ and $1 \text{ min} = 60 \text{ sec}$.

$$\omega = 29.3 \frac{\text{rad}}{\text{sec}} \cdot \frac{1 \text{ rev}}{2\pi \text{ rad}} \cdot \frac{60 \text{ sec}}{1 \text{ min}} \approx 280 \text{ rpm}$$

16.12. (a) Sketch portions of the graphs $y = 3 \sin \theta$ and $y = 3 \cos \theta$.

(b) How do these graphs differ?

(c) What is the period of both functions?

(d) What is the amplitude of both functions?

(a) Same as in Fig. 16-4, but with the ordinates multiplied by 3.

(b) Cosine graph is the sine graph displaced $\pi/2$ radians, or $90°$, to the left.

(c) 2π radians or $360°$

(d) 3

16.13. (a) For what angles does the cosine function have a maximum value?

(b) What is the maximum numerical value of the sine function?

(c) For what angles is the tangent function undefined?

(d) Determine $\sin \theta$ when $\theta = 3\pi/2$ radian.

(a) $0 + n\pi$; $n = 0, \pm 2, \pm 4, \ldots$

(b) 1

(c) $\pi/2 + n\pi$; $n = 0, \pm 1, \pm 2, \ldots$

(d) $3\pi/2 \text{ rad} = 280°$, and $\sin 270° = -\sin 90° = -1$

16.14. *Stellar parallax* (p in Fig. 16-8) is the maximal angle subtended by a radius of the earth's orbit as seen from a star. The star Luyten 726-8 has a parallax of $0.35''$. If a radius of the earth's orbit is $(1.5)(10^8)$ km, what is the approximate distance d to this star?

Express the angle of parallax in radians, using $1° = 3600'' = 0.0174 \text{ rad}$.

$$p = (0.35'') \frac{1°}{3600''} \cdot \frac{0.0174 \text{ rad}}{1°} \approx 0.00000169 \approx (1.7)(10^{-6}) \text{ rad}$$

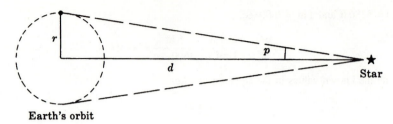

Fig. 16-8

Since p is a small angle, $r \approx S$ = arc length. Substitute the known values in $p = \dfrac{r}{d}$.

$$p \approx \frac{(1.5)(10^8)}{d}$$

$$d \approx \frac{(1.5)(10^8)}{(1.7)(10^{-6})} \approx (8.8)(10^{13}) \text{ km}$$

16.15. What are the period and the amplitude of the function $y = 2 \cos 5\theta$?

$$5\theta + 2\pi = 5\left(\theta + \frac{2\pi}{5}\right)$$

$$\therefore \quad \text{period is} \quad \frac{2\pi^r}{5} = \frac{360°}{5} = 72°$$

The amplitude is equal to the coefficient of $\cos 5\theta$, i.e. 2.

16.16. Construct the graph of $y = \dfrac{1}{3} \cos\left(\dfrac{x}{3}\right)$.

Using a TI-82 graphing calculator use the procedure indicated below.

(1) Set the Mode to Radians.

{MODE}{▼}{▼}<u>Radian</u>{ENTER}

{CLEAR}

(2) Define the function.

{Y =}{(}1{÷}3{)}{×}{COS}{(}{X,T,θ}{÷}3{)}{ENTER}

(3) Set the range.

{WINDOW}{ENTER}

Xmin = 0{ENTER}

Xmax = 12{×}{2nd}{π}{ENTER}

Xscl = 5{ENTER}

Ymin = {(−)}0.4{ENTER}

Ymax = 0.4{ENTER}

Yscl = 0.2{ENTER}

(4)   Plot the graph

{GRAPH}

Figure 16-9 shows the graph.

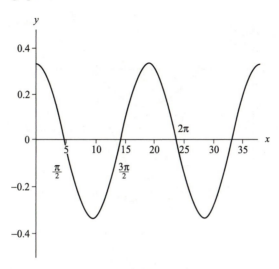

Fig. 16-9

16.17. Use a graphing calculator to compare the graph of the functions $y = 3\cos\theta$ and $y = \cos\theta$.

Using an HP-38G graphing calculator proceed as indicated below.

(1)   Set the Mode to Radians.

{SHIFT}{MODES}

CHOOS{▾}Radians{ENTER}

{HOME}

(2)   Define the functions.

{LIB}FUNCTION{ENTER}

EDIT 3{*}{COS}{X,T,θ}{)}{ENTER}

EDIT{COS}{X,T,θ}{)}{ENTER}

(3)   Define the range.

{SHIFT}{SETUP-PLOT}

XRNG: 0{ENTER}   2{*}{SHIFT}{π}{ENTER}

YRNG:{$-\times$}3{ENTER}   3{ENTER}

XTICK: {SHIFT}{π}{/}2{ENTER}   YTICK: 1{ENTER}

(3)   Plot the graphs.

{PLOT}

Figure 16-10 shows the graphs of both functions.

Notice that both graphs   (a) intersect the horizontal axis at $\theta = (\pi/2)$ and $\theta = (3\pi/2)$;   (b) reach maximum values at $\theta = 0$ and $\theta = 2\pi$ (c) reach minimum value at $\theta = \pi$.

The function $y = 3\cos\theta$ reaches a maximum of 3 and a minimum -3, whereas the function $y = \cos\theta$ reaches maximum and minimum values of 1 and -1, respectively.

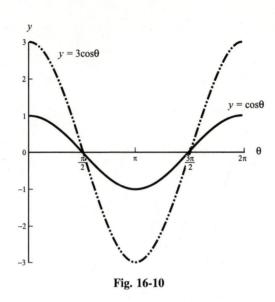

Fig. 16-10

Supplementary Problems

The answers to this set of problems are at the end of this chapter.

16.18. Express the following angles in radians, leaving π as a factor in the answers:

(a) 90°   (b) 120°   (c) −210°   (d) 1080°   (e) 315°   (f) 330°

16.19. Express the following angles in degrees:

(a) $\dfrac{2\pi}{5}$ radian   (c) $\dfrac{\pi}{15}$ radian   (e) 0.4 radian

(b) 3 radian   (d) $-\dfrac{3\pi}{2}$ radian   (f) 12.56 radian

16.20. A central angle of 1.6 radian is inscribed in a circle of radius 25 cm.   What is the length of the intercepted arc?

16.21. Determine the angular speed in radians per second of a motor armature rotating at 1200 rpm.

16.22. Determine the angular speed in radians per minute of the sweep second hand of a watch.

16.23. If the angular speed of a flywheel is 15 radian/sec, determine its speed in revolutions per minute.

16.24. Express $33\frac{1}{3}$ rpm in radians per second.

16.25. The speed of an auto is 84 ft/sec, while the radius of a tire is 14 in.   Determine the angular speed of a point on the circumference of the wheel in   (a) radians per second and   (b) revolutions per minute.

16.26. What is the linear speed of a boy 5 m from the center of a merry-go-round rotating at 12 rpm?

16.27. The parallax of star X is 0.545″, while the parallax of star Y is 0.760″.

(*a*) Which star is nearest to the earth?   (*b*) What is the ratio of their distances?

16.28. The sun's mean horizontal geocentric parallax is 8.80″ (compare Problem 16.8).   The mean distance from the earth to the sun is 9.29×10^7 miles.   Determine the earth's diameter.

16.29. On ordinary graph paper, plot the graph of $y = \sin \theta$.   Let θ vary from $-360°$ to $+360°$ in steps of $15°$.

16.30. On the sheet of graph paper used in Problem 16.29, plot a portion of the graph of $y = \cos \theta$.

16.31. Sketch portions of the graphs of $y = \cos \theta$, $y = 3 \cos \theta$, and $y = 3 \cos 2\theta$.

16.32. Sketch portions of the graph $y = 5 \sin \theta$ and $y = 5 \sin(\theta - \pi/3)$.

16.33. Determine the period and amplitude of the functions:

(*a*) $y = 2 \sin 5\theta$ (*b*) $y = 5 \cos 2\theta$

16.34. What can be said of the graph of equations of the form $y = c \sin k\theta$ when $|k| < 1$?   What happens if $|k| > 1$?

16.35. Graph the equations $y = 1 + \sin x$ and $y = 6 + \cos x$.   How do they differ from the graphs of $y = \sin x$ and $y = \cos x$?

Answers to Supplementary Problems

16.18. (*a*) $\pi/2$ (*b*) $2\pi/3$ (*c*) $-7\pi/6$ (*d*) 6π (*e*) $7\pi/4$ (*f*) $11\pi/6$

16.19. (*a*) $72°$ (*b*) $172°$ (approx.)   (*c*) $12°$ (*d*) $-270°$ (*e*) $22.9°$ (approx.)   (*f*) $720°$ (approx.)

16.20. 40 cm

16.21. 40π or 126 radian/sec approx.

16.22. 2π radian/min

16.23. $450/\pi$ or 143 rpm approx.

16.24. $10\pi/9$ or 3.49 radian/sec approx.

16.25. (*a*) 72 radian/sec   (*b*) $2160/\pi$ or 688 rpm approx.

16.26. 120π or 377 m/min approx.; 2π or 6.28 m/sec approx.

16.27. (*a*) Star Y (*b*) (distance of X)/(distance of Y) $\approx 0.760/0.545 \approx 139/100 \approx 1.39 : 1$

16.28. 7920 miles

16.29. See Fig. 16-11

16.30. See Fig. 16-11

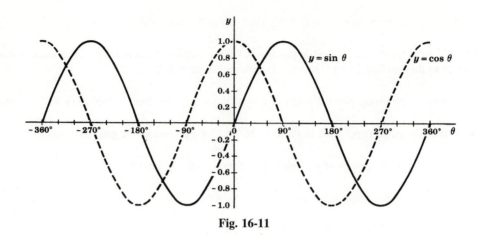

Fig. 16-11

16.31. See Fig. 16-12.

16.32. See Fig. 16-13.

16.33. (*a*) the period is $2\pi/5$, the amplitude is 2;   (*b*) the period is π, the amplitude is 5

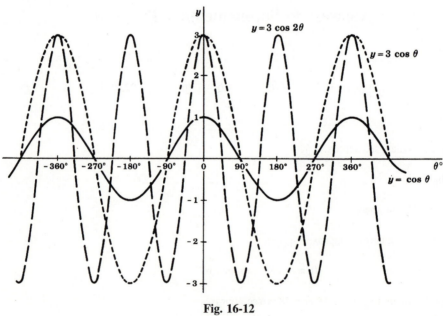

Fig. 16-12

16.34. If $|k| < 1$ the graph stretches out.   If $|k| > 1$ the curves oscillate faster.

16.35. The graph of $y = 1 + \sin x$ is shifted 1 unit up.   The graph of $y = 6 + \cos x$ is shifted 6 units up.

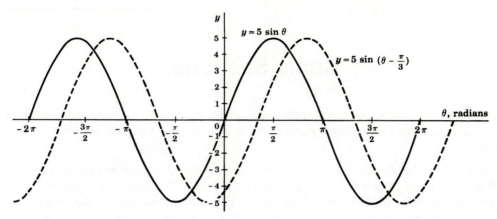

Fig. 16-13

Chapter 17

Conic Sections

Conic sections are the sections of a right circular cone by a plane. Conic sections are of great importance in science and technology, particularly in astronomy. Only the briefest outline can be given here.

The circle, ellipse, parabola, and the hyperbola are often called conic sections, for all of them can be obtained as sections cut from a right circular cone by planes. The cone is thought of as extending indefinitely on both sides of its vertex. The part of the cone on one side of the vertex is called a nappe. Conic sections are of great importance in science and technology, particularly astronomy. Only the briefest outline can be given here.

The Ellipse

17.1 BASIC CONCEPTS

If a plane oblique to the base intersects only the nappe of a right circular cone, the resulting conic section is an *ellipse* (Fig. 17-1).

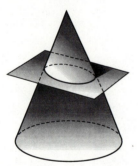

Fig. 17-1. Ellipse as a conic section

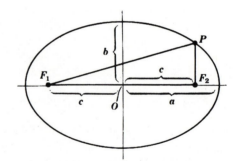

Fig. 17-2. Ellipse and its components

An ellipse is a curve on which all the points have a constant sum of distances from two fixed points.

In Fig. 17-2, the two fixed points F_1 and F_2 are called *foci* (*focuses*). The point O is the *center* of the ellipse, while OF_1 and OF_2 are called *focal distances*, each represented by c. The maximal and minimal chords of an ellipse are called, respectively, its *major* and its *minor* axis. The lengths of the *semimajor* and *semiminor* axes are usually denoted by a and b, as indicated in Fig. 17-2.

In terms of the distances of Fig. 17-2, the definition of the ellipse can be summarized by the equation $FP_1 + PF_2 = 2a$.

It can be shown that in any ellipse $a^2 = b^2 + c^2$. If the origin of rectangular coordinates is established at the center, then the standard equation of the ellipse is

$$\frac{x^2}{a^2} + \frac{y^2}{b^2} = 1$$

The *eccentricity e* of an ellipse is defined as the ratio of the focal distance to the semimajor axis:

$$e = \frac{c}{a}$$

EXAMPLE 17.1. Given an ellipse with a semimajor axis of 15 cm. The distance between the two foci is 18 cm. Determine (a) the eccentricity of the ellipse, (b) the semiminor axis, and (c) the standard equation.

(a) From $2c = 18$ cm, obtain the value of the focal distance.

$$c = \frac{18}{2} = 9$$

The focal distance is 9 cm. Eccentricity: $e = c/a = 9/15 = 0.6$.

(b) Substituting in $b^2 = a^2 - c^2$,

$$b^2 = 15^2 - 9^2 = 225 - 81 = 144 \qquad \text{or} \qquad b = 12$$

The length of the semiminor axis is 12 cm.

(c) The standard equation is

$$\frac{x^2}{225} + \frac{y^2}{144} = 1$$

17.2 HOW TO DRAW AN ELLIPSE

There is a simple way to draw an ellipse. Fasten two pins or thumb tacks on a sheet of paper; these will locate the two foci. Make a loop of thread or string which will slip over the pins. Keeping the loop taut by a pencil point, move the pencil until it gets back to the starting point. The eccentricity of the drawn ellipse will depend on the distance between the foci and the length of the loop. Figure 17-3 illustrates the method.

17.3 HOW TO DETERMINE A TANGENT TO AN ELLIPSE

An interesting property of an elliptical surface acting as a mirror is that it reflects a ray of light or other energy coming from one focus to the other focus. This furnishes a practical method for constructing the tangent at a point on the ellipse. Thus, in Fig. 17-4, if the angle F_1PF_2 is bisected and a perpendicular to the bisector is constructed at P, then this perpendicular is the tangent to the ellipse at point P. The ray F_1P of the Fig. 17-4 is called the *incident ray* and the ray PF_2 is called the *reflected ray*.

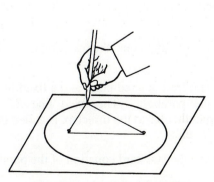

Fig. 17-3. Drawing an ellipse

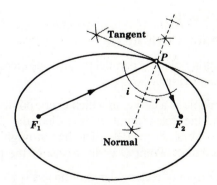

Fig. 17-4. Constructing a tangent
to an ellipse

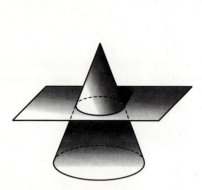

Fig. 17-5.   Circle as a conic section

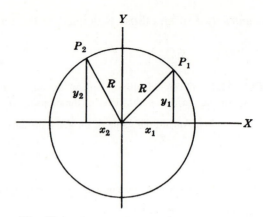

Fig. 17-6.   Rectangular coordinates in a circle

17.4   THE CIRCLE

If a plane cuts a right circular cone parallel to its base, then the resulting section is a *circle* (Fig. 17-5).

If the center of a circle is the origin of rectangular coordinates (Fig. 17-6), then the equation of the circle is

$$x^2 + y^2 = R^2$$

where x and y are the coordinates of any point on the circle and R is its radius.

EXAMPLE 17.2.   The coordinates of a point on a circle with the origin at its center are $x = -5, y = 12$.   (*a*) Write the equation of the circle.   (*b*) Determine the radius of the circle.

(*a*) Substitute the given coordinates in the equation of the circle to determine R^2.

$$(-5)^2 + (12)^2 = R^2$$
$$25 + 144 = R^2$$
$$169 = R^2$$

The equation of the circle is $x^2 + y^2 = 169$.

(*b*) Since R^2 is 169, $R = \sqrt{169} = 13$.   The radius of the circle is 13.   The length of R is in the same units as those for x and y.

The Parabola

17.5   BASIC CONCEPTS

If a cutting plane is parallel to only one element of a right circular cone, the resulting conic section is a *parabola* (Fig. 17-7).

A parabola is a curve on which every point is equidistant from a fixed point and a fixed line.   In Fig. 17-8, the fixed point F is called the *focus* of the parabola.   The fixed line RT is the *directrix*.   Then, $PF = PR$.   The line through F and perpendicular to the directrix is called the *axis* of the parabola.   Point O is the *vertex* of the parabola.

If the origin of rectangular coordinates is taken at O, then the standard equation of the parabola is

$$y^2 = 4px \qquad \text{where} \qquad p = OF = OT = \textit{focal distance}$$

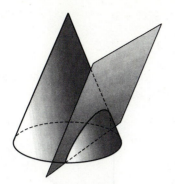

Fig. 17-7. Parabola as a conic
section

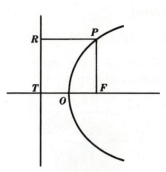

Fig. 17-8. Parabola and its
components

If the same parabola is drawn so that its axis is along the y-axis, then the equation becomes:

$$x^2 = 4py \qquad \text{or} \qquad y = \frac{1}{4p}x^2$$

EXAMPLE 17.3. The coordinates of the focus of a parabola are $(5, 0)$. Write the equation of the parabola if its axis is along the positive x-axis.

Substituting 5 for p in the equation $y^2 = 4px$, we obtain $y^2 = 20x$.

EXAMPLE 17.4. The equation of a parabola is $y = \dfrac{x^2}{6}$.

(*a*) Determine the coordinates of its focus.   (*b*) Sketch the parabola.

(*a*) The focus of this parabola is on the y-axis. Since $4p = 6, p = 6/4 = 1.5$. Therefore, the coordinates of the focus
are $(0, 1.5)$.

(*b*) Make a table of a few values and plot as in Fig. 17-9.

x	0	± 1	± 2	± 3	± 4	± 5
y	0	$1/6$	$2/3$	$1\frac{1}{2}$	$2\frac{2}{3}$	$4\frac{1}{6}$

17.6 FOCUSING PROPERTY OF A PARABOLA

A geometrical property of a parabolic surface is that it reflects to its focus any rays of energy which come in parallel to its axis, as shown in Fig. 17-10. This is the reason for the parabolic shape of the mirror in an astronomical reflecting telescope. Conversely, when a small source of energy is placed at the focus of a parabolic surface, the energy will be reflected as a beam of parallel rays. This is why headlights and radar transmitting antennas are equipped with parabolic reflectors.

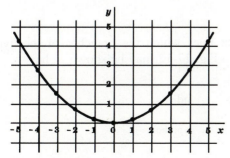

Fig. 17-9. Parabola $y = x^2/6$

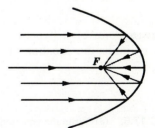

Fig. 17-10. Focusing by a
parabolic surface

The Hyperbola

17.7   BASIC CONCEPTS

If a plane cuts the upper and lower parts (*nappes*) of a right circular cone, the resulting conic section is a *hyperbola* (Fig. 17-11).

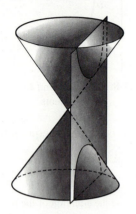

Fig. 17-11.   Hyperbola as a
conic section

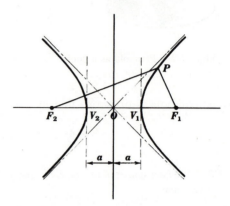

Fig. 17-12.   Hyperbola and its components

A hyperbola is a curve every point on which has a constant difference of distances from two fixed points.

In Fig. 17-12, the fixed points F_1 and F_2 are called *foci* (*focuses*).   The two parts of the hyperbola are its *branches*.   A *vertex* V_1 or V_2 is the intersection of a branch with the line through the foci.   Point O is the *center*.   If the distance between the vertices is $2a$, then from the constant-difference property,

$$|PF_2 - PF_1| = |PF_1 - PF_2| - 2a$$

If the origin of rectangular coordinates be taken at O, and the *focal distances* are $OF_1 = OF_2 = c$, then $b^2 = c^2 - a^2$.   The standard equation of a hyperbola can be shown to be

$$\frac{x^2}{a^2} - \frac{y^2}{b^2} = 1$$

EXAMPLE 17.5.   Write the equation of a hyperbola with $(3, 0)$, $(-3, 0)$ as coordinates of its vertices, and $(5, 0)$, $(-5, 0)$ as coordinates of its foci.

By definition, $c = 5$, $a = 3$.   Since $b^2 = c^2 - a^2$, then $b^2 = 5^2 - 3^2 = 25 - 9 = 16$.   The required equation is: $\dfrac{x^2}{9} - \dfrac{y^2}{16} = 1$.

The *asymptotes* of a hyperbola are two straight lines which the branches of the hyperbola approach as limits; the equation of the asymptotes in Fig. 17-12 may be shown to be:

$$y = \pm \frac{b}{a} x$$

EXAMPLE 17.6.   (*a*) Write the equation of the asymptotes to the hyperbola

$$\frac{x^2}{25} - \frac{y^2}{144} = 1$$

(b) Find the focal distances of the hyperbola in (a).

(a) Since $a^2 = 25$, $a = \sqrt{25} = 5$, and since $b^2 = 144$, $b = 12$. The equations of the asymptotes are:

$$y = (12/5)x \qquad \text{and} \qquad y = -(12/5)x$$

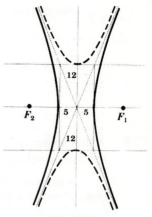

Fig. 17-13

(b) Since $b^2 = c^2 - a^2$,

$$c^2 = b^2 + a^2 = 144 + 25 = 169$$

$$c = \sqrt{169} = 13$$

Therefore, the two focal points F_1 and F_2 have the coordinates $(13, 0)$ and $(-13, 0)$. The hyperbola and its asymptotes are shown in Fig. 17-13 (solid curve).

The hyperbola whose equation is $\dfrac{y^2}{144} - \dfrac{x^2}{25} = 1$ is also shown (dashed curve). These two hyperbolas are said to be *conjugates* of each other.

Solved Problems

17.1. Write the equation of a circle with the center at the origin and with a radius of (a) 7, (b) $\sqrt{2}$.

The equation of a circle with the center at the origin has the form $x^2 + y^2 = R^2$, where R is the radius.

(a) $x^2 + y^2 = 7^2$ or $x^2 + y^2 = 49$

(b) $x^2 + y^2 = (\sqrt{2})^2$ or $x^2 + y^2 = 2$

17.2. What is the diameter of the circle whose equation is $x^2 + y^2 - 81 = 0$?

Put the equation in standard form, $x^2 + y^2 = 81$.   The right member is the square of the radius R.

Thus, $R^2 = 81$ and $R = 9$.

The diameter is $2R = (2)(9) = 18$.

17.3. Which of the following points are on the circle with the equation $x^2 + y^2 = 25$: $(-4, 3)$, $(-3, 4)$, $(2, 3)$, $(4, -3)$, $(0, 5)$, $(\sqrt{21}, 2)$, $(2.5, 2.5)$?

Square the coordinates of each point and add them.   For the point $(-4, 3)$, the result is:

$$(-4)^2 + (3)^2 = 16 + 9 = 25$$

Therefore, the point satisfies the given equation and is on the given circle.

The other points on the circle are: $(-3, 4)$, $(4, -3)$, $(0, 5)$, $(\sqrt{21}, 2)$.

17.4. The equation of circle is $x^2 + y^2 = 289$.   If the x-coordinate of a point on the circle is 15, what is the value of the y-coordinate?

Solve the given equation for y to obtain

$$y^2 = 289 - x^2 \qquad \text{or} \qquad y = \pm\sqrt{289 - x^2}$$

Substitute the value of the x-coordinate in the above equation and simplify.

$$y = \pm\sqrt{289 - 15^2} = \pm\sqrt{289 - 225} = \pm\sqrt{64} = \pm 8$$

17.5. A thin spherical bowl has a diameter of 12 inches and is 2 inches deep.   If the bowl was cut off from a spherical shell what was the radius of the shell?

Let the origin of the coordinate system be placed at the center of the spherical shell, as shown in Fig. 17-14.   Then,

$$x^2 + y^2 = R^2 \qquad (1)$$

We know that $y = 12/2 = 6$ and that $x + 2 = R$.   Therefore, by substitution in equation (1),

$$x^2 + 6^2 = (x + 2)^2$$

from which we obtain

$$\cancel{x^2} + 36 = \cancel{x^2} + 4x + 4$$
$$4x = 32$$
$$x = 8$$
$$R = x + 2 = 8 + 2 = 10$$

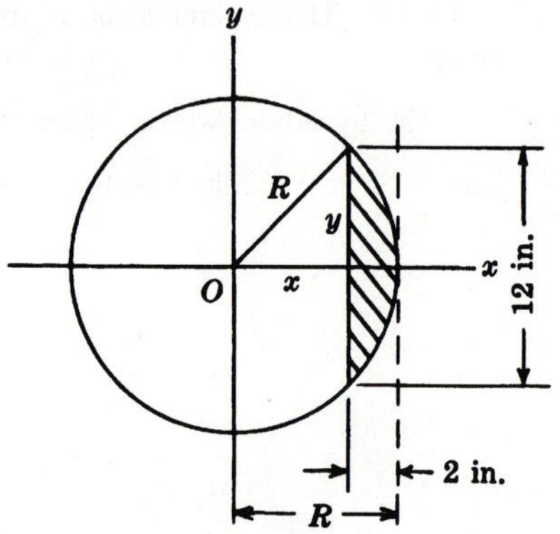

Fig. 17-14

The spherical shell had a radius of 10 inches.

17.6. (a) Write the standard equation for an ellipse with a major axis of 20 and a minor axis of 16.

(b) Determine the focal distance of this ellipse.

(c) Determine the eccentricity of this ellipse.

(a) The semimajor axis a is 20/2 or 10.   The semiminor axis b is 16/2 or 8.

The equation is of the form $x^2/a^2 + y^2/b^2 = 1$, or by substitution of the given values for a and b, the answer is

$$\frac{x^2}{100} + \frac{y^2}{64} = 1$$

(b) In an ellipse, $a^2 = b^2 + c^2$, where c is the focal distance.   Thus, by substitution of the given values, we have:

$$a^2 - b^2 = c^2 = 10^2 - 8^2 = 100 - 64 = 36$$
$$c = \sqrt{36} = 6$$

(c) $e = c/a = 6.10 = 0.6$.

17.7. Given the equation $9x^2 + 25y^2 = 225$.

(a) Rewrite the equation in standard form.

(b) Determine the eccentricity of the ellipse.

(a) Divide both sides of the equation by 225 to obtain the standard form

$$\frac{x^2}{25} + \frac{y^2}{9} = 1$$

(b) From part (a) we have: $a^2 = 25$, $b^2 = 9$, and $c^2 = a^2 - b^2 = 25 - 9 = 16$.

$$e = \frac{c}{a} = \frac{\sqrt{16}}{\sqrt{25}} = \frac{4}{5} = 0.8$$

17.8. The moon's mean distance ($\frac{1}{2}$ (maximum distance + minimum distance) from the earth is 384,000 km. The eccentricity of the moon's orbit is 0.055. What are the nearest and farthest distances of the moon from the earth?

The orbit of the moon as shown in Fig. 17-15 is elliptical with the earth at one of the focuses of the ellipse. The nearest distance is the difference between the semimajor and focal distances; the farthest distance is the sum of these two distances. The mean distance is the length of the semimajor axis. Thus, $a = 384,000$ km; $e = c/a = 0.055$. Solving for c, the focal distance,

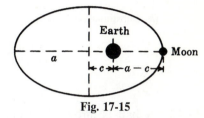

Fig. 17-15

$$c = ea = (0.055)(384,000) \approx 21,120$$

The nearest distance is $384,000 - 21,120 \approx 363,000$ km.

The farthest distance is $384,000 + 21,120 \approx 405,000$ km.

17.9. The equation of a parabola is $x^2 = 10y$. Determine the coordinates of the focus of the parabola.

The equation is of the form $x^2 = 4py$, where $4p = 10$ or $p = 10/4 = 2.5$. The focus of the parabola is on the y-axis. The coordinates of the focus are thus $(0, 2.5)$.

17.10. A point on a parabola with its axis along the x-axis has the coordinates $(18, 12)$. Determine (a) the focal distance and (b) the equation of the parabola.

(a) Substitute the coordinates of the point in the equation $y^2 = 4px$:

$$12^2 = (4)(p)(18)$$

$$144 = 72p$$
$$p = 2$$

The focal distance is 2 units.

(b) The equation of the parabola is $y^2 = (4)(2)x$, or $y^2 = 8x$.

17.11. A building has an arch in the shape of a parabola, with dimensions as shown in Fig. 17-16.

(a) Determine the equation of the parabola.

(b) How high is the arch at point A?

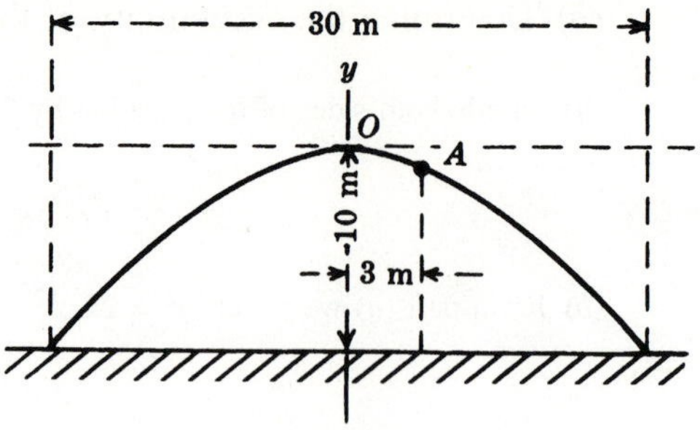

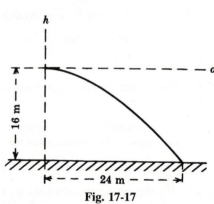

(a) Let the coordinate axes pass through the vertex of the arch. The parabola "opens down" and has an equation of the form $x^2 = -4py$. Substitute the coordinates of one of the points touching the ground, $(15, 0)$, to obtain

$$15^2 = -4p(-10)$$
$$225 = 40p$$

Fig. 17-16

$$p = \frac{225}{40} = \frac{45}{8} \quad \text{and} \quad 4p = 4\left(\frac{45}{8}\right) = \frac{45}{2}$$

The equation of the arch is $x^2 = \dfrac{-45}{2}y$ or $y = -\dfrac{2}{45}x^2$.

(b) Substitute $x = 3$ in the equation to get $y = -\dfrac{2}{45}(3)^2 = -\dfrac{2}{5}$.

This is the vertical distance from the top of the arch which is 10 m above ground. Therefore, point A is $(10 - 2/5)$ m or $9\frac{3}{5}$ m above ground.

17.12. A ball is thrown horizontally from a height h of 16 m. The ball lands at a horizontal distance d of 24 m. If the path of the ball is parabola, write its equation.

Let the point from which the ball was thrown be the origin of the coordinates as shown in Fig. 17-17. The equation of the trajectory is of the form $4ph = -d^2$. Substitute the coordinates of the point on the parabolic path where it intersects the ground to obtain:

$$4p(-16) = -(24)^2$$
$$-64p = -576$$
$$p = 9$$

The equation of the path of the ball is

$$4(9)h = -d^2$$
$$36h = -d^2$$

or $\qquad h = -d^2/36$

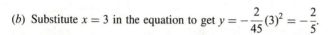

Fig. 17-17

17.13. (a) Find the coordinates of the vertices of the hyperbola whose equation is

$$\frac{x^2}{64} - \frac{y^2}{25} = 1$$

(b) Determine the coordinates of the foci of the hyperbola.

(c) Write the equations of the asymptotes of the hyperbola.

(d) Sketch the hyperbola and its asymptotes.

(a) Since $a^2 = 64$, $a = 8$. Therefore, the vertices are at $(8, 0)$ and at $(-8, 0)$.

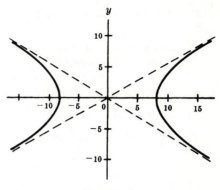

(b) Since $b^2 = 25$, $b = 5$. Thus, $c^2 = a^2 + b^2 = 64 + 25$
= 89 and

$$c = \sqrt{89} \approx 9.43$$

The foci of the hyperbola are at (9.43, 0) and (-9.43, 0).

(c) The equations of the asymptotes of the hyperbola are:

$$y = \pm\frac{6}{a}x \qquad \text{or} \qquad y = \pm\frac{5}{8}x$$

(d) The hyperbola is shown in Fig. 17-18.

Fig. 17-18

17.14. (a) Write an equation for the hyperbola conjugate to that of Problem 17.13.

(b) Determine the coordinates of the hyperbola's vertices and its foci.

(c) Write the equations for the asymptotes of the hyperbola.

(a) The equation of the hyperbola conjugate to

$$\frac{x^2}{64} - \frac{y^2}{25} = 1 \qquad \text{is} \qquad \frac{y^2}{25} - \frac{x^2}{64} = 1$$

(b) Since $a^2 = 25$, $a = 5$; therefore, the vertices of the hyperbola are at (0, 5) and (0, -5). Since $c^2 = a^2 + b^2 = 25 + 64 = 89$, $c = \pm\sqrt{89} \approx \pm9.43$. The foci of the hyperbola are at (0, 9.43) and (0, -9.43).

(c) The equations of the asymptotes of the hyperbola are $x = \pm\frac{8}{5}y$, or $y = \pm\frac{5}{8}x$, which are the same as for the hyperbola in Problem 17.13.

17.15. The eccentricity e of a hyperbola is defined as $e = c/a$. Determine the eccentricity of the hyperbola

$$\frac{x^2}{16} - \frac{y^2}{9} = 1$$

From the equation, $a^2 = 16$, $b^2 = 9$.

$$c^2 = a^2 + b^2 = 16 + 9 = 25$$

$$c = \sqrt{25} = 5$$

$$e = c/a = 5/4 = 1.25$$

The eccentricity of the hyperbola is 1.25. The eccentricity of every hyperbola is greater than 1.

Supplementary Problems

The answers to this set of problems are at the end of this chapter.

17.16. What is the equation of a circle whose radius length is   (a) 8,   (b) $\sqrt{3}$?

17.17. What are the diameter lengths of circles with equations: (a) $x^2 + y^2 = 9$,   (b) $x^2 + y^2 = 15$?

17.18. The equation of a circle is $x^2 + y^2 = 100$.   The y-coordinate of a point on this circle is 6.   What is the value of the x-coordinate of the point?

17.19. The front view of a circular conical surface is shown in Fig. 17-19.

    (a) What figure will result when a plane cuts the surface at P_1? at P_2?

    (b) Sketch the figures which will result when a plane cuts the surface at P_3 and P_4 parallel to a generator of the cone.

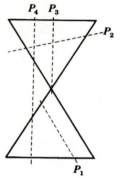

Fig. 17-19

17.20. The focus of a parabola with its axis along the y-axis, and "opening upwards," has the coordinates (0, 2.5).

    (a) Write the equation of the parabola.

    (b) Sketch the parabola.

    (c) Sketch the parabola with the same axis but "opening down," with focus at (0, −2.5).

17.21. A hyperbola has vertices at (4, 0) and (−4, 0), and foci at (5, 0) and (−5, 0).

    (a) Write the equation of the hyperbola.

    (b) Write the equations of the asymptotes of the hyberbola.

    (c) Sketch the hyperbola.

17.22. (a) What are the coordinates of the vertices of the hyperbola whose equation is $\dfrac{x^2}{36} - \dfrac{y^2}{49} = 1$?

    (b) What are the coordinates of the foci for the hyperbola in (a)?

    (c) Write the equations of the asymptotes.

17.23. The equation of an ellipse is $\dfrac{x^2}{169} + \dfrac{y^2}{25} = 1$.

    (a) What is the length of the semiminor axis?

    (b) What is the distance between the two foci?

    (c) Determine the eccentricity of this ellipse.

17.24. What is the eccentricity of   (a) a circle;   (b) a straight line?

    (*Hint*: Think of these two figures as special ellipses.)

17.25. The earth revolves around the sun in an elliptical orbit.   The sun is at one of the foci of the ellipse.   If the nearest distance between the earth and the sun is 91,500,000 miles and the farthest distance is 94,500,000 miles, calculate:

    (a) the approximate distance between the foci of the orbit, and

    (b) the eccentricity of the earth's orbit.

17.26. Given the ellipse shown in Fig. 17-20:

    (a) Using a metric rule, measure the lengths of the major and minor axes in centimeters; measure the distance between the foci.

(b) Calculate the eccentricity of the ellipse.

(c) Write the standard equation of the ellipse.

17.27. The orbit of the asteroid Eros has an eccentricity of 0.22. Its mean distance from the sun is 1.46 astronomical units (AU).

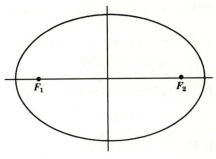

(a) Calculate the nearest distance of Eros from the sun

(b) Calculate the farthest distance of Eros from the sun.

(c) Using a looped string and two pins or thumb tacks, draw to scale a representation of the orbit of Eros.

Fig. 17-20

17.28. The vertices of a hyperbola are at (0, 7) and (0, −7); the focuses of the hyperbola are at (0, 25) and (0, −25).

(a) Write the equation of the hyperbola.

(b) Determine the eccentricity of the hyperbola.

(c) Write the equation of the conjugate hyperbola.

17.29. A projectile is shot horizontally from a height S of 100 ft. The projectile follows a parabolic curve and strikes level ground at a horizontal distance H of 80 ft from launch site. Write the equation of the projectile path.

17.30. A piece of plywood with the dimensions shown in Fig. 17-21 is to be sawed out of a circular piece. What should be the diameter of the circle?

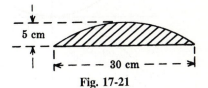

Fig. 17-21

17.31. A reflector has the shape of a dish with a parabolic cross-section. The dish depth is 3 inches; the dish diameter is 2 feet.

(a) Calculate the focal length of the reflector.

(b) Write the equation for the parabola.

17.32. The Oterma comet has an eccentricity of 0.144 and a semimajor axis of 5.92×10^8 km. The sun is at one of the foci of the comet's elliptical orbit. Determine the comet's nearest and farthest distances from the sun.

17.33. In Fig. 17-16 how far from its left end is the arch 4 m high?

Answers to Supplementary Problems

17.16. (a) $x^2 + y^2 = 64$ (b) $x^2 + y^2 = 3$

17.17. (a) 6   (b) $2\sqrt{15} \approx 7.74$

17.18. $x = \pm 8$

17.19. (*a*) parabola; ellipse   (*b*) See Fig. 17-22

17.20. (*a*) $x^2 = 10y$ (*c*) See Fig. 17-23

(*b*) See Fig. 17-23

17.21. (*a*) $x^2/16 - y^2/9 = 1$ (*c*) See Fig. 17-24

(*b*) $y = \pm(3/4)x$

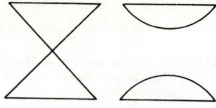

Fig. 17-22

17.22. (*a*) $(6, 0)$, $(-6, 0)$ (*b*) $(\sqrt{85}, 0)$, $(-\sqrt{85}, 0)$ or $(9.2, 0)$, $(-9.2, 0)$ approx.   (*c*) $y = \pm 7x/6$

17.23. (*a*) 5   (*b*) 24   (*c*) 12/13 or 0.923 approx.

17.24. (*a*) 9   (*b*) ∞

17.25. (*a*) 3,000,000 miles   (*b*) 0.0161

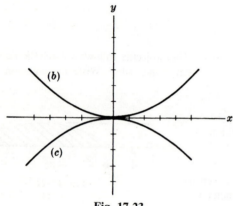

Fig. 17-23

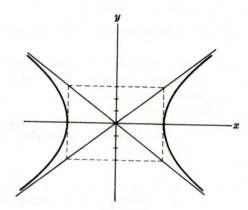

Fig. 17-24

17.26. (*a*) 5.3 cm; 3.4 cm; 4.0 cm approx.   (*b*) 0.75 approx.   (*c*) $x^2/(2.6)^2 + y^2/(1.7)^2 = 1$

17.27. (*a*) 1.14 AU   (*b*) 1.78 AU

17.28. (*a*) $\dfrac{y^2}{49} - \dfrac{x^2}{576} = 1$ (*b*) $e \approx 3.57$ (*c*) $\dfrac{x^2}{576} - \dfrac{y^2}{49} = 1$

17.29. $y = -x^2/64$ or $64y = -x^2$

17.30. $D = 50$ cm

17.31. (*a*) $p = 12$ in.

(*b*) $y^2 = 48x$ (if focus on *x*-axis) or $x^2 = 48y$ or $y = x^2/48$ (if focus on *y*-axis)

17.32. $(5.07)(10^8)$ km; $(6.77)(10^8)$ km

17.33. 3.38 m approx,

Chapter 18

Numbering Systems

18.1   THE BINARY, OCTAL, AND HEXADECIMAL SYSTEMS

The most commonly used numbering system is the decimal system.   It is named this way because it is based on the number 10, hence the prefix "deci." The number 10 itself is called the *basis* or *radix* of the system.   The basis tells us how many different individual symbols may be used to write numbers in the system.   In the decimal system these symbols are called *digits*; they are the familiar 0, 1, 2, 3, 4, 5, 6, 7, 8, and 9.   Note that they range in value from 0 to 9.   This is a particular case of a more general rule that can be stated as follows: *Given any positive basis or radix N, there are N different individual symbols that can be used to write numbers in the system.   The value of these symbols range from 0 to N − 1.*

In addition to the decimal system, there are other numerical systems that are extensively used in the computer and telecommunication fields.   They are the binary system (its basis is 2), the octal system (its basis is 8), and the hexadecimal system (its basis is 16).

EXAMPLE 18.1.   How many different individual symbols do the binary, octal, and hexadecimal systems have?   What is the numerical range of these symbols?

In the binary system the basis is $N = 2$; therefore, there are 2 different individual symbols.   The values of these symbols range from 0 to $(2 − 1)$, that is, from 0 to 1.   These symbols are 0 and 1.   In computer jargon, these symbols are called the 0 bit and 1 bit, respectively.   The word *bit* is an abbreviated form of the words **b**inary dig**it**.

In the octal system the basis is $N = 8$; therefore, there are 8 different individual symbols.   The values of these symbols range from 0 to $(8 − 1)$, that is, from 0 to 7.   These symbols are 0, 1, 2, 3, 4, 5, 6, and 7.   We call all of these symbols by their decimal names.   These are the familiar names one, two, three, and so on.

In the hexadecimal system the basis is $N = 16$; therefore, there are 16 different individual symbols.   The values of these symbols range from 0 to $(16 − 1)$, that is, from 0 to 15.   These symbols are 0, 1, 2, 3, 4, 5, 6, 7, 8, 9, A, B, C, D, E, and F.   Notice that after 9, the symbols are the letters A through F.   In this system, the A stands for 10, the B for 11, the C for 12, the D for 13, the E for 14, and the F for 15.   The reason for choosing letters instead of symbols to represent symbols higher than 9 is to keep all symbols single characters.   We call the symbols 0 through 9 by their decimal names.   The letters are called by their common names.

To explicitly state the basis on which a number is written, we will use a subscript to the lower right side of the number.   The format of the subscript is "(basis." For example, $1010_{(2}$ indicates that the number is binary.   Likewise, $145_{(10}$ is a decimal number, $1A2_{(16}$ is a hexadecimal number, and $145_{(8}$ is an octal number.

Although the word digit generally refers to the individual symbols of the decimal system, it is also common to call the individual symbols of the other numerical systems by this name.

18.2   NUMERICAL VALUES IN POSITIONAL SYSTEMS AND THEIR DECIMAL EQUIVALENTS

All the numerical systems that we have considered so far, including the decimal system, are positional systems.   That is, the value represented by a digit in the numerical representation of a number always depends on its position in the number.   For example, in the decimal system, the 5 in 157 represents the value 50 units; the 5 in 548 represents 500 units.

In the decimal system, we call the rightmost position of a number the ones (10^0) place, the next position from the right the tens (10^1) place, the next position the hundreds (10^2) place, and so on. Notice that the powers increase from the right. The power of the rightmost digit is zero, the next digit has a power of one, the next a power of two, and so on. These powers are sometimes called the *weight* of the digit. To better visualize this we generally write superscripts beginning with 0 (the rightmost digit of the number) and increasing the powers by 1 as we move toward the left digits of the number.

In any positional system, *the decimal value of a digit in the representation of a number is the digit's own value, in decimal, multiplied by a power of the basis in which the number is represented. The sum of all these powers is the decimal equivalent of the number.*

The following example illustrates this.

EXAMPLE 18.2. What is the value of each digit in the number $1554_{(10}$?

Notice that the basis is 10 since we are working in the decimal system. Writing superscripts beginning with 0, on the rightmost digit, and increasing them as we move toward the right, the number would look like this: $1^3 5^2 5^1 4^0$. Using these superscripts as the power of the basis:

$$1554 = (1 \times 10^3) + (5 \times 10^2) + (5 \times 10^1) + (4 \times 10^0) \quad \text{Remember that } 10^0 \text{ is equal to 1.}$$

$$= (1 \times 1000) + (5 \times 100) + (5 \times 10) + (4 \times 1)$$

$$= 1000 + 500 + 50 + 4$$

Observe that reading the number from left to right, the 1 represents 1000 units, the first 5 represents 500 units, the next 5 represents 50 units and the 4 represents the single units.

EXAMPLE 18.3. What is the value of each digit in the number $1554_{(8}$? What is the decimal equivalent of the number?

Notice that the basis is 8 since we are working with the octal system. Using a similar procedure as in the previous example, we can use superscripts and write the number as $1^3 5^2 5^1 4^0$. Using these superscripts as the power of the basis

$$1554_{(8} = (1 \times 8^3) + (5 \times 8^2) + (5 \times 8^1) + (4 \times 8^0) \quad \text{Remember that } 8^0 \text{ is equal to 1.}$$

$$= (1 \times 512) + (5 \times 64) + (5 \times 8) + (4 \times 1)$$

$$= 512 + 320 + 40 + 4$$

$$= 876_{(10}$$

We can observe that reading the octal number from left to right, the 1 represents 512 units, the first 5 represents 320 units, the next 5 represents 40 units and the 4 represents the single units. The decimal equivalent is 876.

EXAMPLE 18.4. What is the value of each digit in the number $1554_{(16}$? What is the decimal equivalent of the number?

Notice that the basis is 16 since we are working with the hexadecimal system. Using superscripts as the power of the basis we can write the number as $1^3 5^2 5^1 4^0$. Therefore, the hexadecimal values of each digit can be calculated as follows:

$$1554_{(16} = (1 \times 16^3) + (5 \times 16^2) + (5 \times 16^1) + (4 \times 16^0) \quad \text{Remember that } 16^0 \text{ is equal to 1}$$

$$= (1 \times 4096) + (5 \times 256) + (5 \times 16) + (4 \times 1)$$

$$= 4096 + 1280 + 80 + 4$$

$$= 5460_{(10}$$

Observe that reading the hexadecimal number from left to right, the 1 represents 4096 units, the first 5 represents 1280 units, the next 5 represents 80 units and the 4 represents the single units. The decimal equivalent is 5460.

Examples 18-2 through 18-4 show that, depending on the basis, numbers with identical representation may have different decimal equivalent values.

EXAMPLE 18.5. What is the value of each digit in the number $11001_{(2}$? What is the decimal equivalent of the number?

The number can be written as $1^4 1^3 0^2 0^1 1^0$ using superscripts. The value of each digit can be calculated as follows:

$$11001_{(2} = (1 \times 2^4) + (1 \times 2^3) + (0 \times 2^2) + (0 \times 2^1) + (1 \times 2^0) \quad \text{Remember that } 2^0 \text{ is equal to 1}$$
$$= (1 \times 16) + (1 \times 8) + (0 \times 4) + (0 \times 2) + (1 \times 1)$$
$$= (16) + (8) + (0) + (0) + 1$$
$$= 25_{(10}$$

Notice that the basis is 2 since we are working with the binary system. Observe that reading the binary number from left to right, the first 1 represents 16 units, the second 1 represents 8 units, and the last 1 represents the single units. The decimal equivalent of the number is 25.

EXAMPLE 18.6. What is the value of each digit in the number $1AB2_{(16}$? What is the decimal equivalent of the number?

Using superscripts the number can be written as $1^3 A^2 B^1 2^0$. The value of each digit can be calculated as follows:

$$1AB2_{(16} = (1 \times 16^3) + (10 \times 16^2) + (11 \times 16^1) + (2 \times 16^0) \quad \text{Remember that } 16^0 \text{ is equal to 1}$$
$$= (1 \times 4096) + (10 \times 256) + (11 \times 16) + (2 \times 1)$$
$$= (4096) + (2560) + (176) + 2$$
$$= 6834_{(10}$$

The basis is 16 since we are working with the hexadecimal system. Notice that we have replaced A and B by their decimal equivalents 10 and 11, respectively. Observe also that reading the hexadecimal number from left to right, the first 1 represents 4096 units, the A represents 2560 units, the B represents 176 units and the 2 represents the single units. The decimal equivalent of the number is 6834.

18.3 CONVERSION OF DECIMAL NUMBERS TO OTHER BASES

So far, we have seen how to obtain the equivalent decimal value of numbers expressed in the binary, octal, or hexadecimal systems. In this section we will learn how to express a given decimal number in any of these bases. Before converting decimal numbers to the binary system, remember that when we divide a number (called the dividend) by another number (called the divisor), we will obtain a quotient and a remainder. If the remainder is zero, we say that the division is exact; otherwise we say that the division is not exact. If we call the dividend (D), the divisor (d), the quotient (q) and the remainder (r), the following formula always holds true:

$$D = d \times q + r \qquad \text{where} \qquad 0 \le r < q$$

With this in mind, we can now proceed to describe an algorithm or method to carry out the conversion of decimal numbers to binary. This process consists primarily of dividing a given number by the basis of the system to which we want to convert it until we get a quotient of zero. The remainders of the different divisions (in the opposite order to that in which they were obtained) will give the representation of the decimal number in the new basis. Algorithm 1 formalizes this process.

Algorithm 1 (Conversion from decimal to any nonnegative integer basis)

Step 1. Set the value of $i = 0$. Call N_i the given decimal number.

Step 2. Divide N_i by the new basis. Obtain the quotient q_i and the remainder r_i.

Step 3. Check the value of the quotient q_i. If the quotient q_i is zero continue with step 5 otherwise continue with step 4.

Step 4.   Increase the value of i by 1.   Set N_i equal to $q_{(i-1)}$ and go back to step 2.

Step 5.   The equivalent number (in the new basis) to the given decimal number is obtained by concatenating (writing next to each other) the remainders r_i obtained in the opposite order in which they were obtained.   That is, the equivalent number (in the new basis) is $r_i r_{i-1} r_{i-2} \cdots r_0$.   Notice that the subscripts decrease from left to right.

EXAMPLE 18.7.   What is the binary representation of $25_{(10}$?

The new basis is 2.

Step 1.   Set $i = 0$ and call $N_0 = 25$.

Step 2.   Divide $N_0 = 25$ by 2, the new basis.   Obtain the quotient q_0 and the remainder r_0.   That is, divide 25 by 2.   Obtain $q_0 = 12$ and $r_0 = 1$.

Step 3.   Check that q_0 is not zero.   In this case, $q_0 = 12$, therefore it is not zero.

Step 4.   Set $i = 1$ (i is incremented by 1).

                        $N_1 = q_{i-1} = q_0 = 12$ (N_i is set to $q_{(i-1)}$).

                        Go back to step 2 of the algorithm.

Step 2 (2nd time).   Divide $N_1 = 12$ by 2.   Obtain $q_1 = 6$ and $r_1 = 0$.

Step 3 (2nd time). $q_1 = 6$.   Therefore, q_1 is not zero.

Step 4 (2nd time).   Set $i = 2$ (i is incremented by 1).

                        Set $N_2 = q_{2-1} = q_1 = 6$.

                        Go back to step 2 of the algorithm.

Step 2 (3rd time).   Divide $N_2 = 6$ by 2.   Obtain $q_2 = 3$ and $r_2 = 0$.

Step 3 (3rd time). $q_2 = 3$. Therefore, q_2 is not zero.

Step 4 (3rd time).   Set $i = 3$ (i is incremented by 1).

                        Set $N_3 = q_{3-1} = q_2 = 3$.

                        Go back to step 2 of the algorithm.

Step 2 (4th time).   Divide $N_3 = 3$ by 2.   Obtain $q_3 = 1$ and $r_3 = 1$.

Step 3 (4th time). $q_3 = 1$.   Therefore, q_3 is not zero.

Step 4 (4th time).   Set $i = 4$ (i is incremented by 1).

                        Set $N_4 = q_{4-1} = q_3 = 1$.

                        Go back to step 2 of the algorithm.

Step 2 (5th time).   Divide $N_4 = 1$ by 2.   Obtain $q_4 = 0$ and $r_4 = 1$.

Step 3 (5th time). q_4 is zero.

Step 5.   The equivalent binary number is obtained by concatenating the remainders of the divisions in the opposite order to that in which they were obtained.   The equivalent binary number is $N_0 = 25_{(10} = r_4 r_3 r_2 r_1 r_0 = 11001_{(2}$.

We can verify this result as follows:

$$1^4 1^3 0^2 0^1 1^0_{(2} = (1 \times 2^4) + (1 \times 2^3) + (0 \times 2^2) + (0 \times 2^1) + (1 \times 2^0)$$

$$= (1 \times 16) + (1 \times 8) + (0 \times 4) + (0 \times 2) + (1 \times 1)$$

$$= 16 + 8 + 0 + 0 + 1$$

$$= 25$$

The steps of this algorithm can also be illustrated in an abbreviated form as follows:

Number	Quotient when dividing by 2	Remainder
25	12	1
12	6	0
6	3	0
3	1	1
1	0	1

Writing the remainders in opposite order, we have that $25_{(10} = 11001_{(2}$.

EXAMPLE 18.8. What is the octal representation of decimal 125?

The new basis is 8.

Step 1.	Set $i = 0$ and $N_0 = 125$.
Step 2.	$125 \div 8$. Obtain $q_0 = 15$ and $r_0 = 5$.
Step 3.	q_0 is not zero.
Step 4.	Set $i = 1$ (i is incremented by 1).

$N_1 = q_{i-1} = q_0 = 15$ (N_i is set to $q_{(i-1)}$).

Go back to step 2 of the algorithm.

Step 2 (2nd time).	$15 \div 8$. Obtain $q_1 = 1$ and $r_1 = 7$.
Step 3 (2nd time).	q_1 is not zero.
Step 4.	Set $i = 2$.

Set $N_2 = q_{2-1} = q_1 = 1$.

Go back to step 2 of the algorithm.

Step 2 (3rd time).	$1 \div 8$. Obtain $q_2 = 0$ $r_2 = 1$.
Step 3 (3rd time).	q_2 is zero.
Step 5.	The equivalent octal number is obtained by concatenating the remainders of the divisions in the opposite order to that in which they were obtained. Therefore $N_0 = 125_{(10} = r_2 r_1 r_0 = 175_{(8}$.

We can verify this result as follows:

$$1^2 7^1 5^0{}_{(8} = (1 \times 8^2) + (7 \times 8^1) + (5 \times 8^0)$$

$$= (1 \times 64) + (7 \times 8) + (5 \times 1)$$

$$= 64 + 56 + 5$$

$$= 125_{(10}$$

The steps of the algorithm can be illustrated in an abbreviated form as follows:

Number	Quotient when dividing by 8	Remainder
125	15	5
15	1	7
1	0	1

Writing the remainders in opposite order, we have that $125_{(10} = 175_{(8}$.

EXAMPLE 18.9. What is the hexadecimal representation of $48_{(10}$?

The new basis is 16.

Step 1.   Set $i = 0$ and $N_0 = 48$.

Step 2. $48 \div 16$.   Obtain $q_0 = 3$ and $r_0 = 0$.

Step 3. q_0 is not zero.

Step 4.   Set $i = 1$ (i is incremented by 1).

                          Set $N_1 = q_{i-1} = q_0 = 3$ (N_i is set to $q_{(i-1)}$).

                          Go back to step 2 of the algorithm.

Step 2 (2nd time). $3 \div 16$.   Obtain $q_1 = 0$ and $r_1 = 3$.

Step 3 (2nd time). q_1 is zero.

Step 5.   The equivalent hexadecimal number is obtained by concatenating the remainders of the divisions in the opposite order to that in which they were obtained.   The equivalent hexadecimal number is $N_0 = 48_{(10} = r_1 r_0 = 30_{(16}$.

We can verify this result as follows:

$$3^1 0^0_{(16} = (3 \times 16^1) + (0 \times 16^0)$$
$$= (3 \times 16) + (0 \times 1)$$
$$= 48 + 0$$
$$= 48_{(10}$$

The steps of the algorithm can be illustrated in an abbreviated form as follows:

| | Quotient when | |
Number	dividing by 16	Remainder
48	3	0
3	0	3

Writing the remainders in opposite order, we have that $48_{(10} = 30_{(8}$.

18.4 CONVERSION BETWEEN HEXADECIMAL AND BINARY NUMBERS

Conversion between these two systems can be accomplished as a two-step process.   If the number is hexadecimal we first convert it to decimal and then to binary.   If the number is binary we first convert it to decimal and then to hexadecimal.   However, there is an even faster process for converting numbers between these two bases.   As Table 18.1 shows, with four binary digits we can represent the numbers 0 through 15.   Since each four-bit binary number corresponds to one and only one hexadecimal digit and vice versa, we can view the hexadecimal system as a shorthand notation of the binary system.

Algorithm 2 gives the process for converting binary numbers to their hexadecimal equivalents.

Algorithm 2 (Conversion from binary to hexadecimal)

Step 1.   Form four-bit groups beginning from the rightmost bit of the number.   If the last group (at the leftmost position) has fewer than four bits, add extra zeros to the left of the bits in this group to make it a four-bit group.

Step 2.   Replace each four-bit group by its hexadecimal equivalent.

EXAMPLE 18.10.   What is the hexadecimal equivalent of the binary number 1010111100001011110100011?

**Table 18.1.   Decimal, Binary, Octal and
Hexadecimal Equivalents**

Decimal	Binary	Octal	Hexadecimal
0	0000	0	0
1	0001	1	1
2	0010	2	2
3	0011	3	3
4	0100	4	4
5	0101	5	5
6	0110	6	6
7	0111	7	7
8	1000	10	8
9	1001	11	9
10	1010	12	A
11	1011	13	B
12	1100	14	C
13	1101	15	D
14	1110	16	E
15	1111	17	F

Step 1.   Forming four-bit groups beginning from the rightmost bit we obtain

$$10 \ 1011 \ 1100 \ 0010 \ 1111 \ 0100 \ 0111$$

Since the last group (the leftmost) has only two bits we add two extra zeros to the left of these bits to make it a four-bit group.   The number now looks like this:

0010 1011 1100 0010 1111 0100 0111   ← Notice the extra zeros added to the leftmost group to make it a 4-bit group.

Step 2.   Replacing each group by its hexadecimal equivalent (see Table 18.1) we obtain 2BC2F47.

The procedure to convert hexadecimals to binary is basically the reverse of the previous algorithm.   However, there is no need to add extra zeros since each hexadecimal number will always convert to a group with four binary bits.

EXAMPLE 18.11.   What is the binary equivalent to the hexadecimal number 3ACBD2?

Replacing each hexadecimal number by its binary equivalent (see Table 18.1) we will obtain

$$0011 \ 1010 \ 1100 \ 1011 \ 1101 \ 0010$$

We have separated the binary number in groups of four bits to facilitate the reading of the resulting binary number.

18.5   RULES FOR FORMING NUMBERS IN ANY SYSTEM

Table 18.1 shows the decimal equivalent of some numbers in binary, octal, and hexadecimal.   In any of these systems how do we form consecutive numbers higher than their largest individual symbol?   Let us examine the decimal system.   First, notice that with a single digit we can only represent the numbers 0 through 9.   To represent numbers exceeding nine we use more than one digit as indicated below:

0 1 2 3 4 5 6 7 8 9 **10 11 12 13 14 15 16 17 18 19 20 21 22 23 24 25 26 27 28 29 30 31 32 33 34 35 36 37 38 39** ...   **90 91 92 93 94 95 96 97 98 99 100 101 102 103 104 105 106 107 108 109 110**

Beginning at the left of the top row, we have written all single digits. Since there is no single digit to represent numbers higher than 9, we form all the two-digit combinations of numbers beginning with 1. Then we form all two-digit combinations that begin with 2 and so on until we reach 99. After exhausting all two-digit combinations we proceed to form combination of three digits. Once we have exhausted all three-digit combinations we proceed with four digits. This process can be continued forever. This rule can be applied to any other numerical system. Table 18.1 shows that we have followed a similar procedure for representing numbers higher than 7 in the octal system.

EXAMPLE 18.12. In the hexadecimal system, the sequence of numbers that exceed F (or decimal 15) is

10 11 12 13 14 15 16 17 18 19 1A 1B 1C 1D 1E 1F 20 21 22 23 24 25 26 27 28 29 2A 2B 2C 2D 2E 2F 30 31 32 33 34 35 36 37 38 39 3A 3B 3C 3D 3E 3F 40 41 and so on.

Solved Problems

18.1. If 0, 1, 2, 3, 4 are the single digits of a numbering system, what is the smallest basis to have such digits?

Notice that the range of numbers goes from 0 to 4. Since 4 is the largest, the minimum basis is 5. Remember that for a given basis N, there are $N-1$ different symbols with a range from 0 to $(N-1)$.

18.2. What is the decimal equivalent of $1631_{(8}$?

$$1631_{(8} = (1 \times 8^3) + (6 \times 8^2) + (3 \times 8^1) + (1 \times 8^0)$$
$$= (1 \times 512) + (6 \times 64) + (3 \times 8) + (1 \times 1)$$
$$= 512 + 384 + 24 + 1$$
$$= 921_{(10}$$

18.3. What is the decimal equivalent of $333_{(5}$?

$$333_{(5} = (3 \times 5^2) + (3 \times 5^1) + (3 \times 5^0)$$
$$= (3 \times 25) + (3 \times 5) + (3 \times 1)$$
$$= 75 + 15 + 3$$
$$= 93_{(10.}$$

18.4. Is 13481 the representation of a valid octal number?

No! This number cannot be an octal number since the largest single digit in this base is a 7 and this number has an 8.

18.5. What is the decimal representation of $1100111_{(2}$?

$$1100111_{(2} = (1 \times 2^6) + (1 \times 2^5) + (0 \times 2^4) + (0 \times 2^3) + (1 \times 2^2) + (1 \times 2^1) + (1 \times 2^0)$$
$$= (1 \times 64) + (1 \times 32) + (0 \times 16) + (0 \times 8) + (1 \times 4) + (1 \times 2) + (1 \times 1)$$
$$= 64 + 32 + 0 + 0 + 4 + 2 + 1$$
$$= 103_{(10}$$

18.6. What is the binary equivalent of $FC2B_{(16}$?

Replacing each decimal number by its binary equivalent (see Table 18.1) we obtain

F	C	2	B
1111	1100	0010	1011

Therefore, $FC2B_{(16} = 1111110000101011_{(2}$.

18.7. What is the hexadecimal equivalent of $1011101001110101111_{(2}$?

Forming four-bit groups beginning with the rightmost bit and adding extra zeros to the leftmost group, if necessary, to make it a 4-bit group, we obtain

$1011101001110101111_{(2} = 0101 \quad 1101 \quad 0011 \quad 1010 \quad 1111 \leftarrow$ An extra zero was added to the leftmost group.

Replacing each group by its hexadecimal equivalent (see Table 18.1) we obtain that

$$1011101001110101111_{(2} = 5D3AF_{(16}$$

18.8. What's the octal equivalent to $11011000101_{(2}$?

Since we are working with the octal basis, we can follow a similar algorithm to the one used in Example 18.9. However, instead of forming groups of four bits, we form three-bit groups beginning with the rightmost bit and add extra digits to the leftmost group if necessary. Proceeding this way, we can form groups of 3-bits as follows:

$11011000101_{(2} = 011 \quad 011 \quad 000 \quad 101 \leftarrow$ A zero was added to the leftmost group.

Replacing each group by its octal equivalent (see Table 18.1), we obtain

$$11011000101_{(2} = 011 \quad 011 \quad 000 \quad 101$$
$$= 3305_{(8}$$

18.9. What is the binary equivalent of $31_{(10}$?

Since we want to express the number in binary we need to apply Algorithm 1 with 2 as the basis.

Step 1.   Set $i = 0$ and $N_0 = 31$.

Step 2. $31 \div 2$.   Obtain $q_0 = 15$ and $r_0 = 1$.

Step 3. q_0 is not zero.

Step 4.   Set $i = 1$.

                   Set $N_1 = q_{i-1} = q_0 = 15$.

                   Go back to step 2 of the algorithm.

Step 2 (2nd time). $15 \div 2$.   Obtain $q_1 = 7$ and $r_1 = 1$.

Step 3 (2nd time). q_1 is not zero.

Set $i = 2$.

Step 4 (2nd time). $N_2 = q_{i-1} = q_1 = 7$.

Go back to step 2 of the algorithm.

Step 2 (3rd time). $7 \div 2$.   Obtain $q_2 = 3$ and $r_2 = 1$.

Step 3 (3rd time). q_2 is not zero.

Step 4 (3rd time).   Set $i = 3$.

$N_3 = q_{i-1} = q_2 = 3$.

Go back to step 2 of the algorithm.

Step 2 (4th time). $3 \div 2$.   Obtain $q_3 = 1$ and $r_3 = 1$.

Step 3 (4th time). q_3 is not zero.

Step 4 (4th time).   Set $i = 4$.

$N_4 = q_{i-1} = q_3 = 1$.

Go back to step 2 of the algorithm.

Step 2 (5th time). $1 \div 2$.   Obtain $q_4 = 0$ and $r_4 = 1$.

Step 3 (5th time). q_4 is zero.

Step 5.   The equivalent binary number is obtained by concatenating the remainders of the divisions in the opposite order to that in which they were obtained. The equivalent binary number is $N_0 = 31_{(10} = r_4 r_3 r_2 r_1 r_0 = 11111_{(2}$.

We can verify this result as follows:

$$1^4 1^3 1^2 1^1 1^0_{(2} = (1 \times 2^4) + (1 \times 2^3) + (1 \times 2^2) + (1 \times 2^1) + (1 \times 2^0)$$

$$= (1 \times 16) + (1 \times 8) + (1 \times 4) + (1 \times 2) + (1 \times 1)$$

$$= 16 + 8 + 4 + 2 + 1$$

$$= 31_{(10}.$$

The conversion process using the abbreviated form of the algorithm is shown below.

Number	Quotient when dividing by 2	Remainder
31	15	1
15	7	1
7	3	1
3	1	1
1	0	1

Writing the remainders in opposite order, we have that $31_{(10} = 11111_{(2}$.

18.10. In any numerical system, what does a 10 represent?

This question seems trivial since we "think" in decimal. However, the answer depends upon the numbering system being considered. Looking at Table 18.1, observe that under each of the columns labeled binary and octal there is a 10. The binary 10 represents the decimal value 2. The octal 10 represents the decimal value 8. Notice that in all cases 10 represents the basis of the system in consideration. This is true in any other numbering system including the decimal and hexadecimal system.

18.11. How many different binary numbers can be represented with four bits?   What are their decimal equivalents?

We can begin by forming all possible combinations using all four bits and then finding their decimal equivalent.   Looking at Table 18.1, we can see that there are 16 different binary numbers.   There is a rule that can give use the maximum number of different numbers that can be represented with N bits.   We can state it as follows: *Given N bits, it is possible to form 2^N different binary numbers.   The range of these values is from 0 to $2^N - 1$.*

18.12. Given a sequence of seven bits, what is the range of values that can be represented in the binary system?   How many different numbers are there?

Since $N = 7$, it is possible to form 2^7 or 128 different binary numbers.   The range of these values will vary from 0 to $2^7 - 1$.   That is, from 0 to 127.

18.13. What is the binary equivalent of $113_{(10}$?

Using the abbreviated form of Algorithm 1, we have

Number	Quotient when dividing by 2	Remainder
113	56	1
56	28	0
28	14	0
14	7	0
7	3	1
3	1	1
1	0	1

Writing the remainders in the opposite order to that in which they were obtained, we have that $113_{(10} = 1110001_{(2}$.

This can be verified as follows:

$$1110001_{(2} = (1 \times 2^6) + (1 \times 2^5) + (1 \times 2^4) + (0 \times 2^3) + (0 \times 2^2) + (0 \times 2^1) + (1 \times 2^0)$$

$$= (1 \times 64) + (1 \times 32) + (1 \times 16) + (0 \times 8) + (0 \times 4) + (0 \times 2) + (1 \times 1)$$

$$= 64 + 32 + 16 + 0 + 0 + 0 + 1$$

$$= 113_{(10}$$

18.14. What is octal representation of $111_{(10}$?

Using the abbreviated form of Algorithm 1, we have

Number	Quotient when dividing by 8	Remainder
111	13	7
13	1	5
1	0	1

Writing the remainders in the opposite order to that in which they were obtained, we have that $111_{(10} = 157_{(8}$.

We can verify this result as follows:

$$157_{(8} = (1 \times 8^2) + (5 \times 8^1) + (7 \times 8^0)$$
$$= (1 \times 64) + (5 \times 8) + (7 \times 1)$$
$$= 64 + 40 + 7$$
$$= 111_{(10}$$

18.15. What is the hexadecimal representation of $142_{(10}$?

Using the abbreviated form of Algorithm 1, we have

Number	Quotient when dividing by 16	Remainder
142	8	14
8	0	8

Writing the remainders in the opposite order to that in which they were obtained, and keeping in mind that E stands for 14 in the hexadecimal system, we have that $142_{(10} = 8E_{(16}$.

This result can be verified as follows:

$$8E_{(16} = (8 \times 16^1) + (14 \times 16^0)$$
$$= (8 \times 16) + (14 \times 1)$$
$$= 142_{(10}$$

Supplementary Problems

The answers to this set of problems are at the end of this chapter.

18.16. Given the decimal equivalent of the following binary numbers:

(a) 1100011101   (b) 11001100111   (c) 100010001   (d) 11110001111   (e) 1111111

18.17. Give the decimal equivalent of the following hexadecimal numbers:

(a) F1AB   (b) ABCD   (c) 5942   (d) 1A2B   (e) 6DCE

18.18. Convert the following decimal numbers to binary:

(a) 1949   (b) 750   (c) 117   (d) 432   (e) 89

18.19. Convert the following hexadecimal numbers to binary:

(a) B52   (b) FAFA   (c) 512   (d) CAFE

18.20. Convert the following hexadecimal numbers to octal:

(a) 35AB   (b) F2CA   (c) 853F   (d) CAB   (e) 7A8B

18.21. Convert the following octal number to decimal:

    (*a*) 7714   (*b*) 3532   (*c*) 6320   (*d*) 1515   (*e*) 4050

18.22. Given a sequence of bits representing a number, how can you tell without making any conversion that the decimal equivalent is even or odd?

Answers to Supplementary problems

18.16.   (*a*) 797   (*b*) 1639   (*c*) 273   (*d*) 1935   (*e*) 127

18.17.   (*a*) 61867   (*b*) 43981   (*c*) 22850   (*d*) 6699   (*e*) 28110

18.18.   (*a*) 11110011101   (*b*) 1011101110   (*c*) 1110101   (*d*) 110110000   (*e*) 1011001

18.19.   (*a*) 101101010010   (*b*) 1111101011111010   (*c*) 10100010010   (*d*) 1100101011111110

18.20.   (*a*) 32653   (*b*) 171312   (*c*) 102477   (*d*) 6253   (*e*) 75213

18.21.   (*a*) 4044   (*b*) 1882   (*c*) 3280   (*d*) 845   (*e*) 2088

18.22. If the rightmost bit of the binary sequence is 1 its decimal equivalent is odd. Otherwise, its decimal equivalent is even.

Chapter 19

Arithmetic Operations in a Computer

19.1 REVIEW OF BASIC CONCEPTS OF ARITHMETIC

Before considering arithmetic operations in the binary, octal, or hexadecimal systems, let us review these operations in the decimal system. We will begin with a simple addition as shown below. Although the explanation may seem trivial, some generalization can be made to all other numerical systems to facilitate their understanding.

EXAMPLE 19.1.   The result of adding the decimal numbers 194 and 271 is 465. How did we obtain this result?

We began by aligning the numbers by their rightmost digits. We then added the numbers beginning with the rightmost digits.

$$194 \ +$$
$$271$$
$$\overline{}$$

5 ← 4 plus 1 is 5. Notice that there is a symbol for the number five.

1   ← carry of 1
$$194 \ +$$
$$271$$
$$\overline{}$$

65 ← 9 plus 7 is 16. Notice that there is no single symbol for the number sixteen. Write 6 and carry 1 since we know that $16 = 10 + 6$.

At this moment we may ask, Why do we write 6? Why do we carry 1? First remember that the single digits go from 0 to 9. Since we have exceeded 9 we need to use two digits. The rules for the formation of numbers of the previous chapter show that 6 "accompanies" the tens digit that we carry. Notice also that we can write 16 as $16 = \mathbf{1} \times 10 + \mathbf{6}$.   Therefore, we write 6 and carry 1. Finally,

1   ← carry of 1.
$$194 \ +$$
$$271$$
$$\overline{}$$
465 ← $1 + 1$ is 2. Then 2 plus 2 is 4. There is a single symbol for the number four.

EXAMPLE 19.2.   Subtract decimal 105 from decimal 4002.

We start by aligning the rightmost digits of the numbers.   The following sequence explains the process.

$$\overset{12}{400\cancel{2}} \ -$$
$$105$$
$$\overline{}$$

7 ← Since 2 is less than 5 we "borrow" 1 unit from the digit to the left of the 2. The borrowed unit is equivalent to "borrowing 10." Therefore, write 7 since $12 - 5 = 7$.

$$\overset{9}{40\cancel{0}2} \ - \quad \leftarrow \text{The 0 to the left of the 2 became a 9 (the basis minus 1).}$$
$$105$$
$$\overline{}$$
97 ← Write 9 since $9 - 0 = 9$.

384

9
4̶0̶02 −
105
‾‾‾‾‾
897 ← Write 8 since 9 − 1 = 8.

3
4̶0̶02 − ← The 4 "paid" the 10 that was initially borrowed by the rightmost 2. The 4 decreases to 3.
  105
‾‾‾‾‾
3897 ← Write 3 since 3 − 0 = 3.

19.2 ADDITION AND SUBTRACTION OF BINARY NUMBERS

When adding or subtracting binary numbers, we can follow a process similar to the one used for adding or subtracting decimal numbers. Using the rules of addition shown in Table 19-1, let us illustrate this with an example.

Table 19-1. Addition of Binary Numbers

$$0 + 0 = 0$$
$$0 + 1 = 1$$
$$1 + 0 = 1$$
$$1 + 1 = 0 \text{ with a carry of 1}$$

EXAMPLE 19.3.   Add the binary numbers 01010 and 10111.

We begin by aligning the rightmost digits of the number; therefore,

0101**0** +
1011**1**
‾‾‾‾‾‾
    1 ← 0 plus 1 is 1. There is a single symbol to represent one in binary.

010**1**0 + ← "Thinking in decimal" we have that 1 plus 1 is 2. There is not a single symbol for the number two in binary.
10**1**11   Two is written as 10 in binary. Therefore, we write 0 and carry 1 as indicated in Table 19.1. Notice that
‾‾‾‾‾   in "decimal" $2 = \mathbf{1} \times 2 + \mathbf{0}$. Therefore, write 0 and carry 1.
   01 ← Write 0 and carry 1.

       1
01**0**10 + ← 1 (the carry) plus 0 is 1. This 1 plus 1 is 2. Therefore, write 0 and carry 1.
10**1**11
‾‾‾‾‾
  001 ← Write 0 and carry 1.

     1
0**1**010 + ← 1 (the carry) plus 1 is 2. This 2 plus 0 is 2. Therefore, write 0 and carry 1.
1**0**111
‾‾‾‾‾
 0001

1

01010 $+$ $\leftarrow$ 1 (the carry) plus 0 is 1. This 1 plus 1 is 2. Therefore, write 0 and carry 1.

10111
___
100001 $\leftarrow$ Write 0 and the carry 1.

Subtraction of binary numbers is similar to addition, except that we may generate a "borrow" of "2" from the higher digits (those to the left of a particular digit). Example 19.4 illustrates this.

EXAMPLE 19.4.   Subtract 111 from 10001.

After aligning the rightmost digits of the number as shown below, we have

10001 $-$

111
___
     0 $\leftarrow$ Write 0 since $1-1=0$.

"10"

10001 $-$ $\leftarrow$ Since we cannot subtract 1 from 0 we "borrow" 1 unit from the next higher unit. This is equivalent to

  111   borrowing "2." Notice that if we were working with the decimal system, in a similar situation, we say that
___
      we borrow "ten" from the next higher place.

     10 $\leftarrow$ Write 1 since $2-1=1$.

   1

10001 $-$ $\leftarrow$ The next 0 becomes 1 (the basis minus 1).

  111
___
   010 $\leftarrow$ Write 0 since $1-1=0$.

  1

10001 $-$ $\leftarrow$ The next 0 also becomes 1 (the basis minus 1).

  111
___
  1010 $\leftarrow$ Write 1 since $1-0=1$. Notice that the 0 in the subtrahend is not shown but assumed.

0

10001 $-$ $\leftarrow$ The 1 "paid" the unit that was borrowed by the rightmost 1. This 1 becomes 0.

  111
___
01010 $\leftarrow$ Write 0 since $0-0=0$.

19.3   ADDITION AND SUBTRACTION OF HEXADECIMAL NUMBERS

Operating with a long string of 0s and 1s is an error-prone task. For this reason it is preferable to carry out arithmetic operations in the hexadecimal system. As we have seen, we can easily transform numbers between these two bases. Adding hexadecimal numbers is a process similar to that of adding decimal or binary numbers. However, it is necessary to remember that there is a value carried whenever the sum exceeds F hexadecimal or 15 decimal. In subtraction operations, whenever necessary, we may borrow "16" from the higher units "to the left." To simplify the process it is convenient to think in "decimal." Therefore, in our heads we may translate letters into their decimal equivalent, do the operations in decimal, and translate the results back to hexadecimal. The following examples illustrate this process.

EXAMPLE 19.5.   What is the sum of the following hexadecimal numbers: 4C1A and 12B3?

After aligning the numbers on their rightmost digit we have

4C1A +

12B**3**

$\overline{}$

   D ← A is equivalent to 10. 10 plus 3 is 13. In hexadecimal 13 is written D. Therefore, write D.

4C1A +

12**B**3

$\overline{}$

  CD ← B is equivalent to 11. 1 plus 11 is 12. In hexadecimal 12 is written C. Therefore, write C.

4C1A +

12B3

$\overline{}$

 ECD ← C is equivalent to 12. 12 plus 2 is 14. In hexadecimal 14 is written E. Therefore, write E.

4C1A +

12B3

$\overline{}$

5ECD ← 4 plus 1 is 5. Therefore, write 5.

EXAMPLE 19.6.   Add $EDF4_{(16}$ and $223_{(16}$.

After aligning the numbers on their rightmost digit we have that

EDF4 +

 223

$\overline{}$

   7 ← 4 plus 3 is 7.

EDF4 +

 22**3**

$\overline{}$

  17 ← F is equivalent to 15. 15 plus 2 is 17. Since $17 = 1 \times 16 + \mathbf{1}$ write 1 and carry 1.

1   ← carry

EDF4 +

 223

$\overline{}$

 017 ← D is equivalent to 13. 13 plus 1 (the carry) is 14. 14 plus 2 is 16. Since $16 = 1 \times 16 + \mathbf{0}$. Write 0 and carry 1.

1   ← carry

EDF4 +

 223

$\overline{}$

F017 ← E is equivalent to 14. 1 (the carry) plus 14 is 15. Write F since its decimal equivalent is 15.

EXAMPLE 19.7.   What is the result of subtracting $51AF_{(16}$ from $F3002_{(16}$?

   "18"

F300~~2~~ −

51AF

$\overline{}$

   3 ← F is equivalent to 15. We cannot subtract 15 from 2. Borrow "16" from the higher order unit to the left
       of the 2. This 16 plus 2 is 18. Write 3 since $18 − 15 = 3$.

  F

F3~~0~~02 −

51AF

$\overline{}$

  53 ← The 0 next to the rightmost 2 became an F (the basis minus 1). A is equivalent to 10. Write 5 since $15 − 10 = 5$.

     F
F3θ02 −

51AF
___

  E53 ← The next 0 also became an F (the basis minus 1). Since $15 - 1 = 14$, write E since its equivalent is decimal 14.

    2
F3002 −

5 1 AF
___

  DE5 3 ← The 3 "paid" the "16" that was "borrowed" by the rightmost 2. The 3 decreases to a 2. Since we cannot
                subtract 5 from 2, we need to "borrow" 16 from the higher unit to the left. This 16 plus 2 is 18. Write D since
                $18 - 5 = 13$. D in hexadecimal is 13.

    E
$\cancel{F}$3002 −

51AF
___

  EDE53 ← The next F "paid" the "16" that was "borrowed" by the 2 to its right. The F became an E. Since $14 - 0 = 14$,
                write E since in hexadecimal E is 14.

19.4 REPRESENTING NONNEGATIVE INTEGERS IN A COMPUTER

Numerical data in a computer is written in basic "units" of storage made up of a fixed number of consecutive bits. The most commonly used units throughout the computer and communication industries are shown in Table 19-2. A number is represented in each of these units by setting the bits according to the binary representation of the number. By convention the bits in a *byte* are numbered right to left, beginning with zero. Thus, the rightmost bit is bit number 0 and the leftmost bit is number 7. The rightmost bit is called the least significant bit. Likewise, the leftmost bit is called the most significant bit. Higher units are numbered also from right to left. In general, the rightmost bit is labeled 0, the leftmost bit is labeled $(n - 1)$ where n is the number of bits available.

Since each bit may have one of two values, 0 or 1, there are 2^8 configurations in a byte. That is, it is possible to represent 256 different binary numbers. The range of these nonnegative numbers varies from 0 to 255. In general, given n bits there are 2^n different numbers that can be represented. The range of these nonnegative integers varies from 0 to $2^n - 1$.

Table 19-2.

Units	Number of bits
A byte	8 consecutive bits
A word	16 consecutive bits
A double word	32 consecutive bits

EXAMPLE 19.8. How many different binary numbers can be represented in a word or a double word? What is the range of these nonnegative values?

Since a word has 16 bits, $n = 16$. Therefore, there are 2^{16} or 65,536 different representations of binary numbers. The range of nonnegative numbers is from 0 to $2^{16} - 1$. That is, from 0 to 65,535. Since a double word has 32 bits, $n = 32$. Thus, it is possible to represent 2^{32} or 4,294,967,296 different binary numbers. The range of these values is from 0 to 4,294,967,295.

19.5 COMPUTER ADDITION

Addition of binary numbers in a computer is carried out with a fixed amount of storage. Therefore, when adding two n-bit numbers the result may have $n + 1$ bits. Since this number is too big to fit in n bits we say that an *overflow* condition has occurred. In the next sections we will consider a more general condition for detecting overflows.

EXAMPLE 19.9. Add the binary numbers 10010101 and 11001011 assuming that the numbers are represented in a byte. Does this addition result in an overflow condition?

$$10010101 \; +$$
$$\underline{11001011}$$

101100000 $\leftarrow$ This number requires 9 bits. An overflow condition has occurred.

The result of this addition has more than 8 bits. This number is too big to fit in a byte. Therefore, there is an overflow. We can verify this result by considering the decimal equivalent of the binary result. In this case $101100000_{(2} = 352_{(10}$. This value falls outside the range that can be represented with a byte.

19.6 REPRESENTING NEGATIVE NUMBERS IN A COMPUTER

So far we have considered nonnegative values. However, the operation of subtraction raises the possibility of having negative results. When representing negative numbers one of the bits is chosen as the *sign bit*. By convention, the leftmost bit (or most significant bit) is considered the sign bit (Fig. 19-1). This convention is illustrated below for a byte. A value of *0* in the sign bit indicates a positive number, whereas a value of *1* indicates a negative number. A similar convention is also followed for higher storage units including words and double words.

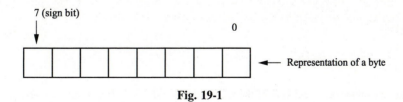

Fig. 19-1

Conventions

19.7 THE SIGN-MAGNITUDE

In this convention the leftmost bit is used exclusively to hold the sign bit. To represent negative numbers using n bits, the sign bit is set to 1 and the remaining $n - 1$ bits are used to represent the absolute value of the number (its magnitude). Positive numbers are represented in a similar way except that the sign bit is set to 0. The range of integer values that can be represented with this convention using n bits varies from $-2^{n-1} + 1$ to $2^{n-1} - 1$.

EXAMPLE 19.10. What is the sign-magnitude representation of decimal 75 and -75 if these values are represented in a byte?

Let us consider the representation of $+75$. The sign bit will be set to 0 since the number is positive. The remaining 7 bits will be used to represent the magnitude. The binary representation of $75_{(10} = 1001011_{(2}$. Therefore, the sign-magnitude representation of $+75$ is $01001011_{(2}$.

Since the magnitude of -75 is the same as that of $+75$, by changing the sign bit from 0 to 1, we will have that the representation of -75 is $11001011_{(2}$. The sign bit is set to 1 since the number is negative.

EXAMPLE 19.11. What is the sign magnitude representation of decimal 0? Assume you are working with bytes. The magnitude of 0 using 7 bits, in binary, is $0000000_{(2}$. However, notice that we can set the sign bit to either 0 or 1, therefore, representing 0 in two different forms: one as "positive" 0 and the other as "negative 0." This double representation of 0 has been one of the biggest drawbacks of this convention.

In sign-magnitude, arithmetic operations are carried out in binary following the usual convention. When adding numbers that have the same sign, we add their magnitudes and copy the common sign into the sign bit. If the signs are different, we first determine the number that has the largest magnitude and then subtract the other number from it. The sign of the result is the sign of the number that has the largest magnitude.

EXAMPLE 19.12. Using the sign-magnitude convention find the result of adding the decimal numbers 50 and 12 if these values are represented in a byte. The sign-magnitude representation of $50_{(10}$ is $00110010_{(2}$; the sign-magnitude representation of $12_{(10}$ is $00001100_{(2}$. Since both numbers are positive the result is positive. Adding their magnitudes we have that

```
0110010 +
0001100
───────
0111110
```

The decimal equivalent of $0111110_{(2}$ is $62_{(10}$. The sign-magnitude representation of the result is $00111110_{(2}$.

EXAMPLE 19.13. What is the sign-magnitude representation of the addition of the decimal numbers -55 and 27 if both values are represented in a byte?

First, let us find the sign-magnitude of both numbers. The representation of $-55_{(10}$ is $10110111_{(2}$. The representation of $27_{(10}$ is $00011011_{(2}$. Since the number with the largest magnitude is -55 the result is negative. That is, the sign-bit of the result is 1. The magnitude is calculated by subtracting the binary representation of 27 from that of 55. That is,

```
0110111 −
0011011
───────
0011100
```

Therefore, the sign-magnitude representation of the result of this addition is $10011100_{(2} = -28_{(10}$.

In the sign-magnitude convention an overflow occurs when the results fall outside the range from $-2^{n-1} + 1$ to $2^{n-1} - 1$.

19.8 ONE'S COMPLEMENT

This convention has the advantage of making the addition of two numbers with different signs the same as for two numbers with the same sign. Positive numbers are represented in the usual way. To represent a negative number in this convention, find the binary representation of the absolute value of the number and then complement all the bits including the sign bit. *To complement the bits of a binary number*, change all 0s to 1s and all 1s to 0s. Changing 0s to 1s and vice versa is equivalent to subtracting from 1 each of the individual bits of the number.

EXAMPLE 19.14.   What is the 1's complement representation of decimal 15 and -15 if these values are represented in a word?

The binary representation of $15_{(10}$ is $0000000000001111_{(2}$.   To represent -15, change 0s to 1s and 1s to 0s.   The resulting number is $1111111111110000_{(2}$.   Notice that the sign bit changed from 0 (positive) to 1 (negative).

EXAMPLE 19.15.   What is the decimal equivalent of the 1's complement number $1111111111000011_{(2}$?

First notice that the number is negative since its sign bit is 1.   The absolute value of this number is found by complementing all its bits.   This results in $0000000000111100_{(2} = 60_{(10}$.   Therefore, the original sequence of bits is the 1's complement representation of decimal -60.

Arithmetic operations in 1's complement are carried out in the usual way, without regard to the sign bits, which are treated as any other bit.   However, when performing an addition, if there is a carry out of the leftmost position, this carry is added to the rightmost bit.   This is known as an *end-around carry*.

An overflow condition will occur if the result of an operation falls outside the range $-2^{n-1} + 1$ to $+2^{n-1} - 1$, where n is the number of bits available to represent the number.   This condition will be detected when operands of the same sign produce a result with opposite sign.   An overflow will occur if either of the following conditions is satisfied: (1) There is a carry into the sign bit and no carry out of the sign bit.   (2) There is no carry into the sign bit and there is a carry out of the sign bit.

EXAMPLE 19.16.   Add the following 1's complement numbers 00110010 and 11100001?   Is there an overflow?

Adding the binary numbers in the usual manner and remembering to add the end-around carry to the rightmost bit, if there is one, we have that:

```
    11        ← carries
  00110010 +
  11100001
  ─────────
1 00010011 +
        1 ← end-around carry
  ─────────
  00010100 ← final result.
```

There is no overflow since there was a carry into the sign bit and a carry out of the sign bit.   To verify this result, let us find first the decimal equivalent of the given numbers.   The decimal equivalent of 00110010 is $50_{(10}$, and the decimal equivalent of 11100001 is $-30_{(10}$.   The decimal equivalent of the result 00010100 is $+20_{(10}$ as it should be.

One's complement also suffers from the "negative and positive" zero malady.   However, because of the end-around carry, the "negative zero" behaves as the "positive" zero.   When performing computations with a computer, the hardware needs to check for both representations of zero.   This extra check for zero and the required sum whenever there is an end-around carry has made the one's complement convention an unpopular choice for hardware manufacturers.

19.9   TWO'S COMPLEMENT

The most popular convention for representing negative numbers is the *2's complement*.   This variation of the 1's complement method corrects the problems of the negative zero and the end-around carry.   The representation of positive numbers in this new convention is similar to that of the 1's complement. Negative numbers are represented using a two-step process: First, complement all bits of the representation of the absolute value of the number.   Second, add 1 to the binary configuration of the number obtained in the previous step.

EXAMPLE 19.17. What is the 2's complement representation of $-45_{(10}$ if the value is represented in a word?

The binary representation of decimal 45 is $0000000000101101_{(2}$.

Complementing all bits of this number we obtain: 1111111111010010.

Adding 1 to this configuration of bits we have

$$
\begin{array}{r}
1111111111010010 \ + \\
1 \\
\hline
1111111111010011
\end{array}
$$

Therefore, the 2's complement of decimal -45 is $1111111111010011_{(2}$.

There is an alternate algorithm to obtain the 2's complement representation of a integer value. Scan the binary representation of the number from right to left until the first 1-bit is encountered. Leave this bit and all the 0-bits to its right, if any, unchanged. Complement all bits to the left of this first 1-bit.

EXAMPLE 19.18. Find the 2's complement of 00010100 using the alternate algorithm.

Scanning the number from right to left we find that the first 1-bit occupies the bit position number 2. Leaving this 1-bit and all the 0-bits to its right unchanged we have

00010**100**

     ↑ First 1-bit. This 1-bit and all 0-bits to its right are left unchanged.

Complementing all bits to the left of the first 1-bit, we have that

11101100 ← Complement all bits to the left of the first 1-bit.

Therefore, the 2's complement representation of 00010100 is 11101100.

The range of numbers that can be represented in two's complement with n bits is -2^{n-1} to $2^{n-1} - 1$. Notice that the range of negative numbers is one unit larger than the range of positive values. Overflows are detected using the similar convention to that of the 1's complement notation.

To add 2's complement integers, follow the usual rules for adding binary numbers and ignore the carry out of the leftmost bit if there is one.

EXAMPLE 19.19. What is the result of adding 01100010 and 00110001 as 2's complement numbers if these values are represented in a byte? Does the result overflow?

Adding the two binary numbers in the usual way we obtain

$$
\begin{array}{l}
\mathbf{11} \qquad\qquad \leftarrow \text{carries} \\
01100010 \ + \\
00110001 \\
\hline
10010011 \ \leftarrow \text{The result is negative since its sign is 1.}
\end{array}
$$

Notice that an overflow has occurred because the operands are both positive but the result is negative. Another way to check that an overflow has occurred is by noticing that there is a carry into the leftmost bit but there is no carry out of it.

EXAMPLE 19.20. What is the result of adding 00100001 and 11101011 as 2's complement numbers if these values are represented in a byte? Does an overflow occur?

Adding the two binary numbers in the usual way we have that

11   11   ← carries
00100001 +
11101011

1 00001100 ← The end-around carry is ignored.

In this case there is an end-around carry which is ignored.   There is no overflow since there is a carry into the sign bit and a carry out of the sign bit.

We can verify the result by noticing that $00100001_{(2} = 33_{(10}$ and that $11101011_{(2} = -21_{(10}$.   The sum is $00001100_{(2} = 12_{(10}$ as it should be.

Subtraction of 2's complement numbers can be reduced to a simpler integer addition as illustrated in the following example.

EXAMPLE 19.21.   What is the result of subtracting 00010110 from 00111100 if both numbers are represented in 2's complement notation?

Subtracting 00010110 from 00111100 is equivalent to adding the negative of 00010110 to 00111100.   Since the 2's complement of 00010110 is 11101010 we have that

00111100 +
11101010

00100110 ← Remember to ignore the carry out of the sign bit.

An alternative way to subtract 2's complement numbers is the following: Subtract one number from the other as if both were unsigned binary numbers.   If a borrow is needed in the leftmost place, borrow as if there were another bit to the left.   This is illustrated in the following example.

EXAMPLE 19.22.   Subtract 00111001 from 00100011.   Assume that both numbers are in 2's complement representation.

00100011 −
00111001

0 ← Write 0 since 1 minus 1 is 0.

00100011 −
00111001

10 ← Write 1 since 1 minus 0 is 1.

00100011 −
00111001

010 ← Write 0 since 0 minus 0 is 0.

10
00100011 −
00111001

1010 ← We cannot subtract 1 from 0. Borrow "2" from higher unit to left. Write 1 since $2 - 1 = 1$.

1
00100011 −
00111001

01010 ← The 0 became 1 (the basis minus 1). Write 0 since 1 minus 1 is 0.

10
0
001̶00011 −

00111001
_____

101010 ← 1 paid the "2" that was borrowed. The 1 became 0. We cannot subtract 1 from 0. Borrow
            "2" from higher unit to the left. Write 1 since 2 minus 1 is 1.

1
0̶0̶100011 −

00111001
_____

1101010 ← The 0 to the left became 1 (the basis minus 1). Write 0 since 1 minus 1 is 0.

    1
1   0̶0100011 −

   00111001
   _____

11101010 ← The 0 to the left became 1 (the basis minus 1). Write 0 since 1 minus 1 is 0.

At this moment we can ask, "Who pays the last '2' that was borrowed?" The answer is simple; the 1 that is assumed to be at the leftmost position (shown here in bold italics).

The user can verify the answer using the method shown in Example 19-21. In fact, the 2's complements of 00111001 is 11000111. If we add this value to 00100011 we have that

   111 ← carries
   00100011 +

   11000111
   _____

   11101010

The result of this addition is identical to the one we obtained previously.

19.10 MULTIPLICATION AND DIVISION OF BINARY NUMBERS

Multiplication and division of binary numbers are carried out in the same way that both operations are carried out in decimal. A multiplication, like $A \times B$, is equivalent to the addition of A to itself B times. Therefore, we can multiply two numbers using successive additions. Likewise, a division such as A/B can be carried out by successive subtraction. Since the computers have special hardware to do additions and subtractions very rapidly, any multiplication or division is carried out as an addition or subtraction, respectively.

Solved Problems

19.1.   What is the result of the following binary addition?   Assume that no overflow can occur.

          101011 +

          011010
          _____

          1 ← 0 + 1 = 1.

101011 +

011010
_____

  01 ← 1 + 1 = 0 and carry 1. Thinking in "decimal" notice that $2 = 1 \times 2 + 0$. Therefore, write 0
        and carry 1.

   1   ← carry

101011 +

011010
_____

  101 ← 1 (the carry) plus 0 is 1. This 1 plus 0 is 1. Therefore, write 1.

101011 +

011010
_____

  0101 ← 1 + 1 = 0 and carry 1. Thinking in "decimal" notice that $2 = 1 \times 2 + 0$. Therefore, write 0
          and carry 1.

  1

101011 +

011010
_____

  00101 ← 1 (the carry) plus 0 is 1. This 1 plus 1 is 2. Therefore, write 0 and carry 1.

1

101011 +

011010
_____

1000101 ← 1 (the carry) plus 1 is 2. This 2 plus 0 is 2. Therefore, write 0 and carry 1.

19.2.   Subtract 100101 from 10010110 in binary.

        "10"

10010110 −

  100101
_____

       1 ← We cannot subtract 1 from 0. Borrow "2" from the higher unit to the left. Since 2−1=1, write 1.

       0

10010110 −

  100101
_____

      01 ← 1 "paid" the "2" that the 0 to its left borrowed. 1 decreased to 0. Since $0 - 0 = 0$, write 0.

10010110 −

  100101
_____

     001 ← Since $1 - 1 = 0$, write 0.

10010110 −

  100101
_____

    0001 ← Since $0 - 0 = 0$, write 0.

10010110 −

  100101
_____

   10001 ← Since $1 - 0 = 1$, write 1.

"10"
100̶10110 −
  100101
─────────────
  110001 ← We cannot subtract 1 from 0. Borrow "2" from the higher unit to the left.   Since
             $2 - 1 = 1$, write 1.

   1
10̶010110 −
  100101
─────────────
 1110001 ← The 0 became 1 (the basis minus 1). Write 1 since $1 - 0 = 0$.

  0
1̶0̶010110 −
  100101
─────────────
01110001 ← 1 "paid" the "2" that the 0 had borrowed. 1 decreased to 0. Since $0 - 0 = 0$, write 0.

19.3.   What is the result of subtracting 1 from 10000000 in binary?

       "10" ← carry
1000000̶0 −
       1
─────────────
       1 ← We cannot subtract 1 from 0. Borrow "2" from the higher unit to the left.   Since
             $2 - 1 = 1$, write 1.

      1 ← carry
100000̶0̶0 −
       1
─────────────
      11 ← The next zero became 1 (the basis minus 1). Write 1 since $1 - 0 = 1$.

      1 ← carry
10000̶000 −
       1
─────────────
     111 ← The next zero also became 1 (the basis minus 1). Write 1 since $1 - 0 = 1$.

     1 ← carry
1000̶0000 −
       1
─────────────
    1111 ← The next zero also became 1 (the basis minus 1). Write 1 since $1 - 0 = 1$.

    1 ← carry
100̶00000 −
       1
─────────────
   11111 ← The next zero also became 1 (the basis minus 1). Write 1 since $1 - 0 = 1$.

    1 ← carry
10̶000000 −
       1
─────────────
  111111 ← The next zero also became 1 (the basis minus 1). Write 1 since $1 - 0 = 1$.

1   ← carry

10000000 −

　　　　　1
———————————

1111111 ← The next zero also became 1 (the basis minus 1). Write 1 since $1 - 0 = 1$.

0   ← carry

10000000 −

　　　　　1
———————————

01111111 ← 1 "paid" the "2" that the 0 had borrowed. 1 decreased to 0. Since $0 - 0 = 0$, write 0.

19.4. What is the result of the following addition in binary?   Assume that no overflow can occur.

10101111 +

11111010
——————————

　　　　　1 ← Write 1 since $1 + 0 = 1$.

10101111 +

11111010
——————————

　　　　　01 ← Write 0 since $1 + 1 = 0$ and carry 1. In decimal, $2 = 1 \times 2 + 0$. Thus, write 0 and carry 1.

　　　1   ← carry

10101111 +

11111010
——————————

　　　　001 ← 1 (the carry) plus 1 is 2. This 2 plus 0 is 2. Therefore, write 0 and carry 1.

　　　1   ← carry

10101111 +

11111010
——————————

　　　1001 ← In "decimal," 1 (the carry) plus 1 is 2. This 2 plus 1 is 3. Since $3 = 1 \times 2 + 1$, write 1 and
　　　　　　　carry 1. In "binary," 1 (the carry) plus 1 is 10. Write 0 and carry 1. The 0 plus 1 is 1. Write
　　　　　　　1 and remember the previous carry.

　　　1   ← carry

10101111 +

11111010
——————————

　　　01001 ← 1 (the carry) plus 0 is 1. This 1 plus 1 is 2. Therefore, write 0 and carry 1.

　　　1   ← carry

10101111 +

11111010
——————————

　　101001 ← In "decimal," 1 (the carry) plus 1 is 2. This 2 plus 1 is 3. Since $3 = 1 \times 2 + 1$, write 1 and
　　　　　　　carry 1. In "binary," 1 (the carry) plus 1 is 10. Write 0 and carry 1. The 0 plus 1 is 1. Write 1
　　　　　　　and remember the previous carry.

　1   ← carry

10101111 +

11111010
——————————

0101001 ← 1 (the carry) plus 0 is 1. This 1 plus 1 is 2. Therefore, write 0 and carry 1 since $2 = 1 \times 2 + 0$.

```
1           ← carry
10101111 +
111111010
```

110101001 ← In "decimal," 1 (the carry) plus 1 is 2. This 2 plus 1 is 3. Since $3 = 1 \times 2 + 1$, write 1 and
carry 1. In "binary," 1 (the carry) plus 1 is 10. Write 0 and carry 1. This 0 plus 1 is 1.
Write 1 and remember the previous carry.

19.5. What is the result of adding the following hexadecimal numbers?   Assume that no overflow
can occur.

```
A1BF3 +
ECD1B
```

E ← B is decimal 11. 3 plus 11 is 14. Write E since E is decimal 14.

```
A1BF3 +
ECD1B
```

0E ← F is decimal 15. 15 plus 1 is 16. Since $16 = 1 \times 16 + 0$, write 0 and carry 1.

```
   1        ← carry
A1BF3 +
ECD1B
```

9 0E ← 1 (the carry) plus B (=11) is 12. This 12 plus D (=13) is 25. Since $25 = 1 \times 16 + 9$, write 9 and
carry 1.

```
   1       ← carry
A1BF3 +
ECD1B
```

E 9 0E ← 1 (the carry) plus 1 is 2. This 2 plus C (=12) is 14. Write E since E is decimal 14.

```
A1BF3 +
ECD1B
```

18 E90E ← A (=10) plus E (=14) is 24. Since $24 = 1 \times 16 + 8$, write 8 and carry 1.

19.6. What is the result of the following hexadecimal subtraction?

```
F300 5 −
   1AD
```

8 ← We cannot subtract D (=13) from 5. Borrow "16," from higher unit to the left. 16 plus 5 is 21. This
21 minus D (=13) is 8. Write 8.

```
   F
F300 5 −
   1AD
```

5 8 ← The 0 became an F (the basis minus 1). F (=15) minus A (=10) is 5. Therefore, write 5.

```
  F
F300 5 −
  1 AD
```

E5 8 ← The next 0 became also an F (the basis minus 1). F (=15) minus 1 is 14.
Write E since E is decimal 14.

      2
F**3**00 5 −

  1AD
_____

2E5 8 ← The 3 "paid" the "16" that was borrowed by the rightmost 5. 3 decreases to 2. Write 2 since
        $2 - 0 = 2$.

F300 5 −

  1AD
_____

F2E5 8 ← Write F since F (=15) minus 0 is 15. Hexadecimal F is decimal 15.

19.7. What is the result of the following hexadecimal addition? Assume that no overflow occur.

  ABC +

  DAB
_____

    7 ← C (=12) plus B (=11) is 23. Since $23 = \mathbf{1} \times 16 + \mathbf{7}$, write 7 and carry 1.

  1   ← carry
  ABC +

  DAB
_____

    67 ← 1 (the carry) plus B (=11) is 12. This 12 plus A (=10) is 22. Since $22 = \mathbf{1} \times 16 + \mathbf{6}$, write 6 and
        carry 1.

  1   ← carry
  ABC +

  DAB
_____

18 6 7 ← 1 (the carry) plus A (=10) is 11. This 11 plus D (=13) is 24. Since $24 = \mathbf{1} \times 16 + \mathbf{8}$, write 8 and
        carry 1.

19.8. What is the result of the following hexadecimal subtraction?

  FFEA −

  EEFF
_____

    B ← Since F (=15) cannot be subtracted from A (=10), borrow "16" from next higher unit. A (=10) plus
      16 is 26. Since 26 minus F (=15) is 11, write B.

    D
  FF**E**A −

  EEFF
_____

    EB ← The E "paid" the "16" borrowed by the A. The E (=14) decreases to D (=13). Since F (=15)
      cannot be subtracted from D borrow "16" from the next higher unit. D (=13) plus $16 = 29$. Since
      29 minus F (=15) is 14, write E.

    E
  F**F**EA −

  EEFF
_____

  0EB ← The F "paid" the "16" borrowed by the D. The F (=15) decreases to E (=14). Since E (=14) minus
      E (=14) is 0, write 0.

FFEA −
EEFF
──────
10EB ← F (=15) minus E (=14) is 1. Therefore, write 1.

19.9. What is the sign-magnitude representation of decimal 77? Use a byte to represent the number.

The sign bit is 0 since the number is positive. The binary representation of 77 is $1001101_{(2}$. Therefore, the sign magnitude is 01001101.

19.10. What is the decimal equivalent of the sign-magnitude binary sequence 10011101?

The number is negative since its sign bit is 1. The magnitude is given by the decimal equivalent of the remaining bits. In this case, $0011101_{(2} = 29_{(10}$. Therefore, the decimal equivalent of the original binary sequence is decimal −29.

19.11. What is the sign magnitude representation of −45? Use a byte to represent the number.

Since the number is negative, the sign bit is 1. The magnitude is given by the binary representation of 45. Since $45_{(10} = 101101_{(2}$, the sign-magnitude of −45 is $1101101_{(2}$.

19.12. What is the decimal equivalent of the sign-magnitude representation given by the binary sequence 01011101?

The number is positive since the sign bit is 0. The magnitude is given by the decimal equivalent of the remaining bits. In this case, $1011101_{(2} = 93_{(10}$. Therefore, the decimal equivalent of the original binary sequence is +93.

19.13. What is the result of adding the sign-magnitude binary numbers 00101011 and 00010001?

Since both numbers are positive, the result is positive. Therefore, the sign bit is 0. To find the magnitude of the result, add the magnitude of the given numbers. That is, add

0101011 +
0010001
──────
         0 ← Write 0 and carry 1.

        1 ← carry
0101011 +
0010001
──────
       00 ← Write 0 and carry 1.

      1   ← carry
0101011 +
0010001
──────
     100 ← Write 1 since 1 + 0 = 1.

0101011 +
0010001

1100 ← Write 1 since 0 plus 1 is 1.

0101011 +
0010001

11100 ← Write 1 since 0 plus 1 is 1.

0101011 +
0010001

111100 ← Write 1 since 0 plus 1 is 1.

0101011 +
0010001

0111100 ← Write 0 since 0 plus 0 is 0.

We can verify the correctness of the result by noting that the decimal equivalents of the operands are $00101011_{(2} = +43_{(10}$ and $00010001_{(2} = +17_{(10}$. Therefore, the result is $00111100_{(2} = +60_{(10}$ as it should be.

19.14. Add the sign-magnitude binary numbers 11001000 and 00110000.

Since the numbers have opposite signs, the sign of the addition is the sign of the number with the largest magnitude. Therefore, it is necessary to determine which of the two numbers has the largest magnitude.

The number 11001000 has a negative sign. Its magnitude is $1001000_{(2} = 72_{(10}$.

The number 00110000 has a positive sign. Its magnitude is $0110000_{(2} = 48_{(10}$.

Since the number with the largest magnitude is -72, the sign of the addition will be negative. The magnitude is calculated by subtracting the absolute value of the smaller number from the absolute value of the largest. That is,

1001000 −
0110000
------
0 ← Write 0 since 0 minus 0 is 0.

1001000 −
0110000
------
00 ← Write 0 since 0 minus 0 is 0.

1001000 −
0110000
------
000 ← Write 0 since 0 minus 0 is 0.

1001000 −
0110000

1000 ← Write 1 since 1 minus 0 is 1.

"10"
1001000 −
0110000

11000 ← 1 cannot be subtracted from 0. Borrow "2" from higher unit to the left. Write 1 since $2 - 1 = 1$.

1
1̶001000 $-$
0̶110000
_____
011000 $\leftarrow$ The next 0 became a 1 (the basis minus 1). Write 1 since 1 minus 0 is 1.

0
1̶001000 $-$
0̶110000
_____
0011000 $\leftarrow$ 1 "paid" the 2 that was borrowed before. It decreases to 0. Write 0 since 0 minus 0 is 0.

We can verify the correctness of this result by noticing that $0011000_{(2} = 24_{(10}$. Thus, the result of this addition, expressed in sign-magnitude, is $10011000_{(2}$.

19.15. What is the 1's complement representation of decimal -27? Use a byte to represent the number.

Since the number is negative, we need to find the binary representation of $+27$. In this case we obtain that $27_{(10} = 00011011_{(2}$. To find the 1's complement of this number change all 0s to 1s and vice versa. The result is $11100100_{(2}$.

19.16. What is the decimal equivalent of the 1's complement sequence 10101010?

The number is negative since its sign bit is 1. Complementing all of its bits we obtain 01010101. The decimal equivalent of this binary sequence is the absolute value of the number. In this case, $01010101_{(2} = 85_{(10}$. Therefore, the original 1's complement binary sequence is equivalent to $-85_{(10}$.

19.17. Add the binary 1's complement numbers shown below. Is there an overflow?

11111100 $+$
00000110
_____
        0 $\leftarrow$ Write 0 since $0 + 0 = 0$.

11111100 $+$
00000110
_____
      10 $\leftarrow$ Write 1 since $1 + 0 = 1$.

11111100 $+$
00000110
_____
    010 $\leftarrow$ In "decimal" we have that $2 = \mathbf{1} \times 2 + \mathbf{0}$. Thus, write 0 and carry 1.

       1 $\leftarrow$ carry
11111100 $+$
00000110
_____
   0010 $\leftarrow$ Working in "decimal," 1 (the carry) plus 1 is 2. This 2 plus 0 is 2. Since $2 = \mathbf{1} \times 2 + \mathbf{0}$, write 0 and carry 1.

$$1 \quad \leftarrow \text{carry}$$
$$11111100 \; +$$
$$00000110$$

00010 $\leftarrow$ Working in "decimal" 1 (the carry) plus 1 is 2. This 2 plus 0 is 2. Since $2 = 1 \times 2 + 0$, write 0 and carry 1.

$$1 \quad \leftarrow \text{carry}$$
$$11111100 \; +$$
$$00000110$$

000010 $\leftarrow$ Working in "decimal" 1 (the carry) plus 1 is 2. This 2 plus 0 is 2. Since $2 = 1 \times 2 + 0$, write 0 and carry 1.

$$1 \quad \leftarrow \text{carry}$$
$$11111100 \; +$$
$$00000110$$

0000010 $\leftarrow$ Working in "decimal" 1 (the carry) plus 1 is 2. This 2 plus 0 is 2. Since $2 = 1 \times 2 + 0$, write 0 and carry 1.

$$1 \quad \leftarrow \text{carry}$$
$$11111100 \; +$$
$$00000110$$

1 00000100 + $\leftarrow$ Write 0 and add the end-around carry to the rightmost bit.

1 $\leftarrow$ end-around carry.

00000011 $\leftarrow$ Final result.

We can verify the correctness of this operation by noticing that $11111100_{(2} = -3_{(10}$ and $00000110_{(2} = +6_{(10}$. The result is $00000011_{(2} = +3_{(10}$ as it should be.

There is no overflow since there is a carry into the sign bit and a carry out of the sign bit.

19.18. What is the decimal equivalent of the 2's complement binary representation 10100101?

Since the sign bit of the given number is 1, the number is negative. Following the 2's complement convention to determine the absolute value of the number, all we need to do is complement all its bits and then add 1 to the resulting number. Complementing all bits produces 01011010. Adding 1 to this number we have

$$01011010 \; +$$
$$1$$
$$01011011$$

Since $01011011_{(2} = 91_{(10}$. The decimal equivalent of the given number is -91.

19.19. What is the 2's complement representation of decimal 87 and -87? Assume that the value is represented in a byte.

The binary equivalent of $87_{(10}$ is $01010111_{(2}$. Since the number is positive, this representation is also its 2's complement representation.

The 2's complement representation of $-87_{(10}$ is found by complementing all bits of the binary equivalent of $87_{(10}$ and then adding 1 to it. This two-step process is shown below.

Complementing all bits of $01010111_{(2}$ produces $10101000_{(2}$.

Adding 1 to this number we obtain

$$
\begin{array}{r}
10101000\ + \\
1 \\
\hline
10101001
\end{array}
$$

Therefore the 2's complement representation of -87 is 10101001.

This result can be verified by noticing that $87_{(10}$ plus $-87_{(10} = 0$.   A similar result is obtained when we add their binary equivalents as indicated below.

$$
\begin{array}{ll}
-87_{(2} & 10101001\ + \\
+87_{(2} & 01010111 \\
\hline
& 1\ 00000000
\end{array}
$$

$1\ 00000000 \leftarrow$ Remember that in 2's complement the carry out of the sign bit is ignored.

$\uparrow$ ignore this carry.

19.20. What is the result of adding the following 2's complement binary numbers?   Does the result overflow?

$$
\begin{array}{r}
00100111\ + \\
11011011 \\
\hline
0
\end{array}
$$

$$
\begin{array}{r}
1 \\
00100111\ + \\
11011011 \\
\hline
10
\end{array}
$$

$$
\begin{array}{r}
1 \\
00100111\ + \\
11011011 \\
\hline
010
\end{array}
$$

$$
\begin{array}{r}
1 \\
00100111\ + \\
11011011 \\
\hline
0010
\end{array}
$$

$$
\begin{array}{r}
1 \\
00100111\ + \\
11011011 \\
\hline
00010
\end{array}
$$

$$
\begin{array}{r}
1 \\
00100111\ + \\
11011011 \\
\hline
000010
\end{array}
$$

$$
\begin{array}{r}
1 \\
00100111\ + \\
11011011 \\
\hline
0000010
\end{array}
$$

1

00100111 +

11011011

———————

100000010 ← Final result.

↑   Ignore this carry.

There is no overflow since there is a carry into the sign bit and a carry out of it.

That this result is correct can be verified by adding the decimal equivalent of the numbers and comparing the sum against the final result that we just obtained.

The decimal equivalent of $00100111_{(2} = 39_{(10}$.

The other operand, 11011011, is negative according to its sign bit. Therefore, to calculate its decimal equivalent we need to find its 2's complement. Changing all 1s to 0s and vice versa, we obtain 00100100. Adding 1 to this number we obtain

00100100 +

1

———————

00100101

In consequence, the decimal equivalent of $11011011_{(2}$ is $-37_{(10}$. Since $39_{(10} - 37_{(10} = 2_{(10}$ we confirm that our final result is correct because $00000010_{(2} = 2_{(10}$.

Supplementary Problems

The answers to this set of problems are at the end of this chapter.

19.21. Perform the following calculations in the binary number system. Assume that no overflow can occur.

(a) 10110111 +
     11011001
     ————————

(c) 10110101 +
     10110101
     ————————

(e) 01110001 −
     01000101
     ————————

(b) 11001100 +
     10010111
     ————————

(d) 01110011 −
     1010010
     ————————

(f) 01111110 −
     00110000
     ————————

19.22. Perform the following calculations in the given number system. Assume that no overflow can occur.

(a) A12B +
     3C54
     ————

(c) ED21 +
     25DF
     ————

(e) BE2A−
     9CFE
     ————

(b) 48A3 +
     C759
     ————

(d) F4CD −
     21DE
     ————

(f) FE25−
     3ACD
     ————

19.23. Represent the following decimal numbers in sign-magnitude. Use a byte to represent the numbers.

(a) 93   (b) −26   (c) 127   (d) −4

19.24. Represent the following decimal numbers in 1's complement representation. Use a word to represent the numbers.

(a) −133   (b) 157   (c) −168   (d) −555

19.25. Represent the following decimal numbers in 2's complement representation. Use a byte to represent the numbers.

    (*a*) -106 (*b*) -95 (*c*) -87 (*d*) -68 (*e*) -145

19.26. Perform the following calculations. Assume that the numbers are represented in sign-magnitude. If an overflow occurs explain why. Use a byte to represent the numbers.

    (*a*) 10110101 +   (*c*) 11010101 +   (*e*) 00111100+
         01111011   00110101   01010101

    (*b*) 01011010 +   (*d*) 10110011 +
         01010101   11011100

19.27. Perform the following calculations. Assume that the numbers are represented in 1's complement. If an overflow occurs explain why. Use a byte to represent the numbers.

    (*a*) 01100100 +   (*c*) 00100001 +   (*e*) 10110101 +
         00011100   11101010   01111011

    (*b*) 10011011 +   (*d*) 11011110 +
         11100011   11101010

19.28. Perform the following calculations. Assume that the numbers are represented in 2's complement. If an overflow occurs explain why. Use a byte to represent the numbers.

    (*a*) 01011100 +   (*c*) 01010101 +   (*e*) 00101000 +
         11010110   00110111   01001101

    (*b*) 10011010 +   (*d*) 01111111 +
         11000110   11001110

Answers to Supplementary Problems

19.21. (*a*) 110010000, (*b*) 101100011, (*c*) 101101010, (*d*) 00100001, (*e*) 00101100, (*f*) 01001110

19.22. (*a*) DD7F, (*b*) 10FFFC, (*c*) 11300, (*d*) D2EF, (*e*) 212C, (*f*) C358

19.23. (*a*) 01011101, (*b*) 10011010, (*c*) 01111111, (*d*) 10000100

19.24. (*a*) 1111111101111010, (*b*) 0000000010011101, (*c*) 1111111101010111, (*d*) 1111110111010100

19.25. (*a*) 10010110, (*b*) 10100001, (*c*) 10101001, (*d*) 10111100, (*e*) It cannot be represented since this number is outside the range of values that can be represented with a byte.

19.26. (*a*) 01000110. There is no overflow.

    (*b*) Overflow! The sum is 0101111 (using 7 bits). There is a carry out of the 7th bit. The number is too big to fit in 7 digits.

    (*c*) 10100000. There is no overflow.

(*d*) Overflow!   The sum is 0001111 (using 7 bits).   There is a carry out of the 7th bit.   The number is too big to fit in 7 digits.

(*e*) Overflow!   The sum is 0010001 (using 7 bits).   There is a carry out of the 7th bit.   The number is too big to fit in 7 digits.

19.27.   (*a*) Overflow!   The sum, 10000000, is negative whereas both operands are positive.   Notice that there is a carry into the sign bit but no carry out of the sign bit.

(*b*) Overflow!   The sum, 01111111, is positive whereas both operands are negative.   Notice that there is no carry into the sign bit but there is a carry out of the sign bit.

(*c*) The sum is 00001100.   No overflow occurs.

(*d*) The sum is 11001001.   No overflow occurs.

(*e*) The sum is 00110001.   No overflow occurs.

19.28.   (*a*) The result is 00110010.   No overflow occurs.

(*b*) Overflow!   The result, 01100000 (using 8 bits), is positive whereas both operands are negative.   Notice that there is carry out of the sign bit but no carry into it.

(*c*) Overflow!   The result is 10001100 (using 8 bits), is negative whereas both operands are positive.   Notice that there is a carry into the sign bit but no carry out of it.

(*d*) The result is 01001101.   No overflow occurs.

(*e*) The result is 01110101.   No overflow occurs.

Chapter 20

Counting Methods

20.1 FUNDAMENTAL COUNTING PRINCIPLE

Assume that a given task can be subdivided in N parts. If the first part can be done in n_1 manners, the second in n_2 manners, the third in n_3 manners, and so on through the Nth, which can be done in n_N manners, then the total number of possible results for completing the task is given by the product.

$$n_1 \times n_2 \times n_3 \times \cdots \times n_N$$

EXAMPLE 20.1. How many different two-digit numbers can we form with the digits 1 through 9?

We need to design numbers of the form **DD** where each D can be any number between 1 and 9. In this case, there are 9 different ways to select the first digit. The second digit can also be selected in 9 different ways. Therefore, there are $9 \times 9 = 81$ different ways to form two-digit numbers.

EXAMPLE 20.2. In computer terminology, a byte is a sequence of 8 consecutive bits or binary digits. Since a binary digit can only be either 0 or 1, how many possible 8-bit numbers can we form?

A byte is of the form **BBBBBBBB** where each B can only be either 0 or 1. Therefore, according to the Fundamental Counting Principle, there are: $2 \times 2 \times 2 \times 2 \times 2 \times 2 \times 2 \times 2 = 2^8$ or 256 different 8-bit bytes that can be formed.

20.2 FACTORIAL OF A NUMBER

Given any integer n greater than zero, the expression $n!$ (n factorial) is defined as follows:

$$n! = n \times (n - 1) \times (n - 2) \times (n - 3) \times \cdots \times 1$$

Notice that the factorial is formed by the product of all counting numbers from n down to 1. Mathematicians by definition assume that $0! = 1$.

EXAMPLE 20.3. Evaluate the expression 5!

We form 5! by calculating the product of all counting numbers from 5 down to 1. That is,

$$5! = 5 \times (5 - 1) \times (5 - 2) \times (5 - 3) \times (5 - 4) = 5 \times 4 \times 3 \times 2 \times 1 = 120$$

The factorial of a number can be calculated directly with a calculator if there is a key for doing so. However, this is not the case for most of the graphing calculators.

EXAMPLE 20.4. Calculate 8! using a graphing calculator.

Using an HP-38G you may proceed as follows:

8 {MATH}

Select the category <u>Prob</u>. You may have to use the directional keys to move to this category.

Select the option ! and press {ENTER}

The display shows 8!

{ENTER}

The display shows 40320.

Using a TI-82 proceed as follows:

8 {MATH}

Highlight the PRB category.   You may have to use the directional keys to move to PRB.

Use the directional keys to select the option ! and press ENTER.

The display shows 8!

{ENTER}

The display shows 40320.

20.3 PERMUTATIONS

Given m different objects, the number of different arrangements (or permutations) that can be formed by taking n ($n \leq m$) objects at a time, denoted by $P(m, n)$, is defined as

$$P(m, n) = n \times (n - 1) \times (n - 2) \times \cdots \times [n - (n - 1)]$$

Notice that in the product defined by $P(m, n)$ there are n decreasing factors beginning with m.

When $n = m$, $P(m, m) = m!$.   That is, the number of different arrangements that can be formed taking into account *all m* elements is $m!$.   Observe that the groups formed are different because (a) their elements are different or   (b) the ordering of their elements is different.

Another way of calculating $P(m, n)$ is given by the formula

$$P(m, n) = \frac{m!}{(m - n)!}$$

EXAMPLE 20.5.   Evaluate $P(7, 3)$ and interpret the result.

The product that defines $P(7, 3)$ has 3 decreasing factors beginning with 7. That is, $P(7, 3) = 7 \times 6 \times 5 = 210$.   This result indicates that, given 7 different objects, we can form 210 different groups (or arrangements) of 3 elements each.

We can verify this result using the formula

$$P(7, 3) = \frac{7!}{(7 - 3)!} = \frac{7!}{4!} = 210$$

EXAMPLE 20.6.   Evaluate $P(4, 4)$ and interpret the result.

According to the definition, $P(4, 4) = 4 \times 3 \times 2 \times 1 = 4! = 24$.   This result indicates that, given 4 different objects, we can form 24 different groups of 4 objects each.   The ordering of elements in each group is different.

EXAMPLE 20.7.   System administrators control access to their computer systems using passwords.   If passwords consist of a sequence of six different letters from the English alphabet, how many different passwords can be formed?

Since there are 26 letters in the English alphabet, there are $P(26, 6) = 26 \times 25 \times 24 \times 23 \times 22 \times 21 = 165,765,600$ different passwords.

Permutations can be calculated with the help of a calculator. The calculator makes use of the formula

$$P(m, n) = \frac{m!}{(m - n)!}$$

EXAMPLE 20.8. Calculate $P(5, 3)$ with the help of a calculator.

Some calculators have a special key to calculate the permutations directly. However, this is not the case for graphing calculators.

Using an HP-38G you may proceed as follows:

{MATH}

Use the directional keys {►▲▼◄} to move to the category <u>Prob</u>. Select the <u>PERM</u> option and press {ENTER}.

The display shows PERM(

Key in the numbers, separate them with a comma, and press ENTER. That is, key in

5{,}3{)}{ENTER}

The display shows 60.

To calculate P(5, 3) with a TI-82 you may proceed as shown below.

Enter the first number

5

{MATH}

Use the directional keys {►▲▼◄} to move to the category <u>PRB</u>. Select the <u>nPr</u> option and press ENTER.

The display shows 5 nPr

Enter the second number and press ENTER.

3 {ENTER}

The display shows 60.

20.4 ARRANGEMENTS WITH DUPLICATE ELEMENTS

If within a given collection of m objects, there are n_1 objects that are equal to one another, n_2 that are equal to one another and so on up to n_k, then the total number of different arrangements that can be formed using all m objects is defined as

$$\frac{m!}{n_1! \, n_2! \, n_3! \cdots n_n!}$$

EXAMPLE 20.9. If we select the word TENNESSEE, how many different passwords of 9 letters can be formed from this word?

Since there 9 letters in the word, $m = 9$.   The subgroups of repeated letters are:

$$n_1 = 4 \text{ (there are 4 Es)} \qquad n_2 = 2 \text{ (there are 2 Ns)} \qquad n_3 = 2 \text{ (there are 2 Ss)}$$

The total number of different passwords is

$$\frac{9!}{4! \times 2! \times 2!} = \frac{362,880}{24 \times 2 \times 2} = \frac{362,880}{96} = 3780$$

20.5   CIRCULAR PERMUTATIONS

If m objects are arranged in a circle, there are $(m - 1)!$ permutations of the objects around the circle.

EXAMPLE 20.10.   In some secured installations, electronic entry locks are designed in a "circular fashion." Each lock has a "fixed" programmable code that lasts only a week.   Users are given entry codes like ACDEB this week, CBDAE next week and so on.   Assume that the "keys" of the locks are labeled A through B.   Using keys A through D, how many different codes can be formed every week?

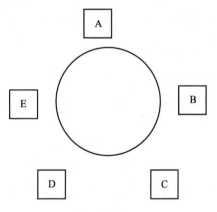

Fig. 20-1

Programming the "keys" in this case is the equivalent of "sitting" the labels at a round table.   One possible initial arrangement is shown above.   There are five initial possible arrangements.   Each initial arrangement can be changed by rotating the circle 1/5 of a revolution.   If we continue rotating the labels some of these "sitting positions" will repeat.   In this case, the number of distinct arrangements around a table is 1/5 of the total number of arrangements in a line.   Altogether, we have $(5 - 1)! = 4! = 24$ possible arrangements every week.

20.6   COMBINATIONS

Given m different objects, the number of subsets (or combinations) that can be formed by taking n ($n \leq m$) objects at a time, denoted by $C(m, n)$, is defined as

$$C(m, n) = \frac{m!}{n!(m - n)!}$$

When forming subsets of a set, the order of the elements within the sets is immaterial.

EXAMPLE 20.11.   Given a collection of 100 light bulbs, in how many ways can a sample of 10 bulbs be selected?

The number of samples that can be formed is given by

$$C(100, 10) = \frac{100!}{10!(100-10)!} = \frac{100!}{10 \times 90!}$$

EXAMPLE 20.12. Given the integer numbers 1 through 6, how many additions of three digits each can be formed?

Each group will have the format XYZ. If we form the sum of these digits we see that $X + Y + Z = Y + Z + X = Z + X + Y$. Notice that the order of the numbers does not alter the sum. Since the order of the elements within each group is immaterial, the total number of 3-digit additions is given by

$$C(6, 3) = \frac{6!}{(6-3)!} = \frac{6!}{3!} = 20$$

Therefore, we can form 20 different additions.

The number of combinations can also be found with the help of a calculator. However, most calculators do not have a special key to do the calculation directly. Example 20.13 shows how to calculate combinations using graphing calculators like the HP-38G or the TI-82.

EXAMPLE 20.13. Find $C(6, 3)$ using a graphing calculator.

Using an HP-38G you may proceed as follows:

(1) Find the COMB built-in application within the MATH menu.

{MATH}

Use the directional keys {▶▲▼◀} to move to the category <u>Prob</u>. Select the <u>COMB</u> option and press ENTER.

The display shows COMB(

(2) Enter the numbers separated by a comma and press ENTER. That is, key in

6{,}3{)}{ENTER}

The display shows 20.

Using a TI-82:

(1) Enter the first number.

6

(2) Find the built-in application <u>nCr</u> within the PRB menu.

{MATH}

Use the directional keys {▶▲▼◀} to move to the category <u>PRB</u>. Select the <u>nCr</u> option and press ENTER.

(3) Enter the second number.

3{ENTER}

The display shows 20.

Solved Problems

20.1. Telephones in the United States have the following format

<table>
<tr><td>**Area Code**</td><td>**Local Number**</td></tr>
<tr><td>(DDD)</td><td>(DDD – DDDD)</td></tr>
</table>

Knowing that local numbers cannot have 0 or 1 as their first two digits, what is the maximum number of local telephone numbers that can be formed within a single area code?

Using the Fundamental Counting Principle there are only 8 choices for the first two digits. For the other digits there are 10 choices. Therefore, there can only be $8 \times 8 \times 10 \times 10 \times 10 \times 10 \times 10 = 6,400,000$ different telephone numbers.

20.2. A computer monitor produces color by blending colors on palettes. If a computer monitor has four palettes and each palette has 4 colors, how many blended colors can be formed?

There are 4 palettes: P P P P. Since there are four choices for each palette, according to the Fundamental Counting Principle, the number of colors that can be formed is $4 \times 4 \times 4 \times 4 = 256$ colors.

20.3. Suppose that four pistons need to be installed in a new engine block. There is a production run of 100 pistons. If the new pistons to be installed in the engine block must come from this production run, how many different possible ways are there to install these pistons?

Assume that the pistons will be arranged as P P P P. There are 100 choices for the first piston; there are 99 choices for the second piston; there are 98 choices for the third piston and 97 choices for the fourth piston. Therefore, the total number of ways in which these pistons can be installed is $100 \times 99 \times 98 \times 97 = 94,109,400$.

20.4. Assume that you are trying to transmit secret codes using 3-letters words. If only 5 letters of the alphabet can be used and no letter can be repeated within the words, how many different secret codes can you possibly send? If words could be 4 letters long, how many letters can be sent?

Using 3 letter words we can only form

$$P(5, 3) = \frac{5!}{(5-3)!} = \frac{5!}{2!} = \frac{120}{2} = 60$$

With 4 letter words, we can form

$$P(5, 4) = \frac{5!}{(5-4)!} = \frac{5!}{1!} = 5! = 120$$

20.5. Assume that passwords in a computer system can be formed using only letters of the English alphabet. If passwords are limited to 8 characters, how many different passwords can be formed if the ordering of the letters is important?

We can form $P(26, 8)$ different passwords. The total number is given by

$$P(26, 8) = \frac{26!}{(26-8)!} = \frac{26!}{18!} = 62,990,928,000$$

20.6. To identify legitimate users, a system analyst has assigned IDs to all registered users. All IDs have the format EEEOO where E stands for an even digit and O for an odd digit. If 0 is not a valid option and no digit can be used more than once, how many different codes can be formed?

Entry codes have the format EEEOO. The total number of 3-digit groups that can be formed with the even digits {2, 4, 6, 8} is given by

$$P(4, 3) = \frac{4!}{(4-3)!} = \frac{24}{1!} = 24$$

With the odd digits {1, 3, 5, 7, 9}, the total number of 2-digit groups that can be formed is given by

$$P(5, 2) = \frac{5!}{(5-2)!} = \frac{5!}{3!} = 20$$

According to the Fundamental Counting Principle the total number of entry codes that can be formed is given by $20 \times 24 = 480$.

20.7. Assume that the users connect to a computer using terminals that are polled by the computer in a "circular" fashion, in how many ways can the computer "visit" these eight users?

We can think of these users as "sitting" at a round table. The computer can visit them in as many as $(8 - 1)! = 7! = 5040$ different ways.

20.8. How many different arrangements of 11 letters can be formed with the letters of the word Mississippi?

The word has 11 letters, thus $n = 11$. The letter i occurs 4 times, thus $n_1 = 4$. The letter s occurs 4 times also, thus, $n_2 = 4$. Since the letter p occurs 2 times, $n_3 = 2$. The total number of distinguishable arrangements that can be formed is

$$\frac{m!}{n_1 \times n_2 \times n_3} = \frac{11!}{4! \times 4! \times 2!} = \frac{39,916,800}{24 \times 24 \times 2} = 34,650$$

20.9. How many different even 4-digit numbers can be formed with the digits of the number 41,852?

The format of 4-digit numbers is DDDD. There are three even digits in the given number. Thus, the possible choices are: DDD2, DDD4 and DDD8. If we consider DDD2, there are $P(4, 3) = 24$ possibilities to fill the remaining digits DDD. Similar reasoning can be used with the other two cases. Therefore, the total number of 4-digit numbers that can be formed is $24 + 24 + 24 = 72$.

20.10. Assume that you are to compose messages of 4 given letters and 4 given digits. The messages can begin with either a letter or a digit; however, in any message letters and digits must alternate and all letters and digits must be used. How many different messages can be composed?

The general form of the groups is $L_1 D_1 L_2 D_2 L_3 D_3 L_4 D_4$ or $D_1 L_1 D_2 L_2 D_3 L_3 D_4 L_4$. Consider the first group, $L_1 D_1 L_2 D_2 L_3 D_3 L_4 D_4$. If we leave the letters "fixed," with the digits {D_1, D_2, D_3, D_4} we can form $P(4, 4) = 4! = 24$ different groups. If we leave the digits fixed, with the letters {L_1, L_2, L_3, L_4}, we can form $P(4, 4) = 4! = 24$ different groups. According with the Fundamental Counting Principle, we can have $P(4, 4) \times P(4, 4) = 4! \times 4! = 24 \times 24 = 576$ different groups. However, since the groups can begin with either a letter or a number, the actual total number of groups is twice that many; therefore, we can form $2 \times 576 = 1152$ different messages.

Supplementary Problems

The answers to this set of problems are at the end of this chapter.

20.11. A computer manufacturer offers customized computer systems. Customers can choose from the options shown below. How many different computer systems can a customer buy, assuming that the individual components are compatible with each other.

Color monitors models (screen size in inches): 100sx (13"), 200sx (15"), 300sx (17"), 400sx (21")

Disk drives sizes: 1.2 Gbytes, 2.1 Gbytes, 3.1 Gbytes, 4.3 Gbytes, 6.4 Gbytes, 9.0 Gbytes

Keyboards: Extended 105, Ergo 0906, Design 041, Extended II, Extended-Ergo 0786

Memory boards: 16 Mbytes, 24 Mbytes, 32 Mbytes, 48 Mbytes, 64 Mbytes

20.12. Assume that you are to generate computer password codes using 10 consonants and 5 vowels. How many different passwords can be generated if each password is made up of 3 consonants and 2 vowels?

20.13. Assume that you have to transmit a message made up of a sequence of 8 pieces of information. There are only three choices for each of the different pieces. If we assume that the order of the individual pieces is important, how many different messages can you send?

20.14. A chemist wants to mix 4 different coloring substances. Assuming that the substances are all mixed in the same proportion, in how many different ways can these colors be mixed?

20.15. If the number of keys of the "circular lock" of Example 20.10 is increased to 10. How many different codes can be formed?

20.16. Storage units in a computer are formed as sequences of binary digits (bits). Some of these units have 16 bits (sometimes called words) or 32 bits (sometimes called double words). How many different sequences of binary digits can be formed in a word? How many can be formed with a double word?

20.17. Assume that you want to assign unique IDs to a set of computer users. If you can only use four given vowels and four given consonants and all vowels and consonants need to be used in every ID, how many different IDs can be formed if the consonants and vowels must be alternated within the IDs?

20.18. Morse code is a system of dots, dashes and spaces frequently used to send messages. How many different messages can be sent with four dots and two dashes?

20.19. Some computer monitors can display any of 4096 different colors. If only 12 colors can be displayed at any one time, how many groups of 12 colors can be displayed?

20.20. How many different 11-letter words of can be formed with the letters of the word ABRACADABRA?

Answers to Supplementary Problems

20.11. 600 different systems

20.12. 1,728,000 different passwords

20.13. 6561 different messages

20.14. $C(4, 4) = 1$

20.15. $9! = 362,880$

20.16. 65,536 (word) or 4,294,967,296 (double word)

20.17. 1152

20.18. 15

20.19. $\dfrac{4096!}{12! \times 4084!}$

20.20. 83,160

Chapter 21

Probability and Odds

21.1 PROBABILITY AND SAMPLE SPACES

When studying probability we call any process for which the result cannot be predicted in advance an *experiment*. For example, rolling two fair die or tossing two coins into the air are experiments. Each repetition of an experiment is called a *trial*. The results of a trial are called *outcomes*. The set of all possible outcomes is called a *sample space*.

EXAMPLE 21.1. Assume that you roll a fair die twice in row and obtain 2 and 5. In this case, there are two experiments or trials since the die was rolled twice. There are two outcomes: 2 and 5. The sample space is {1, 2, 3, 4, 5, 6} since these are the possible outcomes of rolling a die.

An event is any subset of the possible outcomes of a trial. For example, when we roll a pair of dice, there are 36 possible outcomes (see Problem 21-7). An event is, for example, obtaining 5 on one die and 3 on the other.

If all outcomes are equally likely and distinct, the probability of an event is defined as

$$P(\text{event}) = \frac{\text{number of favorable outcomes}}{\text{total number of outcomes}}$$

Notice that this definition requires that all outcomes of the sample space be equally likely and distinct. In this case, we say that the sample space is *uniform*.

EXAMPLE 21.2. What is the probability of getting a 4 when we roll a fair die?

Since the die is fair, there are only 6 possible outcomes. There is only one number 4 on a die. The probability of the event of obtaining a 4, which we can indicate by P(4), can be calculated by the formula given above. Therefore, the probability of obtaining a 4 is given by

$$P(4) = \frac{1}{6}$$

EXAMPLE 21.3. What is the probability of obtaining an even number when we roll a fair die?

There are six possible outcomes. Only three outcomes produce even numbers: {2, 4, 6}.

The probability of P(even) = 3/6 = 1/2 = 0.5.

EXAMPLE 21.4. To control the quality of its packaging and shipping department, a manufacturer tracks a package that contains two very sensitive graphic memory cards. The table shown below gives the sample space. Consider that each row of the table is an outcome. If all outcomes are equally likely, what is the probability of both cards arriving undamaged?

Memory card 1	Memory card 2
Undamaged	Undamaged
Damaged	Damaged
Damaged	Undamaged
Undamaged	Damaged

The sample space contains four possible outcomes. There is only one outcome where both memory cards are undamaged. The probability is $1/4 = 0.25$.

This result indicates that only 25 out 100 shipments will arrive undamaged. It is hard to imagine a manufacturing company doing business with such a low packaging quality.

21.2 PROBABILITY OF SUCCESS AND FAILURE

Trials of an experiment can be classified as failures (or undesirable or unfavorable results) and successes (desirable or favorable results). If an event can succeed in s ways ($s \geq 0$) and fail in f ways ($f \geq 0$), then the probability of a success, called P(s), and the probability of a failure, called P(f), can also be defined as follows:

$$P(s) = \frac{s}{f+s} \qquad P(f) = \frac{f}{f+s}$$

An event that cannot fail has a probability of 1. An event that cannot succeed has a probability of 0. The probability of success, P(s), is always a number between 0 and 1. Likewise, the probability of failure, P(f), is also a number between 0 and 1. The sum of the probability of success and the probability of failure is always equal to 1. All these results can be expressed as follows:

$$0 \leq P(s) \leq 1$$
$$0 \leq P(f) \leq 1$$
$$P(s) + P(f) = 1$$

That the sum of the probability of success and the probability of failure is always equal to 1 can be easily proven by noticing that

$$P(s) + P(f) = \frac{s}{s+f} + \frac{f}{s+f} = \frac{s+f}{s+f} = 1$$

From this equality we can deduce that $P(f) = 1 - P(s)$. Since $P(s) + P(f) = 1$, P(s) and P(f) are called *complements*.

EXAMPLE 21.5. Assume that a box contains three different types of computer external drives. There are 3 drives of 1.44 MBytes each, 7 drives of 100 MBytes each, and 11 drives of 1 GBytes each. What is the probability that an external drive selected at random will be a 1 GBytes drive?

Let P(1 GBytes drive) be the probability of selecting a 1 GBytes drive. There are 11 ways of selecting a 1 GBytes drive since there are 11 of them. Since there are 7 drives of 100 MBytes and 3 drives of 1.44 MBytes, there are 10 or $7 + 3$ different ways of not selecting a 1 GBytes drive. Therefore, $s = 11$ and $f = 10$. According to the formula given above,

$$P(1 \text{ GBytes drive}) = \frac{s}{s+f} = \frac{11}{11+10} = \frac{11}{21}$$

21.3 ODDS

The odds of the successful (or desired outcome) of an event are defined as the ratio of the probability of its success to the probability of its failure. That is, Odds (event) $= P(s)/P(f)$. Odds are sometimes expressed as fractions. For example, in sporting events, we may hear that the odds of team A winning the championship game are 7 to 12. This may be expressed as $7/12$ or $7:12$.

EXAMPLE 21.6. The weather bureau forecasts that the probability of rain tomorrow is 65%. What are the odds that it will not rain tomorrow?

In this problem, the desired result is that it *does not* rain.

The probability of rain, P(rain), is 65%. That is,

$$P(\text{rain}) = \frac{65}{100} = \frac{13}{20}$$

Since the desired result is that it does not rain, we can consider $P(\text{rain}) = P(f) = 13/20$.

The probability that it does not rain, P(not rain) can be calculated as $1 - (13/20) = 7/20$.

We can consider $P(\text{not rain}) = P(s)$. Therefore, the odds that it does not rain tomorrow are

$$\text{Odds(not rain)} = \frac{P(s)}{P(f)} = \frac{\dfrac{7}{20}}{\dfrac{13}{20}} = \frac{7}{13}$$

This result can also be expressed as 7/13 or as 7:13.

When working with probabilities, results are sometimes expressed as fractions. You can use a calculator to express the results of ordinary arithmetic operations in this fashion. For instance, to express as a fraction the result of the previous example you may proceed as follows:

Using an HP-38G graphing calculator:

(1) Set the mode of the computer to fraction.

{SHIFT}{MODES}

(2) Use the directional keys to move to the option NUMBER FORMAT and select the option CHOOS. Highlight the option <u>Fraction</u> and press {ENTER}. Press HOME to return to the home screen.

(3) Enter the arithmetic expression.

1{−}{(}13{/}20{)}{ENTER}.   The display shows 7/20.

Using a TI-82 graphing calculator:

(1) Enter the arithmetic expression.

1{−}{(}13{÷}20{)} ← "Do not press ENTER here"

(2) Display the answer as a fraction.

Press {MATH}

Use the directional keys to move to the option <u>Frac</u> and press {ENTER}{ENTER}. The display shows 7/20.

EXAMPLE 21.7. If a die is rolled twice, what are the odds of getting at least one 3?

The possible outcomes are shown below as pairs. The first number of each pair indicates the result of the first throw; the second number indicates the result of the second throw.

(1, 1), (1, 2), (1, 3), (1, 4), (1, 5), (1, 6)	(2, 1), (2, 2), (2, 3), (2, 4), (2, 5), (2, 6)
(3, 1), (3, 2), (3, 3), (3, 4), (3, 5), (3, 6)	(4, 1), (4, 2), (4, 3), (4, 4), (4, 5), (4, 6)
(5, 1), (5, 2), (5, 3), (5, 4), (5, 5), (5, 6)	(6, 1), (6, 2), (6, 3), (6, 4), (6, 5), (6, 6)

There are 11 pairs that contain a 3.   Therefore, there are 11 favorable cases.

There are 25 unfavorable cases.   That is, there are 25 pairs that do not contain a 3.

Therefore, the odds of getting at least one 3 are 11/25.

If we know the probability of an event we can calculate the odds in favor of it.   Likewise, if we know the odds in favor of an event we can calculate its probability.

In fact, if we know, P(E), the probability of an event E, then odds in favor of event E are given by the formula

$$\text{Odds(E)} = \frac{P(E)}{1 - P(E)}$$

Similarly, if we know that the odds in favor of an event E are $a{:}b$, then the probability of an event E is given by the formula

$$P(E) = \frac{a}{a + b}$$

EXAMPLE 21.8.   If a fair die is rolled, what are the odds of getting a 5?

Since the die is fair, all outcomes are equally likely.   The probability of getting a 5 is 1/6.

The odds are

$$\text{Odds} = \frac{\left(\frac{1}{6}\right)}{1 - \left(\frac{1}{6}\right)} = \frac{1}{5}$$

21.4   PROBABILITY OF INDEPENDENT AND DEPENDENT EVENTS

Two events, E and F, are called *independent*, if the occurrence of one of them has no effect on the occurrence of the other.   For example, if we throw a fair die twice in a row, the probability of obtaining a 4 in the first row is 1/6.   The probability of obtaining a 5 in the second throw is also 1/6.   The results of the first throw does not affect the result of the second throw.   These two events are said to be independent.   That events E and F are independent will be indicated as "E and F."

If two events, E and F, are independent, the probability of both events occurring is the product of each individual probability.   That is,

$$P(E \text{ and } F) = P(E) \times P(F)$$

EXAMPLE 21.9.   A fair die is thrown twice, what is the probability of getting two 5s?

The probability of getting a 5 in the first throw is 1/6.   The outcome of the first throw does not affect the outcome of the second one.   The probability of getting a 5 in the second throw is also 1/6.   Therefore, the probability of getting two consecutive 5s is $(1/6) \times (1/6) = 1/36$.

EXAMPLE 21.10.   In a manufacturing plant of electronic equipment, boards currently under construction are placed on conveyor belts that take them to the different assembly stations.   Boards stay in each station for a few seconds and then move on.   In the final processing line there are two stations.   One station is busy 50% of time.   The next station is busy 60% of the time.   Both stations operate independently of each other.   Find the probability that a board will find both stations busy as it moves through this final line.

The probability of finding both lines busy is $0.5 \times 0.6 = 0.3$.

21.5 PROBABILITY OF EXCLUSIVE EVENTS

Two events, E and F, are said to be *mutually exclusive* if they cannot occur simultaneously. That is, when P(E and F) = 0.

If two events, E and F, are mutually exclusive, we will indicate this as "E or F." Given two mutually exclusive events E and F, the probability that either E or F occurs is defined as the sum of their probabilities. That is,

$$P(E \text{ or } F) = P(E) + P(F)$$

EXAMPLE 21.11. If a fair die is rolled once, what is the probability of getting a 5 or a 6?

Every time we roll a die, there is only one outcome. We cannot get both numbers simultaneously. The events are mutually exclusive. The probability of getting a 5 is 1/6. The probability of getting a 6 is 1/6. The probability of getting a 5 or a 6 is $(1/6) + (1/6) = 2/6 = 1/3$.

21.6 PROBABILITY OF INCLUSIVE EVENTS

Two events, E and F, are said to be *inclusive*, when one or the other or both can occur.

If two events, E and F, are inclusive, the probability that either E or F occurs is the sum of their probabilities minus the probability of both occurring. That is,

$$P(E \text{ or } F \text{ or both}) = P(E) + P(F) - P(E \text{ and } F)$$

Notice that in this definition, by subtracting the probability of both events occurring, we do not count some events twice.

EXAMPLE 21.12. If a fair die is rolled, what is the probability of getting an even number or a number greater than 3?

When a fair die is thrown, it is possible to get an even number; it is possible to get a number that is greater than 3, and it is also possible to get a number that is both, an even number and greater than 3. For example, consider the number 4. The probability of obtaining an even number is 1/2 (see Example 21.3). The probability of obtaining a number greater than 3 is 1/2 since there are three favorable outcomes, {4, 5, 6}, out of six possible outcomes. The probability of obtaining a number that is both even and greater than 3 is 1/3, since there are two possible favorable outcomes, {4, 6}, out of six possibilities. Therefore, the possibility of obtaining a number that is even or greater than 3, is given by

$$P(\text{number even or greater than 3 or both}) = \frac{1}{2} + \frac{1}{2} - \frac{1}{3} = \frac{2}{3}$$

21.7 CONDITIONAL PROBABILITY

Given two events, E and F, if the probability of event F is affected by knowledge of the occurrence of event E, then the probability of event F is said to be *conditional* to that of E. In general, the condition that E occurs reduces the entire sample space to the sample space of E; this space is called the *reduced sample space*.

The conditional probability that event F occurs if event E occurs, denoted by P(F/E), is defined as

$$P(F/E) = \frac{P(E \text{ and } F)}{P(E)} \qquad \text{where} \qquad P(E) \neq 0$$

EXAMPLE 21.13.   In a set of 160 personal computers, there are 80 computers with a hard disk of 1.2 GBytes, 60 computers have hard disks of 2.0 GBytes, and 20 computers have both types of hard disks.   If a computer is selected at random, what is the probability that it has a hard disk of 1.2 GBytes?   What is the probability that the computer has both types of hard disks given that it already has one hard disk of 2.0 GBytes?

The probability of having a hard drive of 1.2 GBytes is

$$P(1 \text{ hard drive of } 2 \text{ GBytes}) = 80/120 = 2/3$$

The fact that the computer has already a 2.0 GBytes hard disk reduces the sample space to 60.   Of this group there are 20 computers that have both types of hard disks; therefore the probability of selecting a computer with both drives is $20/60 = 1/3$.

Solved Problems

21.1.   In a given batch of electronic boards that need to be upgraded or repaired there are 10 boards that are in good condition and 4 boards that are defective.   If a board is selected at random, find the odds of picking up one that is defective.

There are 10 outcomes that are favorable since there are 10 boards in good condition.   There are 4 outcomes that are unfavorable since there are 4 defective boards.   The required odds are

$$P(\text{success}) = \frac{10}{10+4} = \frac{10}{14} = \frac{5}{7}$$

21.2.   If a die is thrown twice, what are the odds of getting two 3s?

Consider the sample space of Example 21.7.   Since there is only one pair where both numbers are 3, there is only one favorable outcome.   The number of unfavorable outcomes is 35.   Therefore, the odds are 1/35.

21.3.   Assume that a new electronic calculator has a probability of failure of 0.016.   Find the odds of failing.

Let us call F the event "the machine fails." The odds in favor of failing are

$$\frac{P(F)}{1 - P(F)} = \frac{0.016}{1 - 0.016} = \frac{2}{123}$$

21.4.   A batch of 15 memory boards contains 3 boards that are defective.   If 2 boards are selected at random, what is the probability that at least one of them is not defective?

Knowing that the sum of the probability of the success of an event plus the probability of failure of the same event is equal to 1, we can solve this problem using "complements." In this case, let us begin by noticing that there are 3 defective boards.   Therefore, the complement of selecting at least a board that is not defective is selecting 2 boards that are defective.   Given that there are 3 defective boards, there are $C(3, 2)$ different ways of selecting two defective boards.   There are $C(15, 2)$ different ways of selecting any two boards.   The probability of selecting two boards that are defective is given by

$$P(2 \text{ boards are defective}) = \frac{C(3, 2)}{C(15, 2)} = \frac{\dfrac{3!}{2!(3-2)!}}{\dfrac{15!}{2!(15-2)!}} = \frac{3}{105} = \frac{1}{35}$$

To find the complement of finding at least a board that is not defective, let us calculate the complement of the probability that we just got.   That is,

$$P(1 \text{ board not being defective}) = 1 - \frac{1}{35} = \frac{35-1}{35} = \frac{34}{35}$$

21.5. Based on past experience, a computer technician has constructed a table of probabilities of computer failures.   Assume that the events indicated in the table are not inclusive.

Type of failure	Probability
High memory failure	0.07
Low memory failure	0.10
Hard disk drive failure	0.20
Zip drive failure	0.23
Tape drive failure	0.22
Removable drive failure	0.18

(*a*) What is the probability that the next computer has a memory failure?

(*b*) What is the probability that it has a drive failure?

The probability that is a memory failure is $0.07 + 0.10 = 0.17$.

The probability that is a drive failure is $0.20 + 0.23 + 0.22 + 0.18 = 0.83$.

21.6. Assume that among the computers brought to a certain shop for repairs, the probability of a computer having a hard disk crash is 0.21.   If only 4 computers were brought into a shop for repair, what is the probability that the four computers had a hard disk crash?

We can safely assume that the events are independent.   The failure of one computer in no way depends on the failure of another.   The probability of all four computers having a hard disk crash is the product of their individual probabilities.   That is,

$$P(\text{all 4 computer having a disk crash}) = 0.21 \times 0.21 \times 0.21 \times 0.21 = 0.00194$$

21.7. If a pair of fair dice are rolled, what is the probability of the numbers adding up to 10?

There are only three favorable outcomes (6, 4), (4, 6) and (5, 5).   The total number of outcomes is 36 (see Example 21.7).   The probability that the numbers add up to 10 is calculated by dividing the number of favorable outcomes by the total number of outcomes.   That is, the probability is $3/36 = 1/12 = 0.083$.

21.8. The proportions of employees of a manufacturing plant born during the various months of the year were as tabulated below.

Jan.	Feb.	Mar.	Apr.	May	June	July	Aug.	Sept.	Oct.	Nov.	Dec.
0.09	0.07	0.10	0.09	0.08	0.10	0.09	0.07	0.08	0.07	0.08	0.07

What is the probability that an employee selected at random was born during the first 3 months of the year?

In this case, there are only three mutually exclusive possibilities, since no employee can be born in more than one month. The probability of being born during the first 3 months is equal to $0.09 + 0.07 + 0.10 = 0.26$.

21.9. The State Inspection Agency announced that last semester, of all cars inspected, 12% were rejected for faulty brakes, 8% were rejected for defective tires, and 4% were rejected for faulty brakes and defective tires. Find the percentage of cars rejected due to faulty brakes or defective tires?

Let B stand for faulty brakes; T stands for defective tires and BT for both faulty brakes and defective tires.

The probabilities are:

$$P(B) = 0.12 \qquad P(T) = 0.08 \qquad P(B \text{ and } T) = 0.04$$

Therefore,

$$P(B \text{ or } T) = P(B) + P(T) - P(BT) = 0.12 + 0.08 - 0.04 = 0.16$$

This result indicates that 16% of the cars were rejected due to faulty brakes or defective tires.

21.10. Assume that products in an assembly line have to go through three soldering machines. If the performance of the machines are reported as S (satisfactory) or U (unsatisfactory), what is the sample space for an experiment where the performance of the three machines are inspected? What is the probability that all machines perform satisfactorily if they all have the same probability of failure?

Since each machine has two outcomes, {S, F}, and there are three machines, the sample space will have $2 \times 2 \times 2 = 8$ possible outcomes. They are: {S, S, S}, {S, S, F}, {S, F, S}, {S, F, F}, {F, S, S}, {F, S, F}, {F, F, S}, {F, F, F}. The probability that all systems perform satisfactorily has only one probable outcome. That is, there is only one triple {S, S, S}. The total number of outcomes is 8. Therefore, the probability of all machines working correctly is $1/8 = 0.125$.

21.11. A box contains 3 golf balls, 8 tennis balls, and 10 metal marbles. What is the probability of selecting at random a metal marble provided that each object has the same chance of being chosen?

There are 10 ways to select a metal marble. There are 21 equally possible outcomes. Therefore, the probability of selecting a metal marble is 10/21.

21.12. Assume that within a group of 40 external modems there are 4 that are defective. If 2 modems are selected at random, what is the probability that at least one of them is nondefective?

To solve this problem, let us first determine what the sample space is. The modems can be divided into two groups: nondefective (36) and defective (4).

With the 36 nondefective modems, we can form $C(36, 2)$ or 630 different groups of two modems each.

With the 4 defective modems, we can form $C(4, 2)$ or 6 different groups of two modems each.

With the 36 nondefective modems and the 4 defective modems, we can form 36×4 or 144 different groups of two modems. The pairs of this last group include one defective modem and one nondefective.

The sample space includes all possible outcomes. That is, $630 + 6 + 144 = 780$ possible cases. The favorable outcomes, that is, outcomes that satisfy the conditions of the program are $630 + 144 = 774$.

Therefore, the probability of having a pair of modems with at least one being nondefective is $774/780 = 387/390$.

21.13. The table below illustrates the crossbreeding of "pure breed" peas according to Mendel's experiments on genetics. Mendel assumed that, when plants are crossbred, each individual receives one gene from each of its parents. The combination of these genes determines the characteristics of each individual of the next generation. In this case, there are two types of genes: one wrinkled gene (W), and one smooth gene (S). The peas of a generation are called smooth if at least one of its genes is an S gene, otherwise they are called wrinkled. In the first generation shown in the table, what is the probability of a pea of being smooth? What is the probability of being wrinkled? Interpret the results.

Gene from 1st plant ($\rightarrow$) Gene from 2nd plant ($\downarrow$)	W	W
S	SW	SW
S	SW	SW

According to the table, there are 4 possible outcomes. Four of these are favorable outcomes for being smooth. Therefore, the probability is $4/4 = 1$. This result indicates that all peas must be smooth. There is no favorable outcome for being wrinkled; the probability if being wrinkled is $0/4 = 0$. This result indicates that there is no chance whatsoever of a pea being wrinkled.

21.14. If two peas of the first generation are crossbred, what is the probability of being smooth? What is the probability of being wrinkled?

Gene from 1st plant ($\rightarrow$) Gene from 2nd plant ($\downarrow$)	S	W
S	SS	SW
W	WS	WW

There are 4 possible outcomes. Three of these outcomes contain an S. Therefore, the probability of being smooth is $3/4$. There is only one outcome favorable to being wrinkled. The probability of being wrinkled is $1/4$.

21.15. Assume that you toss a pair of dice twice in a row. Find the probability of getting a sum of 5 in the first toss and a sum of 6 in the second toss.

Tossing a pair of dice may produce 36 possible different outcomes (refer to Example 21.7). There are 4 favorable outcomes for getting a sum of 5. They are $(1, 4), (4, 1), (2, 3), (3, 2)$. Therefore, the probability of getting a sum of 5 during the first toss is equal to $4/36$ or $1/9$.

The second toss of the dice may also produce 36 possible different outcomes. There are 6 favorable outcomes for getting a sum of 6. They are $(1, 5), (5, 1), (2, 4), (4, 2), (3, 3), (3, 3)$. Therefore, the possibility of getting a sum of 6 during the second toss is equal to 6/36 or 1/6.

The tosses are independent since the first toss does not affect the outcomes of the second toss. Call P(E and F) the probability of getting a sum of 5 and 6 in the first and second toss respectively:

$$P(E \text{ and } F) = (1/9) \times (1/6) = (1/54)$$

21.16. What is wrong with the following argument? If we throw a pair of fair dice the probability of getting at least one ace is $(1/6) + (1/6)$.

We forgot to subtract P(A and B) as A and B are not exclusive.

21.17. A box contains 5 blue balls, 3 red balls, and 7 yellow balls. If we select one of these balls at random, what is the probability that the ball is either blue or yellow?

Since a ball cannot be blue and yellow at the same time the events are mutually exclusive. The individual probabilities are

$$P(\text{blue}) = \frac{5}{15} = \frac{1}{3} \qquad P(\text{yellow}) = \frac{7}{15}$$

Therefore, the probability of selecting a blue or yellow ball is given by

$$P(\text{blue or yellow}) = \frac{1}{3} + \frac{7}{15} = \frac{4}{5}$$

Alternatively, there are $(5 + 7)$ ways of choosing a blue or yellow ball. Since the total number of balls is 15, the probability of obtaining a blue or yellow ball is $(5 + 7)/15 = 4/5$.

21.18. When tossing a fair die, what is the probability of not getting a 6?

All outcomes are likely; the probability of getting a 6 is 1/6. Since the sum of the probability of getting a 6 and not getting a 6 must add up to 1, we have that the probability of not getting a 6 is

$$P(\text{not getting a } 6) = 1 - P(\text{getting a } 6)$$
$$P(\text{not getting a } 6) = 1 - \frac{1}{6} = \frac{5}{6}$$

21.19. A computer analyst is building 5-letter passwords using the letters of the English alphabet at random. What is the probability of getting a password in which the letters are all vowels?

The sample space is the number of different passwords that can be formed with 5 letters of the English alphabet. This number is equal to 7,893,600 or the number of permutations of 26 letters taken 5 at a time. The number of favorable outcomes is the number of different passwords that can be formed using all vowels. This number is equal to 120 or the number of permutations of 5 elements taken 5 at a time. Therefore, the probability of having a password in which all letters are vowels is $1/7,893,600 \approx 0.00000013$.

21.20. In a group of 20 computer programs for capturing images from the screen, 8 can save the images in JPG format, 9 can save images in TIF format, and 5 can save them in both formats. If one program is selected at random, what is the probability that the program can save images in at least one of these formats.

Let T stand for "save the image in TIF format," and J for "save the images in JPG format."

According to the data the probabilities are

$$P(T) = \frac{8}{20} \qquad P(J) = \frac{9}{20} \qquad P(T \text{ and } J) = \frac{5}{20}$$

Therefore, the probability of $P(T \text{ or } J) = 8/20 + 9/20 - 5/20 = 3/5 \approx 0.6$.

Supplementary Problems

The answers to this set of problems are at the end of the chapter.

21.21. If the odds of having a communication failure during a satellite transmission are 21 to 20, what is the probability of failure?

21.22. If a die is thrown once, what is the probability of getting an odd number or a 4?

21.23. A fair die is tossed twice in row. What is the probability of the sum of the outcomes being equal to 8?

21.24. Assume that a computer has two hard disks of different capacities labeled D and E respectively. If the probabilities that disk D and E will fail in the next two years are of 0.7 and 0.5 respectively, what is the probability that both disks will fail independently in the next two years?

21.25. A newly wed couple plan to have three children. What is the sample space of the possible sexes of the children? What is the probability of having 3 girls?

21.26. An experiment produces four outcomes {A, B, C, D}. The probabilities of each of these outcomes are $P(A) = 0.35$, $P(B) = 0.17$, $P(C) = 0.22$. What is the probability of $P(D)$?

21.27. Determine whether the following fractions can be probabilities of some undetermined event. Explain your answer: (*a*) 16/15,   (*b*) 36/25,   (*c*) 15/36.

21.28. Assume that for laptop computers the probability of having a hard disk failure during their first year of operation is 0.14. The probability of recovering any data after a disk failure is 0.001.

(*a*) What is the probability of not having a hard disk failure?

(*b*) What is the probability of not recovering any data after a hard disk failure?

21.29. The owner of a print shop plans to buy 3 high photocopy machines of an inventory of 9. If 2 of these machines have defective circuit:

(*a*) What is the probability that none of the machines purchased have a defective circuit?

(*b*) What is the probability that at least one of the machines purchased will have a defective circuit?

21.30. If a die is rolled twice, what is the probability of both numbers being even?

21.31. During the integration phase of a computer system, it was found that 30% of the time the system failed due to a misunderstanding of the specifications, 20% of the time the system failed to improper coding of the instructions, and 15% the system failed due to both causes. What is the probability that the system failed due to either type of error?

21.32. Suppose that A and B are independent events for which P(A) = 1/2 and P(B) = 1/3. Find the following probabilities: (*a*) P(A and B), (*b*) (P(A|B), (*c*) P(A or B).

Answers to Supplementary Problems

21.21. 21/41 ≈ 0.51

21.22. 2/3

21.23. 5/36

21.24. $(0.7) \times (0.5) = 0.35$

21.25. Sample space = {BBB, GGG, BBG, GBB, BGB, BGG, GBG, GGB}. The probability of having 3 girls is 1/8.

21.26. 0.26

21.27. (*a*) No. The value of this fraction is greater than 1. (*b*) No. The value of this fraction is greater than 1. (*c*) Yes. The quotient is greater than 0 and less than 1.

21.28. (*a*) 0.86, (*b*) 0.999

21.29. (*a*) 0.417, (*b*) 0.583

21.30. 3/12 = 1/14

21.31. 0.35

21.32. (*a*) 1/6, (*b*) 1/2, (*c*) 2/3

Chapter 22

Statistics

The term *statistics* is generally understood with two different but somewhat related meanings. In the most general sense it refers to the scientific methods used for collecting, organizing, summarizing, and presenting data. Statistics is also concerned with the task of analyzing the data collected and making valid conclusions or decisions based on the results of such analysis. For this purpose statisticians use several tools like tables, graphs, and some calculated quantities such as averages. In a more restricted sense, the term *statistics* is used to refer to the data collected by these methods.

22.1 DESCRIPTIVE VERSUS INFERENTIAL STATISTICS

The science of statistics is commonly divided into two categories, *descriptive* and *inferential*. The main emphasis of *descriptive* statistics is on the analysis of observed measures usually taken from a group. Descriptive statistics tries to answer questions such as what are the typical measures, or are there values greater or smaller (extreme values) than all the other measures. *Inferential* statistics, on the other hand, tries to make valid conclusions (inferences or guesses) about an entire group based on observations of a representative and small subset of the group. An example of inferential statistics is a presidential election poll. Here, after interviewing a group, say, of 1000 registered voters across the nation, pollsters try to make predictions about the outcome of the next election.

22.2 POPULATION AND SAMPLES. PARAMETERS AND STATISTICS

When collecting data, statisticians are generally interested in some characteristics of a group of individuals or objects. The characteristics being measured may vary from one group to another. In all cases, the total set of measurements that we are interested in studying is called a *population*. For example, we may be interested in the occurrence of intestinal dysentery among the Yanomami tribes of the Amazonian jungle. In this case, the population is made up of all the individual members of these tribes.

Sometimes a population may be very large and difficult to consider as a whole. For instance, for presidential election polls it is very difficult to interview all registered voters of a country or state. In cases like this, statisticians work with smaller subsets extracted from the population. A subset of the entire population is called a *sample*. For example, a group of 1000 registered voters is a sample of the entire population of eligible voters in the United States. The objective of using samples is to gain insight about the larger group or population

Whenever we talk about population, the facts about the population as a whole are called *parameters*. For example, the number of naturalized citizens in the last ten years as a percentage of all U.S. citizens is a parameter of the U.S. population. Since parameters are descriptions of the population, a population can have many parameters. A fact about a sample is called a *statistic*. For example, the percentage of naturalized U.S. citizens born in France, for example, is a statistic. Notice that in this case the statistic is just a simple number. Statistics may vary from sample to sample since there may be many samples that can be drawn from a particular population. Statistics are used as estimates of population parameters. For instance, the number of cases of intestinal dysentery among the Amazonian tribes may vary depending upon which tribe is being considered. The relation between population, sample, parameter and statistic can be summarized as follows:

$$\text{Population} \rightarrow \text{Sample}$$
$$\downarrow \qquad\qquad \downarrow$$
$$\text{Parameter} \leftarrow \text{Statistics}$$

EXAMPLE 22.1. A pharmaceutical company is interested in determining the effectiveness of a new drug for relieving migraines. The drug is currently prescribed out across the nation by more than 1000 doctors. The company selects at random 500 patients for interviews. What is the population and what is the sample. Will the results obtained from this sample be equal to the results obtained from some other sample?

The population is the entire group of persons who are taking the new drug under the care of their doctors. The sample consists of the 500 patients selected at random. In general, the results obtained from this sample will be different from the results obtained from another sample.

22.3   QUANTITATIVE AND QUALITATIVE DATA

Data that can be gathered for purposes of study can be generally divided into two main categories: quantitative or qualitative. *Quantitative* data are observations that can be measured on a numerical scale. This type of data is the most commonly used for descriptive purposes. Examples of quantitative data are: the number of female students in our high school for the current year, or the number of bacteria in a cup of drinking water.

If each observation in a data set falls in one and only one of a set of categories, the data set is called *qualitative*. Examples of these type of data are: the political party with which a person is affiliated, or the sex of the individual members of your family.

EXAMPLE 22.2. Classify the data obtained if you do a study on: (*a*) the number of heartbeats per minute of all the members of your household, (*b*) the marital status of all your professors, (*c*) the largest selling brand of cereal in your hometown.

Part (*a*) will produce a quantitative response. It can be measured on a numerical scale. For instance, at rest, the average human heart will beat approximately 75 times per minute. Part (*b*) will yield a qualitative result. Each person will have only one marital status (single, married, divorced or widowed). Part (*c*) will produce a qualitative result.

22.4   FREQUENCY DISTRIBUTIONS AND GRAPHICAL REPRESENTATION OF THE DATA

A *frequency distribution* is a summary technique that organizes data into classes and provides in a tabular (table-like) form a list of the classes along with the number of observations in each class. When setting up the categories or classes for classification, the categories must be exclusive and exhaustive. That is, the categories should not overlap and should cover all possible values.

EXAMPLE 22.3. To test the durability of a batch of lithium batteries, 30 twelve-volt batteries were selected to power a set of identical flashlights under temperatures close to 32 degrees (Fahrenheit). The flashlights were left on until the batteries died. The following table shows the number of consecutive hours (rounded to the nearest integer value) in operation of the different groups. Make a frequency distribution for the collected data.

65	69	64	69	68	66	69	64	66	67
65	71	71	69	64	65	70	68	67	68
67	68	72	72	65	66	67	66	65	66

The frequency of distribution of this data, ordering it from the shortest to the longest, is as follows:

Hours:	64	65	66	67	68	69	70	71	72
Frequency:	3	5	5	4	4	4	1	2	2

Notice that the frequency tells us how many batteries lasted a particular number of hours. For instance, three batteries lasted for 64 consecutive hours, five batteries lasted for 65 hours, and so on.

EXAMPLE 22.4. The age of a tree can be estimated as a function of the size of the circumference of its trunk. The following measures represent actual circumference of a sample of 14 blue pine trees (rounded to the nearest inch).

$$22 \quad 18 \quad 25 \quad 27 \quad 22$$
$$18 \quad 25 \quad 22 \quad 30 \quad 27$$
$$30 \quad 25 \quad 25 \quad 22$$

The frequency of distribution of this data is as follows:

Trunk size (inches)	Frequency
18	2
22	4
25	4
27	2
30	2

The frequency tells us how many blue pine trees have a particular trunk size. Frequency distributions can be displayed using different graphical representations. The most popular are bar charts, pie charts, histograms, and line charts. In all cases, the objective is to provide a visual summary of the data that is easy to understand.

22.5 BAR CHARTS

Bar charts are used to illustrate a frequency distribution for qualitative data. A bar chart is a graphical representation in which the length of each bar corresponds to the number of observations in each category. There are no fixed rules for representing data in a bar graph. Some authors use the horizontal axis for categories that are numerical and the vertical axis for categories that are descriptively labeled. If the categories are ordered, it is suggested that they maintain that order in the bar chart. Otherwise, the categories may be ordered according to some other criteria such as alphabetical ordering, or increasing or decreasing ordering of values. Keep in mind that all these are just mere suggestions, for instance, if you believe that a vertical bar chart is more descriptive or more effective than a horizontal chart for the data being considered, by all means, use it.

EXAMPLE 22.5. Figure 22-1 shows a bar chart that represents the percentage (rounded to the next integer) of computer attacks via Internet experienced at a university computer center.

Notice that to create this chart it was necessary to select an appropriate "scale" along the vertical axis. Each mark (or unit) along the vertical axis represents a category. In addition, the vertical axis has been properly labeled to indicate what these categories represent. The horizontal axis has been also labeled to indicate the percentages represented along this axis. The "height" of the bar represents how many attacks have that particular percentage. When displaying values in a bar graph not every value can be "exactly" represented by the height of the bar. In this particular case the values have been conveniently rounded to an integer to facilitate their representation.

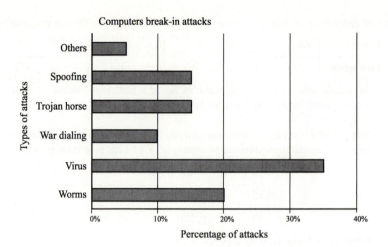

Fig. 22-1

22.6 PIE CHARTS

Bar charts are not the only way of displaying frequency distributions. *Pie charts* can also be used to display similar information. The circle (the pie) represents the complete set of data. The "slices" of the pie are drawn proportional to the frequencies in each category. When drawing a pie chart, each slice represents the percentage of the total that belongs to a particular category.

When calculating the size of the slices, it is important to remember that the angle around the entire circle measures $360°$. Therefore, the relative size of a particular slice can be calculated by multiplying its percentage, in decimal form, by $360°$. The result obtained is the central angle of a sector representing that particular slice.

EXAMPLE 22.6. Table 22.1 shows the percentage of main elements making up earth's crust. Represent this information using a pie chart.

Table 22.1

Category	Percentage
Oxygen	47%
Silicon	28%
Aluminum	8%
Iron	5%
Others	12%

To calculate the size of the individual slices, multiply their corresponding percentages, in decimal form, by $360°$. These results indicate the angle at the center for the different categories under consideration. The sizes of the slices are:

Oxygen $= 0.47 \times 360° = 169.2° \approx 169°$
Silicon $= 0.28 \times 360° = 100.8° \approx 101°$
Aluminum $= 0.08 \times 360° = 28.8° \approx 29°$
Iron $= 0.05 \times 360° = 18°$
Others $= 0.12 \times 360° = 43.2° \approx 43°$

To create the pie chart you may proceed as follows: Draw a circle and an arbitrary radius. That is, draw a circle and a segment joining the center of the circle with an arbitrary point of its circumference. To draw a slice, say the "oxygen slice," use a protractor and draw the angle at the center of 169°. For this, make sure that the vertex of this angle is at the center of the circle and its initial side is the arbitrary radius. Label this slice with its corresponding name and indicate the percentage that it represents. The remaining slices are drawn in a similar way. Figure 22-2 shows the pie chart.

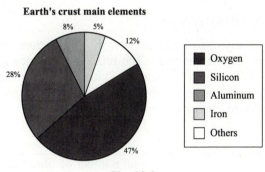

Earth's crust main elements

Fig. 22-2

22.7 FREQUENCY DISTRIBUTION OF LARGE DATA SETS

When working with large data sets, choosing the appropriate number of categories or classes to represent all data is a little bit more elaborate than in the previous examples. In general, when more data is available, more classes can be used. However, if there are too few classes (less than four) or too many classes (more than 20) much of the information that the graphic could convey is lost. Although there are no fixed rules for determining the number of classes, statisticians agree that between 5 and 20 classes are appropriate. When we speak about the width of a class, it is assumed that each class has an upper limit and a lower limit value. The difference between these two limits is called the *width* of the interval, and it should be constant, if possible, for all the classes representing a particular type of data.

22.8 DETERMINING THE CLASS WIDTH

For determining the width of a class for a particular number of classes, use the following formula:

Class width = (largest value in dataset − smallest value in the data set)/number of classes

The difference between the largest value in the data set and the smallest value in the data set is called the *range* of the data. Using this terminology the formula can be rewritten as

$$\text{Class width} = \frac{\text{range}}{\text{number of classes}}$$

Fractional values may be rounded to an integer value, although it is not necessary to do so. However, when determining class widths, the reader should be reasonable since classes that have fractional endpoints may complicate the graph and make it hard to read. That is, fractional intervals may make the graphs a little bit more difficult to draw and understand. Statisticians sometimes prefer widths with odd values. We will adopt this convention for the remainder of the chapter

EXAMPLE 22.7. Assume that in a set of 50 measures the largest value is 180 and the smallest is 70. If there are 8 classes, what is the width of each interval?

According to the formula, the width of each interval is

Class width $= (180 - 70)/8 = 110/8 = 13.75 \approx 14$ (The symbol $\approx$ is read "approximately equal to")

This result indicates that each class has a width of 14 units. Only 8 categories are necessary to account for all observations of this set (see Example 22.8).

22.9 CLASS RELATIVE FREQUENCY

The *relative frequency* of any class is the number of observations or measurements in the class divided by the total number of observations. The relative frequency represents the percentage of the total observations that fall in a given class. Statisticians use the frequency distribution to compare data sets with different numbers of observations. Thus, the class relative frequency serves as a standardizing mechanism to compare data.

$$\text{Relative frequency} = \frac{\text{number in class}}{\text{total number of observations}}$$

EXAMPLE 22.8. A medical team, interested on the effect of lifting moderate weights (less than 50 pounds) on the cardiovascular condition, selected at random a group of 21 middle-aged American men (between 45 and 55 years old). The men were asked to lift a 35-pound bag, carry it for a distance of 20 yards, and drop it at a specified location.

$$120 \quad 130 \quad 144 \quad 132 \quad 140 \quad 160 \quad 170$$
$$145 \quad 152 \quad 130 \quad 127 \quad 148 \quad 130 \quad 146$$
$$135 \quad 151 \quad 155 \quad 121 \quad 151 \quad 135 \quad 167$$

According to the data, the largest value is 170 and the smallest value is 120. Therefore, the range is 50. If we consider only six classes, the width of each class is $50/6 \approx 9$ units. In addition, as a general rule, begin at 0.5 below the smallest measurement and use the second decimal to assure that no measurement falls on a class boundary.

For this particular example, six classes of width 9 will suffice to tabulate all possible values of this data set. Table 22.2 shows the classes, the frequencies and the relative frequency distribution. The first class comprises values from 119.5 (the smallest value minus 0.5) to 128.5 (smallest value minus 0.5 plus the width of the class). Since classes cannot overlap, the next class comprises values from 128.51 to 137.5. Notice how the second decimal is being used to assure that classes cannot overlap. The remaining classes or categories are formed in a similar way. The frequencies are calculated by counting the number of values that fall in each category. The relative frequency, as indicated before, is calculated by dividing the total value in each category by the total number of observations. There are 21 observations in this example.

Table 22.2

Class limits	Frequency	Relative frequency
119.5–128.5	3	0.14
128.51–137.5	6	0.29
137.51–146.5	4	0.19
146.51–155.5	5	0.24
155.51–164.5	1	0.05
164.51–173.5	2	0.1

22.10   CUMULATIVE FREQUENCY

Given a particular class, the *cumulative frequency* of this class is defined as the sum of the frequency of the class and the frequency of all preceding classes.   The cumulative frequency of any given class tells the reader the number of observations that are smaller than or belong to a particular category.   Table 22.3 shows the cumulative frequency of the previous example.   For instance, given the class that comprises values in the range 146.51 to 155.5, we can observe that the cumulative frequency is 18.   This indicates that there are 18 observations with values less or equal to 155.5.   Notice that 18 (in the cumulative frequency column) is obtained by adding $3 + 6 + 4 + 5$ (under the frequency column).

Table 22.3

Class limits	Frequency	Cumulative frequency
119.5–128.5	3	3
128.51–137.5	6	9
137.51–146.5	4	13
146.51–155.5	5	18
155.51–164.5	1	19
164.51–173.5	2	21

22.11   HISTOGRAMS

A *histogram* is the graphical image of a frequency or a relative frequency distribution in which the height of each bar is proportional to the frequency or relative frequency of the class.   Histograms are used when the categories being observed are numerical.   Figure 22-3 shows the histogram for the data of Example 22.8.   Notice that the cumulative frequency is

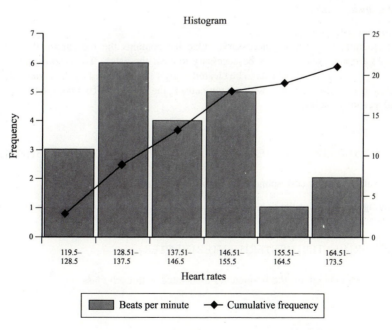

Fig. 22-3

indicated with a superimposed line diagram. It is customary to write the cumulative frequencies in the middle of the bars. The secondary scale at the right shows the values of these cumulative frequencies.

It is possible to plot a histogram with a graphing calculator. Depending upon the type of calculator that you use, there may be more or fewer steps than the ones here indicated. However, the general methodology is the same.

1. Enter the data in a list or data set.

2. Select the statistical calculations.

3. Define the statistical plot.

4. Define the range window.

5. Plot the graph.

EXAMPLE 22.9. Plot a histogram using a graphing calculator and the data of Table 22.4.

Table 22.4

Class limits	Frequency	Cumulative frequency
119.5–128.5	3	3
128.51–137.5	6	9
137.51–146.5	4	13
146.51–155.5	5	18
155.51–164.5	1	19
164.51–173.5	2	21

Using a TI-82 graphing calculator

(1) Enter the histogram data in a list.
To create a histogram, two lists are necessary. One list contains the measurements; the other contains the frequencies. As a previous step, it may be necessary to clear the lists. To do so use the ClrList option of the STAT menu. Enter the name of the lists to be cleared using {2nd} followed by the name of the lists (use the keys L1 through L6); separate the list names with a comma using the key {,}. To enter the list values in the STAT list editor proceed as shown below.

{STAT}EDIT{ENTER}

119.5{ENTER}128.51{ENTER}137.51{ENTER}146.51{ENTER}155.51{ENTER}164.51{ENTER}

Notice that the values entered correspond to the lower end of each class.

After entering the values of the first list use the directional key to move to the top of the second list. Enter the data as indicated below.

3{ENTER}6{ENTER}4{ENTER}5{ENTER}1{ENTER}2{ENTER}

Notice that the values entered are the frequencies associated with each class.

(2) Define the plot.

{2nd}{STAT PLOT}Plot1{ENTER}

You may have to use the directional keys to select the different options.

Set Plot1 on.   To do this highlight the On option and then press {ENTER}.

Move to the rightmost position of the Type option to highlight Histogram and press {ENTER}.

Choose L1 as the Xlist by highlighting L1 and pressing {ENTER}.

Choose L2 as the Freq. list by highlighting L2 and pressing {ENTER}.

(3)   Define the Width of the class and Viewing Window.

Begin by pressing {WINDOW}

Be aware that for histograms Xscl determines the width of each bar, beginning at Xmin.   The ZoomStat option of ZOOM adjusts Xmin and Xmax to include all values, but does not change Xscl.   Finally, the condition (Xmas − Xmin)/Xscl ≤ 47 must be satisfied.   If the top of the histogram is not shown, use TRACE to determine the value for Ymax.

{WINDOW}{ENTER}

Xmin = 119.5 {ENTER} (low end of first class)

Xmax = 173.5 {ENTER} (high end of last class)

Xscl = 9 {ENTER} (width of the class)

Ymin = {(−)}10 {ENTER}

Ymax = 10 {ENTER}

Yscl = 1 {ENTER}

(4)   Draw histogram.

{TRACE}.

You may use the directional keys to move to the different classes.   At the lower left corner of the window you will see the ends of each interval and the frequency for that interval.   Depending on the window settings that you define it may be necessary to choose the option ZoomStat of the Zoom menu to view the histogram.   You get to this menu by pressing the key {ZOOM}.

22.12   MEASUREMENTS OF CENTRAL TENDENCY

When working with a set of values, it is desirable to have a single value around which all values in the set tend to cluster.   This number then serves as a representative for the whole set.   Numbers of this type are called measures of central tendency.   The most common measures are the arithmetic mean, the median, and the mode.

22.13   AVERAGE OR ARITHMETIC MEAN

Given a set of n measurements, say, $x_1, x_2, \ldots, x_n$, the *average* or *arithmetic mean*, $\bar{x}$, is defined as indicated below.   The $\bar{x}$ is read "x bar."

$$\bar{x} = \frac{x_1 + x_2 + \cdots + x_n}{n}$$

EXAMPLE 22.10.   In a manufacturing plant a packaging machine is supposed to fill small bags of marbles with exactly 50 marbles per bag.   However, a random sample of four bags indicates that one bag has 54 marbles, another has 48, and two have 47 marbles.   What is the average content of these bags?

In this case, the measurements are the number of marbles in each bag. There are four measurements. That is $n = 4$. Therefore, we have that the average number of marbles per bag $= (54 + 48 + 47 + 47)/4 = 49$. On average, each bag has about 49 marbles.

EXAMPLE 22.11. Table 22.5 shows the wingbeats per second and the maximum flight speed in miles/hour of some insects with speeds greater than 3 miles/hour. Calculate the average number of wingbeats/second and the average maximum speed in miles/hour.

Table 22.5

Insect	Wingbeats/second	Max. flight speed (mi/h)
White butterfly	12	9
Damselfly	16	3
Dragonfly	40	34
Cockchafer beetle	50	17
Bumblebee	130	7
Housefly	200	4

Since there are six measurements, $n = 6$.

Average number of wingbeats $= (12 + 16 + 40 + 50 + 130 + 200)/6 = 74.66$ wingbeats/second.

The average maximum speed $= (9 + 4 + 34 + 17 + 7 + 4)/6 = 12.5$ miles/hour

To calculate averages with a calculator, you may enter the values in a list or data set and then select the appropriate option. To calculate these averages using an HP-38G we can proceed as follows:

(1)   Select statistics.

    {LIB}Statistics{ENTER}

(2)   Enter values in data set.

    You may have to clear the data set. To do this press {SHIFT}{CLEAR}, highlight the ALL columns option and press {ENTER}

    12{ENTER}16{ENTER}40{ENTER}50{ENTER}130{ENTER}200{ENTER}. ← Wingbeats values.

    Use the directional key {▶} to move to column C2 and enter the speed values. That is, enter

    9{ENTER}4{ENTER}34{ENTER}17{ENTER}7{ENTER}4{ENTER}.

(3)   Calculate the average.

    Before calculating the average of the first data set make sure that you are working with this data set. Press {SYMB} and make sure that there is a √ mark in the column where you entered the data. If the data was entered in column C1, move to that column using the directional keys if you are not already there. Make sure that there is a √ by the C1 column. If not, press √ CHK. Then press {NUM} followed by the option STATS to display the results. The display shows an average of 74.66667. This value is indicated in the calculator as Mean Σ. You may go back to the data sets by pressing the option O.K.

    To calculate the average of the second data set, make sure you are working with the column where you entered the data. Use a process similar to the one explained for C1. If you do not see on the symbolic view display the column C2, you can enter it by pressing the option C followed by the number 2 in any row labeled H2 through H4 that is empty. Then press {NUM} followed by the option STATS to display the results. The display shows an average of 12.5 miles/hour. This value is indicated in the calculator as Mean Σ.

22.14   WEIGHTED MEAN

If a sample consists of n distinct values $x_1, x_2, \ldots x_n$ that occur with frequencies $f_1, f_2, \ldots f_n$, the *weighted mean, w* is defined as follows

$$\bar{w} = \frac{x_1 f_1 + x_2 f_2 + \cdots + x_n f_n}{f_1 + f_2 + \cdots + f_n}$$

EXAMPLE 22.12.   Table 22.6 shows the time in hours (rounded to the nearest hour) that it took to test the performance of a new computer system during the integration phase of the software.   Calculate the mean number of hours to test the system.

Table 22.6

No. of hours	Frequency
5	2
6	5
7	11
8	4

The weighted average is computed as follows:

(1)   Multiply each value by its frequency and add the results of these products:

$$(5 \times 2) + (6 \times 5) + (7 \times 11) + (8 \times 4) = (10) + (30) + (77) + (32) = 149$$

(2)   Divide the sum obtained in the previous step by the sum of the frequencies.   The sum of the frequencies is $2 + 5 + 11 + 4 = 22$.   Therefore, the weighted mean value is $149/22 \approx 6.77$.

That is, it takes about 7 hours to test the system.

22.15   MEDIAN

Given a set of n measurements, say, $x_1, x_2, \ldots, x_n$, arranged in increasing or decreasing order, the *median* is the middle entry, if n is odd.   If n is even, the median is the average of the two middle entries.

EXAMPLE 22.13.   Calculate the median of the following data set: 12, 14, 6, 21, 23, 15, 25, 2, 7, 8.

The measurements in increasing order, are 2, 6, 7, 8, 12, 14, 15, 21, 23, 25.   Since there are 10 numbers, n is even.   The median is the average of the two middle entries.   That is, the median is $(12 + 14)/2 = 13$.   Notice that the median does not have to be one of the given measures.

EXAMPLE 22.14.   Find the median of the following data set: 18, 13, 14, 21, 11, 16, 15, 21, 25.

The data in increasing order 11, 12, 13, 14, 15, 16, 18, 21, 25.   Since there are 9 pieces of data, n is odd.   The median is the middle number; in this case, 15.

22.16   MODE

Given a set of n measurements, say, $x_1, x_2, \ldots, x_n$ the *mode* is the observation that occurs most often.   That is, the mode is the number that occurs with the greatest frequency.   If every measurement occurs only once, then there is no mode.   If only one measurement occurs more often, the data is said to be *unimodal*. If two different measurements occur with the same frequency, then the data set is *bimodal*. It may be the case that in a data set there are two or more modes. Data sets with three modes are called *trimodal*, data sets with four modes are called *quadrimodal*, and so on.

EXAMPLE 22.15. Find the mode for the following data set: 15, 32, 94, 94, 47, 18, 29.

In this sequence, the number that occurs with the greatest frequency is 94. This number occurs twice. Therefore, 94 is the mode. Since there is only one mode, this data set is said to be unimodal. Notice that to find the mode there is no need to order the numbers in any way.

22.17 MEASURES OF DISPERSION

Although the average or arithmetic mean is one of the most commonly used measures of central tendency, it may not always be a good descriptor of the data. For example, the average is affected by outliers. By this name statisticians call those values that are very large or very small relative to the other numbers or values of a data set. Since the average may not provide an accurate description of the data, it may be necessary to complement the information that it provides with a measure of the degree to which the data cluster around or wander away from the average. Consider, for example, the data sets A and B shown below.

Both sets have the same mean and neither has a mode. A careful examination of both data sets will reveal that the data set B is more variable than A. Notice that for the data set A, the value 7 seems to be a "typical" value. The remaining values of the data set are not too far from 7. However, in the data set B, the variations from 7 are much bigger. That is, the values of data set B are more spread. This important information about the nature of the data is not provided by any of the measures of central tendency.

Data set A:	5	6	7	8	9
Data set B:	1	2	7	12	13

22.18 SAMPLE RANGE

The simplest measure of sample variation is the *sample range*. This value, generally denoted by R, is calculated by subtracting the smallest measure from the highest measure. This value is sometimes very helpful when determining quality control procedures of small samples. Although the range is easy to obtain, it may be affected by a single extreme value.

EXAMPLE 22.16. For the data set A, the range is equal to 4. That is, $R = 9 - 5 = 4$. For data set B, the range is 12 since $R = 13 - 1 = 12$.

22.19 VARIANCE

Given a set of n measurements, say $x_1, x_2, \ldots, x_n$, the *sample variance*, denoted by s^2, can be obtained as follows.

(1) Compute the difference of each observation from the arithmetic mean. These differences are called deviations. To check that this operation was done correctly, add the deviations. The sum of the deviations for a set of data is always 0.

(2)   Square each difference.

(3)   Sum the squares of step 2 and divide this sum by $n - 1$, where n is the number of measurements.

Some authors distinguish between the variance for the entire population, denoted by σ^2 and the sample variance just defined.   To find the *population variance*, σ^2, divide the sum of step 3 given above by n instead of $n - 1$.

EXAMPLE 22.17.   Compute the sample variance of 102, 110, 116, 120 and 132.

(1)   Compute the mean of the numbers.

$$\bar{x} = \frac{102 + 110 + 116 + 120 + 132}{5} = \frac{580}{5} = 116$$

(2)   Compute the difference of each value from the arithmetic mean and square the difference.

Data	Deviation from the mean	Square of the deviation
102	$102 - 116 = -14$	196
110	$110 - 116 = -6$	36
116	$116 - 116 = 0$	0
120	$120 - 116 = 4$	16
132	$132 - 116 = 16$	256

The sum of the deviations should be 0.   In this case $(-14 - 6 + 0 + 4 + 16) = 0$.

(3) Sum the squares of step 2 and divide this sum by the number of values minus 1.

$$s^2 = \frac{196 + 36 + 0 + 16 + 256}{4} = \frac{504}{4} = 126$$

22.20   STANDARD DEVIATION

Another measure of dispersion is the *standard deviation*.   This measure is defined as the square root of the variance and has the advantage that it is expressed in the same units as the original data.   The standard deviation indicates the degree to which the numbers in a data set are "spread out."   The larger the measure, the more spread.   Since there are two measures of variance, there are two standard deviations.   One standard deviation for the population data and one for the sample data.   The sample standard deviation is denoted by s.   The population standard deviation is denoted by σ

EXAMPLE 22.18.   Compute the sample standard deviation of the following measurements: 11, 15, 26, 31, 42 and 48.

(1) Compute the arithmetic mean of the numbers.

$$\bar{x} = \frac{11 + 15 + 26 + 31 + 42 + 48}{6} = \frac{173}{6} = 28.83$$

(2)   Compute the difference of each value from the arithmetic mean and square the results.

Data	Deviation from the mean	Square of the deviation
11	$11 - 28.83 = -17.83$	317.90
15	$15 - 28.83 = -13.83$	191.27
26	$26 - 28.83 = -2.83$	8.01
31	$31 - 28.83 = 2.17$	4.71
42	$42 - 28.83 = 13.17$	173.45
48	$48 - 28.83 = 19.17$	367.49

Due to rounding errors the sum of the deviations may not be exactly 0.

(3)   Sum the squares of step 2 and divide this sum by the number of values minus 1.

$$s^2 = \frac{317.90 + 191.27 + 8.01 + 4.71 + 173.45 + 367.49}{6 - 1} = \frac{1062.83}{5} = 212.57$$

(4)   Calculate the square root of the variance.

$$s = \sqrt{212.57} = 14.58$$

It is difficult to describe exactly what the standard deviation measures. However, it can be used to measure how far the data values are from their mean. There are some rules that help us to understand this concept a little bit better. These rules govern the percentage of measurements in a particular data set that fall within a specified number of standard deviations of the mean. Although three rules are introduced, only two (Chebyshev's and Camp-Meidel) are used in this section. These rules are:

Chebyshev's Rule.   For any set of numbers, regardless of how they are distributed, the fraction of them that lie within k ($k > 1$) standard deviations of their mean is at least $100[1 - (1/k^2)]\%$.

Camp-Meidel Inequality.   For $k > 1$, at least $100[1 - (1/2.25k^2)]\%$ of the observations of a unimodal distribution fall within k standard deviations of the mean.

Empirical Rule.   For distributions that are bell-shaped, approximately 68% of the observations are within one standard deviation from the mean, approximately 95% are within two standard deviations from the mean, and almost all observations are within three standards deviations from the mean.

Whenever the three given rules can be applied, in general, Chebyshev's Rule gives the more conservative estimates.

EXAMPLE 22.19.   In a Public Health study to determine the effects of oral contraceptives, a team of physicians examined a total of 1800 pill users (aged between 17 and 40). It was later determined that their average blood pressure was 121 mm, with a standard deviation of 12.5 mm. Describe the sample using the Chebyshev and Camp-Meidel rules within two standard deviations.

In this case $k = 2$. Therefore, according to Chebyshev's rule, *at least* 75% of the women have a blood pressure in the interval from $121 - 2(12.5) = 96$ to $121 + 2(12.5) = 146$. This is equivalent to saying that *at most* 25% will have a blood pressure below 96 and or above 146.

If we assume a unimodal distribution, the Camp-Meidel rule estimates that *at least* 88.9% will have a blood pressure within the same interval from 96 to 146. This is equivalent to saying that *at most* 11.1% of the women have a blood pressure outside that interval.

22.21   RANDOM VARIABLE

Whenever we assign numerical values to the random outcomes of an experiment, a variable is used to represent these values.   Since we assume that the outcomes occur at random, this variable is called a *random variable*.   The collection of all possible values that a random variable can take is called its *space*.   If a random variable can only take certain fixed values, the variable is called a *discrete* variable.   When the variable can take on a continuous set of real values, the variable is said to be *continuous*.

EXAMPLE 22.20.   Classify the type of random variable that can be associated with events like   (*a*) the number of errors in a computer program,   (*b*) the number of babies born yesterday in the local hospital,   (*c*) all possible temperatures between 100°C and 200°C.

The first two events are associated with random variables.   In both situations, the elements of their spaces can be counted using nonnegative integers (counting numbers).   For instance, we could have 500 errors or 100 babies, but never an error and a half or three-quarters of a baby.   The last event is associated with a continuous random variable.   This variable can take any value between any two given measurements.

22.22   NORMAL DISTRIBUTION

A *normal distribution* is a frequency distribution often encountered when there is a large number of values in a data set.   This distribution has a bell-shaped curve, is symmetric about a vertical line through its center, and is completely determined by the mean, and the standard deviation.   A normal distribution is called the *standard normal curve* if the points on the horizontal axis represent standard deviations from the mean.   The area under the curve along a certain interval is numerically equal to the probability that the random variable will have a value in the corresponding interval.   The total area under the curve and above the horizontal axis is 1.   Figure 22-4 shows a standard normal curve.

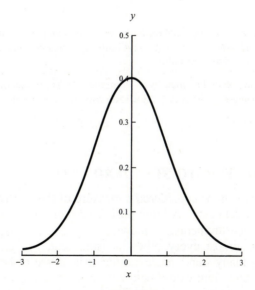

Fig. 22-4

22.23 EMPIRICAL RULE

About 68% of all data values of a normal curve lie within 1 standard deviation of the mean (in both directions), about 95% within 2 standard deviations, and about 99.7% within 3 standard deviations.

EXAMPLE 22.21. Assume that the average lifetime of long-life electric bulbs is 2000 hours with a standard deviation of 150 hours. Sketch a normal curve that represents the frequency of lifetimes of the electric bulbs.

The average or arithmetic mean is 2000, therefore the curve is centered at 2000. Let μ stand for the mean.

To sketch the curve, find the values defined by the standard deviation in a normal curve. That is, find

$$\mu + 1\sigma = 2000 + 1(150) = 2150 \qquad \mu + 2\sigma = 2000 + 2(150) = 2300 \qquad \mu + 3\sigma = 2000 + 3(150) = 2450$$

$$\mu - 1\sigma = 2000 - 1(150) = 1850 \qquad \mu - 2\sigma = 2000 - 2(150) = 1700 \qquad \mu - 3\sigma = 2000 - 3(150) = 1550$$

Figure 22-5 shows the sketch of the normal curve.

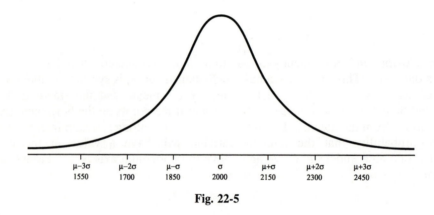

$\mu-3\sigma$	$\mu-2\sigma$	$\mu-\sigma$	σ	$\mu+\sigma$	$\mu+2\sigma$	$\mu+3\sigma$
1550	1700	1850	2000	2150	2300	2450

Fig. 22-5

EXAMPLE 22.22. Using the data of the previous example, what percentage of light bulbs will last between 1850 hours and 2150 hours? Of the total sample, how many light bulbs does this percentage represent? Assume that the data was collected using a sample of 1500 light bulbs.

These two values fall within the interval defined by $\bar{X} - 1\sigma$ and $\bar{X} + 1\sigma$. According to the empirical rule, about 68% of the values fall within this category. That is, $0.68 \times 1500 = 1020$ bulbs will last between 1850 hours and 2150 hours.

22.24 CONVERTING VALUES INTO STANDARD UNITS

The *z-score*, defined below, can be considered a measure of the relative position, with respect to the mean, of the elements of a data set. A large positive *z*-score indicates that the measurement is larger than almost all other measurements. Likewise, a large negative *z*-score indicates that the measurement is smaller than almost every other measurement. A given value is converted to standard units by seeing how many standard deviations it is above or below the average. Values above the average are given a plus sign; values below the average are given a minus sign. This type of conversion is generally known as the *z*-score formula and is given by the formula indicated below.   The z-score is a unit-free measure.   That is, regardless of the units in which the values of the data set are measured, the z-score of any value remains the same.

$$z = \frac{\text{given value-mean}}{\text{standard deviation}}$$

EXAMPLE 22.23.   Assume that the mean blood pressure of a group of 100 individuals is 121 mm with a standard deviation of 12.5 mm.   Express the values 146 mm and 115 mm in standard units and represent the relative positions of these values with respect to the mean on a straight line.

To express the value of 146 mm in standard units, begin by drawing a horizontal line and mark the value of mean, 121, somewhere on this line.   Also mark on this line the value of 146.   Since this value is greater than the mean, write it to the right of the mean.   The line should look like the one shown below.

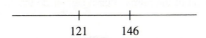

Draw a second line directly below the first line and write 0 directly below the average

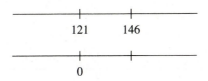

Subtract the mean from the given value and divide this difference by the standard deviation.   Mark this value on the second line directly below the value 146.   That is, $(146 - 121)/12.5 = 2$.   This value is the representation of the given value in standard units.   The two lines should look like this:

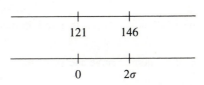

If we consider the standard deviation of 12.5 as a yardstick, we can observe that it fits exactly twice in the space between the mean and the given value.

Repeating this procedure for the value of 115, we find that this value is -0.48 standard units away from the mean.   The minus sign indicates that it is below the mean.   It will appear to the left of the mean.   The two lines should look like this:

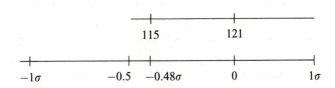

22.25   FINDING AREAS UNDER THE NORMAL CURVE

The table Areas Under the Standard Normal Curve (see Appendix) gives the percentage of all scores in a normal distribution that lie between the mean and z deviations from the mean. The column labeled z represents the number of standard deviations from the mean. The columns represent fractions of standards deviations. The entries in the table represent the probability that a value will be between zero and z standard deviations. Areas under the curve for some particular interval, say between 1 and 2 standard deviations, is generally expressed as $P(0 \leq z \leq 2)$. The following examples illustrate how to find the percentage of all values that lie between the mean and a particular standard deviation.

EXAMPLE 22.24.   Find the percentage of all measurements that lie 1 standard deviation above the mean. That is, find $P(0 \leq z \leq 1)$.

First, let us make a sketch by hand of the normal curve and shade the area that we are interested in (Fig. 22-6).

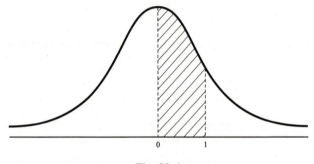

Fig. 22-6

Refer to the table Areas Under the Standard Normal Curve in the Appendix. In this problem $z = 1.00$ (written with two decimals). Under the z column find the value 1.0. The entry at the intersection of the row $z = 1.0$ and the column labeled 0 gives us the value 0.3413. This result indicates that 34.13% of all measurements lie between the mean and one standard deviation above the mean.

EXAMPLE 22.25.   Find the percentage of all measurements that lie 1.45 standard deviations above the mean. That is, find $P(0 \leq z \leq 1.45)$.

First, let us make a sketch by hand of the normal curve and shade the area that we are interested in (Fig. 22-7).

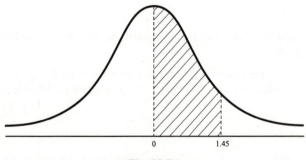

Fig. 22-7

Under the z column of the table Areas Under the Standard Normal Curve in the Appendix, locate the value $z = 1.4$. The entry at the intersection of this row and the column labeled 5 gives us the value 0.4265. This results indicates that 42.65% of all measurements lie between the mean and 1.45 standard deviations above the mean.

EXAMPLE 22.26.   Find the percentage of all measurements that lie 1.73 standard deviations below the mean.   That is, find $P(-1.73 \leq z \leq 0)$.

First, make a sketch by hand of the normal curve and shade the area under consideration (Fig. 22-8).

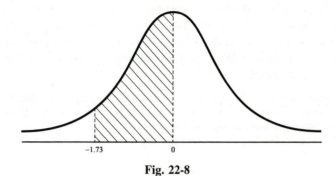

Fig. 22-8

Notice that the table Areas Under the Standard Normal Curve in the Appendix only gives us the percentage of values located to the right of the mean.   That is, between 0 and z.   However, in this case we are required to "go" below the mean.   That is, the area under consideration is to the left of the mean.   To solve this problem, we will use the fact that the normal curve is symmetrical about is mean.   This implies that the percentage of values under the curve between 0 and -1.73 is the same as the percentage of values under the curve between 0 and 1.73.   Therefore, read the value at the intersection of the row $z = 1.7$ and the column labeled 3.   This value is 0.4582.   This indicates that 45.82% of the measurements will lie between the mean and 1.73 standard deviations below the mean.

EXAMPLE 22.27.   Assume that average life span of high-resolution monitors is 24,000 hours with standard deviation of 1800 hours.   If the data has a normal distribution, what is the probability that a monitor will have a life span of at least 21,000 hours?

First, let us find where the value of 21,000 hours lies with respect to the mean.   We do this by finding the value of z.   In this case, $z = (21,000 - 24,000)/1800 = -1.66$.   This result indicates that the value 21,000 lies to the left of the mean and 1.66 deviations from it.   Notice that in this problem we are interested in the area to the right of the mark -1.66 since we want a life span of at least 21,000 hours.   That is, the area we are interested in is $P(z \geq -1.66)$.

The sketch in Fig. 22-9 indicates the area under consideration.

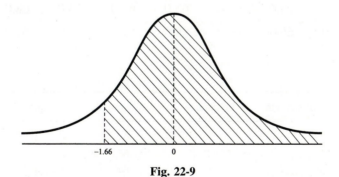

Fig. 22-9

Using the symmetry of the normal curve about its mean, we find that the area under the curve between 0 and 1.66 is 0.4515 (See Example 22.26).   Therefore, the probability of a monitor having a life span of 21,000 hours is the sum of the percentages under the area from 0 to -1.66 plus the percentage under the entire area to the right of 0.   Adding these percentages we obtain $0.4515 + 0.5 = 0.9515$.   Remember that the entire area under the curve is equal to 1.   Therefore the area under each half is 0.5.   This value represents the percentage of the entire area under the curve to the right of the mean.   The result indicates that 95.15% of all monitors will have a life span of at least 2100 hours.

Solved Problems

22.1. A distributor of computer diskettes has decided that if less than 2% of the diskettes produced by one of its suppliers are defective, the company will continue buying diskettes from this supplier. Otherwise, the distributor will buy diskettes from some other sources that meet the standards. Since testing every diskette is time-consuming and not practical, management has decided to choose 50 diskettes at random from the shipment they will receive on Monday from the supplier. What is the population of interest to the manufacturer? How big is its size of the sample?

   The population consists of all diskettes that the distributor will received on Monday from the supplier. The sample consists of the 50 diskettes that will be selected at random from all the diskettes from the supplier that the distributor will received on Monday.

22.2. Geologists can sometimes determine the slope of a river from the average size of the diameter of the stones found in its delta. If a technical group collects a random sample of 150 stones from the river delta, what is the population of interest in this study? What is the sample?

   The population of interest is the diameter of all stones found in the delta region. In this case, the sample consists of the 150 stones selected at random.

22.3. A manufacturer of printed wiring boards used in the assembly of personal computers is testing the quality of its soldering process. To test the strength of the solder joints, some of the soldered leads are pulled until they either break or come off the board. The pull strengths (measured in pounds) for 30 tests are shown below. Make a frequency distribution and draw a bar chart for this data.

   Pull strengths (rounded to the nearest pound):

59.0	58.0	55.0	63.0	62.0	64.0	62.0	60.0	64.0	60.0
55.0	68.0	66.0	65.0	65.0	64.0	53.0	68.0	66.0	66.0
69.0	59.0	66.0	66.0	68.0	61.0	68.0	62.0	64.0	65.0

The frequency distribution is as follows:

Pull strength	Frequency	Pull strength	Frequency
53	1	62	3
55	2	63	1
58	1	64	4
59	2	65	3
60	2	66	5
61	1	68	4
		69	1

Figure 22-10 shows the bar chart representation for this data.

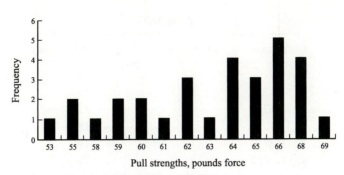

Fig. 22-10

22.4. A chemical compound with the empirical formula $K_2Cr_2O_7$ has the following percentage composition by weight: K = 26.57%, Cr = 35.36%, and O = 38.07%. Represent the composition of this compound using a pie chart.

To draw a pie chart it is necessary to determine the size of each of its "slices." To do this, multiply the percentage given by 360°. These results indicate the angle at the center for each slice. The sizes of the slices are:

Potassium = 26.57% × 360° = 95.652° ≈ 96°
Chromium = 35.36% × 360° = 127.296° ≈ 127°
Oxygen = 38.07% × 360° = 137.052° ≈ 137°

Figure 22-11 shows the percentage composition of the chemical compound.

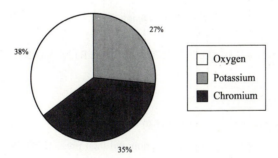

Fig. 22-11

22.5. Construct a frequency histogram using the data shown below. Explain how to generate the diagram using a graphing calculator.

23	13	82	13	67	52	25	17	15	60
32	49	37	55	81	39	99	88	24	30
13	19	53	50	18	16	40	51	61	35

Class width = (largest value in the data set − smallest value in data set)/number of classes. If there are 6 classes, then the class width = (99 − 13)/6 ≈ 14.33. Let us choose 15, an odd number, as the width of the class following the convention adopted earlier in the chapter.

Let us subtract 0.5 from the lowest value in the data set and make the result the initial boundary of the first class. In addition, let us use the second-digit convention of the initial boundary (the left end) of each class to

Table 22.7

Class limits	Frequency	Cumulative frequency
12.5–27.5	3	3
27.51–42.5	6	9
42.51–57.5	4	13
57.51–72.5	5	18
72.51–87.5	1	19
87.51–102.5	2	21

assure that no measurement falls in a class boundary.   The classes, frequencies, and cumulative frequencies are shown in Table 22.7.

To draw the histogram for the non-users group using a TI-82 follow the steps indicated below.

(1)   Enter the data in a data list.

To draw a histogram two lists are always necessary.   One list contains the measurements; the other contains the frequencies.   Clear the lists before entering the data.   To clear lists L_1 and L_2 key in the following sequence: {STAT}4{ENTER}{2^{nd}}{L_1}{,}{2^{nd}}L_2{ENTER}.

Enter the data for list 1.

{STAT}EDIT{ENTER}

12.5{ENTER}27.51{ENTER}42.51{ENTER}57.51{ENTER}72.51{ENTER}87.51{ENTER}

Notice that in this list we entered the lower boundary of each class.

To enter data in L2, navigate to that list using the directional keys {◄ ► ▲ ▼}.   Enter the frequency as follows:

11{ENTER}6{ENTER}6{ENTER}3{ENTER}2{ENTER}2{ENTER}.

(2)   Define the plot

{2^{nd}}{STAT PLOT}Plot1{ENTER}.

You may have to use the directional keys to select the plot.

Set Plot1 on.   To do this highlight this option and then press {ENTER}.

Select the histogram as the type of graph.   Highlight the rightmost option in the option Type and press {ENTER}.

Choose L1 as the Xlist.   To do this highlight L1 and press {ENTER}.

Choose L2 as the Freq. list.   To do this highlight L2 and press {ENTER}.

(3)   Define the range

{WINDOW}{ENTER}

Xmin = 12.5 {ENTER} (smallest frequency)

Xmax = 102.51 {ENTER} (highest frequency)

Xscl = 15 {ENTER} (the width of the class)

Ymin = {(−)}10 {ENTER}

Ymax = 10 {ENTER}

Yscl = 1 {ENTER}

(4)   Draw the histogram

{TRACE}

You may have to ZOOM to see the histogram.   The option ZoomStat allows you to do this.   The histogram for this data set is shown in Fig. 22-12.

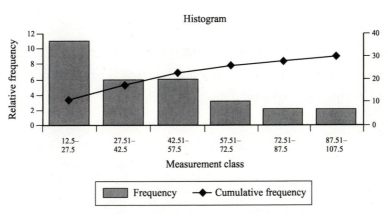

Fig. 22-12

22.6.   To determine the impact strengths (measured in ft-lb) of sheets of an insulating material, the sheets are cut lengthwise and crosswise.   Calculate the mean, median, range, sample variance, and sample standard deviation of the two data sets.   Explain how to calculate these values for the lengthwise set using a calculator.

Lengthwise	Crosswise	Lenthwise	Crosswise
1.15	0.89	0.90	0.69
0.86	0.71	0.93	0.79
0.89	0.47	0.89	0.79
0.93	0.87	0.94	0.81
0.87	0.74	0.95	0.79

There are 10 measures in each group.

The arithmetic mean of the lengthwise group is calculated as follows:

$$\bar{x} = \frac{1.15 + 0.86 + 0.89 + 0.93 + 0.87 + 0.90 + 0.93 + 0.89 + 0.94 + 0.95}{10} = 0.931$$

Arranging the values in ascending order we have

0.86   0.87   0.89   0.89   0.90   0.93   0.93   0.94   0.95   1.15

Since the number of values is even, the median is the mean of the two middle entries.   In this case, the median is $(0.90 + 0.93)/2 = (1.83/2) = 0.915$.

The highest value is 1.15.   The smallest value is 0.86.   The range is $R = 1.15 - 0.86 = 0.29$.

The sample variance is calculated as follows:

(1)   Compute the difference of each value from the arithmetic mean.

$$0.86 - 0.931 = -0.071 \qquad 0.89 - 0.931 = -0.041 \qquad 0.90 - 0.931 = -0.031$$
$$0.87 - 0.931 = -0.061 \qquad 0.89 - 0.931 = -0.041 \qquad 0.93 - 0.931 = -0.001$$
$$0.93 - 0.931 = -0.001 \qquad 0.94 - 0.931 = 0.009 \qquad 0.95 - 0.931 = 0.019$$
$$1.15 - 0.931 = 0.219$$

(2)   Square each difference.

$$(-0.071)^2 = 0.005041 \quad (-0.061)^2 = 0.003721 \quad (-0.001)^2 = 0.000001 \quad (0.219)^2 = 0.047961$$
$$(-0.041)^2 = 0.001681 \quad (-0.041)^2 = 0.001681 \quad (0.009)^2 = 0.000081 \quad (-0.031)^2 = 0.000961$$
$$(-0.001)^2 = 0.000001 \quad (0.019)^2 = 0.000361$$

(3)   Sum the squares of the previous step and divide the sum by 9.

$$0.005041 + 0.001681 + 0.00001 + 0.003721 + 0.001681 + 0.000361$$
$$+ 0.000001 + 0.000081 + 0.047961 + 0.000961 = 0.06149$$

Dividing the previous sum by 9, we have that the sample variance, s^2, is equal to 0.0068322.

The sample standard deviation, s, is the square root of the sample variance; therefore,

$$s = \sqrt{0.0068322} = 0.826573$$

This result can be rounded to 0.83.

Similar procedures can be followed to calculate the values of the *crosswise* data set. The reader can verify that the results are:

Sum of all values $= 7.55$

Arithmetic mean $= 0.755$

Median $= 0.79$

Sample variance $= 0.0139833$

Sample standard deviation $= 0.1182511$

To obtain these results using a calculator we can proceed as follows.

Using an HP-38G

(1)   Select statistics.

{LIB}Statistics{ENTER}

(2)   Enter values in the data set.

You may have to clear the data set. To do this, press {SHIFT}{CLEAR}; highlight the ALL columns option and press {ENTER}.

0.89 {ENTER}0.71{ENTER}0.47{ENTER}0.87{ENTER}0.74{ENTER}0.69{ENTER}

0.79 {ENTER}0.79{ENTER}0.81{ENTER}0.79{ENTER}.

(3)   Calculate the statistics.

Before calculating the statistics make sure you are working with the correct data set. Refer to example 22.11 on how to do this.

STATS

You can read the results directly from the display.

22.7. Line graphs can also be used to represent statistical data. In this case, the line graph shown in Fig. 22-13 indicates the monthly precipitation, in inches, for the cities of Duluth (Minnesota) and Miami (Florida). From the graph, answer the following questions: (*a*) Which city receives the greatest precipitation in one month? (*b*) In which city does the least amount of precipitation occur in one month? (*c*) In which city does the greatest change in precipitation occur from one moth to the next?

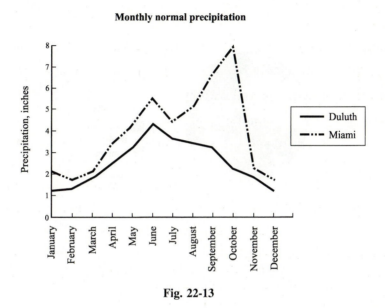

Fig. 22-13

To Answer part (*a*) of the question, notice that Miami receives about 8 inches during the month of October, reaching its maximum value during this month. To answer part (*b*) notice that Duluth receives the least amount of precipitation during the months of January and December, approximately 1.2 inches of rain. Notice that this graph reaches its lowest point during this month. For the answer to part (*c*) notice that Miami experiences the greatest change in precipitation, from one month to the next, between the months of October and November.

22.8. Table 22.8 shows the work load (measured in lines of code generated) of the different divisions of the ACME Company. Construct a pie chart for the following breakdown of work loads. The full pie represents the company total effort.

Table 22.8

Division	Lines of code
Electronic games	4,945
Graphic analyzers	2,735
Database applications	1,445
Service	1,368
Total	10,493

To draw the pie chart we need to calculate the percentages (relative to the total number of lines of code) for each of the company program areas. That is, we divided the program size in each area by the total number of lines.   We then multiply by 100 to obtain the percentage.   Rounding to integers according to the rules of Chapter 1, we obtain

Electronic games	$(4945/10{,}493) \times 100 \approx 47\%$
Graphic analysers	$(2735/10{,}493) \times 100 \approx 26\%$
Database applications	$(1445/10{,}493) \times 100 \approx 14\%$
Service	$(1368/10{,}493) \times 100 \approx 13\%$

The resulting pie chart is shown in Fig. 22-14.   The angle at the center for each slice can be obtained by multiplying the appropriate percentage by 360°.

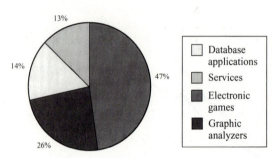

Fig. 22-14

22.9.   The numbers of hours between failures of 30 computer programs recently installed are listed below.   Construct a relative frequency histogram for this data using six classes.

		Number of hours between failures			
320	265	300	421	342	220
163	278	199	248	175	323
220	187	193	297	250	432
188	196	295	245	190	237
425	279	305	367	284	310

The range is $432 - 163 = 269$.   The width of the class is $269/6 \approx 45$.   There are 6 classes.   Table 22.9 shows the classes, the frequency and the cumulative frequency.   Figure 22-15 shows the histogram.

Table 22.9

Class limits	Frequency	Cumulative frequency
12.5–27.5	3	3
27.51–42.5	6	9
42.51–57.5	4	13
57.51–72.5	5	18
72.51–87.5	1	19
87.51–102.5	2	21

Histogram

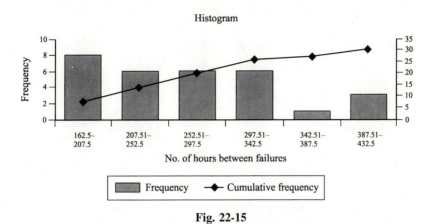

Fig. 22-15

22.10. Assume that the telephone connections via modems to a university computer center for the purpose of downloading graphic files last, on the average, 6 minutes with a standard deviation of 2 minutes. What is the probability that a telephone connection randomly selected from the computer center logs lasted at least 10 minutes?

To answer this question, determine how many standard deviations from the mean is the value of 10 minutes. This particular value is 2 standard deviation from the mean since $(10 - 6)/2 = 4/2 = 2$. The probability that the selected call lasted for at least 10 minutes is equal to the value of the area indicated in Fig. 22-16.

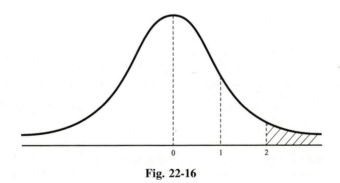

Fig. 22-16

To calculate this value refer to the table Areas under the Standard Normal Curve in the appendix. From this table we observe that the area under the curve between the mean and 2 standard deviations is 0.4772. To calculate the value indicated in the sketch, we need to subtract the value just obtained from the value corresponding to the entire area to the right of the mean. Since the latter value is equal to 0.5, the area to the right of 2 standard deviation is $0.5 - 0.4772 = 0.0228$. This result indicates that the probability of a telephone call lasted 10 minutes or more is very low.

22.11. What is the probability that a value in a normal distribution falls between 1.0 and 2.0 standard deviations from the mean?

Draw a sketch of a normal curve and shade in the area in consideration as shown in Fig. 22-17.

This problem cannot be solved directly by looking up the values in a table. First notice that the table only gives the values between the mean and a particular standard deviation. In this case, values in the intervals 0

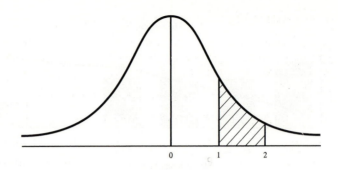

Fig. 22-17

through 1 and 0 through 2. The table does not give directly any z value that falls in the interval 1 to 2. To find the value in this area proceed as follows: (*a*) Find the area under the curve of the interval 0–1. (*b*) Find the area under the curve of the interval 0–2. (*c*) Subtract the value for the interval 0–1 from the value for the interval 0–2.

Looking at the table on page 475, we have that $P(9 \leq z \leq 1) = 0.3413$. and $P(1 \leq z \leq 2) = 0.4772$.

Therefore, the area in the interval 1–2 is $(0.4772 - 0.3413) = 0.1359$.

22.12. Given a normal distribution with mean of 100 and a standard deviation of 200, what is the probability that a value will be greater than 300?

To solve this problem it is better to express the value in standard units. Doing this changes the nature of the problem from one of finding the probability of a value being greater than 300 to that of finding the probability of a random variable falling in an interval past 1 standard deviation above the mean. Therefore,

(*a*) Express the given value in standard units.

$$z = (300 - 10)/200 = 1$$

That is, the given value is above the average and exactly 1 standard deviation away from the mean

(*b*) Find the probability that a random variable lies in the interval starting at 1 and extending its right indefinitely.

The area under the curve from the mean and extending to the right indefinitely has a value of 0.5. Remember that the entire area under the curve is equal to 1. Subtracting from this area the area under the curve in the interval 0–1 will give the value of the area to the right of 1. That is, the probability of a random variable falling in this interval.

$$P(z > 1) = P(0 < z < \infty) - P(0 < z < 1) = 0.5 - 0.3413 = 0.1587$$

Figure 22-18 shows the normal curve and the area under consideration.

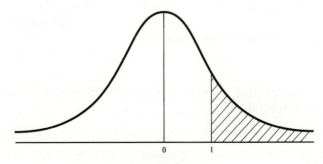

Fig. 22-18

22.13. Find the area of Problem 22.11 with the help of a calculator.

Some graphing calculators allow us to solve this problem by subtracting areas under the curve. The calculator finds, given the mean, the variance, and a particular value, the probability (the area under the curve) that a normal random variable is greater than the given value.

To solve this problem using an HP-38G proceed as indicated below.

(*a*) Find the area greater than 1 standard deviations from the mean.

{MATH}

Find the probability function UTPN (upper tail normal probability) under the Prob menu of the math functions. You may have to use the directional keys to select this option. Assuming that you are already at the MATH menu, you may press the following keys to select the UTPN function.

{▲}{►}{▲}{▲}{ENTER}

The displays shows UTPN(

Key in the following values:

0{,}1{,}1{)}{ENTER}

(these values represent, in order, a mean of 0, a variance of 1, and a value of 1).

The display shows UTPN(0,1,1) = 0.158655253931.

Save this value in a memory variable, say, A. To do this press

STO►{A...Z}{A}{ENTER}

(*b*) Find the area greater than 2 standard deviations from the mean.

{MATH}

{▲}{►}{▲}{▲}{ENTER}

The displays shows UTPN(

Key in the following values:

0{,}1{,}2{)}{ENTER}

(these values represent, in order, a mean of 0, a variance of 1, and a value of 2).

The display shows UTPN(0,1,2) = 2.27501319482E-2.

Save this value in a memory variable, say, B. To do this press

STO►{A...Z}{B}{ENTER}

(*c*) Subtract the area of step (*b*) from the area of step (*a*).

{A...Z}{A}{−}{A...Z}{B}{ENTER}

The display shows 0.135905121983. This value can be rounded to 0.1359.

22.14. Find the total value under the curve for the region indicated in Fig. 22-19.

Due to the symmetry of the normal curve, the region from −1 to 0 is a mirror image of the region from 0 to 1. The table only gives values for the region from 0 to 1. This value is 0.3413. Therefore, for the region in consideration, the total area is twice that value. That is, 0.6826. Notice that this result agrees with the empirical rule that 68% of the total observations fall within 1 standard deviation (in both directions) of the mean.

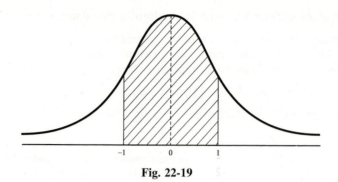

Fig. 22-19

22.15. Given a set of 1500 measures, find the fraction that will fall within 1.5 standard deviation of the mean using (a) Chebyshev's rule (b) the Camp-Meidel inequality.

According to the Chebyshev's rule at least $1 - 1/(1.5)^2 = 0.56$ or 56% of the total measures will fall within 1.5 standard deviation of the mean.   In this case, at least 840 measures will lie within that interval.

If we assume that the distribution is unimodal, the Camp-Meidel inequality states that $1 - (1/2.25k^2) = 0.80$, or 80% of the measurements will fall within this interval.

Supplementary Problems

The answers to this set of problems are at the end of this chapter.

22.16. The reliability of personal computers is sometimes determined by the average life span of some of its hardware components such as hard disks, tape drives, and memory boards.   In order to study the reliability of some of these components, a set of 200 computers were selected at random and then tested continually until one of the components failed.   The life span of each component that failed was recorded.   What is the population of interest?   What is the sample of this problem?

22.17. An insurance company would like to determine the percentage of doctors who have been involved in malpractice suits in the South of the United States.   The company selects 1500 doctors at random and determines the number of doctors who have been involved in a malpractice suit.   Describe the population and the sample.   If another sample of the same size is selected, do you expect the results obtained to be similar?

22.18. Classify the following data as quantitative or qualitative: (a) the ages of the students in Physics 101;   (b) the types of payment either cash, check, or credit made by customers at a hardware store;   (c) the ranking of the faculty as Instructor, Assistant Professor, Associate Professor, and Full Professor in a university;   (d) the capacity of a full tank in the latest U.S.-made cars;   (e) the color of the eyes of your classmates.

22.19. A large university in the South has been keeping track of the number of man-months of their computer projects.   A man-month is defined as the number of months that it would take a full-time programmer to write the program working by himself.   Make a frequency distribution diagram from the data below.

27	18	20	23	19	24
22	16	25	24	18	21
19	30	23	26	22	19
24	24	19	25	20	23
16	18	23	18	20	22

22.20. Construct a line graph for efficiency of the American car.   Use the data of the Table 22.10.   Efficiency is measured in miles per gallon.

Table 22.10

Model	Efficiency	Model	Efficiency
1978	18.5	1985	28
1979	19	1986	26.2
1980	20.1	1987	26
1981	21.5	1988	26.4
1982	23	1989	26.7
1983	25	1990	27.5
1984	26.3	1991	28

22.21. Classify the following variables as discrete or continuous: (*a*) the number of planes that take off from a regional airport;   (*b*) the weights of new born babies at the local hospital;   (*c*) summer temperatures in the South;   (*d*) the number of car crashes at a particular city intersection.

22.22. A pharmaceutical company uses a new breed of rabbits to study the effectiveness of its new drugs.   The data shown below represents the weight gain (in grams) for 20 randomly selected rabbits over a period of 3 months.   Find the range of the data.   If you were to use five classes, what are the different classes and the frequency of each class?

17.5	18.4	17.3	11.6	12.5	11.3	19.6	18.5	13.6	19.4
19.3	12.4	17.6	14.1	12.6	14.3	17.4	16.5	17	15

22.23. The data shown below corresponds to pull strength tests (in pounds force) for pins soldered to the wings of military airplanes used as battlefield surveyors.   Find the population and sample means, median, standard deviation, and sample and population variance.

33.5	28.9	27.5	31.35	30.8	27.1	29.7
25.5	29.05	29.9	32.85	29.05	34.45	33.5

22.24. A survey of the number of times a telephone rings before being answered is shown below.   Find the weighted mean of the number of rings.

Number of rings	Frequency	Number of rings	Frequency
0	0	4	17
1	19	5	10
2	32	6	2
3	25	7	3

22.25. Find the area under the standard normal curve to the left of $z = 0.57$.

22.26. Find the area under the normal curve between $z = 0$ and $z = 1.2$.

22.27. Find the area under the curve to the right of $z = -1.28$.

Answers to Supplementary Problems

22.16. The population of interest consists of the hardware components of the PCs (hard disks, tape drives, and memory boards). The sample is the set of 200 computers selected at random.

22.17. The population consists of all medical doctors who practice in the South. The sample in this case consists of the 1500 doctors selected at random. If another sample is selected, we should expect the percentage of doctors involved in a malpractice suit to be different.

22.18. (*a*) quantitative (age is generally recorded in years);   (*b*) qualitative (these are not numerical quantities);   (*c*) qualitative (these are not numerical quantities);   (*d*) quantitative (capacity is generally measured in numbers of gallons);   (*e*) qualitative

22.20. Figure 20-20 shows the line graph of the efficiency of the American cars.

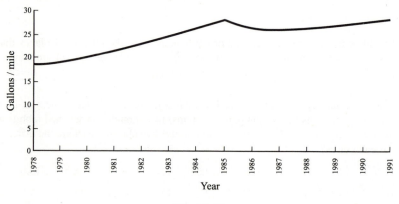

Fig. 22-20

22.21. (*a*) discrete   (*b*) continuous   (*c*) continuous   (*d*) discrete

22.22. Range $= 8.3$ (maximum $= 19.6$; minimum $= 11.3$). The classes and their frequencies are 11.3 through 12.96 (5), 12.97 through 14.63 (3), 14.64 through 16.3 (1), 16.31 through 17.97 (6), 17.98 through 19.64 (5).

22.23. mean $= 30.23$ median $= 29.8$ Population standard deviation $= 2.57$ population variance $= 6.59$ sample variance $= 7.09$ Sample standard deviation $= 2.66$

22.24. weighted mean $= 2.86$

22.25. 0.7157

22.26. 0.3849

22.27. 0.8997

Appendix

MATHEMATICAL SYMBOLS

$+$	plus		$\parallel$	parallel; parallel to
$-$	minus		$\parallel_s$	parallels
$\pm$	plus or minus		$\perp$	perpendicular; perpendicular to
$\times, \cdot$	times		$\perp_s$	perpendiculars
$\div, /$	divided by		$\triangle$	triangle
$:$	ratio sign		$\triangle\!\!\!\triangle$	triangles
$=$	equals		$\square$	square
$\neq$	not equal		$\square$	rectangle
$\approx$	approximately equal		$\square$	parallelogram
$\equiv$	identically equal		$\odot$	circle
$>$	greater than		$\odot$	circles
$<$	less than		$\overline{AB}$	line segment AB
$\geq$	greater than or equal to		$\overarc{CD}$	arc CD
$\leq$	less than or equal to		$\sim$	similar
$\gg$	much greater than		$\cong$	congruent
$\ll$	much less than		$\therefore$	therefore
$\rightarrow$	approaches		$^\circ$	degree
$\propto$	varies as; proportional to		$'$	minute
∞	infinity		$''$	second
$\| \|$	absolute value; magnitude		$\sum$	summation
$\sqrt{}$	radical sign; square root		$(\ )$	parentheses
$\sqrt[3]{}$	cube root		$[\ ]$	brackets
$\angle$	angle		$\{\ \}$	braces
$\angle\!\!\!\angle$	angles		$!$	factorial

METRIC UNIT PREFIXES

Name	Abbreviation	Multiple
Exa	E	10^{18}
Peta	P	10^{15}
Tera	T	10^{12}
Giga	G	10^{9}
Mega	M	10^{6}
Kilo	k	10^{3}
Hecto	h	10^{2}
Deka	da	10
Deci	d	10^{-1}
Centi	c	10^{-2}
Milli	m	10^{-3}
Micro	μ	10^{-6}
Nano	n	10^{-9}
Pico	p	10^{-12}
Femto	f	10^{-15}
Atto	a	10^{-18}

MATHEMATICAL CONSTANTS

$\pi = 3.142$ $1/\pi = 0.3183$ $\log \pi = 0.4971$

$2\pi = 6.283$ $1/2\pi = 0.1592$ $\ln \pi = 1.145$

$4\pi = 12.57$ $1/4\pi = 0.07958$ $\log \pi^2 = 0.9943$

$\pi/2 = 1.571$ $\pi^2 = 9.870$ $\log \pi/2 = 0.1961$

$\pi/4 = 0.7854$ $1/\pi^2 = 0.1013$

$\pi/180 = 0.01745$ $\sqrt{\pi} = 1.772$

$4\pi/3 = 4.189$

$e = 2.718$ $\log 2 = 0.3010$ $\sqrt{2} = 1.414$

$e^2 = 7.389$ $\ln 2 = 0.6931$ $\sqrt{3} = 1.732$

$1/e = 0.3679$ $\ln 10 = 2.303$ $1/\sqrt{2} = 0.7071$

$\log e = 0.4343$ $1/\sqrt{2} = 0.5774$

GREEK ALPHABET

A, α	Alpha	H, η	Eta	N, ν	Nu	T, τ	Tau
B, β	Beta	Θ, θ, ϑ	Theta	Ξ, ξ	Xi	Y, υ	Upsilon
Γ, γ	Gamma	I, ι	Iota	O, o	Omicron	Φ, ϕ, φ	Phi
Δ, δ	Delta	K, κ	Kappa	Π, π	Pi	X, χ	Chi
E, ϵ	Epsilon	Λ, λ	Lambda	P, ρ	Rho	Ψ, ψ	Psi
Z, ζ	Zeta	M, μ	Mu	Σ, σ	Sigma	Ω, ω	Omega

APPENDIX

SQUARES AND SQUARE ROOTS

Number	Square	Square Root	Number	Square	Square Root	Number	Square	Square Root
1	1	1.000	51	26 01	7.141	101	1 02 01	10.050
2	4	1.414	52	27 04	7.211	102	1 04 04	10.100
3	9	1.732	53	28 09	7.280	103	1 06 09	10.149
4	16	2.000	54	29 16	7.348	104	1 08 16	10.198
5	25	2.236	55	30 25	7.416	105	1 10 25	10.247
6	36	2.449	56	31 36	7.483	106	1 12 36	10.296
7	49	2.646	57	32 49	7.550	107	1 14 49	10.344
8	64	2.828	58	33 64	7.616	108	1 16 64	10.392
9	81	3.000	59	34 81	7.681	109	1 18 81	10.440
10	1 00	3.162	60	36 00	7.746	110	1 21 00	10.488
11	1 21	3.317	61	37 21	7.810	111	1 23 21	10.536
12	1 44	3.464	62	38 44	7.874	112	1 25 44	10.583
13	1 69	3.606	63	39 69	7.937	113	1 27 69	10.630
14	1 96	3.742	64	40 96	8.000	114	1 29 96	10.677
15	2 25	3.873	65	42 25	8.062	115	1 32 25	10.724
16	2 56	4.000	66	43 56	8.124	116	1 34 56	10.770
17	2 89	4.123	67	44 89	8.185	117	1 36 89	10.817
18	3 24	4.243	68	46 24	8.246	118	1 39 24	10.863
19	3 61	4.359	69	47 61	8.307	119	1 41 61	10.909
20	4 00	4.472	70	49 00	8.367	120	1 44 00	10.954
21	4 41	4.583	71	50 41	8.426	121	1 46 41	11.000
22	4 84	4.690	72	51 84	8.485	122	1 48 84	11.045
23	5 29	4.796	73	53 29	8.544	123	1 51 29	11.091
24	5 76	4.899	74	54 76	8.602	124	1 53 76	11.136
25	6 25	5.000	75	56 25	8.660	125	1 56 25	11.180
26	6 76	5.099	76	57 76	8.718	126	1 58 76	11.225
27	7 29	5.196	77	59 29	8.775	127	1 61 29	11.269
28	7 84	5.292	78	60 84	8.832	128	1 63 84	11.314
29	8 41	5.385	79	62 41	8.888	129	1 66 41	11.358
30	9 00	5.477	80	64 00	8.944	130	1 69 00	11.402
31	9 61	5.568	81	65 61	9.000	131	1 71 61	11.446
32	10 24	5.657	82	67 24	9.055	132	1 74 24	11.489
33	10 89	5.745	83	68 89	9.110	133	1 76 89	11.533
34	11 56	5.831	84	70 56	9.165	134	1 79 56	11 576
35	12 25	5.916	85	72 25	9.220	135	1 82 25	11.619
36	12 96	6.000	86	73 96	9.274	136	1 84 96	11.662
37	13 69	6.083	87	75 69	9.327	137	1 87 69	11.705
38	14 44	6.164	88	77 44	9.381	138	1 90 44	11.747
39	15 21	6.245	89	79 21	9.434	139	1 93 21	11.790
40	16 00	6.325	90	81 00	9.487	140	1 96 00	11.832
41	16 81	6.403	91	82 81	9.539	141	1 98 81	11.874
42	17 64	6.481	92	84 64	9.592	142	2 01 64	11.916
43	18 49	6.557	93	86 49	9.644	143	2 04 49	11.958
44	19 36	6.633	94	88 36	9.695	144	2 07 36	12.000
45	20 25	6.708	95	90 25	9.747	145	2 10 25	12.042
46	21 16	6.782	96	92 16	9.798	146	2 13 16	12.083
47	22 09	6.856	97	94 09	9.849	147	2 16 09	12.124
48	23 04	6.928	98	96 04	9.899	148	2 19 04	12.166
49	24 01	7.000	99	98 01	9.950	149	2 22 01	12.207
50	25 00	7.071	100	1 00 00	10.000	150	2 25 00	12.247

COMMON LOGARITHMS

No.	0	1	2	3	4	5	6	7	8	9	Proportional Parts								
											1	2	3	4	5	6	7	8	9
10	0000	0043	0086	0128	0170	0212	0253	0294	0334	0374	4	8	12	17	21	25	29	33	37
11	0414	0453	0492	0531	0569	0607	0645	0682	0719	0755	4	8	11	15	19	23	26	30	34
12	0792	0828	0864	0899	0934	0969	1004	1038	1072	1106	3	7	10	14	17	21	24	28	31
13	1139	1173	1206	1239	1271	1303	1335	1367	1399	1430	3	6	10	13	16	19	23	26	29
14	1461	1492	1523	1553	1584	1614	1644	1673	1703	1732	3	6	9	12	15	18	21	24	27
15	1761	1790	1818	1847	1875	1903	1931	1959	1987	2014	3	6	8	11	14	17	20	22	25
16	2041	2068	2095	2122	2148	2175	2201	2227	2253	2279	3	5	8	11	13	16	18	21	24
17	2304	2330	2355	2380	2405	2430	2455	2480	2504	2529	2	5	7	10	12	15	17	20	22
18	2553	2577	2601	2625	2648	2672	2695	2718	2742	2765	2	5	7	9	12	14	16	19	21
19	2788	2810	2833	2856	2878	2900	2923	2945	2967	2989	2	4	7	9	11	13	16	18	20
20	3010	3032	3054	3075	3096	3118	3139	3160	3181	3201	2	4	6	8	11	13	15	17	19
21	3222	3243	3263	3284	3304	3324	3345	3365	3385	3404	2	4	6	8	10	12	14	16	18
22	3424	3444	3464	3483	3502	3522	3541	3560	3579	3598	2	4	6	8	10	12	14	15	17
23	3617	3636	3655	3674	3692	3711	3729	3747	3766	3784	2	4	6	7	9	11	13	15	17
24	3802	3820	3838	3856	3874	3892	3909	3927	3945	3962	2	4	5	7	9	11	12	14	16
25	3979	3997	4014	4031	4048	4065	4082	4099	4116	4133	2	3	5	7	9	10	12	14	15
26	4150	4166	4183	4200	4216	4232	4249	4265	4281	4298	2	3	5	7	8	10	11	13	15
27	4314	4330	4346	4362	4378	4393	4409	4425	4440	4456	2	3	5	6	8	9	11	13	14
28	4472	4487	4502	4518	4533	4548	4564	4579	4594	4609	2	3	5	6	8	9	11	12	14
29	4624	4639	4654	4669	4683	4698	4713	4728	4742	4757	1	3	4	6	7	9	10	12	13
30	4771	4786	4800	4814	4829	4843	4857	4871	4886	4900	1	3	4	6	7	9	10	11	13
31	4914	4928	4942	4955	4969	4983	4997	5011	5024	5038	1	3	4	6	7	8	10	11	12
32	5051	5065	5079	5092	5105	5119	5132	5145	5159	5172	1	3	4	5	7	8	9	11	12
33	5185	5198	5211	5224	5237	5250	5263	5276	5289	5302	1	3	4	5	6	8	9	10	12
34	5315	5328	5340	5353	5366	5378	5391	5403	5416	5428	1	3	4	5	6	8	9	10	11
35	5441	5453	5465	5478	5490	5502	5514	5527	5539	5551	1	2	4	5	6	7	9	10	11
36	5563	5575	5587	5599	5611	5623	5635	5647	5658	5670	1	2	4	5	6	7	8	10	11
37	5682	5694	5705	5717	5729	5740	5752	5763	5775	5786	1	2	3	5	6	7	8	9	10
38	5798	5809	5821	5832	5843	5855	5866	5877	5888	5899	1	2	3	5	6	7	8	9	10
39	5911	5922	5933	5944	5955	5966	5977	5988	5999	6010	1	2	3	4	5	7	8	9	10
40	6021	6031	6042	6053	6064	6075	6085	6096	6107	6117	1	2	3	4	5	6	8	9	10
41	6128	6138	6149	6160	6170	6180	6191	6201	6212	6222	1	2	3	4	5	6	7	8	9
42	6232	6243	6253	6263	6274	6284	6294	6304	6314	6325	1	2	3	4	5	6	7	8	9
43	6335	6345	6355	6365	6375	6385	6395	6405	6415	6425	1	2	3	4	5	6	7	8	9
44	6435	6444	6454	6464	6474	6484	6493	6503	6513	6522	1	2	3	4	5	6	7	8	9
45	6532	6542	6551	6561	6571	6580	6590	6599	6609	6618	1	2	3	4	5	6	7	8	9
46	6628	6637	6646	6656	6665	6675	6684	6693	6702	6712	1	2	3	4	5	6	7	7	8
47	6721	6730	6739	6749	6758	6767	6776	6785	6794	6803	1	2	3	4	5	5	6	7	8
48	6812	6821	6830	6839	6848	6857	6866	6875	6884	6893	1	2	3	4	4	5	6	7	8
49	6902	6911	6920	6928	6937	6946	6955	6964	6972	6981	1	2	3	4	4	5	6	7	8
50	6990	6998	7007	7016	7024	7033	7042	7050	7059	7067	1	2	3	3	4	5	6	7	8
51	7076	7084	7093	7101	7110	7118	7126	7135	7143	7152	1	2	3	3	4	5	6	7	8
52	7160	7168	7177	7185	7193	7202	7210	7218	7226	7235	1	2	2	3	4	5	6	7	7
53	7243	7251	7259	7267	7275	7284	7292	7300	7308	7316	1	2	2	3	4	5	6	6	7
54	7324	7332	7340	7348	7356	7364	7372	7380	7388	7396	1	2	2	3	4	5	6	6	7

COMMON LOGARITHMS (cont.)

No.	0	1	2	3	4	5	6	7	8	9	1	2	3	4	5	6	7	8	9
											Proportional Parts								
55	7404	7412	7419	7427	7435	7443	7451	7459	7466	7474	1	2	2	3	4	5	5	6	7
56	7482	7490	7497	7505	7513	7520	7528	7536	7543	7551	1	2	2	3	4	5	5	6	7
57	7559	7566	7574	7582	7589	7597	7604	7612	7619	7627	1	2	2	3	4	5	5	6	7
58	7634	7642	7649	7657	7664	7672	7679	7686	7694	7701	1	1	2	3	4	4	5	6	7
59	7709	7716	7723	7731	7738	7745	7752	7760	7767	7774	1	1	2	3	4	4	5	6	7
60	7782	7789	7796	7803	7810	7818	7825	7832	7839	7846	1	1	2	3	4	4	5	6	6
61	7853	7860	7868	7875	7882	7889	7896	7903	7910	7917	1	1	2	3	4	4	5	6	6
62	7924	7931	7938	7945	7952	7959	7966	7973	7980	7987	1	1	2	3	3	4	5	6	6
63	7993	8000	8007	8014	8021	8028	8035	8041	8048	8055	1	1	2	3	3	4	5	5	6
64	8062	8069	8075	8082	8089	8096	8102	8109	8116	8122	1	1	2	3	3	4	5	5	6
65	8129	8136	8142	8149	8156	8162	8169	8176	8182	8189	1	1	2	3	3	4	5	5	6
66	8195	8202	8209	8215	8222	8228	8235	8241	8248	8254	1	1	2	3	3	4	5	5	6
67	8261	8267	8274	8280	8287	8293	8299	8306	8312	8319	1	1	2	3	3	4	5	5	6
68	8325	8331	8338	8344	8351	8357	8363	8370	8376	8382	1	1	2	3	3	4	4	5	6
69	8388	8395	8401	8407	8414	8420	8426	8432	8439	8445	1	1	2	2	3	4	4	5	6
70	8451	8457	8463	8470	8476	8482	8488	8494	8500	8506	1	1	2	2	3	4	4	5	6
71	8513	8519	8525	8531	8537	8543	8549	8555	8561	8567	1	1	2	2	3	4	4	5	5
72	8573	8579	8585	8591	8597	8603	8609	8615	8621	8627	1	1	2	2	3	4	4	5	5
73	8633	8639	8645	8651	8657	8663	8669	8675	8681	8686	1	1	2	2	3	4	4	5	5
74	8692	8698	8704	8710	8716	8722	8727	8733	8739	8745	1	1	2	2	3	4	4	5	5
75	8751	8756	8762	8768	8774	8779	8785	8791	8797	8802	1	1	2	2	3	3	4	5	5
76	8808	8814	8820	8825	8831	8837	8842	8848	8854	8859	1	1	2	2	3	3	4	5	5
77	8865	8871	8876	8882	8887	8893	8899	8904	8910	8915	1	1	2	2	3	3	4	4	5
78	8921	8927	8932	8938	8943	8949	8954	8960	8965	8971	1	1	2	2	3	3	4	4	5
79	8976	8982	8987	8993	8998	9004	9009	9015	9020	9025	1	1	2	2	3	3	4	4	5
80	9031	9036	9042	9047	9053	9058	9063	9069	9074	9079	1	1	2	2	3	3	4	4	5
81	9085	9090	9096	9101	9106	9112	9117	9122	9128	9133	1	1	2	2	3	3	4	4	5
82	9138	9143	9149	9154	9159	9165	9170	9175	9180	9186	1	1	2	2	3	3	4	4	5
83	9191	9196	9201	9206	9212	9217	9222	9227	9232	9238	1	1	2	2	3	3	4	4	5
84	9243	9248	9253	9258	9263	9269	9274	9279	9284	9289	1	1	2	2	3	3	4	4	5
85	9294	9299	9304	9309	9315	9320	9325	9330	9335	9340	1	1	2	2	3	3	4	4	5
86	9345	9350	9355	9360	9365	9370	9375	9380	9385	9390	1	1	2	2	3	3	4	4	5
87	9395	9400	9405	9410	9415	9420	9425	9430	9435	9440	0	1	1	2	2	3	3	4	4
88	9445	9450	9455	9460	9465	9469	9474	9479	9484	9489	0	1	1	2	2	3	3	4	4
89	9494	9499	9504	9509	9513	9518	9523	9528	9533	9538	0	1	1	2	2	3	3	4	4
90	9542	9547	9552	9557	9562	9566	9571	9576	9581	9586	0	1	1	2	2	3	3	4	4
91	9590	9595	9600	9605	9609	9614	9619	9624	9628	9633	0	1	1	2	2	3	3	4	4
92	9638	9643	9647	9652	9657	9661	9666	9671	9675	9680	0	1	1	2	2	3	3	4	4
93	9685	9689	9694	9699	9703	9708	9713	9717	9722	9727	0	1	1	2	2	3	3	4	4
94	9731	9736	9741	9745	9750	9754	9759	9763	9768	9773	0	1	1	2	2	3	3	4	4
95	9777	9782	9786	9791	9795	9800	9805	9809	9814	9818	0	1	1	2	2	3	3	4	4
96	9823	9827	9832	9836	9841	9845	9850	9854	9859	9863	0	1	1	2	2	3	3	4	4
97	9868	9872	9877	9881	9886	9890	9894	9899	9903	9908	0	1	1	2	2	3	3	4	4
98	9912	9917	9921	9926	9930	9934	9939	9943	9948	9952	0	1	1	2	2	3	3	4	4
99	9956	9961	9965	9969	9974	9978	9983	9987	9991	9996	0	1	1	2	2	3	3	3	4

NATURAL TRIGONOMETRIC FUNCTIONS

Angle		Values of Functions			Angle		Values of Functions		
Degrees	Radians	sin	cos	tan	Degrees	Radians	sin	cos	tan
0	0.0000	0.0000	1.000	0.0000	45	0.7854	0.7071	0.7071	1.000
1	0.01745	0.01745	0.9998	0.01746	46	0.8028	0.7193	0.6947	1.036
2	0.03491	0.03490	0.9994	0.03492	47	0.8203	0.7314	0.6820	1.072
3	0.05236	0.05234	0.9986	0.05241	48	0.8378	0.7431	0.6691	1.111
4	0.06981	0.06976	0.9976	0.06993	49	0.8552	0.7547	0.6561	1.150
5	0.08727	0.08716	0.9962	0.08749	50	0.8727	0.7660	0.6428	1.192
6	0.1047	0.1045	0.9945	0.1051	51	0.8901	0.7772	0.6293	1.235
7	0.1222	0.1219	0.9926	0.1228	52	0.9076	0.7880	0.6157	1.280
8	0.1396	0.1392	0.9903	0.1405	53	0.9250	0.7986	0.6018	1.327
9	0.1571	0.1564	0.9877	0.1584	54	0.9425	0.8090	0.5878	1.376
10	0.1745	0.1736	0.9848	0.1763	55	0.9599	0.8192	0.5736	1.428
11	0.1920	0.1908	0.9816	0.1944	56	0.9774	0.8290	0.5592	1.483
12	0.2094	0.2079	0.9782	0.2126	57	0.9948	0.8387	0.5446	1.540
13	0.2269	0.2250	0.9744	0.2309	58	1.012	0.8480	0.5299	1.600
14	0.2444	0.2419	0.9703	0.2493	59	1.030	0.8572	0.5150	1.664
15	0.2618	0.2588	0.9659	0.2680	60	1.047	0.8660	0.5000	1.732
16	0.2792	0.2756	0.9613	0.2868	61	1.065	0.8746	0.4848	1.804
17	0.2967	0.2924	0.9563	0.3057	62	1.082	0.8830	0.4695	1.881
18	0.3142	0.3090	0.9511	0.3249	63	1.100	0.8910	0.4540	1.963
19	0.3316	0.3256	0.9455	0.3443	64	1.117	0.8988	0.4384	2.050
20	0.3491	0.3420	0.9397	0.3640	65	1.134	0.9063	0.4226	2.144
21	0.3665	0.3584	0.9336	0.3839	66	1.152	0.9136	0.4067	2.246
22	0.3840	0.3746	0.9272	0.4040	67	1.169	0.9205	0.3907	2.356
23	0.4014	0.3907	0.9205	0.4245	68	1.187	0.9272	0.3746	2.475
24	0.4189	0.4067	0.9136	0.4452	69	1.204	0.9336	0.3584	2.605
25	0.4363	0.4226	0.9063	0.4663	70	1.222	0.9397	0.3420	2.748
26	0.4538	0.4384	0.8988	0.4877	71	1.239	0.9455	0.3256	2.904
27	0.4712	0.4540	0.8910	0.5095	72	1.257	0.9511	0.3090	3.078
28	0.4887	0.4695	0.8830	0.5317	73	1.274	0.9563	0.2924	3.271
29	0.5062	0.4848	0.8746	0.5543	74	1.292	0.9613	0.2756	3.487
30	0.5236	0.5000	0.8660	0.5774	75	1.309	0.9659	0.2588	3.732
31	0.5410	0.5150	0.8572	0.6009	76	1.326	0.9703	0.2419	4.011
32	0.5585	0.5299	0.8480	0.6249	77	1.344	0.9744	0.2250	4.332
33	0.5760	0.5446	0.8387	0.6494	78	1.361	0.9782	0.2079	4.705
34	0.5934	0.5592	0.8290	0.6745	79	1.379	0.9816	0.1908	5.145
35	0.6109	0.5736	0.8192	0.7002	80	1.396	0.9848	0.1736	5.671
36	0.6283	0.5878	0.8090	0.7265	81	1.414	0.9877	0.1564	6.314
37	0.6458	0.6018	0.7986	0.7536	82	1.431	0.9903	0.1392	7.115
38	0.6632	0.6157	0.7880	0.7813	83	1.449	0.9926	0.1219	8.144
39	0.6807	0.6293	0.7772	0.8098	84	1.466	0.9945	0.1045	9.514
40	0.6981	0.6428	0.7660	0.8391	85	1.484	0.9962	0.08716	11.43
41	0.7156	0.6561	0.7547	0.8693	86	1.501	0.9976	0.06976	14.30
42	0.7330	0.6691	0.7431	0.9004	87	1.518	0.9986	0.05234	19.08
43	0.7505	0.6820	0.7314	0.9325	88	1.536	0.9994	0.03490	28.64
44	0.7679	0.6947	0.7193	0.9657	89	1.553	0.9998	0.01745	57.29
					90	1.571	1.000	0.0000	∞

UNITS AND CONVERSION FACTORS

LENGTH

1 centimeter (cm) $= 10^{-2}$ meter $= 0.3937$ inch

1 meter (m) $= 100$ centimeters $= 39.37$ inch $= 1.094$ yard $= 1000$ millimeters (mm)

1 kilometer $= 10^3$ meters $= 0.6214$ statute mile

1 light-year $= 9.464 \times 10^{12}$ km

1 micrometer (μ) $= 10^{-6}$ meter $= 10^{-4}$ centimeter

1 nanometer (mμ) $= 10^{-9}$ meter $= 10^{-7}$ centimeter

1 Angstrom (Å) $= 10^{-10}$ meter $= 10^{-8}$ centimeter

1 inch (in.) $= 2.540$ centimeters

1 yard (yd) $= 0.9144$ meter

1 mile (mi) $= 1.609$ kilometer

MASS

1 kilogram (kg) $= 1000$ grams $= 2.205$ lb (avoirdupois)

1 pound (avoirdupois) (lb) $= 0.4536$ kilogram $= 453.6$ grams

1 ounce (avoirdupois) (oz) $= 28.35$ grams

1 slug $= 14.59$ kilograms $= 32.17$ pounds (avoirdupois)

1 atomic mass unit (amu) $= 1.660 \times 10^{-24}$ gram $= 1.660 \times 10^{-27}$ kilogram

TIME

1 second (s, sec) $= 1/60$ minute (min) $= 1/3600$ hour (h, hr) $= 1/86,400$ day (d)

FORCE

1 newton (N) $= 10^5$ dyne $= 0.2248$ pound force $= 7.233$ poundals

1 pound force $= 4.448$ newtons $= 4.448 \times 10^5$ dyne $= 32.17$ poundals

1 dyne (dyn) $= 10^{-5}$ newton $= 2.248 \times 10^{-6}$ pound force $= 7.233 \times 10^{-5}$ poundal

ENERGY

1 joule (J) $= 1$ newton-meter $= 10^7$ ergs $= 0.7376$ foot-pound force $= 9.481 \times 10^{-4}$ Btu
$= 2.778 \times 10^{-7}$ kilowatt-hours $= 0.2390$ calorie $= 2.390 \times 10^{-4}$ kilocalorie

1 erg $= 10^{-7}$ joule $= 1$ dyne-centimeter

1 foot-pound force (ft-lb-f) $= 1.356$ joule $= 1.285 \times 10^{-3}$ Btu $= 3.766 \times 10^{-7}$ kilowatt-hours $= 3.240 \times 10^{-4}$ kcals

1 kilogram-calorie or large calorie (kcal) $= 4.186 \times 10^3$ joules $= 1.163 \times 10^{-3}$ kilowatt-hour

1 gram-calorie (cal) $= 4.186$ joules

1 Btu (60 °F) $= 1.055 \times 10^3$ joule $= 2.52 \times 10^2$ calories

1 kilowatt-hour (kWh) $= 3.600 \times 10^6$ joules $= 3.413 \times 10^3$ Btu $= 8.600 \times 10^2$ kcal

1 electron volt (eV) $= 1.602 \times 10^{-12}$ erg $= 1.602 \times 10^{-19}$ joule

1 atomic mass unit (amu) $= 1.492 \times 10^{-3}$ erg $= 1.492 \times 10^{-10}$ joule $= 9.315 \times 10^8$ eV $= 931.5$ MeV

POWER

1 watt (W) $= 1$ joule/sec

1 kilowatt (kW) $= 100^3$ joule/sec $= 7.376 \times 10^2$ foot-pound force/sec $= 1.341$ horsepower
$= 0.2389$ kcal/sec $= 0.9483$ Btu/sec

1 horsepower (hp) $= 745.7$ watts $= 0.7457$ kilowatt $= 550$ foot-pound force/sec $= 0.7070$ Btu/sec

SPEED

1 m/s $= 3.281$ ft/sec $= 2.237$ mi/hr $= 3600$ km/h

1 ft/sec $= 0.3048$ m/sec $= 0.6818$ mi/hr $= 1.097$ km/h

1 mph $= 1.467$ ft/sec $= 0.4470$ m/sec

MISCELLANEOUS USEFUL CONSTANTS

DENSITY

air, 760 mm Hg, 20 °C:	$1.205 \, \text{kg/m}^3$; $1.205 \times 10^{-3} \, \text{g/cm}^3$
water, 760 mm Hg, 20 °C:	$1.000 \times 10^3 \, \text{kg/m}^3$; $1.000 \, \text{g/cm}^3$
mercury, 760 mm Hg, 20 °C:	$1.355 \times 10^4 \, \text{kg/m}^3$; $13.55 \, \text{g/cm}^3$
aluminum, hard-drawn, 760 mm Hg, 20 °C:	$2.699 \times 10^3 \, \text{kg/m}^3$; $2.699 \, \text{g/cm}^3$
copper, hard-drawn, 760 mm Hg, 20 °C:	$8.89 \times 10^3 \, \text{kg/m}^3$; $8.89 \, \text{g/cm}^3$

WEIGHT

One cubic foot of water weighs 62.4 pounds.

One U.S. gallon of water weighs 8.33 pounds.

PRESSURE

Standard atmosphere $= 76 \, \text{cm Hg} = 760 \, \text{mm Hg} = 1.013 \times 10^6 \, \text{dyne/cm}^2 = 1.013 \times 10^5 \, \text{newton/m}^2 = 1.013 \, \text{bar} = 1013 \, \text{millibar} = 29.92 \, \text{in. Hg} = 33.90 \, \text{feet of water} = 14.70 \, \text{lb/in}^2$

HEAT AND TEMPERATURE

Freezing point of water, $0 \, °\text{C} = 32 \, °\text{F} = 273.15 \, \text{K} = 491.67 \, °\text{R}$

Heat of fusion of water, $0 \, °\text{C} = 79.71 \, \text{cal/g}$

Heat of vaporization of water $= 539.6 \, \text{cal/g}$ at $100 \, °\text{C}$

Specific heat of copper, $25 \, °\text{C} = 0.924 \, \text{cal/(g} \, °\text{C)}$

Specific heat of aluminum, $25 \, °\text{C} = 0.215 \, \text{cal/(g} \, °\text{C)}$

Specific heat of lead, $25 \, °\text{C} = 0.0305 \, \text{cal/(g} \, °\text{C)}$

SOUND

Velocity of sound in air, $0 \, °\text{C} = 331.36 \, \text{m/sec} = 1087 \, \text{ft/sec} \approx 742 \, \text{mi/hr}$

Velocity of sound in water, $25 \, °\text{C} = 1498 \, \text{m/s}$

Velocity of sound in brass $= 4700 \, \text{m/sec}$

Velocity of sound in iron $= 5950 \, \text{m/sec}$

FORMULA SUMMARY

Pythagorean Theorem

$a^2 + b^2 = c^2$

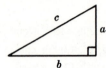

Quadratic Equation

$ax^2 + bx + c = 0$

Solution, $x = \dfrac{-b \pm \sqrt{b^2 - 4ac}}{2a}$

If $b^2 - 4ac$ is positive, the roots are real and unequal.

If $b^2 - 4ac$ is zero, the roots are real and equal.

If $b^2 - 4ac$ is negative, the roots are imaginary.

Plane Figures

RECTANGLE

    Perimeter, $P = 2a + 2b$

    Area, $A = ab$

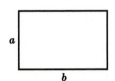

TRIANGLE

    Perimeter, $P = a + b + c$

    Area, $A = \dfrac{hc}{2}$

$$A = \sqrt{s(s - a)(s - b)(s - c)} \qquad s = \frac{1}{2}(a + b + c)$$

TRAPEZOID

    Area, $= \frac{1}{2}(a + b)h$

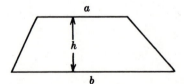

CIRCLE

    Diameter, $d = 2r$

    Circumference, $C = 2\pi r = \pi d$

    Area, $A = \pi r^2 = \dfrac{\pi d^2}{4} \approx 0.785\, d^2$

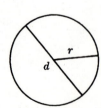

FORMULA SUMARY (cont.)

Solid Figures

RECTANGULAR SOLID

Volume, $V = abc$

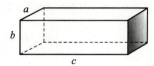

SPHERE

Surface area, $A = 4\pi r^2 = \pi d^2$

Volume, $V = \dfrac{4}{3}\pi r^3 = \dfrac{\pi d^3}{6}$

RIGHT CIRCULAR CYLINDER

Lateral surface area, $S = 2\pi rh$

Total surface area, $A = 2\pi rh + 2\pi r^2$

Volume, $V = \pi r^2 h$

RIGHT CIRCULAR CONE

Lateral surface area, $S = \pi rL$

Total surface area, $A = \pi rL + \pi r^2$

Volume, $V = \dfrac{\pi r^2 h}{3}$

Trigonometric Functions

$$\sin B = \frac{b}{c} \qquad \csc B = \frac{c}{b}$$

$$\cos B = \frac{a}{c} \qquad \sec B = \frac{c}{a}$$

$$\tan B = \frac{b}{a} \qquad \cot B = \frac{a}{b}$$

Sine Law

$$\frac{a}{\sin A} = \frac{b}{\sin B} = \frac{c}{\sin C}$$

Cosine Law

$$c^2 = a^2 + b^2 - 2ab\cos C$$

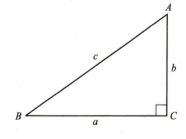

Temperature Conversion

F—Fahrenheit;   C—Celsius or centigrade;   K—Kelvin or absolute;   R—Rankine

$$^\circ C = \frac{5}{9}(^\circ F - 32) \qquad ^\circ F = \frac{9}{5}(^\circ C + 32) \qquad K = {}^\circ C + 273.16 \qquad ^\circ R = {}^\circ F + 459.6$$

BINOMIAL THEOREM

If n is a positive integer, then the nth power of a binomial $a + b$ can be expanded by the formula

$$(a + b)^n = a^n + \frac{n!}{1!(n - 1)!} a^{n-1} b + \frac{n!}{2!(n - 2)!} a^{n-2} b^2 + \cdots + b^n$$

The above expression can be reduced to a form more convenient for computations:

$$(a + b)^n = a^n + n a^{n-1} b + \frac{n(n - 1)}{1 \cdot 2} a^{n-2} b^2 + \cdots + b^n$$

Either of the above equations is called the binomial expansion.

EXAMPLE 1.   Expand $(x + y)^4$.

Since $n = 4$, application of the binomial expansion results in the equation

$$(x + y)^4 = x^4 + 4x^{4-1} y + \frac{4 \cdot 3}{1 \cdot 2} x^{4-2} y^2 + \frac{4 \cdot 3 \cdot 2}{1 \cdot 2 \cdot 3} x^{4-3} y + \frac{4 \cdot 3 \cdot 2 \cdot 1}{1 \cdot 2 \cdot 3 \cdot 4} x^{4-4} y^4$$
$$= x^4 + 4x^3 y + 6x^2 y + 4xy^3 + y^4$$

If the binomial has a fractional or negative exponent, then the expansion of the binomial results in an infinite series of terms.   However, the coefficients of the series are computed exactly as in the second form of the binomial expansion above.

If one of the two terms in the binomial is much smaller than the other, then the infinite series can be used to make some useful approximations.   Thus, the value of $(1 \pm x)^n$, where $|x|$ is much smaller than 1, is approximated by $1 \pm nx$, or the first two terms of the binomial series.

EXAMPLE 2.   Determine the approximate numerical value of $1/1.02$.

Rewrite the given ratio as $\dfrac{1}{1 + 0.02} = (1 + 0.02)^{-1}$.

With $n = -1$ and $x = 0.02$, the approximate value of the ratio is

$$1 + (-1)(0.02) = 1 - 0.02 = 0.98$$

PHYSICAL CONSTANTS

Symbol	Name	Numerical Value and Units	
		SI or mks	cgs
g	Acceleration of gravity (sea level, 45°N)	$9.806 \, \text{m/s}^2$	$980.6 \, \text{cm/s}^2$
amu	Atomic mass unit	$1.661 \times 10^{-27} \, \text{kg}$	$1.661 \times 10^{-24} \, \text{g}$
$\mathcal{N}$	Avogadro's number	$6.022 \times 10^{26}/\text{kmole}$	$6.022 \times 10^{23}/\text{g-mole}$
k	Boltzmann's constant	$1.381 \times 10^{-23} \, \text{J/K}$	$1.381 \times 10^{-16} \, \text{erg}/°\text{K}$
e/m_e	Charge-to-mass ratio for electron	$1.759 \times 10^{11} \, \text{C/kg}$	$5.273 \times 10^{17} \, \text{esu/g}$
e	Electron charge	$1.602 \times 10^{-19} \, \text{C}$	$4.803 \times 10^{-10} \, \text{esu}$
m_e	Electron rest mass	$9.110 \times 10^{-31} \, \text{kg}$	$9.110 \times 10^{-28} \, \text{g}$
F	Faraday constant	$9.649 \times 10^{7} \, \text{C/kmole}$	$2.893 \times 10^{14} \, \text{esu/mole}$
G	Gravitational constant	$6.673 \times 10^{-11} \, \text{N} \cdot \text{m}^2/\text{kg}^2$	$6.673 \times 10^{-8} \, \text{dyne} \cdot \text{cm}^2/\text{g}^2$
J	Mechanical equivalent of heat (15°C)	$4.186 \, \text{J/cal}$	$4.186 \times 10^{7} \, \text{erg/cal}$
M_n	Neutron rest mass	$1.675 \times 10^{-27} \, \text{kg}$	$1.675 \times 10^{-24} \, \text{g}$
h	Planck's constant	$6.626 \times 10^{-34} \, \text{J} \cdot \text{s}$	$6.626 \times 10^{-27} \, \text{erg} \cdot \text{s}$
M_p	Proton rest mass	$1.673 \times 10^{-27} \, \text{kg}$	$1.673 \times 10^{-24} \, \text{g}$
R_∞	Rydberg constant	$1.097 \times 10^{7}/\text{m}$	$1.097 \times 10^{5}/\text{cm}$
V_0	Standard volume of an ideal gas	$2.241 \times 10 \, \text{m}^3/\text{kmole}$	$2.241 \times 10^{4} \, \text{cm}^3/\text{g mole}$
σ	Stefan–Boltzmann constant	$5.670 \times 10^{-8} \, \text{J}/(\text{m}^2 \cdot \text{K}^4 \cdot \text{s})$	$5.670 \times 10^{-5} \, \text{erg/cm}^2 \cdot \text{K}^4 \cdot \text{s}$
R_0	Universal gas constant	$8.314 \, \text{J}/(\text{kmole} \cdot \text{K})$	$8.314 \times 10^{7} \, \text{erg}/(\text{g-mole} \cdot \text{K})$
c	Velocity of light in vacuum	$2.998 \times 10^{8} \, \text{m/s}$	$2.998 \times 10^{10} \, \text{cm/s}$
$\lambda_{\max} T; \, \omega$	Wien's displacement law constant	$2.898 \times 10^{-3} \, \text{m} \cdot \text{K}$	$2.898 \times 10^{-1} \, \text{cm} \cdot \text{K}$

ASTRONOMICAL CONSTANTS

Symbol or Abbreviation	Name	Numerical Value and Units	
AU	Astronomical unit	1.496×10^{11} m	1.496×10^{13} cm
LY	Light-year	9.460×10^{15} m	9.460×10^{17} cm
pc	Parsec	3.086×10^{16} m	3.086×10^{18} cm
$R_\odot$	Radius of sun	6.960×10^{8} m	6.960×10^{10} cm
$R_\oplus$	Radius of earth (equatorial)	6.378×10^{6} m	6.378×10^{8} cm
$M_\odot$	Mass of sun	1.991×10^{30} kg	1.991×10^{33} g
$M_\oplus$	Mass of earth	5.977×10^{24} kg	5.977×10^{27} g
$L_\odot$	Luminosity of sun	3.86×10^{26} joule/s	3.86×10^{33} erg/s
S	Solar constant	1.36×10^{3} joule/m$^2 \cdot$ s	1.36×10^{6} erg/cm$^2 \cdot$ s
	Constant of aberration	20.496″	
	Solar parallax	8.794″	
	Radius of moon	1.738×10^{6} m	1.080×10^{3} mi
	Mean distance of moon	3.844×10^{10} m	2.389×10^{5} mi
	Mass of moon	7.349×10^{22} kg	7.349×10^{25} g
	Sidereal year	$365^{d}6^{h}9^{m}10^{s}$	365.2564 mean solar days
	Tropical year	$365^{d}5^{h}48^{m}46^{s}$	365.2422 mean solar days $= 3.1558 \times 10^{7}$ s
	Sidereal month	$27^{d}7^{h}43^{m}12^{s}$	27.3217 mean solar days
	Synodic month	$29^{d}12^{h}44^{m}3^{s}$	29.5306 mean solar days

$$1 \text{ pc} = 3.262 \text{ LY} = 206{,}265 \text{ AU} = 3.086 \times 10^{13} \text{ km} = 1.920 \times 10^{13} \text{ mi}$$

$$1 \text{ LY} = 0.3068 \text{ pc} = 6.324 \times 10^{4} \text{ AU} = 9.460 \times 10^{12} \text{ km} = 5.879 \times 10^{12} \text{ mi}$$

AREAS UNDER THE STANDARD NORMAL CURVE FROM 0 TO z

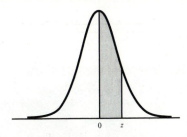

z	0	1	2	3	4	5	6	7	8	9
0.0	0.0000	0.0040	0.0080	0.0120	0.0160	0.0199	0.0239	0.0279	0.0319	0.0359
0.1	0.0398	0.0438	0.0478	0.0517	0.0557	0.0596	0.0636	0.0675	0.0714	0.0754
0.2	0.0793	0.0832	0.0871	0.0910	0.0948	0.0987	0.1026	0.1064	0.1103	0.1141
0.3	0.1179	0.1217	0.1255	0.1293	0.1331	0.1368	0.1406	0.1443	0.1480	0.1517
0.4	0.1554	0.1591	0.1628	0.1664	0.1700	0.1736	0.1772	0.1808	0.1844	0.1879
0.5	0.1915	0.1950	0.1985	0.2019	0.2054	0.2088	0.2123	0.2157	0.2190	0.2224
0.6	0.2258	0.2291	0.2324	0.2357	0.2389	0.2422	0.2454	0.2486	0.2518	0.2549
0.7	0.2580	0.2612	0.2642	0.2673	0.2704	0.2734	0.2764	0.2794	0.2823	0.2852
0.8	0.2881	0.2910	0.2939	0.2967	0.2996	0.3023	0.3051	0.3078	0.3106	0.3133
0.9	0.3159	0.3186	0.3212	0.3238	0.3264	0.3289	0.3315	0.3340	0.3365	0.3389
1.0	0.3413	0.3438	0.3461	0.3485	0.3508	0.3531	0.3554	0.3577	0.3599	0.3621
1.1	0.3643	0.3665	0.3686	0.3708	0.3729	0.3749	0.3770	0.3790	0.3810	0.3830
1.2	0.3849	0.3869	0.3888	0.3907	0.3925	0.3944	0.3962	0.3980	0.3997	0.4015
1.3	0.4032	0.4049	0.4066	0.4082	0.4099	0.4115	0.4131	0.4147	0.4162	0.4177
1.4	0.4192	0.4207	0.4222	0.4236	0.4251	0.4265	0.4279	0.4292	0.4306	0.4319
1.5	0.4332	0.4345	0.4357	0.4370	0.4382	0.4394	0.4406	0.4418	0.4429	0.4441
1.6	0.4452	0.4463	0.4474	0.4484	0.4495	0.4505	0.4515	0.4525	0.4535	0.4545
1.7	0.4554	0.4564	0.4573	0.4582	0.4591	0.4599	0.4608	0.4616	0.4625	0.4633
1.8	0.4641	0.4649	0.4656	0.4664	0.4671	0.4678	0.4686	0.4693	0.4699	0.4706
1.9	0.4713	0.4719	0.4726	0.4732	0.4738	0.4744	0.4750	0.4756	0.4761	0.4767
2.0	0.4772	0.4778	0.4783	0.4788	0.4793	0.4798	0.4803	0.4808	0.4812	0.4817
2.1	0.4821	0.4826	0.4830	0.4834	0.4838	0.4842	0.4846	0.4850	0.4854	0.4857
2.2	0.4861	0.4864	0.4868	0.4871	0.4875	0.4878	0.4881	0.4884	0.4887	0.4890
2.3	0.4893	0.4896	0.4898	0.4901	0.4904	0.4906	0.4909	0.4911	0.4913	0.4916
2.4	0.4918	0.4920	0.4922	0.4925	0.4927	0.4929	0.4931	0.4932	0.4934	0.4936
2.5	0.4938	0.4940	0.4941	0.4943	0.4945	0.4946	0.4948	0.4949	0.4951	0.4952
2.6	0.4953	0.4955	0.4956	0.4957	0.4959	0.4960	0.4961	0.4962	0.4963	0.4964
2.7	0.4965	0.4966	0.4967	0.4968	0.4969	0.4970	0.4971	0.4972	0.4973	0.4974
2.8	0.4974	0.4975	0.4976	0.4977	0.4977	0.4978	0.4979	0.4979	0.4980	0.4981
2.9	0.4981	0.4982	0.4982	0.4983	0.4984	0.4984	0.4985	0.4985	0.4986	0.4986
3.0	0.4987	0.4987	0.4987	0.4988	0.4988	0.4989	0.4989	0.4989	0.4990	0.4990
3.1	0.4990	0.4991	0.4991	0.4991	0.4992	0.4992	0.4992	0.4992	0.4993	0.4993
3.2	0.4993	0.4993	0.4994	0.4994	0.4994	0.4994	0.4994	0.4995	0.4995	0.4995
3.3	0.4995	0.4995	0.4995	0.4996	0.4996	0.4996	0.4996	0.4996	0.4996	0.4997
3.4	0.4997	0.4997	0.4997	0.4997	0.4997	0.4997	0.4997	0.4997	0.4997	0.4998
3.5	0.4998	0.4998	0.4998	0.4998	0.4998	0.4998	0.4998	0.4998	0.4998	0.4998
3.6	0.4998	0.4998	0.4999	0.4999	0.4999	0.4999	0.4999	0.4999	0.4999	0.4999
3.7	0.4999	0.4999	0.4999	0.4999	0.4999	0.4999	0.4999	0.4999	0.4999	0.4999
3.8	0.4999	0.4999	0.4999	0.4999	0.4999	0.4999	0.4999	0.4999	0.4999	0.4999
3.9	0.5000	0.5000	0.5000	0.5000	0.5000	0.5000	0.5000	0.5000	0.5000	0.5000

(*From Schaum's Outline of Theory and Problems of Statistics, Second Edition, Murray R. Spiegel, McGraw-Hill, 1988*)

HEWLETT-PACKARD GRAPHING CALCULATOR 38G

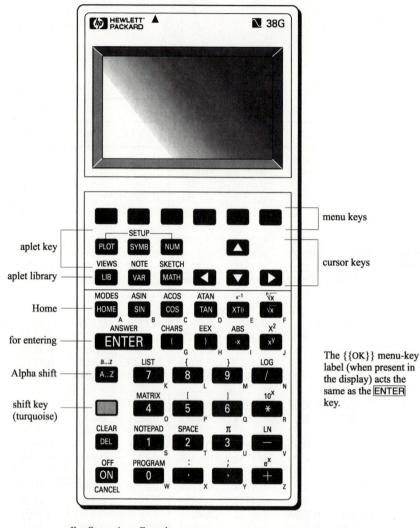

Key Conventions - Examples:

{SHIFT} {ANSWER} Means press the shift key followed by the ENTER key.

A...Z {A} Means press the Alpha-shift key followed by the A key.

Reprinted with permission of the Hewlett-Packard Company.

TEXAS INSTRUMENT GRAPHING CALCULATOR TI-82

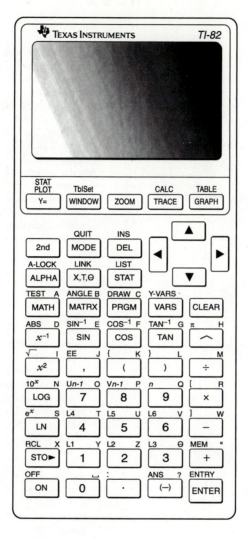

Key Conventions - Example:

{2nd} {L4} Means press the [2nd] key followed by the {L4} key.

Illustration of TI-82 Graphing Calculator courtesy of Texas Instruments. Reprinted with permission of Texas Instruments.

Index

478